Nonlinear Circuits and Systems with Memristors

Fernando Corinto • Mauro Forti • Leon O. Chua

Nonlinear Circuits and Systems with Memristors

Nonlinear Dynamics and Analogue Computing via the Flux-Charge Analysis Method

Springer

Fernando Corinto
Department of Electronics &
Telecommunications
Politecnico di Torino
Torino, Italy

Mauro Forti
Department of Information
Engineering and Mathematics
University of Siena
Siena, Italy

Leon O. Chua
University of California
Berkeley, CA, USA

ISBN 978-3-030-55650-1 ISBN 978-3-030-55651-8 (eBook)
https://doi.org/10.1007/978-3-030-55651-8

This Springer imprint is published by the registered company Springer Nature Switzerland AG
The registered company address is: Gewerbestrasse 11, 6330 Cham, Switzerland

To Novella, Giorgia, and Sara
and in memory of Antonio—F.C.
To Emiliana and in memory of Silvano—M.F.
To Diana—L.O.C.

Foreword by Sung Mo (Steve) Kang

Nonlinear Circuits and Systems with Memristors is a timely contribution to the field of nanoelectronic circuits and systems. Personally, the contents of this book are dear to me, especially since I was one of the early graduate students of Professor Leon Chua at the UC Berkeley who at that time introduced memristors and memristive systems. The seminal paper on memristor by Leon Chua was published in 1971, followed by the Chua and Kang paper on memristive devices and systems in 1976. In microelectronics industry, CMOS Very Large-Scale Integrated (VLSI) circuits have been the dominant workhorse, the development of which has followed the Moore's law. However, as the downscaling faced its limitations and because of the volatility of charge storage in ultrasmall capacitors in VLSI chips, nonvolatile resistance has become critically important as a new state variable. The demands for Resistive RAMs (RRAMs) and other memory devices such as MRAM, PCRAM, STT-RAM, which do not depend on charge storage, have increased. Analog computing has also become of increasing importance for ultralow energy computing as in neuromorphic computing. Almost four decades later, when Stan Williams and his associates in HP announced nanoscale memristors in the May 2008 Nature paper, a new epoch for integrated memristor circuits was established. HP's memristor was the first solid-state realization of the "two-terminal memristor." It was my honor and pleasure to organize the first symposium on memristors and memristive systems in November 2008 under sponsorship of NSF and HP at the Berkeley campus, the fountain of the memristor research. Since then, memristor electronics has been pursued globally at a phenomenal growth rate. In particular, the application of crossbar arrays of memristors for analog neuromorphic computing has been pursued actively as one of the most promising approaches to the mimicry of brain functions. Two-terminal memristors are used in crossbar arrays with programmable memductances as learning weights. Coincidentally in their Proceedings of the IEEE paper (1976), Chua and Kang identified the conductive channels of the Hodgkin–Huxley model for neuromorphic signal (action potential) generation to be memristive. A few years ago, after giving a seminar on memristors at the Rowan University in New Jersey, its faculty members asked me when a textbook would be published on the subject matter for education of electronic circuits including memristors. Although this book is better suited for graduate students, I would expect that undergraduate textbooks on electronic circuits would include basic memristor circuits in the not too distant future.

In the beginning of this book, the authors take the axiomatic approach to describe the four basic circuit elements, namely R, L, C, and M (memristor), with focus on interesting cases involving nonlinear characteristics of elements. Through systematic formulation of circuit equations and rigorous analysis, the authors illustrate peculiar circuit behaviors of nonlinear RLC circuits, including oscillations and bifurcations. Theoretical discussions become even more interesting after including memristors for most general nonlinear RLCM circuits. The introduction of the flux-charge analysis method (FCAM) is not only novel but also uniquely revealing. The introduction of Kirchhoff's charge Law (KqL) and Kirchhoff's flux Law (KφL) in lieu of KCL and KVL is enlightening and prepares the readers for systematic formulation and analysis of RLCM circuits. The application of FCAM to neuromorphic systems uncovers numerous peculiar dynamic behaviors. Undoubtedly, this book is unique and provides a rich set of interesting and intriguing topics for both interested graduate students and researchers. I congratulate the authors for the publication of this book and hope that the readers will benefit from studying the contents to enrich the field of memristive nanoelectronics.

Baskin School of Engineering
University of California
Santa Cruz, CA, USA

Sung Mo (Steve) Kang
SOE Distinguished Chair Professor
Chancellor Emeritus, UC Merced
President Emeritus, KAIST

Foreword by Ronald Tetzlaff

An increasing number of different two-terminal devices manufactured in distinct technologies can be classified as memristors with most popular developments so far, in technologies implementing dense, low-power nonvolatile memories and neuromorphic systems operating according to biological principles. Especially, recent investigations show a strong interest in new computing architectures required for future digital systems that are often based on communicating intelligent sensor–processor architectures. The development of new memristor computing concepts in order to overcome the limits of classical technology caused by the so-called von-Neumann bottleneck is based on principles in which processing and data storage are carried out in the same physical location. These technologies are ranging from crossbar arrays to artificial and bio-inspired neural networks, including models of biological computation in a human brain. The availability and application of these so-called in-memory computing systems would obviate the need for energy-expensive data transfer operations on the memory-CPU in future IOT networks. While different authors have shown in an increasing number of publications that certain mathematical operations can be performed rather efficiently on crossbar arrays, attractive concepts of universal computation are based on the emergence of complex behavior in strongly nonlinear dynamical arrays. By investigating the so-called reaction–diffusion systems, Leon Chua has derived the fundamental results that prove that the emergence of complexity in these structures is based on local activity and in particular on a parameter subset called the "Edge of Chaos." Typically, the treatment of such highly nonlinear, memristive spatiotemporal systems, which exhibit oscillations, pattern formation, and wave propagation phenomena, is based on the availability of compact device models and on new mathematical strategies to gain a deep understanding of the mechanisms of such systems in order to enable the derivation of highly efficient in-memory computing methods. Although mem-elements and memristive circuits have been addressed in a bulk of publications, an in-depth mathematical treatment and understanding based on circuit theory has been provided in only a few investigations. Mostly, present model-based simulators operating in the voltage-current domain make qualitatively incorrect predictions, especially when applied to the problems of nonlinear dynamics. There is a lack of accurate but stable compact device models that allow the simulation and design of future memristive dynamic circuits, which are needed for the digitalization of today's technology.

Starting from an axiomatic approach to introduce basic circuit elements and fundamental concepts of circuit theory, including the treatment of oscillatory RLC circuits, the authors continue with their original development, the flux-charge analysis method (FCAM), which is a unique approach to gain a deep understanding of nonlinear dynamics in memristive circuits. The circuit theoretic concept which is based on Kirchhoff's charge law and Kirchhoff's flux law is clearly outlined and prepares the reader for the following deep insight into the theory of nonlinear dynamics, including the bifurcation without parameters in memristive RLCM circuits. The application of the FCAM method to different circuits considering cellular neural networks completes this unique book, which offers a comprehensive insight into the concepts that allow an in-depth treatment of circuits with mem-elements.

It is my pleasure to congratulate the authors on this important contribution to the theory of memristors, dedicated to graduate students and a broad class of researchers.

Institute of Circuits and Systems
Faculty of Electrical and Computer Engineering
Faculty of Computer Science
TU Dresden, Dresden, Germany

Ronald Tetzlaff

Foreword by R. Stanley Williams

I first became aware of the seminal papers on memristors by Leon Chua and the extension of memristive systems by Chua and Kang in 2004. I was struck by the similarity of the "pinched hysteresis loop" characteristic of the axiomatically defined memristor and the experimental data we had been collecting in my group for nearly 6 years, and I spent a couple of frustrating years trying to figure out if and how they were connected. After suffering an auto accident in 2006 and being incapacitated for the month of August, I spent several weeks downloading and reading about 300 papers. Somehow, my subconscious brain made a connection among them and I had a genuine eureka! moment, where I understood how and why our resistance switching devices were memristors. I spent the next year primarily in writing invention disclosures and patent applications, and in 2008 with my HP colleagues we published our "Missing Memristor Found" paper in Nature. The next couple of years were spent on expanding our understanding of the physical mechanisms behind memristance in transition metal oxides, developing improved compact models that we could use for circuit simulations, improving our devices, and exploring a wide range of potential applications for the "fourth fundamental circuit element." During this period, I read many more papers by Chua and collaborators, and I realized that the memristor was the literal "tip of the iceberg" for an enormous and comprehensive body of work on nonlinear dynamical circuits and networks. I have spent much of the past decade studying and learning how to utilize nonlinear dynamics and chaos for computation in anticipation of the saturation of Moore's scaling for CMOS circuits. I could see that for some functions, one or two memristors could emulate the properties of a neuron with better fidelity than a circuit with several hundred to thousands of transistors, thus my goal was not a one-off replacement for a transistor, but rather dramatic simplification and efficiency improvements in complete circuits for neuromorphic computing. However, the existing papers on nonlinear dynamical circuits were necessarily very formal mathematical developments of unfamiliar (to me) concepts that required extreme effort on my part to understand. In order to provide guidance for me and others, I invited Leon Chua in 2015 to present a series of lectures sponsored by HP that gave him the opportunity to provide an overview of his entire body of work. He gracefully accepted and devoted a huge amount of work into organizing and presenting the twelve-part series "The Chua Lectures: From Memristors and Cellular Nonlinear Networks to the Edge of Chaos," which is available on YouTube. During the lectures, the interesting topic of a Hamiltonian approach to electronic circuits that was based on flux and charge came up, which I found to be fascinating but had no bandwidth to pursue myself.

I was thus delighted when Fernando Corinto presented a copy of this book and asked me to provide some remarks for a preface. I started to read it and felt a warm glow through the first four introductory chapters. I had just finished teaching a graduate-level class for electrical engineers on nonlinear dynamics and chaos, and those chapters with their excellent figures cemented home the concepts that I had been trying to help my students learn. Just one example is the discussion of impasse points and how (and why) to break them. I wish I had the book before I taught the class, because the choice of topics and the presentation were ideal—much better than what I have found elsewhere in texts or papers and certainly better than my lecture notes. Then I began with chapter five, which is the inception of the core of the book. I was both challenged and intrigued as I continued to read. It took me quite a long time to finish the book, not because of lack of interest but rather I kept coming up with completely new ideas for experiments that I wanted to develop, which included designing special-purpose circuits and apparatus for performing measurements. Building a charge and flux meter, musing about the implications of Noether's theorems for electronic circuits, and analyzing a coupled oscillator circuit for computing were just a few of the projects I investigated. I spent a lot of time on these pursuits and had great fun in the process—at this stage, it is hard to know if any of my concepts will work, but I am excited by the prospects. As I write this, I am in a mode where my lab is closed because of the COVID-19 epidemic, but I hope to start work on the ideas inspired by this book as soon as possible. *Nonlinear Circuits and Systems with Memristors* has reinforced much that I have previously learned and taught, and it has also challenged me to consider the new ways of thinking about how to understand and even measure the properties of nonlinear dynamical systems. It is not the final elucidation of the field, but rather a portal and invitation to explore new opportunities.

Department of Electrical and Computer Engineering
Texas A&M University
Bryan, College Station, TX, USA
May 13, 2020

R. Stanley Williams
Hewlett Packard Enterprise
Company Chair Professor

Preface

Conventional Von Neumann computing architectures based on CMOS technology are currently facing challenges termed "the heat and memory wall" in addition to the advent of Moore's Law slowdown [1–3]. Von Neumann bottleneck originates in particular from the speed limitations due to the constant data movements between the memory (e.g., the Random Access Memory—RAM) and the Central Processing Unit (CPU), since RAM and CPU have physically distinct locations. Going beyond CMOS and overcoming Von Neumann restrictions are long-term visions aimed at developing completely new nanoscale components with unconventional functions and dynamics that are capable of outperforming similar CMOS implementations to sustain the growth of the electronics industry at the end of Moore's Law.

The memristor (a shorthand for memory-resistor) is one of the most promising candidate information processing devices for beyond CMOS and more than Moore semiconductor technology. The memristor has been theoretically envisioned by L. O. Chua in 1971 [4] as the fourth basic passive circuit element, in addition to the resistor, inductor, and capacitor, using an axiomatic approach on device modeling and symmetry arguments on the basic electric quantities. A memristor is a state-dependent resistor, where the resistance (also named memristance) is not fixed but rather depends on the history of the voltage or current. A memristor is endowed with a number of new peculiar features that are not shared by the other basic circuit elements.

For a long time, the memristor remained basically an object of academic interest since no passive physical device was known behaving as a memristor. Almost four decades after the publication of the seminal paper [4], the researchers at Hewlett and Packard headed by R. Stanley Williams first announced and presented to the world a nanoscale device displaying memristive features [5]. This work has sprung a huge worldwide cross-disciplinary interest on memristor and its applications ranging from tunable electronics, neuromorphic/in-memory computing, biosensors, data storage, and complex nonlinear systems [6–8].

On-chip memory, biologically inspired computing and in-memory computing, i.e., the integration of storage and computation in the same physical location [9], are categories that are expected to significantly benefit from memristor developments.

This is in turn particularly relevant to future computing needs such as cognitive processing, big-data analysis, and low-power intelligent systems based on the Internet of Things.

Neuromorphic/synaptic electronics is an emerging field of research aiming to overcome Von Neumann platforms by building artificial neuronal systems that mimic the extremely energy-efficient biological synapses. Neuromorphic memristive architectures integrated into edge computing devices are expected to increase the data processing capability at lower power requirements and reduce several overheads for cloud computing solutions [10]. The introduction of photovoltaic/photonic aspects into neuromorphic architectures could produce self-powered adaptive electronics and open new possibilities in artificial neuroscience, neural communications, sensing, and machine learning. This would enable, in turn, a new era for computational systems owing to the possibility of attaining high bandwidths with much reduced power consumption [11].

Memristor devices operated as a nonvolatile memory (NVM) are emerging for data storage and unconventional computing systems. In this case, a memristor should display two (or more) largely different values of memristance and be a nonvolatile device. The memristor is driven from one memristance to others via a suitable current (or voltage) pulse, an operation that is often referred to as set/reset. This mechanism permits to exploit nonvolatile resistive states of memristor to encode information bits. A continuous range of resistive states enables the use of memristor devices as *analogue programmable resistors*. Such devices, whose resistance can be precisely modulated electronically, and which can support important synaptic functions (e.g., Spike-Timing-Dependent Plasticity—STDP), pave the way to denser low-power analogue circuits, multi-state memory, and large-scale synapse implementation in neuromorphic systems and cellular neural networks with adaptation capabilities [10, 12].

Memristor devices embedded in nonlinear circuits can generate a complex nonlinear dynamical evolution of their memristance that can be exploited to build nanoscale oscillators potentially described by low-order mathematical/circuit models. Networks of interconnected and interacting oscillators can develop cooperative and collective dynamics, e.g., phase synchronization and other self-organizing spatiotemporal phenomena for alternative computing schemes overpassing the limits of conventional digital and Boolean computation. A memristor-based nonlinear oscillator is proposed in [13] as a source of tunable chaotic behavior that can be incorporated into a Hopfield computing network to improve the efficiency and accuracy of converging to a solution for computationally hard problems. The breakthrough is the development of networks of interacting nanoscale memristor oscillators and their computational schemes based on cooperative and collective dynamics. It is then crucial to develop a technological platform comprehensive of functional materials and hardware memristors including physical/circuit models of their operation as a key in hand tool for the large-scale industrial exploitation.

Aims and Scope

The main goal of this book is to develop systematic theoretical methodologies to analyze nonlinear circuits including memristors as nonlinear dynamic elements. Such dynamical nonlinear circuits are referred to as *memristor circuits*. These methods are essential for understanding the peculiar dynamic properties and computational capabilities of memristors/mem-elements and exploit their dynamics to implement future unconventional computing systems.

It is known from circuit theory that it is natural and effective to analyze a nonlinear RLC circuit, i.e., a circuit with linear/nonlinear resistors (R), inductors (L), and capacitors (C) (but without memristors), in the traditional voltage-current (v, i)-domain, i.e., using constitutive relations of circuit elements and Kirchhoff laws expressed in terms of voltages v and currents i. However, since the definition of a memristor involves a link between flux φ (the integral of voltage or voltage momentum) and charge q (the integral of current or current momentum), it is natural to ask the following:

1. Is there a more effective domain, with respect to the traditional (v, i)-domain, to analyze the nonlinear dynamics of memristor circuits?
2. Do we expect to observe new and peculiar nonlinear dynamic phenomena when adding one (or more) memristors to a classical RLC circuit?

An extended literature is available on the analysis of dynamic memristor circuits. Several papers point out via experimental or numerical means that, generally speaking, including a memristor in an RLC circuit greatly enriches the dynamics. In particular, it appears that due to the memristor there can coexist different dynamics and regimes (e.g., convergent, oscillatory, complex dynamics) for the same set of circuit parameters. Several basic aspects of the observed behaviors are however elusive and remain unclear and in large part unexplained. It is not even clear from a mathematical viewpoint whether the order of the dynamics of an RLC circuit increases or not when a memristor is added to it.

The chief aim of the book is to answer these fundamental questions and provide an analytic treatment and clear explanation of dynamic phenomena reported in the literature via numerical or experimental means. The treatment is rigorous and based on tools and techniques with foundation in nonlinear circuit theory. Whenever possible, we try however to keep mathematics at the minimum indispensable level for an accurate description.

In this book, we identify and select progressively relevant classes of memristor circuits, widely investigated in the literature and used in the applications, and describe a new method for their dynamic analysis. The new analysis method introduced in this book is based on suitable forms of Kirchhoff laws and constitutive relations of circuit elements expressed in the *flux-charge* (φ, q)*-domain* rather than in the traditional (v, i)-domain. Thus the name *Flux-Charge Analysis Method*, or FCAM, in short. FCAM has been mainly developed in a series of recent articles [14–20].

The authors' aim is to make clear to the readers that there are several main advantages when using FCAM with respect to the traditional analysis in the voltage-current domain. Two key advantages are related to the principle of *reduction of order for the dynamics* and the possibility to deal with a *smoother dynamics*, in the (φ, q)-domain, with respect to the (v, i)-domain. In addition, any fundamental property of a memristor circuit proved via FCAM in the (φ, q)-domain has a corresponding one for an RLC circuit in the (v, i)-domain. Via this analogy, systematic methods for writing the dynamic equations of memristor circuits in the form of a Differential Algebraic Equation (DAE), or a State Equation (SE), are developed.

Moreover, FCAM permits to highlight and rigorously show the existence of new peculiar dynamic behaviors displayed by memristor circuits. These include the presence of *invariant manifolds* and the *coexistence of different nonlinear dynamics, attractors, and regimes*, a complex dynamic scenario that is sometimes referred to in the literature as *extreme multistability*. The coexistence of different attractors is shown to be related to a new type of bifurcations, due to changing the initial conditions for a fixed set of circuit parameters, which are named *bifurcations without parameters*.

Given the extremely rich dynamic scenario in memristor circuits, it is natural to wonder if there is the possibility to control and programme different attractors and regimes in an effective way. This book shows that there is a positive answer to this question by developing a simple programming procedure using impulsive voltage or current sources (a natural way for transmitting signals from a neuromorphic viewpoint).

Applications of the results are considered to oscillatory and chaotic memristor circuits and also to arrays of memristor oscillators and neuromorphic architectures. Finally, FCAM is generalized to higher-order elements with memory (also named *mem-elements*) as memcapacitors and meminductors and also to some classes of extended memristors.

Next, we discuss in some more detail the organization of the book, the content of each single chapter, and the prerequisites and the audience for which the book is intended.

We then conclude the preface by reporting a slightly abridged version of an article by Chua [21] devoted to a reminiscence of the genesis and the thought process he followed to theoretically introduce the fourth circuit element.

Organization of the Book

The book is mainly organized in three parts:

- Foundation of Nonlinear Circuit Theory
- Flux-Charge Analysis Method (FCAM)
- Applications and Extension of FCAM

The first part is a summary of nonlinear circuit theory pillars, so that readers are gently introduced to concepts underlying the FCAM via a self-contained book. Although expert researchers might pass over this part, the holistic approach used reveals the generality of the FCAM and poses the basis for its extension to nonlinear circuits with mem-elements. The FCAM is the core of the second part, whereas the third part is devoted to FCAM applications. A brief summary of the chapters included in each part is reported in the following.

Foundation of Nonlinear Circuit Theory
Chapter 1 provides the reader with an axiomatic definition of the basic nonlinear circuit elements that could be used to model a wide variety of nonlinear devices. A black-box approach is used, independent of the internal composition, material, geometry, and architecture of each device. The axiomatic approach leads naturally to the definition of the fourth basic circuit element, i.e., the memristor, and also higher-order circuit elements, such as the memcapacitor and meminductor.

Chapter 2 deals in more depth with the behavior of a memristor by examining some of its main properties and signatures. A brief account of the HP memristor is provided, as well as a discussion on extended memristors and a classification of memristive devices. The technological realization of real memristive devices as well as materials and physics phenomena underlying the memristor behavior are briefly discussed.

Chapters 3 and 4 are devoted to synthetically give the needed theoretic background on nonlinear RLC circuits (without memristors). Especially, Chap. 3 deals with a synthetic description of the main methods to analyze nonlinear RLC circuits, i.e., tableau, nodal, and mesh analysis. Special emphasis is then paid to give conditions for the existence of the state equation (SE) representation and the techniques to write the SEs of nonlinear RLC circuits.

Chapter 4 briefly discusses some main dynamical phenomena that can be observed in autonomous nonlinear RLC circuits of first, second, and third order, including convergence of solutions, oscillations, and complex dynamic behavior. The chapter ends with some basic considerations on local bifurcations of equilibrium points and global bifurcations of limit cycles in autonomous RLC nonlinear circuits depending on parameters.

Flux-Charge Analysis Method (FCAM)
Chapters 5–7 are at the core of the book, and they are devoted to develop FCAM for the dynamic analysis of nonlinear circuits containing memristors.

In particular, Chap. 5 starts with a critical review of Kirchhoff Laws in the (φ, q)-domain, and, on this basis, it then proceeds to develop the new method of analysis in the (φ, q)-domain (FCAM) for wide classes of memristor circuits. The chapter also establishes an analogy between RLC circuits in the (v, i)-domain and memristor circuits in the (φ, q)-domain and, via this analogy, gives a general formulation of memristor circuit equations.

Chapter 6 discusses the applications of FCAM to some basic low-order autonomous memristor circuits highlighting the main advantages of FCAM related to the reduction of order, and smoother dynamics, in the (φ, q)-domain.

New peculiar and intriguing dynamical features of memristor circuits, such as the presence of invariant manifolds, coexisting dynamics and attractors (extreme multistability), and bifurcations without parameters, are addressed and characterized analytically.

Finally, Chap. 7 develops a systematic analytic procedure for programming different dynamics and regimes in non-autonomous memristor circuits by means of impulsive voltage or current sources.

Applications and Extension of the FCAM

Chapter 8 describes the application of FCAM to study synchronization phenomena in arrays of locally connected oscillators, while Chap. 9 exploits FCAM to design a class of memristor neural networks that are able to process signals in the (φ, q)-domain according to the principle of in-memory computing.

Chapter 10 develops a circuit model of a class of extended memristors by interconnecting basic circuit elements as an ideal memristor and a nonlinear resistor. FCAM is then used to analyze the dynamics in the case the nonlinear resistor has a piecewise-linear voltage-current characteristic. Finally, Chap. 11 discusses the extension of FCAM to nonlinear circuits containing also higher-order elements such as memcapacitors and meminductors, showing that such circuits display even richer dynamic features with respect to memristor circuits.

Prerequisites and Audience

From a didactic viewpoint, the present book is intended for a graduate-level course in engineering on memristors including relevant aspects of nonlinear circuit theory connected with the emerging nanoscale devices. It may also be used for self-study or as a reference book by engineers, physicists, and applied mathematicians. From a research viewpoint, the book is an account of the state of the art on some main research directions in the analysis of memristor circuits. As such it may be used as a solid basis to start and pursue researches on the challenging and rapidly evolving field of memristors and nanoscale devices.

The prerequisite of this book is a graduate-level course in circuit theory introducing basic aspects of the analysis of nonlinear circuits at the level of classical textbooks (e.g., [22]). The needed mathematical background corresponds to the essential level of calculus, ordinary differential equations, and algebra that are part of the graduate student curricula in engineering, physics, and mathematics. Whenever possible, mathematics is kept to a minimum without losing rigor and accuracy in the description. Ad hoc mathematical textbooks are adequately referred to in order to give the reader the possibility to further elaborate on some of the presented topics.

Acknowledgments

This book has benefited from the interaction and collaboration of literally hundreds of people, thanks to the highly multi-disciplinary discussion process adopted in the European COST Action *"Memristors: Devices, Models, Circuits, Systems and Applications (MemoCiS)."* The result is that there are many people to thank and in special manner the Action Chair Julius Georgiou and the Working Group leaders Sabina Spiga, Themis Prodromakis, Dalibor Biolek, Ronald Tetzlaff, Elisabetta Chicca, Sandro Carrara, and Bernabe Linares-Barranco. We have also enjoyed the cross-fertilizing exchange with Abu Sebastian, Thomas Mikolajick, Gianaurelio Cuniberti, Giacomo Indiveri, Stravos Stavrinides, Rodrigo Picos, Enrique Ponce, Enrique Miranda, Nikolay V. Kuznetsov, Georgios Sirakoulis, Said Hamdioui, J. Joshua Yang, Qiangfei Xia, Wei Lu, Yuchao Yang, Shahar Kvatinsky, Carlo Ricciardi, Mattia Frasca, Daniele Ielmini, Stefano Brivio, and Marius Orlowski.

We have received invaluable assistance through day-by-day conversations with our colleagues. It is our pleasure to mention in particular the stimulating discussions and constructive comments by Pier Paolo Civalleri, Alon Ascoli, Alberto Tesi, Mauro Di Marco, Luca Pancioni, Giacomo Innocenti, Valentina Lanza, Jacopo Secco, Francesco Marrone, and Gianluca Zoppo.

The *Ministero dell'Istruzione, dell'Universitá e della Ricerca (MIUR)* is acknowledged for the support of the project *"Analogue COmputing with Dynamic Switching Memristor Oscillators: Theory, Devices and Applications (COSMO)."* The support of the Politecnico di Torino under the Visiting Professors Program 2018 is also acknowledged.

We are especially indebted to Sung Mo (Steve) Kang, Ronald Tetzlaff, and R. Stanley Williams for their prolusion to the book, for thought-provoking criticisms of many chapters, and for suggestions to include relevant contents and examples. We are also grateful to our departments for providing an environment to write the book. Thanks to the staff of Springer for being efficient and cooperative during the production process. We especially enjoyed working with Maria Bellantone and Barbara Amorese displaying just the right combination of patience and gentle pressure.

Finally and most importantly, we thank our families for their encouragement. Your faith in us is a powerful driving force.

Memristor: Remembrance of Things Past

This part is mainly based on the work [21] written by L. O. Chua to present his personal scientific and historical account of memristor concept.

> Postulated in 1971, the memristor did not see the light of day until a serendipitous discovery at HP nearly four decades later. Here, I reminisce on the crisis that inspired me to develop an axiomatic nonlinear circuit theory where the memristor emerges naturally as the fourth basic circuit element.

The dawn of nonlinear electronics was ushered during the 1960s by a Cambrian-esque explosion of newly minted two-terminal electronic devices, bearing such intimidating monikers as Esaki diode, Josephson junction, varactor diode, thyristor, impact ionization avalanche transit time (IMPATT) diode, Gunn diode, and ovonic threshold switch. Trained in the old school of linear circuit analysis, hordes of electronics engineers were awed and shocked upon witnessing a parade of such strongly nonlinear and dynamical electronic devices unfolding at such a breathless rate. To many, the surreal proliferation of exotic devices would conjure the opening scene of Dickens's tale of yore: "It was the best of times, it was the worst of times, it was the age of wisdom, it was the age of foolishness, it was the epoch of belief, it was the epoch of incredulity, it was the season of Light, it was the season of Darkness, it was the spring of hope, it was the winter of despair, we had everything before us, we had nothing before...."

The surreal proliferation of these exotic devices was an exciting time full of opportunity and challenge. At this time, I was working on my PhD research to make sense of the cornucopia of exotic nonlinear devices. Rather than charting a taxonomy to pigeonhole them, I opted for an axiomatic definition of a few basic nonlinear circuit elements that could be used to model a broad variety of nonlinear devices. I joined the Purdue University upon graduation in 1964 and was assigned to revamp its outdated circuit analysis curriculum, thereby providing me an ideal launching pad for teaching nonlinear circuit theory through my device-independent black-box approach. The axioms that would predict the memristor had made their debut in the world's first textbook on nonlinear circuit theory [23] in 1969. It took a year for me to derive and prove mathematically the unique circuit-theoretic properties and memory attributes of this yet unnamed device, earning its accolade as the fourth circuit element [24, 25] and its justification for submission to the *IEEE Transactions on Circuit Theory* on November 25, 1970 [4]. Its publication in the following year coincided with my move to the University of California, Berkeley, to spearhead research in a new frontier dubbed nonlinear circuits and systems. The memristor was soon relegated to the back burner due to lack of research funding, where, like Rip Van Winkle, it would slumber until awakened. This was despite recognition by the IEEE of the potential of the memristor back in 1973, when they awarded me the prestigious IEEE W.R.G. Baker Prize Paper Award for the most outstanding paper reporting the original work in all IEEE publications.

To analyze circuits made of strongly nonlinear and dynamical electronic devices, it is necessary to have realistic device models made of well-defined nonlinear circuit elements as building blocks [23, 26], which did not exist then, because the electrical engineers from that bygone epoch were taught to overcome nonlinearities by expanding them in a Taylor series and then retaining only the linear term that neatly maps into a linear circuit model. But just like the ancient parable about the blind men and the elephant, such models invariably gave rise to grossly inaccurate and misleading results. Unfortunately, the "linearize then analyze" culture endowed upon the electronic engineers of the day had made it impossible to devise such generalizations due in part to the lack of a circuit-theoretic foundation for basic nonlinear circuit elements. Even Richard Feynman would sloppily use the old-fashioned name "condenser"—instead of capacitor—in his 1963 classic, "The Feynman Lectures on Physics," when he wrote on pages 6–12, vol. II, "The coefficient of proportionality is called the Capacity, and such a system of two conductors is called a Condenser" [27]. Electronic circuit theory is concerned only with the prediction of the voltage $v(t)$ and current $i(t)$ associated with the external terminals sticking out of an enclosing cocoon (henceforth called a black box) whose interior may contain some newly minted nonlinear electronic device or an interconnection of various electronic devices, batteries, and so on. And thus, basic nonlinear circuit elements must be defined from a black-box perspective, independent of their internal composition, material, geometry, and architecture.

From the circuit-theoretic point of view, the three basic two-terminal circuit elements are defined in terms of a relationship between two of the four fundamental circuit variables, namely, the current i, the voltage v, the charge q, and the flux φ. Out of the six possible combinations of these four variables, five have led to well-known relationships [23]. Two of these relationships are already given by $q(t) = \int_{-\infty}^{t} i(\tau)d\tau$ and $\varphi(t) = \int_{-\infty}^{t} v(\tau)d\tau$. Three other relationships are given, respectively, by the axiomatic definition of the three classical circuit elements, namely, the resistor (defined by a relationship between v and i), the inductor (defined by a relationship between φ and i), and the capacitor (defined by a relationship between q and v). Only one relationship remains undefined: the relationship between φ and q. From the logical as well as axiomatic points of view, it is necessary for the sake of completeness to postulate the existence of a fourth basic two-terminal circuit element, which is characterized by a $\varphi - q$ relationship (see [21, Fig. 1(a)]). This element is christened the memristor because it behaves like a nonlinear resistor with memory, remembering things past.

As a proof of principle, three memristors were built in 1969 using operational amplifiers and off-the-shelf electronic components. They demonstrated three distinct $\varphi - q$ curves on a bespoke memristor curve tracer that I had designed for this purpose [4]. However, the challenge of fabricating a passive monolithic memristor remained unfulfilled for 37 years until a team of scientists from the HP lab, under the leadership of Dr. R. Stanley Williams, reported in May 1, 2008, issue of *Nature* the world's first operational memristor made by sandwiching a thin film of titanium dioxide between platinum electrodes [5, 28]. The pinched hysteresis loop fingerprint of this seminal HP memristor is reproduced below the four-element quartet in [21, Fig. 1(a)]. It is truly remarkable that the pinched hysteresis loop is in fact an endearing signature that nature endowed upon all basic nonlinear circuit elements [29, 30] capable of remembering their past, including the memcapacitor and the meminductor that I identified to be lossless at the opening lecture of the first UC Berkeley Memristor and Memristive Systems Symposium in November 2008 [31].

Prior to Dr. Williams's *Nature* paper, no one understood how certain experimental two-terminal solid-state devices could remember and switch between two or more values of resistance without a power supply. The seminal *Nature* publication has provided a unifying foundation for all nonvolatile memory devices, which go by such names as ReRAM, PCRAM, STT-RAM, RRAM, MRAM, FRAM, PCM, and the Atomic Switch [32], where the device's high and low resistance states are used to code the 0 and 1 binary bits, instead of the conventional voltage or current, which collapses to zero when power is interrupted.

News of the nanoscale HP memristor has breathed new life into a device of antiquity that had been relegated to the dustbin of history. Overnight, it has triggered a torrent of research and development activities on memristors worldwide in both industry and academia (see Figure 1b, c in [21]), in anticipation of its disrupting potential in AI and neuromorphic computing. Companies such as Panasonic and Fujitsu have sold several hundred million chips with embedded memristor memory based on tungsten oxide since 2013, and Taiwan Semiconductor Manufacturing Company (TSMC) has announced that it has developed a 22-nm memristor crossbar process for ASIC embedded memories that will be available for mass production in 2019. Aside from its predicted eventual replacement of the over-extended flash memories, DRAMs, and even hard drives [2] with disrupting nonvolatile memristor technology, memristors can emulate synapses and ion channels in neurons and muscle fibers [33], sweat ducts in human skin, and even the primitive amoeba's amazing counting ability [29]. Moreover, the memristor's scalable diminutive physical size makes it the right stuff for building brain-like intelligent machines. Furthermore, because even plants can remember events and communicate through memristors [34], could memristors in fact be the *sine qua non* for emulating life itself?

I close my above reminiscence (which is an embellishment on a previous work [35]) by reproducing the last paragraph of my 1971 article, "Memristor-the missing circuit element" [4]. In hindsight, this passage foreshadowed the pinched hysteresis loop fingerprints not only of the ideal memristor predicted in the article but also of all memristors cited in my later works [29–31, 33, 36, 37]. *Although no physical memristor has yet been discovered in the form of a physical device without internal power supply, the circuit-theoretic and quasi-static electromagnetic analyses presented in Sections III and IV make plausible the notion that a memristor device with a monotonically increasing* $\varphi - q$ *curve could be invented, if not discovered accidentally. It is perhaps not unreasonable to suppose that such a device might already have been fabricated as a laboratory curiosity but was improperly identified! After all, a memristor with a simple* $\varphi - q$ *curve will give rise to a rather peculiar—if not complicated hysteretic—*$v - i$ *curve when erroneously traced in the current-versus-voltage plane. (Moreover, such a curve will change with frequency as well as with the tracing waveform.) Perhaps, our perennial habit of tracing the* $v - i$ *curve of any new two-terminal device has already misled some of our device-oriented colleagues and prevented them from discovering the true essence of some new device, which could very well be the missing memristor.*

Torino, Italy
Siena, Italy
Berkeley, CA, USA

Fernando Corinto
Mauro Forti
Leon O. Chua

References

1. M.M. Waldrop, The chips are down for Moore's law. Nat. News **530**(7589), 144 (2016)
2. R. Stanley Williams, What's next? [The end of Moore's law]. Comput. Sci. Eng. **19**(2), 7–13 (2017)
3. M.A. Zidan, J.P. Strachan, W.D. Lu, The future of electronics based on memristive systems. Nat. Electron. **1**(1), 22 (2018)
4. L.O. Chua, Memristor-The missing circuit element. IEEE Trans. Circuit Theory **18**(5), 507–519 (1971)
5. D.B. Strukov, G.S. Snider, D.R. Stewart, R.S. Williams, The missing memristor found. Nature **453**(7191), 80–83 (2008)
6. P. Mazumder, S.M. Kang, Waser, R. (Eds.), Special issue on memristors: devices, models, and applications. Proc. IEEE **100**(6), (2012)
7. R. Tetzlaff (Ed.)., *Memristors and Memristive Systems* (Springer, New York, 2014)
8. L. Chua, G. Sirakoulis, A. Adamatzky (Eds.), *Handbook of Memristor Networks. Vol. 1 and 2* (Springer, New York, 2019)
9. C. Li, Z. Wang, M. Rao, D. Belkin, W. Song, H. Jiang, P. Yan, Y. Li, P. Lin, M. Hu, et al., Long short-term memory networks in memristor crossbar arrays. Nat. Mach. Intell. **1**(1), 49 (2019)
10. O. Krestinskaya, A.P. James, L.O. Chua, Neuromemristive circuits for edge computing: a review. IEEE Trans. Neural Netw. Learn. Syst. 1–20 (2019). https://doi.org/10.1109/TNNLS.2019.2899262
11. A. Pérez-Tomás, Functional oxides for photoneuromorphic engineering: toward a solar brain. Adv. Mater. Interfaces 1900471 (2019).
12. S.H. Jo, T. Chang, I. Ebong, B.B. Bhadviya, P. Mazumder, W. Lu, Nanoscale memristor device as synapse in neuromorphic systems. Nano Lett. **10**(4), 1297–1301 (2010)
13. S. Kumar, J.P. Strachan, R. Stanley Williams, Chaotic dynamics in nanoscale NbO_2 Mott memristors for analogue computing. Nature **548**(7667), 318 (2017)
14. F. Corinto, M. Forti, Memristor circuits: flux–charge analysis method. IEEE Trans. Circuits Syst. I Regul. Pap. **63**(11), 1997–2009 (2016)
15. F. Corinto, M. Forti, Memristor circuits: bifurcations without parameters. IEEE Trans. Circuits Syst. I Regul. Pap. **64**(6), 1540–1551 (2017)
16. F. Corinto, M. Forti, Memristor circuits: pulse programming via invariant manifolds. IEEE Trans. Circuits Syst. I Regul. Pap. **65**(4), 1327–1339 (2018)
17. F. Corinto, M. Forti, Complex dynamics in arrays of memristor oscillators via the flux–charge method. IEEE Trans. Circuits Syst. I Regul. Pap. **65**(3), 1040–1050 (2017)
18. F. Corinto, M. Gilli, M. Forti, Flux-charge description of circuits with non-volatile switching memristor devices. IEEE Trans. Circuits Syst. II Expr. Briefs **65**(5), 642–646 (2018)

19. F. Corinto, M. Di Marco, M. Forti, L. Chua, Nonlinear networks with mem-elements: complex dynamics via flux-charge analysis method. IEEE Trans. Cybern. 1–14 (2019). https://doi.org/10.1109/TCYB.2019.2904903
20. M. Di Marco, M. Forti, L. Pancioni, Memristor standard cellular neural networks computing in the flux–charge domain. Neural Netw. **93**, 152–164 (2017)
21. L. Chua, Memristor: remembrance of things past. IEEE Micro **38**(5), 7–12 (2018)
22. L.O. Chua, C.A. Desoer, E.S. Kuh, *Linear and Nonlinear Circuits* (McGraw-Hill, New York, 1987)
23. L.O. Chua, *Introduction to Nonlinear Network Theory* (McGraw-Hill, New York, 1969)
24. J.M. Tour, T. He, The fourth element. Nature **453**(7191), 42–43 (2008)
25. L.O. Chua, The fourth element. Proc. IEEE **100**(6), 1920–1927 (2012)
26. L.O. Chua, Nonlinear circuit foundations for nanodevices. I. The four-element torus. Proc. IEEE **91**(11), 1830–1859 (2003)
27. R. Feynman, R.B. Leighton, M. Sands, *The Feynman Lectures on Physics* (Addison-Wesley, Boston, 1963)
28. R. Stanley Williams, How we found the missing memristor. IEEE Spectr. **45**(12), 28–35 (2008)
29. L. Chua, If it's pinched it's a memristor. Semiconduct. Sci. Technol. **29**(10), 104001 (2014)
30. L. Chua, Everything you wish to know about memristors but are afraid to ask. Radioengineering **24**(2), 319–368 (2015)
31. L. Chua, Introduction to memristors (2009). IEEE, New York. https://ieeexplore.ieee.org/xpl/articleDetails.jsp?arnumber=EDP091$&$contentType=Education+$\delimiter"026E30F%26$+Learning
32. K. Terabe, T. Hasegawa, T. Nakayama, M. Aono, Quantized conductance atomic switch. Nature **433**(7021), 47 (2005)
33. L. Chua, V. Sbitnev, H. Kim, Hodgkin–Huxley axon is made of memristors. Int. J. Bifurc. Chaos **22**(3), 1230011 (2012)
34. A.G. Volkov, Memristors and electrical memory in plants. Memory Learn. Plants, 139–161 (2018)
35. L.O. Chua, How we predicted the memristor. Nat. Electron. **1**, 322 (2018)
36. L. Chua, Resistance switching memories are memristors. Appl. Phys. A **102**(4), 765–783 (2011)
37. L. Chua, Memristor, Hodgkin–Huxley, and edge of chaos. Nanotechnology **24**(38). 383001 (2013)

Contents

List of Symbols

$\mathbb{R}$	Set of real numbers
$\mathbb{R}^n$	Set of n-dimensional vectors
$\mathbf{x}$	$= (x_1, x_2, \ldots, x_n)^T \in \mathbb{R}^n$, vector in $\mathbb{R}^n$
$\mathbf{x}^T$	Transpose of vector $\mathbf{x}$
$\mathbf{M}$	$= (m_{ij}) \in \mathbb{R}^{n \times m}$, $n \times m$ matrix
$\mathbf{M}^T$	Transpose of matrix
$\mathbf{M}^{-1}$	Inverse of matrix
$\det \mathbf{M}$	Determinant of matrix
$\mathrm{tr}\mathbf{M}$	Trace of matrix
$f(x)$	$: \mathbb{R} \rightarrow \mathbb{R}$, function of scalar variable x
$f'(x)$	Derivative of f with respect to its argument x
$f^{-1}(x)$	Inverse function
$\mathbf{f}(\mathbf{x})$	$: \mathbb{R}^n \rightarrow \mathbb{R}^m$, vector function
$\mathbf{J}_\mathbf{f}$	$= (\partial f_i / \partial x_j)$, Jacobian of $\mathbf{f}$
∇f	$= (\partial f/\partial x_1, \partial f/\partial x_2 \ldots, \partial f/\partial x_n)^T$, gradient of $f(\mathbf{x}) : \mathbb{R}^n \rightarrow \mathbb{R}$
j	$= \sqrt{-1}$
Re	Real part
Im	Imaginary part
$\forall$	For all
$\in$	Is a member of
$\cap$	Intersection
$\doteq$	Definition
$\dot{x}(t)$	Derivative with respect to time t. The explicit notation $dx(t)/dt$ is also used
$\ddot{x}(t)$	$= d^2x(t)/dt^2$
ODE	Ordinary differential equation
DAE	Differential algebraic equation
SE	State equation
IVP	Initial value problem
EP	Equilibrium point

IC	Initial condition
BWP	Bifurcation without parameters
1D	One dimensional
i	Current
A	Ampere
v	Voltage
V	Volts
$p(t)$	Instantaneous power at t
W	Watt
q	Charge, or current momentum
C	Coulomb
φ	Flux, or voltage momentum
Wb	Weber
Ω	Ohm
S	Siemens
F	Farad
H	Henry
J	Joule
ac	Alternate current
dc	Direct current
σ	Charge momentum
ρ	Flux momentum
$v^{(\alpha)}$	where α is an integer, α-derivative (or integral) of voltage v
$i^{(\beta)}$	where β is an integer, β-derivative (or integral) of current i
KCL	Kirchhoff current law
KVL	Kirchhoff voltage law
KqL	Kirchhoff charge law
KφL	Kirchhoff flux law
$\mathcal{G}$	Graph
$\mathcal{C}$	Cut-set
$\mathcal{T}$	Tree
Q	Fundamental cut-set matrix
A	Reduced incidence matrix
B	Fundamental loop matrix
(v, i)-domain	Voltage-current domain
(φ, q)-domain	Flux-charge domain
t_0	Initial instant for transient analysis
$q(t; t_0)$	$= q(t) - q(t_0)$, incremental charge
$\varphi(t; t_0)$	$= \varphi(t) - \varphi(t_0)$, incremental flux
CR	Constitutive relation
FCAM	Flux-charge analysis method
CMOS	Complementary metal-oxide-semiconductor
ASIC	Application-specific integrated circuit
CPU	Central processing unit
RAM	Random-access memory

DRAM	Dynamic random-access memory
PCM	Phase-change memory
NVM	Non-volatile memory
STDP	Spike-Timing-Dependent Plasticity
CNN	Cellular neural network
MIM	Metal-Insulator-Metal structure
IoT	Internet of Things
HP	Hewlett Packard

Part I
Foundation of Nonlinear Circuit Theory

Chapter 1
Device Modeling and Circuit Elements

This chapter discusses a device-independent black-box approach to model a broad variety of physical devices. For the purpose of network theory, a circuit element can be considered as a black-box, whose electrical behavior is defined in terms of a mathematical model (i.e., a set of algebraic and/or differential and/or integral equations), relating currents and voltages at various terminals of the device. The physical means required to implement the black-box are irrelevant. The circuit element is therefore an ideal entity, corresponding to the best abstraction and to the most suited description of a physical device.

1.1 Axiomatic Approach to Device Modeling

Device modeling is more of an "art" than science. Since the aim of any model is to "mimic" as accurately as possible a physical device $\mathcal{D}$, it is crucial to identify all important properties and behaviors of $\mathcal{D}$. Most existing models of physical devices have been derived via two basic approaches: the *physical approach* and the *black-box approach*. The physical approach relies on a careful study of *physics* and *operating mechanisms* of the device (e.g., solid-state devices as Gunn diode, Josephson junction, etc.). On the other hand, the black-box approach is fundamental when the device physics and operating mechanisms are so complex that the physical approach results to be impractical (e.g., bio-physical systems as membranes, nerves, etc.). In either approach, a mathematical description (e.g., a system of nonlinear *algebraic* and *differential* equations) which is capable of simulating most of the observed properties of a device is essential. This crucial step is where most of the "art" is involved.

Once the mathematical description of $\mathcal{D}$ is obtained, a *unified theory of device modeling* can be developed by an axiomatic approach based on the concept of basic

F. Corinto et al., *Nonlinear Circuits and Systems with Memristors*,
https://doi.org/10.1007/978-3-030-55651-8_1

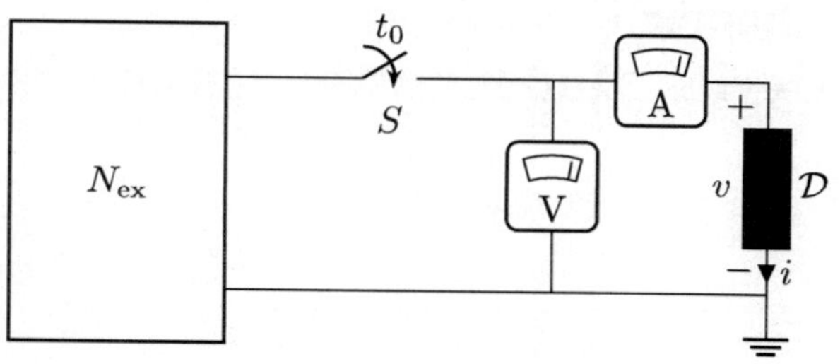

Fig. 1.1 *Gedanken experiment* for the axiomatic definition of a two-terminal circuit element

circuit elements [1]. In the following, the fundamental principles of device modeling via nonlinear circuit elements are summarized.

1.1.1 Device Model and Circuit Elements

Let us consider a two-terminal physical device $\mathcal{D}$ (e.g., an electric contrivance, a solid-state nanoscale device, a biological system, etc.) having two accessible terminals where the voltage $v(t)$ can be measured via a voltmeter and the current $i(t)$ via an ammeter.[1] Such a physical device $\mathcal{D}$ is also referred to as a black-box since its interior is not necessarily known.

The chief issue is how to characterize a two-terminal black-box $\mathcal{D}$ such that its response due to any electrical signal can be estimated via a *device model*. To this aim let us think a *Gedanken Experiment* as in Fig. 1.1 where $\mathcal{D}$ is connected to any *excitation network* N_{ex} (also named Gedanken Probing Circuit) made of an arbitrary interconnection of devices, batteries or time-dependent sources. The switch S is closed at some initial instant t_0 and the voltage $v(t)$ and current $i(t)$ are measured for all times $t \geq t_0$. Any pair of measured waveforms $(v(t), i(t))$ for $t \geq t_0$ is named an *Admissible* (v, i) *Signal* (AVIS) pair associated with the considered device $\mathcal{D}$ subject to an excitation network.

The collection of all possible AVIS pairs associated with $\mathcal{D}$ obtained by repeating the *Gedanken Experiment* for any possible excitation network and any initial instant t_0 is named AVIS-pad, or simply A-pad, and is denoted by

$$\mathcal{F}(\mathcal{D}) = \{(v_1(t), i_1(t)), (v_2(t), i_2(t)), \ldots, (v_n(t), i_n(t)), \ldots\}.$$

In principle, $\mathcal{F}(\mathcal{D})$ completely characterizes the behavior of $\mathcal{D}$ in the sense that, given any excitation $v(t)$ (resp., $i(t)$) and t_0, the corresponding $i(t)$ (resp., $v(t)$) can

[1]Note that the voltage $v(t)$ and the current $i(t)$ are defined axiomatically via two instruments called voltmeter and ammeter, without invoking any physical concepts such as electric field, magnetic field, charge, flux linkage, etc. One does not even have to know how a voltmeter, or an ammeter, works. They are just names assigned to the instruments.

be retrieved via $\mathcal{F}(\mathcal{D})$. In other words, $\mathcal{F}(\mathcal{D})$ acts as an analog memory bank, thus a lookup table containing the results of all possible experiments on $\mathcal{D}$.

Since it is impossible to measure and then store the results of an infinite number of experiments, it is clearly impractical to use $\mathcal{F}(\mathcal{D})$ to characterize $\mathcal{D}$, i.e., the above *Gedanken Experiment* is only a thought experiment. However, a large number of real-world two-terminal devices $\mathcal{D}$ can be described via mathematical equations, i.e., a *mathematical model* of $\mathcal{D}$, obtained either by a physical approach or a black-box approach, consisting of one or more (algebraic, differential, or integral) equations that provide an accurate approximation to the corresponding measured admissible signal pairs $(v(t), i(t))$. This leads to the conceptual definition of an "ideal" *circuit element* whose associated AVIS pairs are *generated* (rather than measured) via some prescribed mathematical model referred to as Constitutive Relation (or CR for short).[2]

Example 1.1 (Linear Resistor) A particular class of physical devices $\mathcal{R}$ are described as a black-box with a mathematical equation (Ohm's law)

$$v(t) = Ri(t)$$

or in implicit form

$$f_R(v, i) = v - Ri = 0 \tag{1.1}$$

where R in Ohm (Ω) is a parameter named *resistance*. In this case the mathematical model (1.1) permits to generate the A-pad

$$\mathcal{F}(\mathcal{R}) = \{(Ri_1(t), i_1(t)), (Ri_2(t), i_2(t)), \ldots, (Ri_n(t), i_n(t)), \ldots\}.$$

In circuit theory, (1.1) is the CR of an (ideal) circuit element referred to as *linear resistor*. Along this line of reasoning the CR (1.1) can be thought of as a compact mathematical model for generating the infinite set $\mathcal{F}(\mathcal{R})$ of a physical resistor.

1.1.2 Remarks on the Concept of a Device Model

A model of a device is not an equivalent circuit (i.e., an arbitrary interconnection of circuit elements) of $\mathcal{D}$ because no real device can be exactly mimicked by a circuit or mathematical model. In fact, depending on the application (e.g., amplitude and frequency of operation), a given device may have many distinct models. As reported in [1]: *"There is no "best model" for all occasions. The best model in a given situation is the simplest model capable of yielding realistic solutions."*

---

[2]From now on the term device will be used only when referring to a physical device, while any "ideal" device will be referred to as a circuit element, or simply an element.

A circuit element defined mathematically by a CR should ideally display the same qualitative behavior that a device $\mathcal{D}$ exhibits when connected to an excitation network and it should have *predictive ability*, i.e., it should be capable of predicting previously unknown operating modes through computer simulation. Readers interested in a detailed discussion of the main qualitative properties (e.g., well-posedness, simulation capability, qualitative similarity, predictive ability, and structural stability) that a circuit model of a device $\mathcal{D}$ should possess can refer to the fundamental article [1].

The mathematical equations describing a circuit element are in general nonlinear and may include a combination of algebraic equations, ordinary differential equations, partial differential equations, and integral equations. A two-terminal circuit element is said to be *lumped* if and only if its CR can be expressed by a finite number of equations involving only algebraic, ordinary differentiation, and integration operations on the instantaneous values of the terminal variables $\{v, i\}$ and/or a finite number of additional internal variables $\{x_1, x_2, \ldots, x_n\}$. Otherwise, the two-terminal circuit element is said to be *distributed*.

From now on the book is concerned with *lumped circuit elements* only. The CR of a lumped circuit element may involve not only $v(t)$ and $i(t)$ but also their higher-order derivatives and integrals defined recursively as follows for any integers α and β.

- If $\alpha > 0$ and $\beta > 0$

$$v^{(\alpha)}(t) = \frac{d}{dt}\left(v^{(\alpha-1)}(t)\right), \quad \alpha = 1, 2, \ldots, \tag{1.2a}$$

$$i^{(\beta)}(t) = \frac{d}{dt}\left(i^{(\beta-1)}(t)\right), \quad \beta = 1, 2, \ldots, \tag{1.2b}$$

- if $\alpha < 0$ and $\beta < 0$

$$v^{(\alpha)}(t) = \int_{-\infty}^{t} v^{(\alpha+1)}(\tau)d\tau, \quad \alpha = -1, -2, \ldots \tag{1.3a}$$

$$i^{(\beta)}(t) = \int_{-\infty}^{t} i^{(\beta+1)}(\tau)d\tau, \quad \beta = -1, -2, \ldots \tag{1.3b}$$

where $v^{(0)}(t) = v(t)$ and $i^{(0)}(t) = i(t)$ are the voltage and the current, respectively.

Important subclasses of circuit elements are described by CRs that also include the terminal variables $v^{(\alpha)}$ and $i^{(\beta)}$ with $\alpha = -1$ and $\beta = -1$, thus

$$\begin{aligned} v^{(-1)}(t) = \varphi(t) &= \int_{-\infty}^{t} v(\tau)d\tau = \int_{-\infty}^{t_0} v(\tau)d\tau + \int_{t_0}^{t} v(\tau)d\tau \\ &= \varphi(t_0) + \int_{t_0}^{t} v(\tau)d\tau \end{aligned} \tag{1.4}$$

and

$$i^{(-1)}(t) = q(t) = \int_{-\infty}^{t} i(\tau)d\tau = \int_{-\infty}^{t_0} i(\tau)d\tau + \int_{t_0}^{t} i(\tau)d\tau = q(t_0) + \int_{t_0}^{t} i(\tau)d\tau \tag{1.5}$$

where $\varphi(t)$ and $q(t)$ are the associated *flux* and *charge*, respectively. Flux and charge, a.k.a. *voltage momentum* and *current momentum* [2], respectively, can in principle be measured via suitable ballistic instruments called a flux-meter and a charge-meter, provided the initial values $\varphi(t_0)$ and $q(t_0)$ are known.[3]

It is worth to note that although $q(t)$ and $\varphi(t)$ are given the names charge and flux, respectively, they need not be associated with a real physical charge as in the case of a classical capacitor built by sandwiching a pair of parallel plates between an insulator, or a real physical flux as in the case of a classical inductor built by winding a copper wire around an iron core.

In addition, $q(t)$ and $\varphi(t)$ are related to $i(t)$ and $v(t)$, respectively, by (1.2) with $\alpha = 0$ and $\beta = 0$, thus

$$v(t) = \frac{d}{dt}\left(v^{(-1)}(t)\right) = \frac{d\,\varphi(t)}{dt} \tag{1.6}$$

and

$$i(t) = \frac{d}{dt}\left(i^{(-1)}(t)\right) = \frac{d\,q(t)}{dt}. \tag{1.7}$$

In the next section the terminal variables $\{v, i, \varphi, q\}$ allow us to introduce an axiomatic definition of the basic circuit elements.

1.2 Four Basic Two-Terminal Circuit Elements

Let us consider the four fundamental electric variables $\{v, i, \varphi, q\}$ associated with a two-terminal circuit element. There are six distinct pairwise combinations of $\{v, i, \varphi, q\}$, but two of them, namely, $\{v, \varphi\}$ and $\{i, q\}$, are by definition dependent relationships according to (1.4) and (1.5) (or (1.6) and (1.7)). The remaining four combinations $\{v, i\}$, $\{q, v\}$, $\{\varphi, i\}$, and $\{\varphi, q\}$ are instead *a priori* unrelated. Clearly, any mathematical relationship between each pair of electrical variables needs to be established solely by the considered two-terminal circuit element.

Example 1.2 (Nonlinear Resistor) An important class of devices is modeled by a circuit element, called *nonlinear resistor*, defined by a CR given in implicit form by

[3] In practice one can never know an AVIS pair over the infinite past. Measurements are set up to begin at some initial time t_0. Consequently, $\varphi(t_0)$ and $q(t_0)$ represent a summary of the history of v' and i, respectively, measured at $t = t_0$.

$f_R(v, i) = 0$, where, differently from the linear resistor in Example 1.1, f_R is now a nonlinear function of voltage v and current i. The nonlinear resistor is said to be current-controlled (resp., voltage controlled) in the case the CR is given in explicit form as $v = \hat{v}(i)$ (resp., $i = \hat{i}(v)$). In the current-controlled case the corresponding A-pad results to be

$$\mathcal{F}(\mathcal{R}) = \{(\hat{v}(i_1(t)), i_1(t)), (\hat{v}(i_2(t)), i_2(t)), \ldots, (\hat{v}(i_n(t)), i_n(t)), \ldots\}.$$

Since both Examples 1.1 and 1.2 involve the same pair of circuit variables (v, i), all two-terminal devices $\mathcal{R}$ modeled by a circuit element with CR $f_R(v, i) = 0$ are named *resistors*.

This notwithstanding, most two-terminal devices cannot be described by a CR between the variable pair $(v(t), i(t))$. Important subclasses can be expressed by a relationship between the variable pairs $(q(t), v(t))$ or $(\varphi(t), i(t))$.

Example 1.3 (Nonlinear Capacitor) A subclass of two-terminal devices can be characterized by an A-pad involving the pair $(q(t), v(t))$, namely

$$\mathcal{F}(\mathcal{C}) = \{(\hat{q}(v_1(t)), v_1(t)), (\hat{q}(v_2(t)), v_2(t)), \ldots, (\hat{q}(v_n(t)), v_n(t)), \ldots\}.$$

Such $\mathcal{F}(\mathcal{C})$ permits to define an ideal circuit element called voltage-controlled *nonlinear capacitor* whose CR is given by the algebraic equation $f_C(q(t), v(t)) = q(t) - \hat{q}(v(t)) = 0$, or $q(t) = \hat{q}(v(t))$.

Example 1.4 (Nonlinear Inductor) Another important subclass of two-terminal devices can be characterized by an A-pad involving the electric variable pair $(\varphi(t), i(t))$, namely

$$\mathcal{F}(\mathcal{L}) = \{(\hat{\varphi}(i_1(t)), i_1(t)), (\hat{\varphi}(i_2(t)), i_2(t)), \ldots, (\hat{\varphi}(i_n(t)), i_n(t)), \ldots\}.$$

Such $\mathcal{F}(\mathcal{L})$ permits to define an ideal circuit element called current-controlled *nonlinear inductor* whose CR is given by the algebraic equation $f_L(\varphi(t), i(t)) = \varphi(t) - \hat{\varphi}(i(t)) = 0$, or $\varphi(t) = \hat{\varphi}(i(t))$.

The subclass of two-terminal devices characterized in terms of the variable pair (v, q) (resp., (φ, i)) is modeled by a circuit element called *capacitor* (resp., *inductor*) with CR $f_C(v, q) = 0$ (resp., $f_L(\varphi, i) = 0$). For *logical consistency*, and *symmetry considerations*, it is necessary to define a fourth circuit element via the CR

$$f_M(\varphi(t), q(t)) = 0 \tag{1.8}$$

between the variable pair (φ, q). This element was postulated and named the *memristor* (acronym for memory-resistor) by Chua in the seminal paper [3]. A physical device $\mathcal{D}$ described by such a circuit element has been fabricated in 2008 as a $\mathrm{TiO_2}$ nanodevice at HP laboratories [4] (see also Chap. 2).

Summing up, the above axiomatic approach allows to give a systematic definition of the four basic circuit elements.

Definition 1.1 (Resistor, Capacitor, Inductor, and Memristor) A two-terminal circuit element is called resistor, capacitor, inductor, and memristor if and only if its CR can be expressed by an algebraic relationship involving only the variable pairs (v, i), (φ, i), (q, v), and (φ, q), respectively. In particular, the CR

- $f_R(v, i) = 0$ defines a resistor
- $f_C(q, v) = 0$ defines a capacitor
- $f_L(\varphi, i) = 0$ defines an inductor
- $f_M(\varphi, q) = 0$ defines a memristor.

Recall that an algebraic relation is any equation involving strictly algebraic operations only. In particular, no differentiation, integration, or time delay operation may be involved.

The four basic circuit elements are graphically represented in Fig. 1.2, along with their respective symbols [5]. Note that the standard symbols for resistor, capacitor, and inductor are enclosed by a thin rectangle with a dark band at the bottom because it is essential to distinguish the reference polarity of each nonlinear element if its CR is not odd symmetric. The reference direction for each terminal current and reference voltage polarity of each terminal voltage can be chosen arbitrarily. However, it is a standard procedure in circuit theory to coordinate them as indicated in Fig. 1.1. This choice is referred to as the passive convention since the product $p(t) = v(t)i(t)$ yields the instantaneous electric power delivered to $\mathcal{D}$. The opposite choice to coordinate v and i is named active convention.

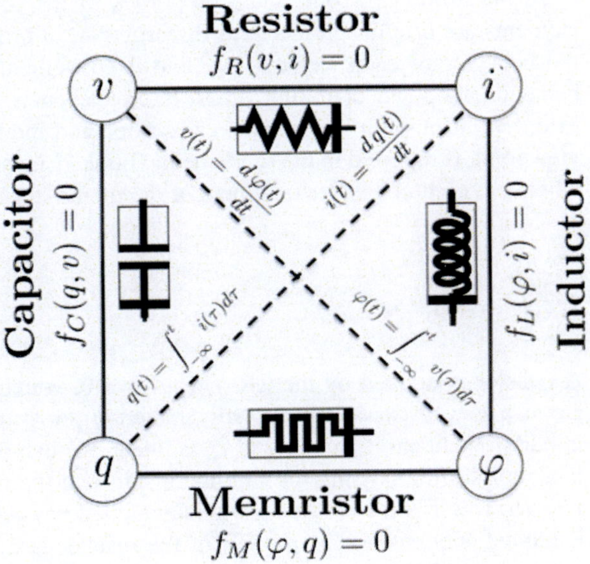

Fig. 1.2 Diagram of four basic circuit elements including the corresponding CRs and symbols

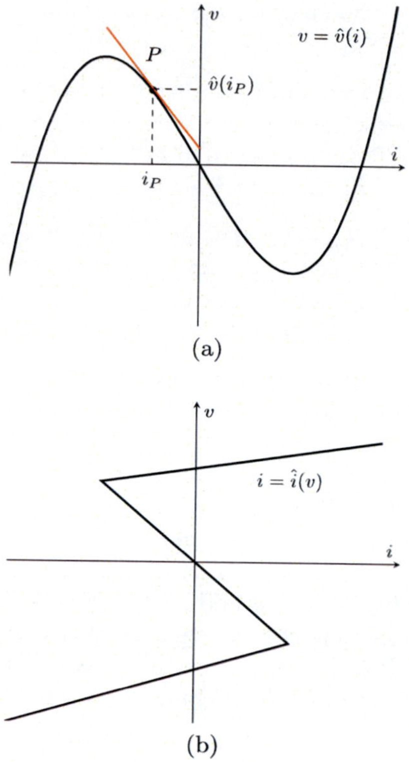

Fig. 1.3 (**a**) Current-controlled characteristic of a nonlinear resistor in the v-i plane. Point P on the characteristic and tangent straight line at P. The slope of the line corresponds to the differential resistance at P. (**b**) Voltage-controlled characteristic of a nonlinear resistor

In the following, some main properties of the four basic two-terminal circuit elements are briefly summarized. In particular, in order to gain physical insight in the behavior of each element, we find it convenient to examine its *small-signal* behavior about an operating point P on the curve associated with its CR. An extensive discussion of resistors, capacitors, and inductors, also from an energetic viewpoint, is reported in the fundamental book "Linear and Nonlinear Circuits" [5], whereas a comprehensive treatment of the memristor is reported in Chap. 2.

1.2.1 Resistor

A resistor is defined by the CR $f_R(v, i) = 0$, which in general corresponds to a curve, a.k.a., *resistor characteristic*, in the v–i (or i–v) plane (Fig. 1.3). The resistor is said to be linear if and only if f_R is linear. In such case the resistor characteristic is a straight line passing through the origin and the resistor obeys the Ohm's law $v = Ri$, where R is a constant parameter named *resistance*. Its reciprocal, $G = 1/R$, is named conductance. If the CR of the resistor is not a linear function, then the resistor is said to be nonlinear.

As already stated in Example 1.2, a nonlinear resistor is said to be current-controlled if it is possible to explicitly write $v = \hat{v}(i)$, i.e., the voltage is a (single-valued) function of the current (Fig. 1.3a). Similarly, a nonlinear resistor is voltage-controlled if it is possible to explicitly write $i = \hat{i}(v)$, i.e., the current is a (single-valued) function of the voltage (Fig. 1.3b).

The power delivered to the resistor at time t by the remainder of the circuit to which it is connected is given by $p(t) = v(t)i(t)$. Obviously, $p(t) \geq 0$ if and only if $v(t)$ and $i(t)$ have the same sign for all t. We call a two-terminal resistor *passive* if and only if its characteristic lies in the closed first and third quadrants of the $v - i$ plane or the $i - v$ plane. A resistor is said to be *active* if it is not passive.

Consider a current-controlled nonlinear resistor $v = \hat{v}(i)$. The resistor is passive if and only if $i\hat{v}(i) \geq 0$ for any i. It is said to be *eventually passive* if it absorbs power for all large values of the current, i.e., there exists $\tilde{i} > 0$ such that $i\hat{v}(i) \geq 0$ for any $|i| \geq \tilde{i}$.

Let us consider an operating point $P = (i_P, \hat{v}(i_P))$, i.e., a point on the characteristic of a current-controlled nonlinear resistor. The slope $R(i_P)$ of the characteristic at P is named *small-signal* or *differential* resistance of the resistor at P, i.e.,

$$R(i_P) = \hat{v}'(i_P) = \left.\frac{d\hat{v}(i)}{di}\right|_{i_P}.$$

The resistor is said to be locally passive at P if we have $R(i_P) \geq 0$. Otherwise, if $R(i_P) < 0$, it is said to be locally active at P.

In the particular case where the CR is a horizontal line, i.e., $v(t) = E = \text{const}$, or a vertical line, i.e., $i(t) = I = \text{const}$, the resistor coincides with a (constant) voltage source or a current source, respectively. A voltage source such that $E = 0$ is called a short circuit, whereas a current source such that $I = 0$ is called an open circuit.

It is remarkable that several commercially available two-terminal devices are described by a nonlinear function of voltage and current, e.g., pn-junction diodes, zener diodes, varistors, tunnel diodes, etc. Such devices can be realistically modeled in a broad range of voltages, currents, and frequencies, as two-terminal nonlinear resistors.

Example 1.5 (pn-Junction Diode) The pn-junction diode is modeled by a voltage-controlled nonlinear resistor with CR given by the "diode junction law"

$$i(t) = I_s\left(\exp\left(v(t)/V_T\right) - 1\right)$$

where I_s is the *reverse saturation current* (a constant on the order of microamperes) and V_T is the *thermal voltage* ($V_T = 26$ mV at room temperature). The characteristic is reported in Fig. 1.4. Note that the pn-junction diode has a monotone strictly increasing characteristic $i = \hat{i}(v)$, hence the differential resistance is always positive and the pn-junction diode is locally passive at any operating point P. Moreover, the characteristic is also current controlled in the interval $(-I_s, +\infty)$.

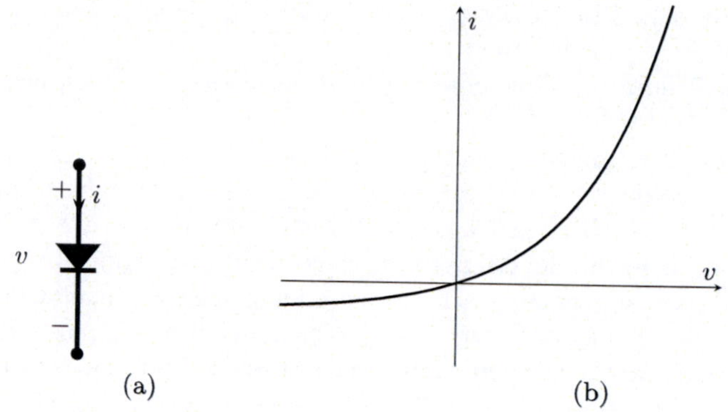

Fig. 1.4 (**a**) Symbol and (**b**) nonlinear characteristic of a pn-junction diode

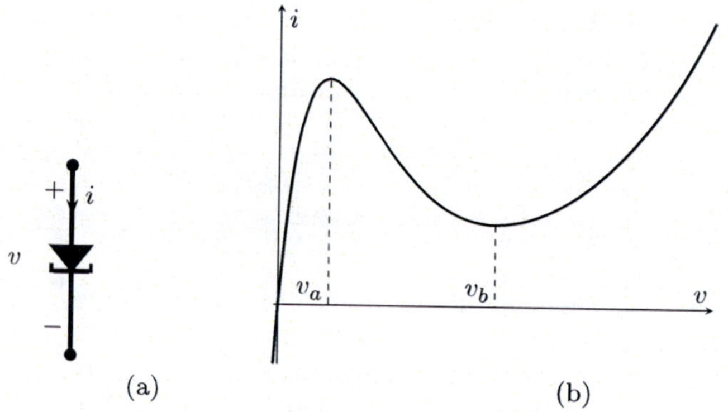

Fig. 1.5 (**a**) Symbol and (**b**) nonlinear characteristic of a tunnel diode

Example 1.6 (Tunnel Diode) A tunnel diode is a semiconductor diode invented in 1957 by Leo Esaki that is based on the electron tunneling effect. It can be modeled by a nonlinear voltage-controlled, but not current-controlled, resistor with a *non-monotonic* characteristic as shown in Fig. 1.5.

One truly peculiar property of a tunnel diode is that there is an interval of voltages, namely $v \in (v_a, v_b)$, for which the characteristic has a negative slope, i.e., the diode displays a negative differential resistance. In other words, it is locally active at any operating point $P = (v, \hat{i}(v))$ such that $v \in (v_a, v_b)$. Local activity makes it possible to use a tunnel diode to implement nonlinear oscillators. We also observe that nonmonotone resistor characteristics might lead to mathematical problems in writing the dynamic equations of circuits containing such resistors and inductor or capacitors (cf. Chap. 4).

1.2.2 Capacitor

A capacitor is defined by the CR $f_C(q, v) = 0$, which corresponds to a curve, a.k.a., *capacitor characteristic*, in the q–v (or v–q) plane. The capacitor is linear if and only if f_C is linear. In such case the characteristic is a straight line passing through the origin and the capacitor satisfies $q = Cv$, where C is a constant parameter named *capacitance*. The description of a capacitor in terms of voltage v and current i results to be $i = C\frac{dv}{dt}$, which is the classical CR of a linear capacitor. If the CR of the capacitor is not a linear function, then the capacitor is said to be nonlinear.

The capacitor is voltage-controlled if it is possible to explicitly write $q = \hat{q}(v)$, i.e., the charge is a (single-valued) function of the voltage. Similarly, it is said to be charge-controlled if it is possible to write $v = \hat{v}(q)$, i.e., the voltage is a (single-valued) function of the charge.

By considering a voltage-controlled capacitor, the slope $C(v_P) = \hat{q}'(v_P)$ of the characteristic at an operating point $P = (v_P, \hat{q}(v_P))$ is named *small-signal* or *differential capacitance* of the capacitor at P. Note that, in terms of current and voltage, a nonlinear capacitor exhibits a CR in differential form

$$i = C(v)\frac{dv}{dt}$$

where $C(v)$ is the small-signal capacitance at v.

The capacitor is said to be locally passive at P if we have $C(v_P) \geq 0$. Otherwise, if $C(v_P) < 0$, it is said to be locally active at P.

Example 1.7 (Varactor Diode) Varactor diodes are widely used in communication systems. The physical structure is a pn-junction diode designed to exploit the capacitive phenomena associated with the depletion layer. The physical approach permits to derive that the charge accumulated on the top layer is equal to

$$q(t) = -K\sqrt{V_0 - v(t)} = \hat{q}(v(t)), \quad \forall v < V_0$$

where $V_0 \in (0.2\,V, 0.9\,V)$ is the contact potential and K is a constant parameter related to the physical and geometrical parameters of the semiconductor structure. The varactor diode results to be a voltage-controlled capacitor for $v < V_0$ with a small-signal capacitance

$$C(v) = \hat{q}'(v) = \frac{1}{2}K\frac{1}{\sqrt{V_0 - v}}, \quad \forall v < V_0.$$

For all values $v > V_0$ the q–v characteristic is not defined and the varactor diode behaves like a nonlinear resistor.

1.2.3 Inductor

An inductor is defined by the CR $f_L(\varphi, i) = 0$, which corresponds to a curve, a.k.a., *inductor characteristic*, in the φ–i (or i-φ) plane. The inductor is linear if and only if f_L is linear. In such case the characteristic is a straight line passing through the origin and the inductor satisfies $\varphi = Li$, where L is a constant parameter named *inductance*. The description of an inductor in terms of voltage v and current i results to be $v = L\frac{di}{dt}$, which is the classical CR of a linear inductor. If the CR of the inductor is not a linear function, then the inductor is said to be nonlinear.

The inductor is current-controlled if it is possible to explicitly write $\varphi = \hat{\varphi}(i)$, i.e., the flux is a (single-valued) function of the current. Similarly, it is said to be flux-controlled if it is possible to explicitly write $i = \hat{i}(\varphi)$, i.e., the current is a (single-valued) function of the flux.

By considering a current-controlled inductor, the slope $L(i_P) = \hat{\varphi}'(i_P)$ of the characteristic at an operating point $P = (i_P, \hat{\varphi}(i_P))$ is named *small-signal* or *differential inductance* of the inductor at P. Note that, in terms of current and voltage, a nonlinear inductor exhibits a CR in differential form

$$v = L(i)\frac{di}{dt}$$

where $L(i) = \hat{\varphi}'(i)$ is the small-signal inductance .

The inductor is said to be locally passive at P if we have $L(i_P) \geq 0$. Otherwise, if $L(i_P) < 0$, it is said to be locally active at P.

Example 1.8 (Josephson Junction) The Josephson junction is made up of two superconductors separated by an insulating layer. Superconductors physics permits to derive the algebraic equation describing the device in terms of the current and the flux

$$i(t) = \hat{i}(\varphi(t)) = I_0 \sin(k\varphi(t)) \tag{1.9}$$

where I_0 is a device parameter and $k = 4\pi(e/h)$ (e = electron charge and h = Planck's constant). The Josephson junction is a flux-controlled (but not current-controlled) inductor with a characteristic as shown in Fig. 1.6 and a reciprocal small-signal inductance (in H^{-1})

$$\Gamma(\varphi) = \hat{i}'(\varphi) = kI_0 \cos(k\varphi).$$

It is important to observe that the physical variable φ is proportional to the difference between two quantum mechanical phases, and is not related to the familiar magnetic flux in a winding coil known from electromagnetic field theory.

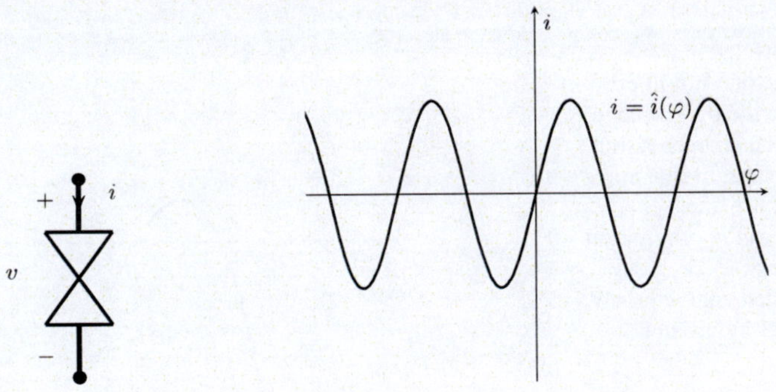

Fig. 1.6 (**a**) Symbol and (**b**) nonlinear characteristic of a Josephson junction

1.2.4 Memristor

A memristor is defined by the CR $f_M(\varphi, q) = 0$, which corresponds to a curve, a.k.a. *memristor characteristic*, in the φ–q (or q-φ) plane. The memristor is linear if and only if f_M is linear; in such case the characteristic is a straight line passing through the origin and the memristor satisfies $\varphi = Mq$, where M is a constant named the *memristance*. Its reciprocal, $W = M^{-1}$, is named memductance. Note that, by differentiating this relationship, the Ohm's law $v = Mi$ is obtained. This means that it is not possible to distinguish a linear memristor from a linear resistor, so that the existence and relevance of a memristor as a new element cannot be predicted from classical *Linear Circuit Theory*. If f_M is a nonlinear function, then the memristor is nonlinear.

The memristor is charge-controlled if it is possible to explicitly write $\varphi = \hat{\varphi}(q)$, i.e., the flux is a (single-valued) function of the charge. Similarly, it is flux-controlled if it is possible to write $q = \hat{q}(\varphi)$, i.e., the charge is a (single-valued) function of the flux (Fig. 1.7).

By considering a charge-controlled memristor, the slope $M(q_P) = \hat{\varphi}'(q_P)$ of the characteristic at an operating point $P = (q_P, \hat{\varphi}(q_P))$ is named *small-signal* or *differential* memristance of the memristor at P and has dimension of Ohm. Note that, in terms of voltage and current, a nonlinear memristor exhibits a CR in differential form

$$v(t) = \hat{\varphi}'(q(t))i(t) = M(q(t))i(t) \tag{1.10}$$

where

$$M(q(t)) = \hat{\varphi}'(q(t)) = \hat{\varphi}'\left(\int_{-\infty}^{t} i(\tau)d\tau\right)$$

is the small-signal memristance at $q(t)$.

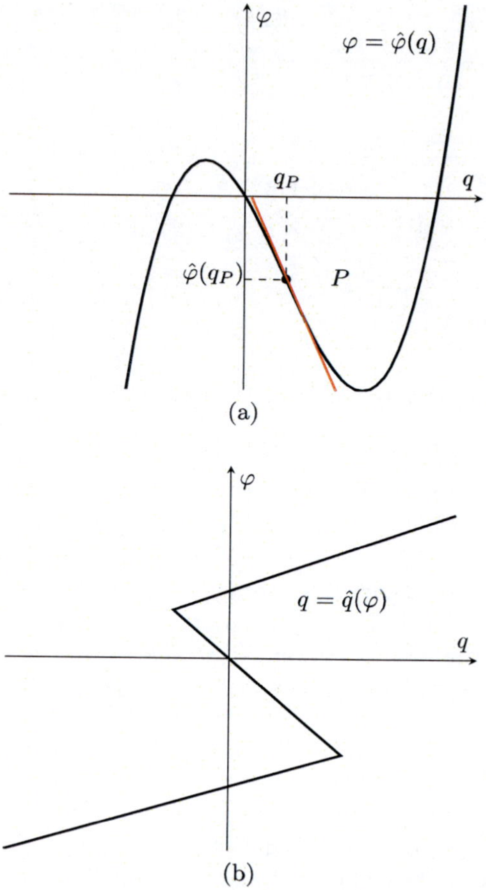

Fig. 1.7 **(a)** Charge-controlled characteristic of a nonlinear memristor in the $q - \varphi$ plane. Point P on the characteristic and tangent straight line at P. The slope of the line corresponds to the differential memristance at P. **(b)** Flux-controlled characteristic of a nonlinear memristor

An examination of (1.10) shows that a two-terminal charge-controlled memristor behaves like a linear resistor described by Ohm's Law; however, its small-signal resistance (memristance) is not a constant, but depends upon the instantaneous value of the charge which book-keeps the history of the current that has flown through the memristor. In other words, the memristance holds a "memory of the (past) current until the instant t." This is the reason why the name "Memory-Resistor," or memristor, for short, was assigned to this heretofore missing fourth basic nonlinear circuit element [3].

Similar considerations hold, *mutatis mutandis*, for a flux-controlled memristor $q = \hat{q}(\varphi)$. The slope $W(\varphi_P) = \hat{q}'(\varphi_P)$ of the characteristic at an operating point $P = (\varphi_P, \hat{q}(\varphi_P))$ is named *small-signal* or *differential* memductance of the memristor at P and has dimension of Ohm^{-1}. In terms of voltage and current, a nonlinear memristor exhibits a CR in differential form

$$i(t) = \hat{q}'(\varphi(t))v(t) = W(\varphi(t))v(t) \tag{1.11}$$

where

$$W(\varphi(t)) = \hat{q}'(\varphi(t)) = \hat{q}'\left(\int_{-\infty}^{t} v(\tau)d\tau\right)$$

is the small-signal memductance at $\varphi(t)$.

Remark 1.1 In the framework of Linear Circuit Theory the terms Resistor, Inductor, and Capacitor are usually used interchangeably without any ambiguity with the respective terms resistance, inductance, and capacitance. In nonlinear network theory, however, this usage becomes ambiguous and only the name of the element (Resistor, Inductor, and Capacitor) should be used.

Example 1.9 (More Realistic Josephson Junction Circuit Model) The classical circuit model of a Josephson junction is made up of the parallel connection of a linear capacitor C, a linear resistor R, and a nonlinear inductor L with the CR in (1.9). A more rigorous quantum mechanical analysis of the Josephson junction dynamics reveals the presence of an additional small current component due to interference among quasi-particle pairs [6]. This heretofore neglected current component is given approximately by

$$i = G\cos(k_0\varphi)v \tag{1.12}$$

where G and k_0 are device constants. One of the reasons why this component had been ignored in the past is due to the lack of a familiar circuit element for its representation. Observe, however, that if we define a flux-controlled memristor with CR

$$q = G_0\sin(k_0\varphi) \tag{1.13}$$

where $G_0 = G/k_0$, then differentiating both sides of (1.13) would give us exactly (1.12). In other words, the neglected current component defined in (1.12) is nothing more than that flowing into a memristor, which can be simply added in parallel to the classical model to obtain the more realistic Josephson junction circuit model shown in Fig. 1.8. It is instructive to note that the Josephson junction provides us with the simplest nonlinear circuit model of a device made from all four basic circuit elements introduced so far.

1.3 Higher-Order Circuit Elements

The axiomatic approach used in Definition 1.1 can be generalized to introduce an infinite variety of higher-order circuit elements as follows.

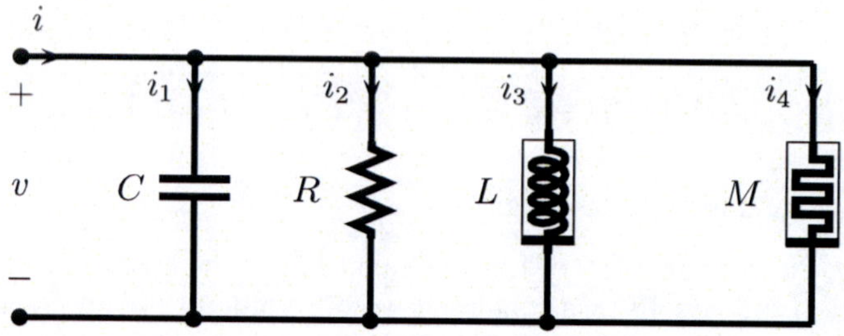

Fig. 1.8 More realistic circuit model of a Josephson junction consisting of a parallel connection of four basic circuit elements, namely, a linear capacitor C, a linear resistor R, a nonlinear inductor L with CR $i_3 = I_0 \sin(k\varphi_3)$, and a nonlinear memristor with CR $q_4 = G_0 \sin(k_0\varphi_4)$ which incorporates the small, hitherto neglected current component due to interference among quasi-particle pairs. Note that the Josephson junction is represented with the symbol of a nonlinear flux-controlled inductor and not with that introduced in Fig. 1.6

Definition 1.2 ((α, β)-Element) A two-terminal circuit element is called an (α, β)-element if and only if it is defined by an algebraic CR involving only the signal pair $v^{(\alpha)}$ and $i^{(\beta)}$ given in (1.2) and (1.3), where α and β are integers.

The electric symbol of an (α, β)-element is reported in Fig. 1.9, while the same figure also depicts a typical curve representing its characteristic in the $v^{(\alpha)} - i^{(\beta)}$ plane (in the case of a $i^{(\beta)}$-controlled element).

Such an infinite family of circuit elements is defined not only for the sake of generality. Rather, these circuit elements are essential for developing a rigorous and comprehensive mathematical theory of nonlinear circuits in the following sense: for any integer k, if one excludes all elements with $|\alpha| > k$ and $|\beta| > k$, then it is possible to construct hypothetical circuits whose dynamic behavior is pathological since solutions are defined up until some finite time instants but cannot be prolonged in time thereafter due to the presence of "singularities" called *impasse points* [5, 7, 8]. Examples of circuits with impasse points shall be discussed in Chaps. 4 and 6.

It is unlikely, however, that (α, β)-elements with $|\alpha| > 2$ and $|\beta| > 2$ will be needed in modeling most real-world devices.[4] In the next section the higher-order elements named *memcapacitor* and *meminductor* are introduced by exploiting the electrical variables defined for $\alpha = -2$ and $\beta = -2$, i.e., $v^{(-2)}$ and $i^{(-2)}$. The reader is referred to Chap. 2 for further properties of these elements.

Finally, it is important to remark that every (α, β)-element can be synthesized via the procedure illustrated in [3, 5] using a family of linear active two ports called *mutators*. They can also be emulated via various off-the-shelf digital components,

[4]It can be proved that any (α, β)-element with $|\alpha| > 2$ and $|\beta| > 2$ is active in the sense that it can be built only with active components, such as transistors and operational amplifiers, which require a power supply.

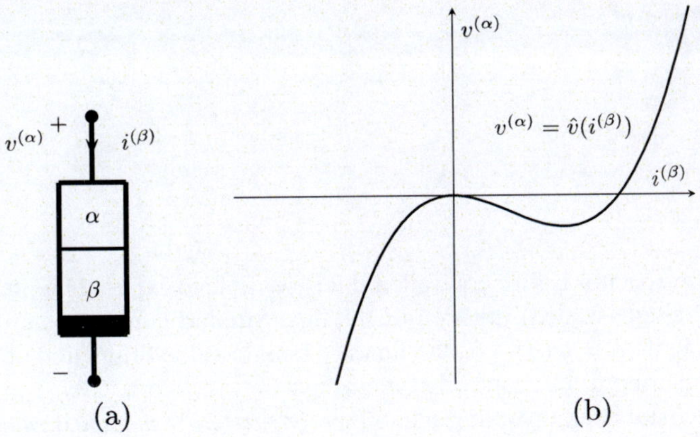

Fig. 1.9 **(a)** Symbol of an (α, β)-element and **(b)** possible nonlinear characteristic in the $v^{(\alpha)}-i^{(\beta)}$ plane

or by programmable software interfaced with analog-to-digital (A/D) and digital-to-analog (D/A) converters.

1.3.1 Memcapacitor

Let us introduce the electric quantity

$$\sigma(t) = \int_{-\infty}^{t} q(\tau)d\tau \tag{1.14}$$

a.k.a. charge momentum.

Definition 1.3 (Memcapacitor) A two-terminal element is called memcapacitor if and only if its CR can be expressed by an algebraic relationship $f_{MC}(\varphi, \sigma) = 0$ involving the variable pair φ, σ.

The CR corresponds to a curve, a.k.a. *memcapacitor characteristic*, in the φ–σ (or σ–φ) plane. The memcapacitor is linear if and only if f_{MC} is linear, in which case the characteristic is a straight line passing through the origin and the memcapacitor satisfies $\sigma = C\varphi$, where C is a constant named *capacitance*. Note that by differentiating this relationship we obtain $q = Cv$ and also $i = Cdv/dt$, i.e., a linear (ideal) capacitor. This means that it is not possible to distinguish a linear memcapacitor from a linear capacitor, so that its existence and relevance as a new element cannot be predicted from classical Linear Circuit Theory. If f_{MC} is a nonlinear function, then the memcapacitor is nonlinear. The symbol of a memcapacitor is given in Fig. 1.10, where the dark band reminds that the nonlinear characteristic may be not symmetric with respect to the origin. Hereinafter only nonlinear memcapacitors are considered.

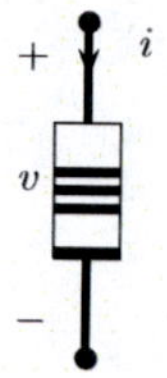

Fig. 1.10 Symbol of a memcapacitor

The memcapacitor is flux-controlled if it is possible to explicitly write $\sigma = \hat{\sigma}(\varphi)$, i.e., σ is a (single-valued) function of the flux. Similarly, it is σ-controlled if it is possible to write $\varphi = \hat{\varphi}(\sigma)$, i.e., the flux is a (single-valued) function of σ.

By considering a flux-controlled memcapacitor, the slope $C(\varphi_P) = \hat{\sigma}'(\varphi_P)$ of the characteristic at an operating point $P = (\varphi_P, \hat{\sigma}(\varphi_P))$ is named *small-signal* or *differential* memory capacitance of the memcapacitor at P and has dimension of Farad. Note that, in terms of voltage and charge, a memcapacitor exhibits a CR in differential form

$$q(t) = \frac{d\sigma(t)}{dt} = \hat{\sigma}'(\varphi(t))\frac{d\varphi(t)}{dt} = C(\varphi(t))v(t) \tag{1.15}$$

where

$$C(\varphi) = \hat{\sigma}'(\varphi)$$

is the small-signal memory capacitance.

An examination of (1.15) shows that a two-terminal flux-controlled memcapacitor behaves like a linear capacitor; however, its small-signal capacitance is not a constant, but depends upon the instantaneous value of the flux which book-keeps the history of the voltage that has been applied to the memcapacitor. In other words, the capacitance has a “memory” and then the name Memory-Capacitor, or memcapacitor, for short.

1.3.2 Meminductor

Let us introduce the electric quantity

$$\rho(t) = \int_{-\infty}^{t} \varphi(\tau)d\tau \tag{1.16}$$

a.k.a. flux momentum.

Definition 1.4 (Meminductor) A two-terminal element is called meminductor if and only if its CR can be expressed by an algebraic relationship $f_{ML}(\rho, q) = 0$ involving the variable pair ρ, q. The symbol of a meminductor is in Fig. 1.11.

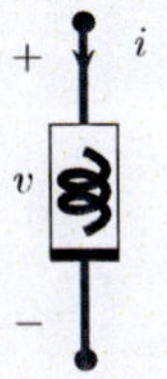

Fig. 1.11 Symbol of a meminductor

The CR corresponds to a curve, a.k.a. *meminductor characteristic* in the ρ–q (or q–ρ) plane. It is readily seen that a linear meminductor acts as a linear inductor with CR $\varphi = Li$, or $v = Ldi/dt$, where L is a constant named *inductance*.

A charge-controlled meminductor is defined by a CR

$$\rho = \hat{\rho}(q)$$

and the corresponding CR in terms of φ and i is obtained by time differentiation as

$$\varphi(t) = \frac{d\rho(t)}{dt} = \hat{\rho}'(q(t))\frac{dq(t)}{dt} = L(q(t))i(t)$$

where

$$L(q) = \hat{\rho}'(q)$$

has dimension of Henry and is called the small-signal memory inductance. Note that $L(q(t))$ depends upon the instantaneous value of the charge and hence it takes into account the history of the current flown through the meminductor.

1.3.3 Periodic Table of Circuit Elements

It is instructive to visualize the (α, β)-elements in a circuit-element-array, named *Periodic Table of Circuit Elements*, shown in Fig. 1.12. Each (α, β)-element is located at the intersection between a vertical line through α and a horizontal line through β. Note in particular the four dots with coordinates $(0, 0)$, $(-1, 0)$, $(0, -1)$, and $(-1, -1)$ that represent the four basic circuit elements named Resistor, Inductor, Capacitor, and Memristor, respectively. We will refer to the other elements as Mixed and Higher-Order Algebraic Elements. Among the latter elements, we have encountered a Memcapacitor, i.e., a $(-1, -2)$-element, and a Meminductor, i.e., a $(-2, -1)$-element.

It can be shown that the small-signal impedance in the frequency domain at an operating point of (α, β)-elements has periodic properties with respect to α and β, thus the name periodic table. The reader is referred to [8] for a detailed discussion.

The circuit elements belonging to the family of the Frequency Dependent Negative Resistors (FDNRs), introduced in the 1960s and synthesized as active

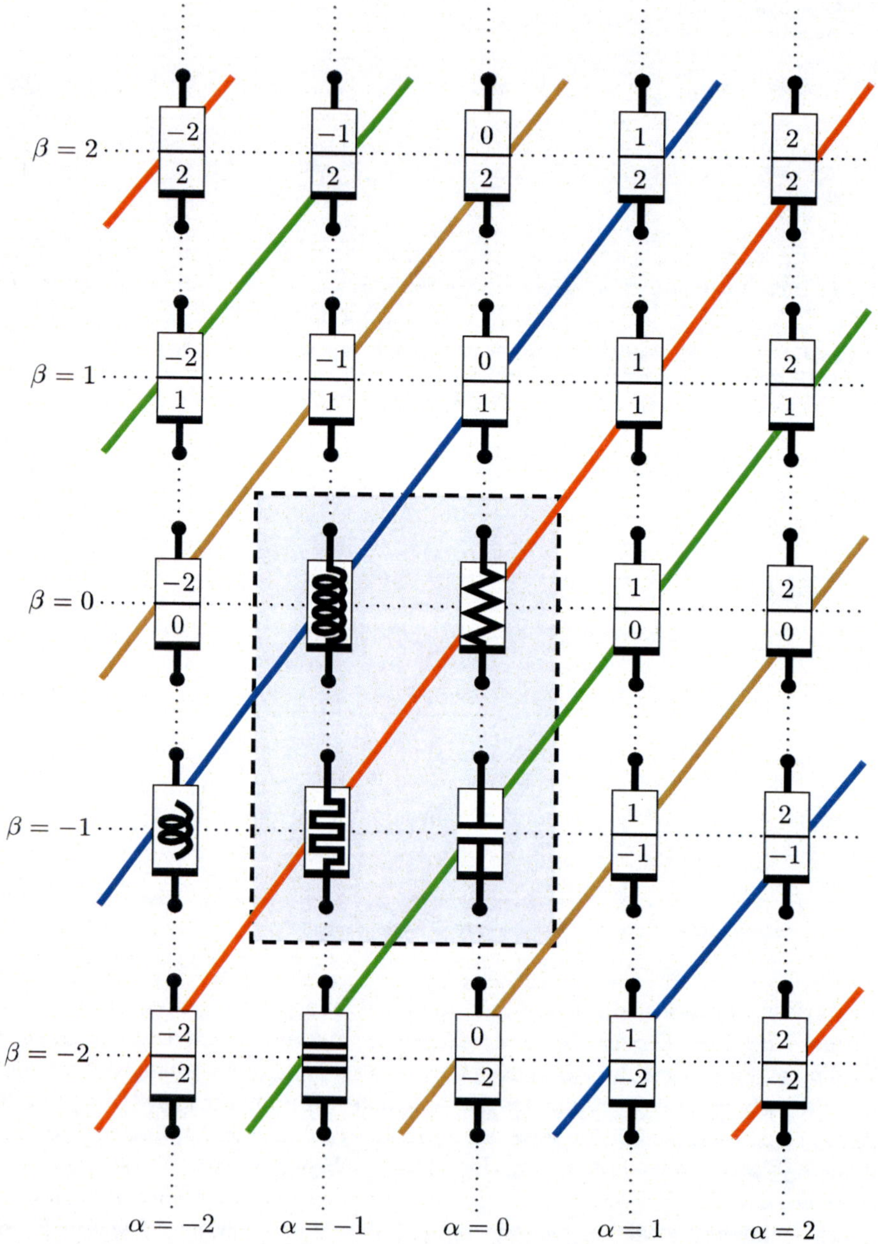

Fig. 1.12 Periodic table of circuit elements. The four basic circuit elements, i.e., resistor, capacitor, inductor, and memristor, are enclosed within a dashed rectangle and they correspond to the four dots with coordinates (0, 0), (0, −1), (−1, 0), and (−1, −1), respectively. The higher-order elements named Memcapacitor and Meminductor correspond to the dots (−1, −2) and (−2, −1), respectively

elements [9, 10], behave as elements from the periodic table. Such resistors with negative resistances show a quadratic dependence of the small-signal conductance or resistance on the frequency and they correspond to the $(1, -1)$ and $(-1, 1)$-elements of the periodic table. The article [11] introduced the FDPC (Frequency Dependent Positive Conductance) whose conductance depends on the fourth power of frequency and thus corresponds the $(2, -2)$-element of the periodic table. It is also worth remarking that the inerter, proposed in 2002 in [12], is the mechanical analog of the $(1, 0)$ circuit element of the periodic table.

The following properties hold [1].

Property 1.1 (Element Independence Property) The (infinitely many) two-terminal elements in the Circuit-Element-Array are all independent of each other in the sense that no element defined by a nonlinear $v^{(\alpha)} - i^{(\beta)}$ curve can be synthesized by a combination of any other elements in the array.

Property 1.2 (Element Closure Property) Arbitrary interconnection of elements of the same type (i.e., same α and β) is equivalent to another element of the same type.

1.3.4 Algebraic and Dynamic Elements

From a circuit-theoretic foundation point of view, it is desirable to classify the universe of all lumped circuit elements into two mutually exclusive categories. The first category is that of the *algebraic elements* and coincides with the (α, β)-elements, i.e., the circuit elements in the periodic table. According to Definition 1.2, a two-terminal element is algebraic if and only if its CR can be expressed by an algebraic relationship involving at most two dynamically independent electric variables $v^{(\alpha)}$ and $i^{(\beta)}$, where α and β are integers. The second category is that of the *dynamic elements*, i.e., all elements that are not algebraic.

Example 1.10 Consider a nonlinear capacitor described by the equation $i = C(v)dv/dt$. Such a CR involves three different variables i, v, and $v^{(1)} = dv/dt$. However, by recasting the equation in the form $q = \hat{q}(v)$, where we let $C(v) = \hat{q}'(v)$, we conclude that the capacitor needs to be classified as an algebraic $(\alpha, \beta) = (0, -1)$-element.

Example 1.11 (ac Dynamic Hysteresis Model of an Inductor) A realistic model of a hysteretic iron-core inductor operating under periodic excitations is given by a two-terminal element with CR

$$v = g(i - f(v^{(-1)}))$$

where $g(\cdot)$ and $f(\cdot)$ are continuous monotone-increasing functions obtained from the hysteresis loops [1]. Here the CR involves three terminal variables, i.e., $\{v, i, v^{(-1)}\}$. In this case, it is *not* possible to find (α, β) such that the CR relation is expressed as an algebraic relationship involving only $v^{(\alpha)}$ and $i^{(\beta)}$, so that we need to classify this element as a dynamic two-terminal element.

Example 1.12 Consider a two-terminal element given by a nonlinear capacitor described by the equation $i = C(v)dv/dt$, i.e., an algebraic $(\alpha', \beta') = (0, -1)$-element, in parallel to a nonlinear voltage-controlled resistor $i = \hat{i}(v)$, i.e., an $(\alpha'', \beta'') = (0, 0)$-element. The parallel obeys to $i = C(v)dv/dt + \hat{i}(v)$. Also in this case it is not possible to find (α, β) such that the CR of this two-terminal element is expressed by an algebraic relationship between $v^{(\alpha)}$ and $i^{(\beta)}$. Hence, such an element is dynamic.

It can be seen that any two-terminal element obtained by interconnecting nonlinear elements with different (α, β) from the circuit-element-array results in a dynamic element. Since, by definition, any lumped element that is not algebraic is dynamic, the class of dynamic elements is therefore much larger than that of algebraic elements.

Most realistic circuit models of devices are made of the interconnection of algebraic elements with different (α, β). Consequently, most realistic circuit models are dynamic elements. These elements are usually described by a CR given by a state equation and an output equation of the following form

$$\begin{aligned} \dot{\mathbf{x}} &= \mathbf{f}(\mathbf{x}, \eta) \\ \xi &= g(\mathbf{x}, \eta) \end{aligned} \tag{1.17}$$

where $\eta, \xi \in \mathbb{R}$, $\mathbf{x} = (x_1, x_2, \ldots, x_n)^T \in \mathbb{R}^n$ is a vector of internal state variables, $\mathbf{f} : \mathbb{R}^{n+1} \to \mathbb{R}^n$ and $g : \mathbb{R}^{n+1} \to \mathbb{R}$.

We can now define four basic classes of dynamic two-terminal elements that are analogous to the four basic algebraic two-terminal elements in Definition 1.1.

Definition 1.5 A two-terminal element is called an R-, L-, C-, or M-dynamic element if and only if it can be described by a CR having the form (1.17), where $\{\xi, \eta\}$ denotes $\{(v, i)\}$, $\{(\varphi, i)\}$, $\{(v, q)\}$, $\{(\varphi, q)\}$, respectively.

As an example, four distinct relevant circuit families of this kind are singled out as follows.

1. Current-Controlled R-dynamic two-terminal element

$$\begin{aligned} \dot{\mathbf{x}} &= \mathbf{f}(\mathbf{x}, i) \\ v &= g(\mathbf{x}, i) \end{aligned} \tag{1.18}$$

2. Current-Controlled L-dynamic two-terminal element

$$\begin{aligned} \dot{\mathbf{x}} &= \mathbf{f}(\mathbf{x}, i) \\ \varphi &= g(\mathbf{x}, i) \end{aligned} \tag{1.19}$$

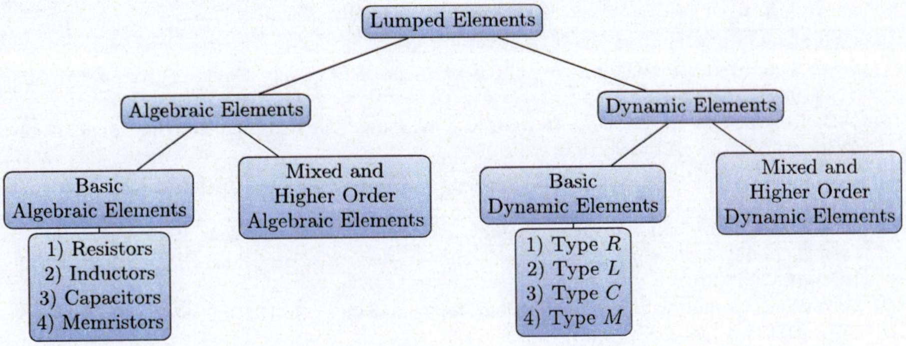

Fig. 1.13 Classification of lumped circuit elements

3. Voltage-Controlled C-dynamic two-terminal element

$$\dot{\mathbf{x}} = \mathbf{f}(\mathbf{x}, v) \qquad (1.20)$$
$$q = g(\mathbf{x}, v)$$

4. Charge-Controlled M-dynamic two-terminal element

$$\dot{\mathbf{x}} = \mathbf{f}(\mathbf{x}, q) \qquad (1.21)$$
$$\varphi = g(\mathbf{x}, q)$$

Along the previous lines it would also be possible to define Mixed and Higher-Order Dynamic Elements. A classification of all lumped circuit elements on the basis of the definitions thus given is provided in Fig. 1.13.

Remark 1.2 (Time-Varying Elements) So far we have considered only time-invariant elements. The treatment can be immediately extended to include time-varying elements. For example, a time-varying (α, β) element is defined by an algebraic CR $f(v^{(\alpha)}, i^{(\beta)}, t) = 0$ including an explicit dependence on time t.

Remark 1.3 (Multiterminal Elements) The treatment in the chapter for two-terminal (or one-port) elements also has a natural extension to multiterminal and multiport elements. The interested reader is referred to [1] for a detailed discussion. In the book, we use multiterminal and multiport elements selectively in specific parts of the book and within examples.

References

1. L. Chua, Device modeling via basic nonlinear circuit elements. IEEE Trans. Circuits Syst. **27**(11), 1014–1044 (1980)

2. F. Corinto, P.P. Civalleri, L.O. Chua, A theoretical approach to memristor devices. IEEE J. Emerg. Sel. Top. Circuits Syst. **5**(2), 123–132 (2015)
3. L.O. Chua, Memristor-The missing circuit element. IEEE Trans. Circuit Theory **18**(5), 507–519 (1971)
4. D.B. Strukov, G.S. Snider, D.R. Stewart, R.S. Williams, The missing memristor found. Nature **453**(7191), 80–83 (2008)
5. L.O. Chua, C.A. Desoer, E.S. Kuh, *Linear and Nonlinear Circuits* (McGraw-Hill, New York, 1987)
6. A.M. Kadin, *Introduction to Superconducting Circuits* (Wiley, Hoboken, 1999)
7. L.O. Chua, Dynamic nonlinear networks: state-of-the-art. IEEE Trans. Circuits Syst. **27**(11), 1059–1087 (1980)
8. L.O. Chua, Nonlinear circuit foundations for nanodevices. I. The four-element torus. Proc. IEEE **91**(11), 1830–1859 (2003)
9. L.T. Bruton, Frequency selectivity using positive impedance converter-type networks. Proc. IEEE **56**(8), 1378–1379 (1968)
10. A. Antoniou, Bandpass transformation and realization using frequency-dependent negative-resistance elements. IEEE Trans. Circuit Theory **18**(2), 297–299 (1971)
11. A.M. Soliman, Realizations of ideal FDNC and FDNR elements using new types of mutators. Int. J. Electron. Theor. Exp. **44**(3), 317–323 (1978)
12. M.C. Smith, Synthesis of mechanical networks: the inerter. IEEE Trans. Autom. Control **47**(10), 1648–1662 (2002)

Chapter 2
Fundamental Properties of Mem-Elements

This chapter is devoted to discuss some basic properties of memristors, memcapacitors, and meminductors, a.k.a. *mem-elements*, that are both of theoretic and practical interest. In the first part (Sects. 2.1 and 2.2), the main focus is on features of a memristor as a $(-1, -1)$-element of the periodic table (cf. Chap. 1), hereinafter also named *ideal memristor*.[1] Such properties include the passivity and the no energy storage property. It is also shown why a memristor is able to store a continuum of stable states and, as such, can implement a nonvolatile memory. In addition, the chapter illustrates some basic signatures of memristors, as the pinched hysteresis loop displayed in the voltage-current (v, i)-domain in response to a zero-mean periodic input (e.g., a sinusoidal signal). In this regard it is however stressed that

Pinched hysteresis loops are not models (i.e., constitutive relations).

In other words, *pinched hysteresis loops don't have any predicting ability because they represent just the device response to a specific input*!

The second part of the chapter discusses the concept of memristive devices and systems, introduced by Chua and Kang [1], as a generalization of an ideal memristor. Memristive systems retain some properties that are analogous to those of an ideal memristor. In particular, any memristive system still displays a pinched hysteresis loop in the (v, i)-domain when subject to a zero-mean periodic input. However, a memristive system may also feature basically different properties with respect to an ideal memristor. For instance, a memristive device may be volatile, as it happens for a thermistor. The concept of memristive devices and a classification

[1] In classic Circuit Theory "ideal circuit elements" are defined by linear CRs characterized by just one single parameter (e.g., the ideal resistor by the resistance R, the ideal capacitor by the capacitance C, the ideal inductor by the inductance L, etc.). On the other hand, the term ideal is also used for nonlinear circuit elements in which the parasitic effects are disregarded. According to this point of view, the (α, β)-elements of the periodic table can be considered as ideal elements.

F. Corinto et al., *Nonlinear Circuits and Systems with Memristors*,
https://doi.org/10.1007/978-3-030-55651-8_2

of memristive devices and systems are discussed in Sects. 2.3 and 2.4, respectively. A brief discussion on fundamental properties of memcapacitors and meminductors is provided in Sect. 2.5.

Finally, Appendix 1 of this chapter is concerned with the technological realization of real passive electronic devices (HP memristor, Resistive RAM, Phase-Change Memories—PCMs, Magnetic RAM) whose behavior can be described in terms of a memristor. Materials and physical phenomena underlying the memristor behavior and the historical links between memristor and resistive switching devices are briefly discussed. The appendix ends with a discussion on the applications of memristor devices in nonlinear complex circuits and systems, including neuromorphic architectures and nonlinear oscillators.

2.1 Ideal Memristor: Basic Properties and Signatures

In this section, and in the next one, we refer to ideal memristors only. In Chap. 1 an ideal memristor has been defined by the CR $f_M(\varphi, q) = 0$ in the flux-charge (φ, q)-domain. In the charge-controlled case, the nonlinear characteristic can be explicitly written as

$$\varphi = \hat{\varphi}(q). \tag{2.1}$$

In the (v, i)-domain a charge-controlled ideal memristor is characterized by the *state-dependent Ohm's Law*

$$v = \hat{\varphi}'(q)i \tag{2.2}$$

where

$$\frac{dq}{dt} = i. \tag{2.3}$$

Dually, in the flux-controlled case we have $q = \hat{q}(\varphi)$ and the ideal memristor obeys

$$i = \hat{q}'(\varphi)v \tag{2.4}$$

where

$$\frac{d\varphi}{dt} = v. \tag{2.5}$$

Recall that $M(q) = \hat{\varphi}'(q)$ and $W(q) = \hat{q}'(\varphi)$ are named memristance and memductance, respectively.

Next, chief properties and peculiar signatures of ideal memristors are summarized. Readers interested in further details can also refer to [1–6].

2.1.1 Passivity

Property 2.1 An ideal memristor is *passive* if and only if $\hat{\varphi}'(q) \geq 0$, for any q, in the charge-controlled case or $\hat{q}'(\varphi) \geq 0$, for any φ, in the flux-controlled case, namely, if and only if its nonlinear characteristic is a *monotone nondecreasing* function.

Proof To verify the sufficiency of this condition, let us consider, without loss of generality, a flux-controlled ideal memristor. Since the memductance $W(\varphi) = \hat{q}'(\varphi) \geq 0$ for all φ, then $p(t) = v(t)i(t) = W(\varphi(t))v^2(t) \geq 0$, i.e., the memristor always absorbs electric power and hence it is passive.

Let us now sketch the proof of necessity. Suppose the memristor has a non-monotone characteristic $\hat{q}(\varphi)$ and it is connected to an (ideal) voltage source $v_s(t)$ as in Fig. 2.1. Suppose the initial flux across the memristor is $\varphi(0) = 0$ at the initial instant $t = 0$. Apply via the voltage source a rectangular voltage pulse in $[t_0, t_0+\Delta T]$ that moves the operating point of the memristor at $Q = (\varphi_Q, \hat{q}(\varphi_Q))$ on the characteristic curve such that $\hat{q}'(\varphi_Q) < 0$. Hence, the memductance is negative in a neighborhood of φ_Q. If we then apply a sinusoidal signal $v_s(t) = \epsilon \sin(\omega t)$, with a sufficiently small amplitude ϵ, it can be seen that the energy $w(t) = \int_0^t p(\tau)d\tau$ tends to $-\infty$ as $t \to +\infty$. Then, the memristor is active since it is able to supply to the voltage source an infinite amount of energy. ■

Example 2.1 (Active Memristor) Given a passive flux-controlled (ideal) memristor with a characteristic $q = b\varphi^3$, where $b > 0$, we can realize an active (ideal) memristor simply by connecting in parallel a negative conductance $G < 0$ obtained for instance by means of a Negative Impedance Converter (NIC) [7, pag. 192]. It is easily seen that such an element is equivalent to an active flux-controlled memristor with a non-monotone characteristic $q = G\varphi + b\varphi^3$.

2.1.2 No Energy Storage Property

Property 2.2 A passive memristor *cannot store or deliver energy.*

Proof Let us consider a simple circuit given by a passive charge-controlled ideal memristor, with an initial charge $q(0) \neq 0$, which is connected to a resistor $R > 0$ at $t = 0$. The governing equation for $t \geq 0$ is

$$\left[R + \hat{\varphi}'(q(t))\right]\frac{dq(t)}{dt} = 0. \tag{2.6}$$

Since the resistor is passive, we have $\varphi'(q) \geq 0$, hence $R + \varphi'(q(t)) > 0$ and then $i(t) = 0$ for any $t \geq 0$. It follows that $q(t) = q(0)$ and $p(t) = v(t)i(t) = 0$ for any $t \geq 0$, i.e., the ideal memristor cannot deliver energy to the resistor. ■

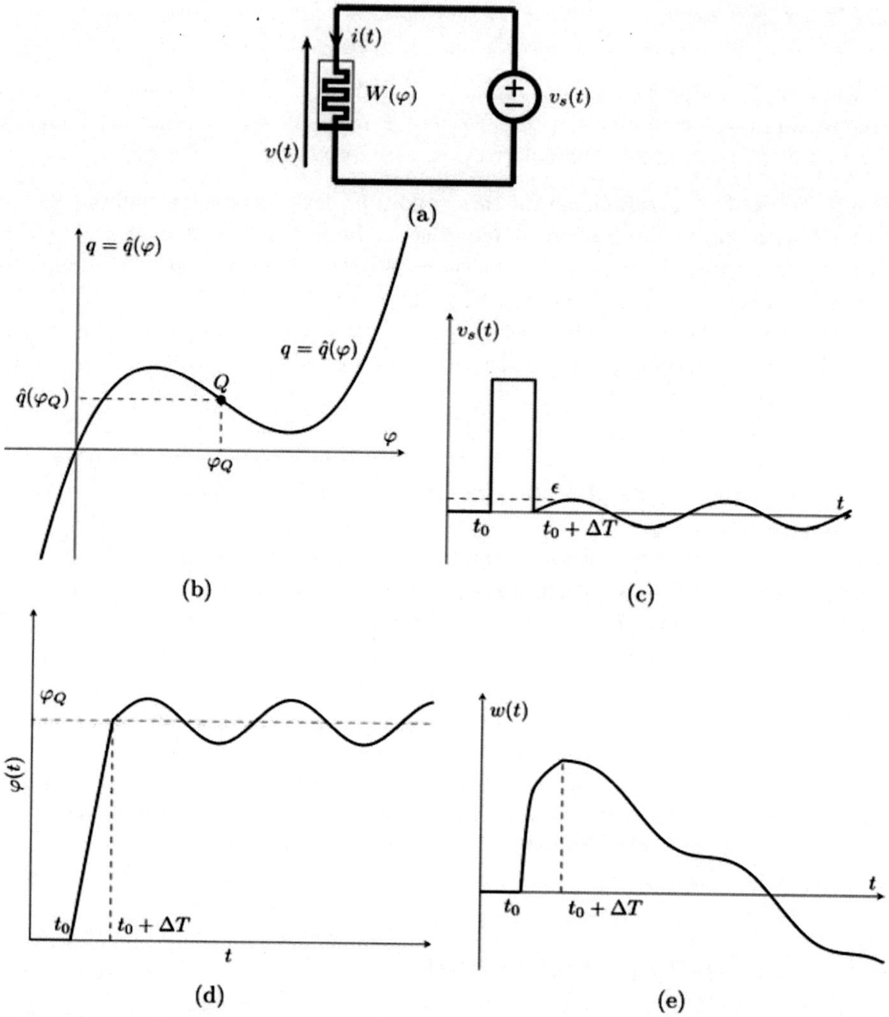

Fig. 2.1 (**a**) Elementary circuit with an ideal memristor having a non-monotonic characteristic as in (**b**). (**c**) Voltage pulse followed by a small-amplitude sinusoid used to show that such a memristor is active. (**d**) Time-domain behavior of $\varphi(\cdot)$ and (**e**) behavior of the energy $w(\cdot)$ absorbed by the memristor

Except for pathological cases, it is always possible to extract stored energy from a passive one-port made of resistors, capacitors, and inductors by simply connecting a load (e.g., a resistor) across it. However, for the case of an ideal passive memristor, such energy discharge is never possible. To emphasize this unique property, we label it the "no energy discharge property."

2.1.3 Zero Crossing Property and Pinched Hysteresis Loop

Property 2.3 An ideal charge-controlled memristor satisfies the following *zero-crossing property*

$$i(t) = 0 \Rightarrow v(t) = 0.$$

Similarly, for an ideal flux-controlled memristor we have

$$v(t) = 0 \Rightarrow i(t) = 0.$$

If a memristor is both charge-controlled and flux-controlled, then both the voltage $v(t)$ and the current $i(t)$ must be zero at the same instants, thus the memristor exhibits the *identical zero-crossing property*

$$i(t) = 0 \Leftrightarrow v(t) = 0.$$

Proof Follows straightforwardly from the state-dependent Ohm's law (2.2) or (2.4). ■

The identical zero-crossing property holds for instance in the case of the HP memristor discussed in Sect. 2.4.2.1. If an ideal memristor is active, in general the identical zero-crossing property no longer holds (see Example 2.6).

The zero-crossing property for passive ideal memristors implies the following *zero phase-shift property*. In any passive ideal memristor the phase shift[2] between a periodic current waveform $i(t)$ (resp., voltage waveform $v(t)$) and its associated periodic voltage waveform $v(t)$ (resp., current waveform $i(t)$) is zero. This property implies that, unlike capacitors and inductors, it is impossible to store energy in a passive memristor (cf. Sect. 2.1.2 for more details).

An additional consequence of the zero-crossing property is that the Lissajous figure obtained combining the memristor voltage $v(t)$ and current $i(t)$ results to be a parametric curve that is pinched in the origin, a.k.a. *pinched hysteresis loop* in the (v, i)-plane. It is in fact a common practice to study the behavior of an ideal flux-controlled memristor by applying a sinusoidal voltage source and measuring the corresponding current, i.e., analyzing the pinched hysteresis loops in the (v, i)-plane exhibited by the memristor.

The next examples illustrate the zero-crossing property and the corresponding pinched hysteresis loops. All the numerical values in the examples are assumed to be normalized with respect to a suitable set of physical quantities, i.e., parameters are given in adimensional form.

[2]The phase shift is considered as the lateral difference between two or more specific points (e.g., zeros, maxima, ...) on waveforms along a common axis.

Example 2.2 (Hysteresis Loops of a Passive Ideal Memristor) Consider a passive ideal flux-controlled memristor with characteristic $q = \varphi + \frac{1}{3}\varphi^3$, suppose it is subject to a voltage $v(t) = \sin(\omega t)$ and the initial condition is $\varphi(0) = 0$. Simple calculations permit to derive that:

$$\varphi(t) = \frac{1}{\omega}(1 - \cos(\omega t)) \tag{2.7}$$

$$q(t) = \frac{1}{\omega}[1 - \cos(\omega t)] + \frac{1}{3\omega^3}[1 - \cos(\omega t)]^3 \tag{2.8}$$

$$i(t) = \left\{1 + \frac{1}{\omega^2}[1 - \cos(\omega t)]^2\right\}\sin(\omega t). \tag{2.9}$$

Figure 2.2 reports the pinched hysteresis loops observed in the (v, i)-plane when $\omega = 1$, $\omega = 3$, and $\omega = 9$, respectively. Note that the hysteresis loops occur because maxima and minima of the sinusoidal input current take place at a different instant with respect to the corresponding voltage. Clearly, hysteresis loops are pinched since $i = 0$ if and only if $v = 0$.

As seen in Example 2.2, hysteresis loops depend upon the frequency of the applied sinusoidal signals and their area *shrink as the frequency increases*. When the frequency tends to infinity, then a pinched hysteresis loop tends to a straight line, i.e., the ideal memristor behaves as a linear resistor whose resistance is independent of the amplitude of the input signal but depends on the initial condition (for the flux

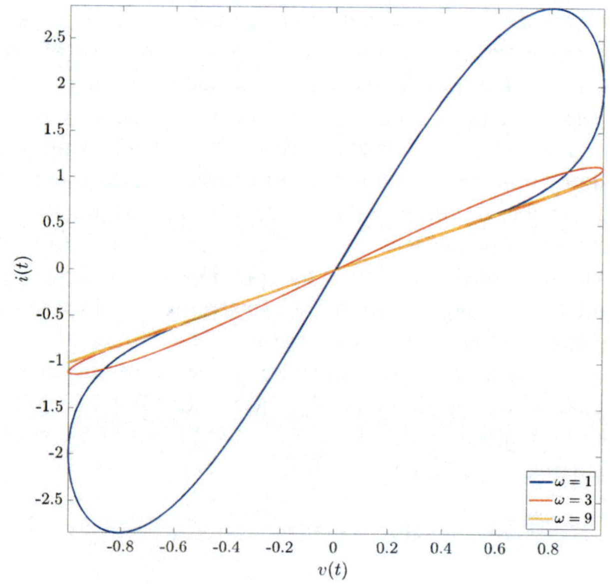

Fig. 2.2 Pinched hysteresis loops of a passive flux-controlled memristor in response to voltage sinusoidal inputs with the same amplitude but different angular frequencies

or charge). To see this, consider a flux-controlled memristor subject to a sinusoidal voltage $v(t) = A\sin(\omega t)$ for $t \geq 0$. We may write

$$\varphi(t) = \varphi(0) + \int_0^t A\sin(\omega\tau)d\tau = \varphi(0) + \frac{A}{\omega}(1 - \cos(\omega t)) \simeq \varphi(0) \tag{2.10}$$

for any t, as $\omega \to +\infty$.

It can also be seen that, for the same sinusoidal input, the *shape of the pinched hysteresis loop is strongly dependent upon the initial condition.*

Example 2.3 (Influence of Initial Conditions on Pinched Hysteresis Loops) Consider a flux-controlled memristor with a piecewise linear characteristic

$$q = \hat{q}(\varphi) = 0.01\varphi + 0.04(|\varphi + 0.25| - |\varphi - 0.25|)$$

as in Fig. 2.3a that is subject to the voltage $v(t) = 1.2\sin(t)$. For two different initial conditions $\varphi(0) = -0.25$ and $\varphi(0) = -0.3$ we observe two quite different pinched hysteresis loops as shown in Fig. 2.3b, c, respectively.

Example 2.4 (Experiment to Characterize an Ideal Memristor) Consider a flux-controlled memristor with CR $q = \sin(\varphi)$ and apply three different voltage signals $v(t) = \sin(\omega t)$ with $\omega = 3.3, \omega = 6.6$ and $\omega = 9.9$, respectively. The corresponding waveforms of $v(t)$, $\varphi(t) = \int_{-\infty}^{t} v(\tau)d\tau = \varphi(0) + \int_0^t v(\tau)d\tau$, $q(t) = \sin\varphi(t)$ and $i(t) = dq(t)/dt$, are plotted in Fig. 2.4, assuming the initial flux $\varphi(0) = 0$.

Suppose we did not know the collection of signals depicted in Fig. 2.4 is derived, or measured, from an ideal memristor characterized by such sinusoidal CR, and we are asked to develop a circuit model which would reproduce with sufficient accuracy the waveforms in Fig. 2.4.

To this end we may depict such curves in the planes v–i, φ–i, v–q, and φ–q, respectively, as shown in Fig. 2.5a–d. Since the input is periodic, each curve is also periodic. However, it is seen that in the planes v–i, φ–i, and v–q the obtained curves are closed loops, hence the two-terminal element cannot be modeled by a nonlinear resistor, inductor, or capacitor (cf. Chap. 1). Only in the plane φ–q we obtain an open-ended loci which is traced back and forth during the application of the sinusoidal signal. We conclude that, based on these experiments, the considered two-terminal element may be reasonably described by a memristor. Other experiments with different signals would of course be needed to confirm this finding.

Example 2.5 (Constitutive Relation of an Ideal Memristor) Let us consider a simple circuit with a voltage source v_s and a linear resistor R connected to an unknown device $\mathcal{D}$ (see Fig. 2.6). The voltage source v_s in series with the resistance R can model a real voltage source supplying the device $\mathcal{D}$. The circuit is subject to different waveforms $v_s(t)$ with zero mean value (i.e., no dc component), namely sinusoids $v_s(t) = A\sin(\omega t)$, a combination of sinusoids and triangular waveforms.

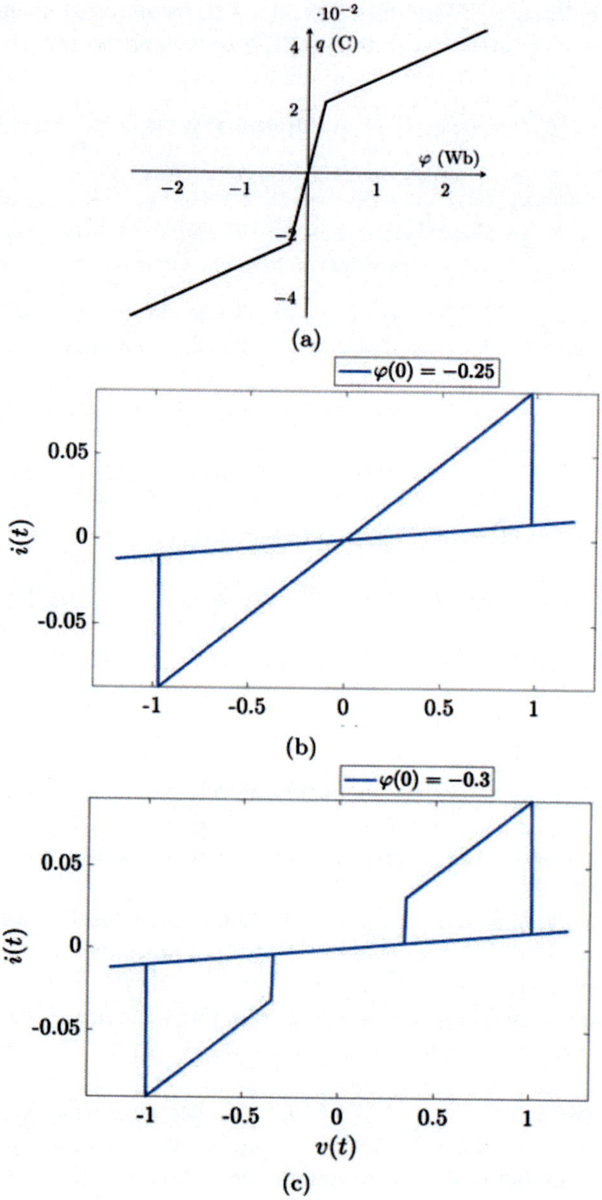

Fig. 2.3 (**a**) Piecewise linear characteristic of a passive memristor. (**b**) Pinched hysteresis loop for an initial memristor flux $\varphi(0) = -0.25$ and (**c**) for an initial flux $\varphi(0) = -0.3$. The example is adapted from [4]

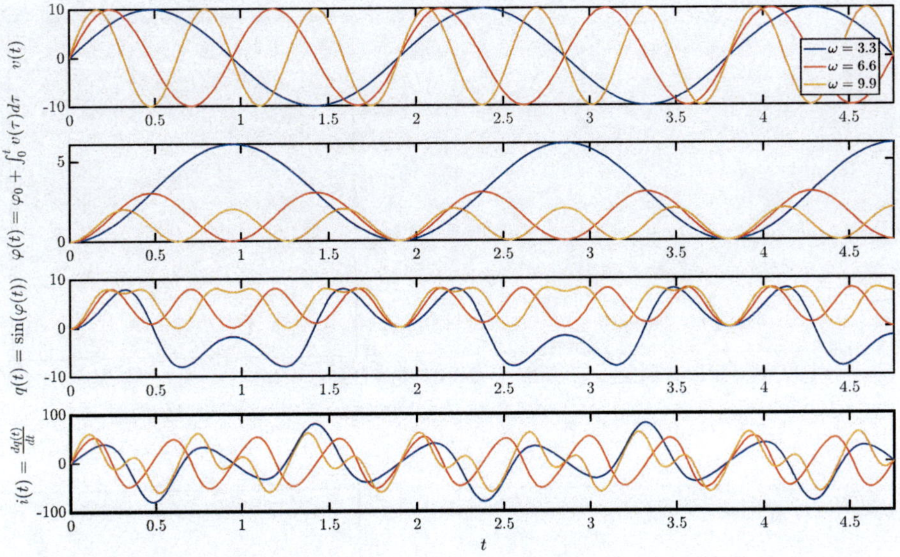

Fig. 2.4 Waveforms of $v(t)$, $\varphi(t)$, $q(t)$, and $i(t)$. The example is adapted from [8]

The aim is to obtain the CR of $\mathcal{D}$ from electrical variables (voltage/flux and current/charge) measured at the terminals of circuit elements. In particular, we use a voltmeter to measure voltage $v(t)$ on the memristor and an ammeter to measure current $i(t)$ through the memristor. By definition, the time integral[3] of the voltages $v_s(t)$ and $v(t)$ gives the fluxes $\varphi_s(t)$ and $\varphi(t)$, respectively; the charge $q(t)$ through the circuit is derived from the time integral of $i(t)$. Finally, for each one of the three circuit elements in Fig. 2.6 the "admissible pairs" can be expressed in terms of either current and voltage or charge and flux.

Let us consider the following case-studies (all variables are expressed in a coherent system of units):

- $v_s(t)$ is a zero-mean triangular waveform with unitary amplitude and period $T = 4$. The corresponding flux $\varphi_s(t)$, assuming $\varphi_s(0) = 0$, is reported in Fig. 2.7a. The current $i(t)$ and its time integral $q(t)$, assuming $q(0) = 0$, are shown in Fig. 2.7b, while Fig. 2.7c reports $v(t)$ and $\varphi(t)$, assuming $\varphi(0) = 0$. Let us focus on the admissible pairs of $\mathcal{D}$, namely (v, i) in the voltage-current domain and (φ, q) in the flux-charge domain, respectively, whose representations in the corresponding domain are shown in Fig. 2.8a, b. It turns out that $(\varphi(t), q(t))$ traces back and forth a unique nonlinear (cubic) characteristic $\varphi = q^3/3$ defining the CR of an ideal (charge-controlled) memristor, whereas $(v(t), i(t))$ produces a pinched hysteresis loop.

[3]The time integral of electrical variables can be also measured via ballistic instruments (see Sect. 1.1.2).

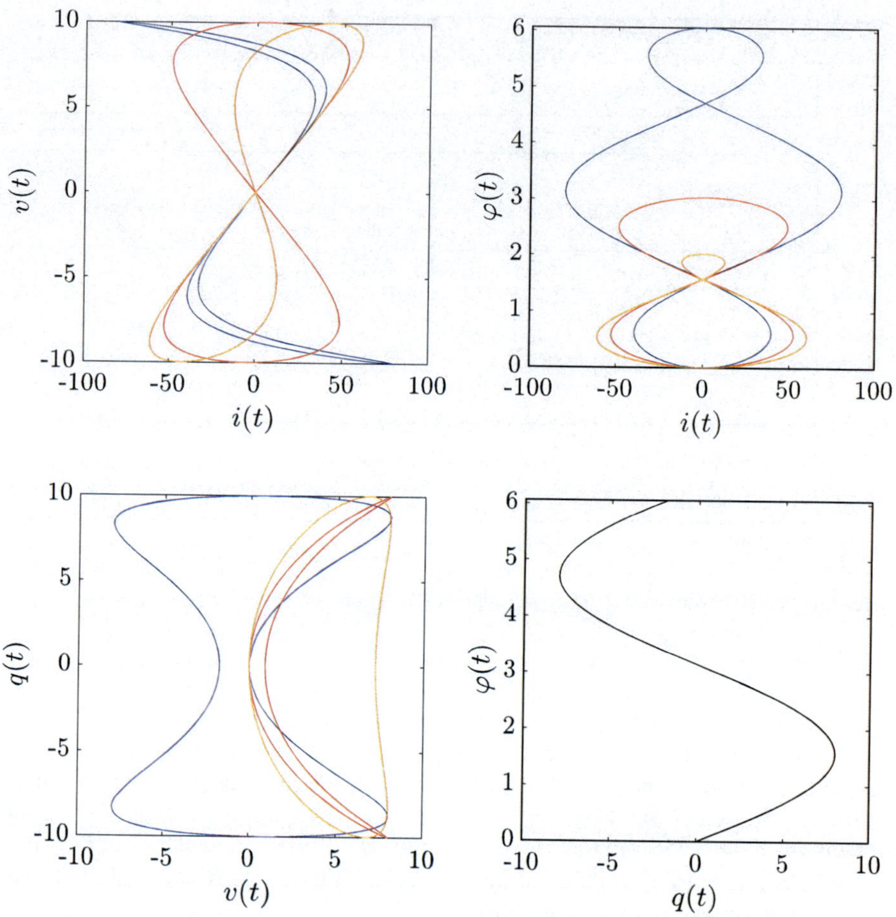

Fig. 2.5 Loci in the v–i, φ–i, v–q, and φ–q planes for the waveforms in Fig. 2.4

Fig. 2.6 Circuit for determining the CR of an unknown device $\mathcal{D}$

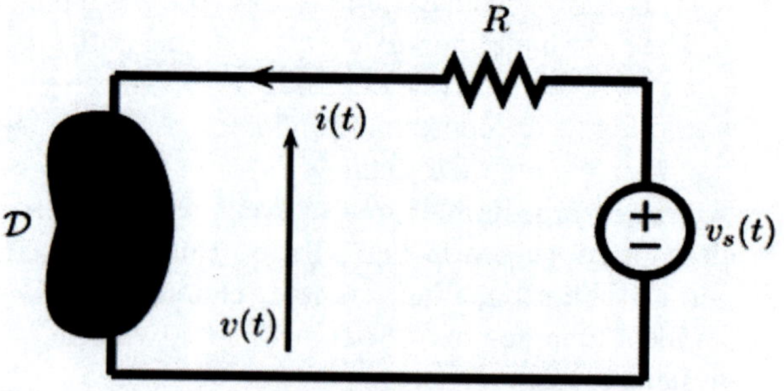

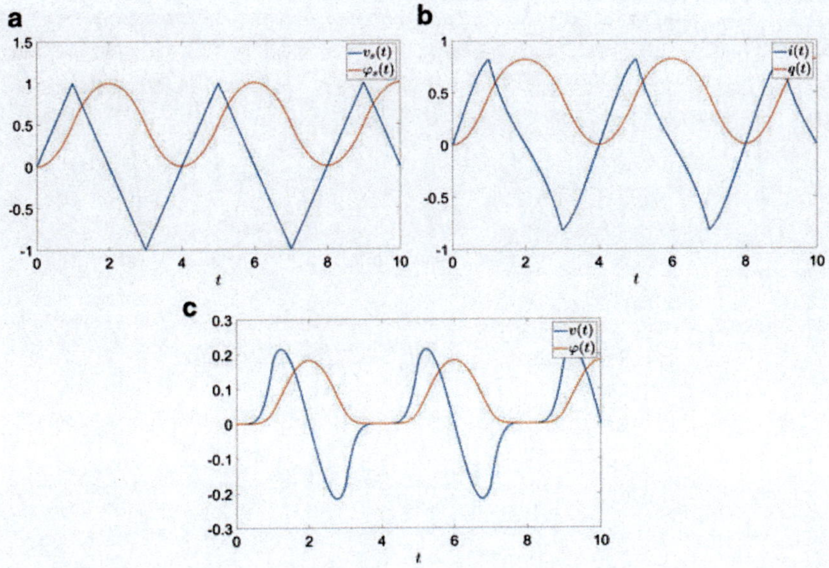

Fig. 2.7 Waveforms of the electrical variables in the circuit in Fig. 2.6. (**a**) Voltage $v_s(t)$ **and flux** $\varphi_s(t)$ applied by the source. (**b**) Current $i(t)$ and charge $q(t)$ through the circuit elements. (**c**) Voltage $v(t)$ **and flux** $\varphi(t)$ across the unknown device $\mathcal{D}$

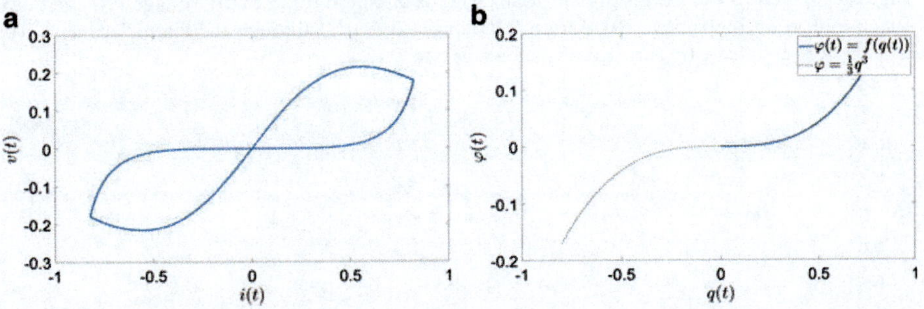

Fig. 2.8 Admissible pairs of $\mathcal{D}$ in the (v, i)- and (φ, q)-domain. (**a**) The voltage and current pair of $\mathcal{D}$ describes a pinched hysteresis loop in the voltage-current domain. (**b**) **The flux and charge** pair of $\mathcal{D}$ **describes a cubic CR in the flux-charge domain**

- $v_s(t)$ is a zero-mean sinusoidal waveform with $A = 2$ and period $T = 2$ (i.e., $\omega = \pi$). Following the same approach of the previous case, the corresponding $v_s(t)$, $\varphi_s(t)$, $i(t)$, $q(t)$, $v(t)$, and $\varphi(t)$ waveforms are given in Fig. 2.9a, b, c, respectively. Then, the admissible pairs (v, i) and (φ, q) of $\mathcal{D}$ are shown in Fig. 2.10a and 2.10b, respectively. It is worth to compare this case with the previous one to spot the differences/analogies due to the different voltage source $v_s(t)$. It is readily seen that the curve traced in the (φ, q)-domain corresponds to the same nonlinear function $\varphi = q^3/3$, whereas the pinched hysteresis loop

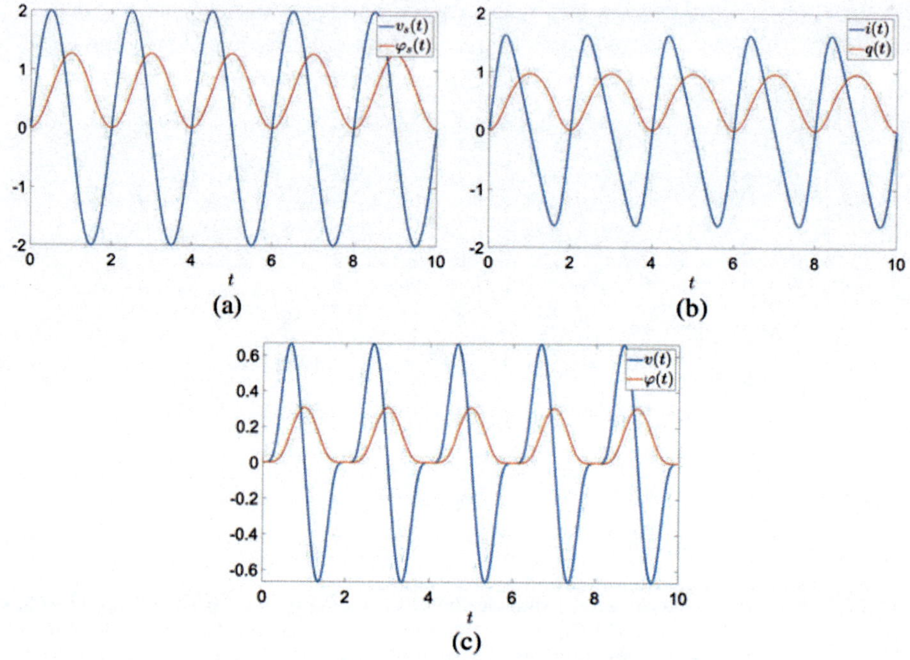

Fig. 2.9 Waveforms of the electrical variables in the circuit in Fig. 2.6. (**a**) Voltage $v_s(t)$ and flux $\varphi_s(t)$ applied by the source. (**b**) Current $i(t)$ and charge $q(t)$ through the circuit elements. (**c**) Voltage $v(t)$ and flux $\varphi(t)$ across the unknown device $\mathcal{D}$

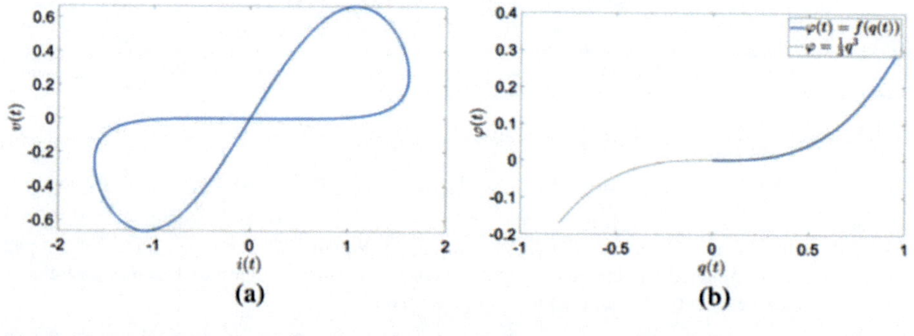

Fig. 2.10 Admissible pairs of $\mathcal{D}$ in the (v, i)- and (φ, q)-domain. (**a**) The voltage and current pair of $\mathcal{D}$ describes a pinched hysteresis loop in the voltage-current domain. (**b**) The flux and charge pair of $\mathcal{D}$ describes a cubic CR in the flux-charge domain

in the (v, i)-domain changes due to the different voltage source. This second experiment confirms that $\mathcal{D}$ results to be an ideal memristor with a cubic CR $\varphi = q^3/3$.

We stress once more that the meaning of the CR of an ideal memristor is as follows (cf. Chap. 1): "admissible pairs" of waveforms $(\varphi(t), q(t))$ (resp., $(q(t), \varphi(t))$) that

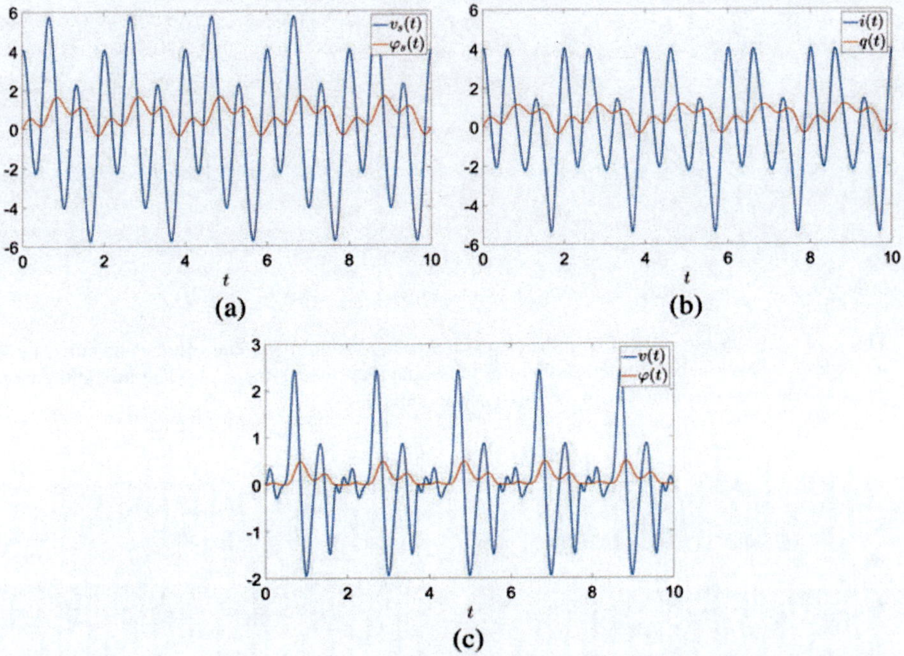

Fig. 2.11 Waveforms of the electrical variables in the circuit in Fig. 2.6. (**a**) Voltage $v_s(t)$ and **flux** $\varphi_s(t)$ applied by the source. (**b**) Current $i(t)$ and charge $q(t)$ through the circuit elements. (**c**) Voltage $v(t)$ **and flux** $\varphi(t)$ across the unknown device $\mathcal{D}$

one measures from a memristor must fall along the curve φ vs. q (resp., q vs. φ). There is no requirement that the input charge $q(t)$ **(resp., flux** $\varphi(t)$) be dc, or be sinusoidal! It is crucial to understand that the CR of a memristor is nothing but a graphical read-out table of all admissible (q vs. φ), or equivalently (φ vs. q), waveforms pairs.

Figures 2.11 and 2.12 present the electrical variables due to a multitone periodic voltage source $v_s(t) = 2\sin(\omega t) + \cos(3\omega t)$ including two commensurable angular frequencies. If $v_s(t)$ includes incommensurable frequencies, e.g., $v_s(t) = 2\sin(\omega t) + \cos(\sqrt{3}\omega t)$, the waveforms in Figs. 2.13 and 2.14 are obtained. Note that the waveforms $(v(t), i(t))$ measured from an ideal memristor with CR $\varphi = q^3/3$ are neither dc nor sinusoidal and give rise to exotic pinched hysteresis curves (not closed when frequencies are incommensurable). However, once such $v(t)$ and $i(t)$ are integrated in time, the corresponding admissible waveform pairs $(q(t), \varphi(t))$ always describe the same cubic $\varphi = q^3/3$ CR.

Different for nonlinear resistors, it may not be possible to choose a nonzero dc voltage (resp., current) as the input signal to the memristor, because the corresponding current (resp., voltage) response may never settle to a dc equilibrium. In fact, an ideal memristor, and all nonvolatile memristors (Sect. 2.3) do not have an EP when subject to a nonzero dc signal.

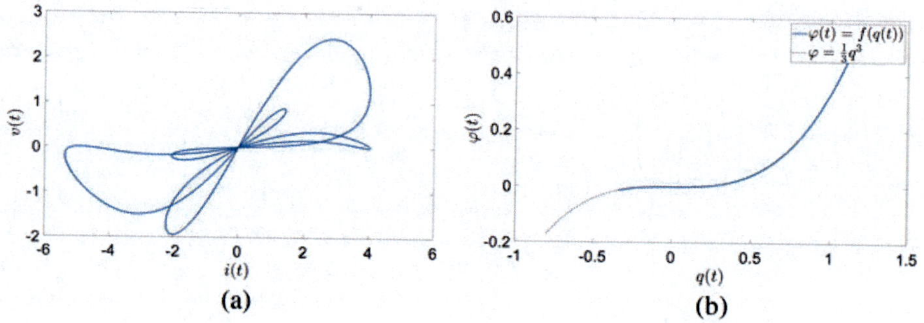

Fig. 2.12 Admissible pairs of $\mathcal{D}$ in the (v, i)- and (φ, q)-domain. (**a**) The voltage and current pair of $\mathcal{D}$ describes a pinched hysteresis loop in the voltage-current domain. (**b**) The flux and charge pair of $\mathcal{D}$ describes a cubic CR in the flux-charge domain

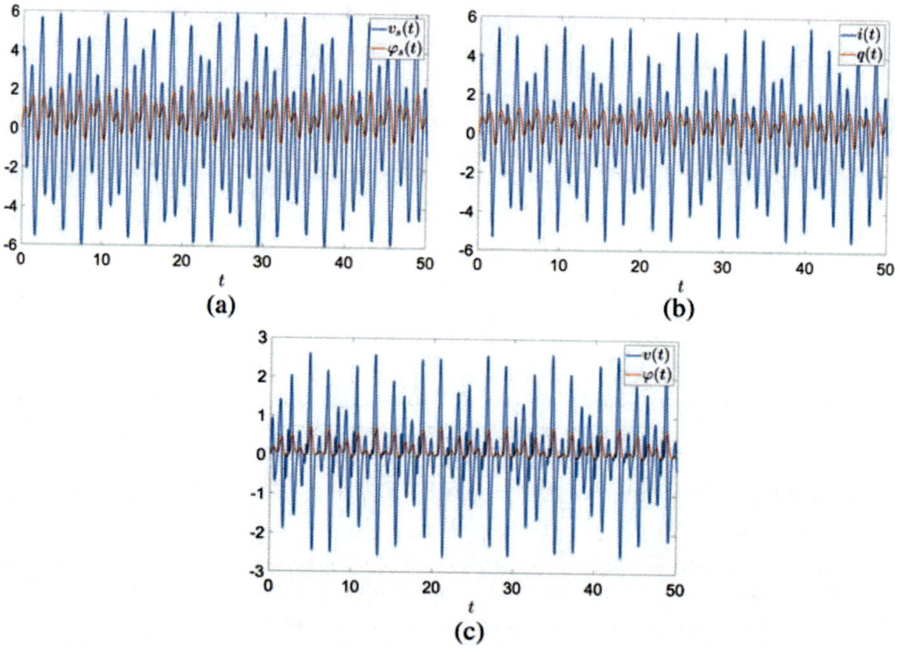

Fig. 2.13 Waveforms of the electrical variables in the circuit in Fig. 2.6. (**a**) Voltage $v_s(t)$ and flux $\varphi_s(t)$ applied by the source. (**b**) Current $i(t)$ and charge $q(t)$ through the circuit elements. (**c**) Voltage $v(t)$ and flux $\varphi(t)$ across the unknown device $\mathcal{D}$

Remark 2.1 The concept of "pinched hysteresis loop" represents one of the main fingerprint for ideal memristors and it is the basis for the experimental definition of memristor devices. On the other hand, the examples reported above show that a pinched hysteresis loop in the (v, i)-domain is just the response of a memristor to a specific zero-mean periodic input. Therefore, the pinched hysteresis loop *is only a signature, but cannot be used as the CR of a memristor. In fact, it doesn't have*

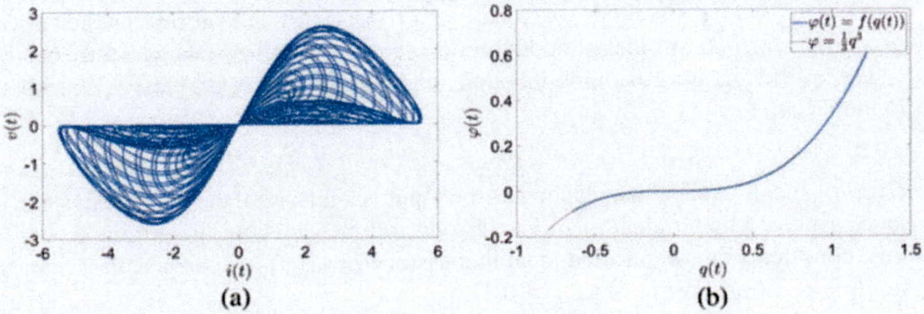

Fig. 2.14 Admissible pairs of $\mathcal{D}$ in the (v, i)- and (φ, q)-domain. (**a**) The voltage and current pair of $\mathcal{D}$ describes a pinched hysteresis loop in the voltage-current domain. (**b**) The flux and charge pair of $\mathcal{D}$ describes a cubic CR in the flux-charge domain

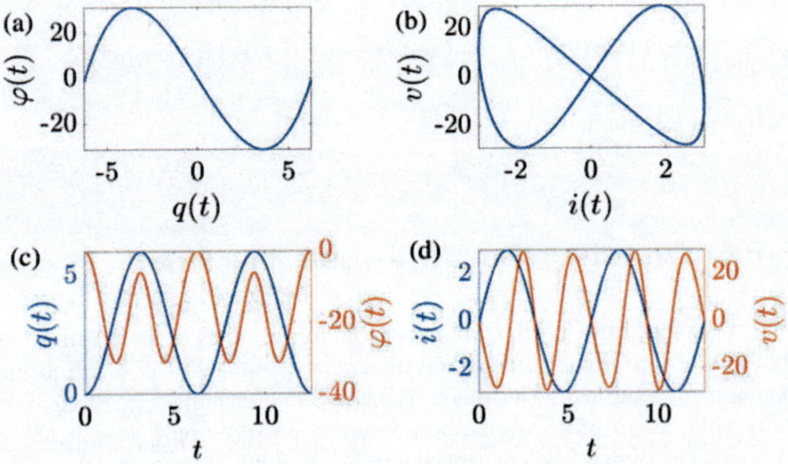

Fig. 2.15 (**a**) Characteristic of an active charge-controlled memristor. (**b**) Hysteresis loop when the memristor is subject to a sinusoidal current. (**c**) Time-domain behavior of charge and flux and (**d**) of current and voltage

predicting ability since different periodic input (voltages) yield different periodic responses (currents) and then different pinched hysteresis loops.

Example 2.6 (Zero-Crossing Property for an Active Ideal Memristor) Let us consider an active charge-controlled memristor with CR $\varphi = -13q + \frac{1}{3}q^3$. Figure 2.15 shows the hysteresis loop displayed when the memristor is subject to a sinusoidal current $i(t) = 3\sin(t)$ and $q(0) = 0$. Note that the loop intersects the second and fourth quadrant of the v–i plane. Moreover, there are instants such that the voltage is 0 but the current does not vanish, thus violating the simultaneous zero crossing property of voltage and current.

Pinched hysteresis loops of ideal memristors present further mathematical properties when sinusoidal/periodic inputs are considered. A detailed study can be found in [9]. The next example illustrates the odd symmetry property with respect to the origin.

Example 2.7 (Pinched Hysteresis Loops with Odd Symmetry with Respect to the Origin) It can be seen that, as long as the input is sinusoidal, the pinched hysteresis loop displayed by an ideal memristor is odd symmetric about the origin. To verify this, consider a flux-controlled ideal memristor $q = \hat{q}(\varphi)$ and suppose it is subject to $v(t) = \sin(\omega t)$. We have for $t \geq 0$

$$i(t) = \hat{q}'(\varphi(t))v(t) = \hat{q}'(\varphi(t)) \sin(\omega t).$$

Since

$$\varphi(t) = \varphi(0) + \int_0^t \sin(\omega \tau) d\tau = \varphi(0) + \frac{1}{\omega}[1 - \cos(\omega t)]$$

then $\varphi(t)$ is even and hence $i(t)$ is odd.

2.2 Ideal Memristors and Non-volatile Memories

Before discussing how ideal memristors can be used as a nonvolatile memory capable to store information in a "physical state variable," it is worth to mention that many electronic memories made of passive two-terminal circuit elements are typically volatile. In other words, a continuous power supply (and hence power consumption) is needed to keep on an electronic circuit in its state and then to realize a nonvolatile memory. This principle is illustrated by the following simple example.

Example 2.8 (Binary Memory with Tunnel Diode) Consider the simple circuit in Fig. 2.16a with an Esaki (tunnel) diode having a non-monotone current-voltage CR $i = \hat{i}(v)$. When power is on, i.e., $E \neq 0$, the intersection between the load line $E = Ri + v$ and the diode characteristic $i = \hat{i}(v)$ shows that the circuit has three different solutions (i.e., operating points). From a real implementation of the circuit it can be observed that two of these solutions (Q_1 and Q_2) are asymptotically stable (Fig. 2.16b), while one solution (Q_0) is unstable (see also Example 4.4 in Chap. 4 for a discussion on stability of these solutions). As such the circuit can implement a binary memory in the states Q_1 and Q_2. However, when power is turned off (i.e., $E = 0$), there is a unique solution where voltage and current vanish and the circuit "forgets Q_1 and Q_2." Then, the circuit acts as a volatile memory (Fig. 2.16c) and the power source E is essential to retain the two states Q_1 and Q_2.

Now, let us discuss the ideal memristor as a nonvolatile memory and how to operate it.

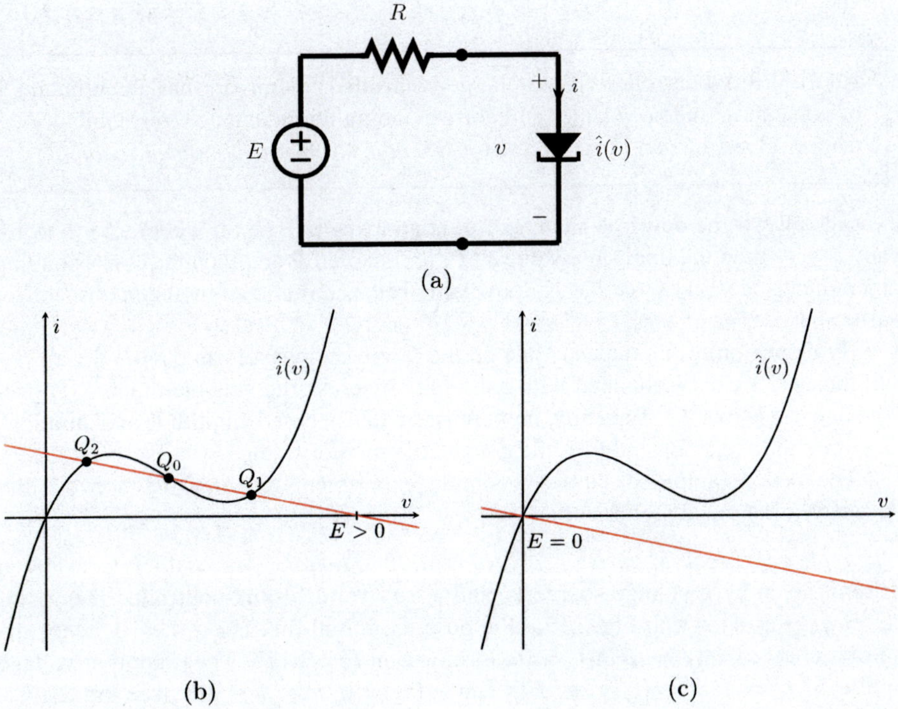

Fig. 2.16 (**a**) Volatile memory implemented by a circuit with tunnel diode. (**b**) The circuit has two stable memory states Q_1 and Q_2 (and an unstable state Q_0) when power is on, i.e., $E > 0$. (c) There is only one equilibrium state with $v = 0$ and $i = 0$ when power is turned off, i.e., $E = 0$

2.2.1 *Ideal Memristor as Nonvolatile Memory Memristor*

Consider the CR of a flux-controlled memristor in the (v, i)-domain

$$i = \hat{q}'(\varphi)v \tag{2.11}$$

where

$$\frac{d\varphi}{dt} = v. \tag{2.12}$$

Suppose power is turned off, i.e., we let $v = 0$. The EPs, i.e., the stationary states of the memristor, are obtained by letting $d\varphi(t)/dt = 0$. It is seen from (2.12) that *any value of the flux is an EP*. Moreover, any EP is stable (but not asymptotically stable). Analogous considerations hold for a charge-controlled memristor.

Then:

> **An ideal flux-controlled (resp., charge-controlled) memristor has a continuum of nonvolatile memory states and corresponding memductance (resp., memristance) values, i.e., an ideal memristor is a *nonvolatile memory memristor*.**

Actually, if the power is switched off at an instant T, i.e., if we let $v(t) = 0$, for any $t \geq T$, then the final state value $\varphi(T)$ is a stable EP and then the corresponding memductance value $\hat{q}'(\varphi(T))$ is nonvolatile because it is kept by the memristor for all $t > T$.

By definition, $\varphi(T)$ depends just on the (arbitrary) waveform of $v(t)$ for all $t \leq T$, thus $\varphi(T)$ can be changed with continuity by choosing suitable inputs $v(t)$ over the interval $(-\infty, T]$. Typically, input pulses with specified amplitude and duration[4] are used to tune $\varphi(T)$ and then the nonvolatile memory state.

The next example discusses a simple experiment to show the nonvolatility property of an ideal memristor.

Example 2.9 (Ideal Memristor as Nonvolatile Memory Memristor) Consider a circuit given by a voltage source v_s and a passive ideal flux-controlled memristor $q = \hat{q}(\varphi)$ as shown in Fig. 2.17a. Suppose the initial flux is $\varphi(t_0) = 0$, hence the initial charge $q(t_0) = Q(t_0) = 0$, as shown in Fig. 2.17b. Then, apply a voltage pulse $v_s(t) = E$, $t \in [t_0, t_0 + \Delta T]$ and $v_s(t) = 0$, $t \geq t_0 + \Delta T$ (see Fig. 2.17c). Although $v_s(t) = 0$ (hence $i(t) = 0$), for all $t \geq t_0 + \Delta T$, i.e., the power supply turns off, the memristor is in a steady state with

$$\varphi(t) = \varphi(t_0 + \Delta T) = \int_{t_0}^{t_0+\Delta T} v_s(t)dt = E\Delta T = \varphi_Q \tag{2.13}$$

for all $t \geq t_0 + \Delta T$ (see Fig. 2.17d). Thus the ideal memristor stores its final memductance value at $t_0 + \Delta T$, i.e., $\hat{q}'(\varphi(t_0 + \Delta T)) = \hat{q}'(\varphi_Q)$, in a permanent way by means of the steady state (flux) $\varphi_Q = E\Delta T$ (see Fig. 2.17b). *The ideal memristor thus acts as a nonvolatile memory.*

2.2.2 Ideal Memristor as Continuum Memory Memristor

A flux-controlled memristor can be used as a nonvolatile analog memory when it can *store a continuum of different memductance values* in correspondence with the stable EPs of the state (flux),[5] as illustrated next.

---

[4] Amplitude and duration of pulses depend on technological processes and materials used to fabricate memristor devices.

[5] Similar considerations hold for charge-controlled memristors.

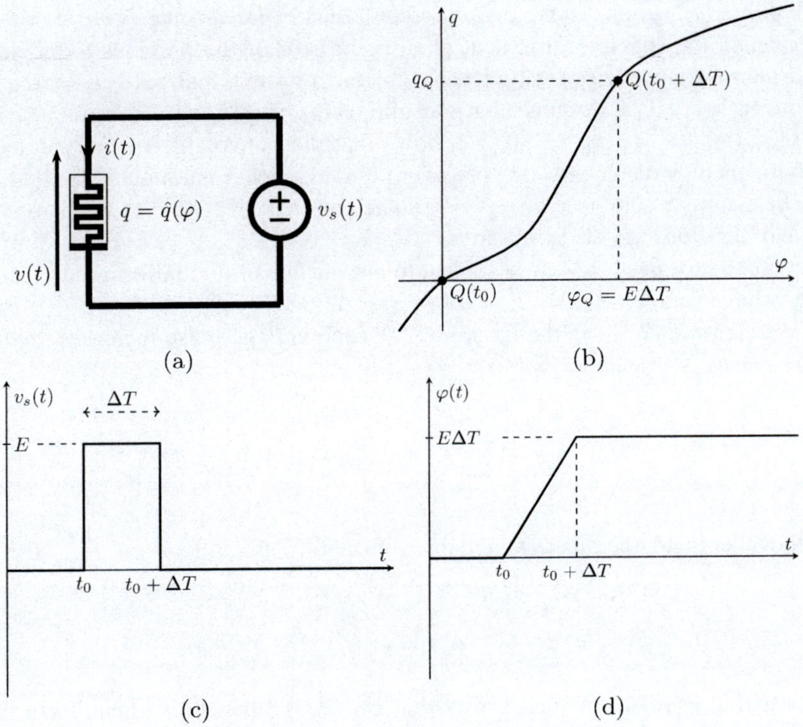

Fig. 2.17 An experiment to show that an ideal memristor acts as a nonvolatile memory. (**a**) Elementary circuit with memristor and voltage source. (**b**) Characteristic of the memristor. (**c**) Pulse applied via a voltage source and (**d**) time-domain behavior of memristor flux

Example 2.10 (Memristor as an Analog Memory) Suppose that the CR of a passive flux-controlled memristor is given by a monotonically increasing cubic function

$$\hat{q}(\varphi) = \frac{1}{3}\varphi^3.$$

The corresponding memductance

$$\hat{q}'(\varphi) = \varphi^2$$

assumes a continuum of values in the interval $[0, +\infty)$. The memristance $\hat{q}'(\varphi)$ vs. state φ map is shown in Fig. 2.18; this allows one to tune the memductance continuously by using short voltage impulses (cf. Example 2.11). In conclusion, such an ideal memristor is nonvolatile and continuously tunable via pulses and basically corresponds to an analogue, variable, and programmable resistor.

Example 2.11 (Programming a Continuum Memory Ideal Memristor) Let us consider again the (passive) flux-controlled memristor of the previous Example 2.10 with memductance $\hat{q}'(\varphi) = \varphi^2$ and subject to an external voltage source v_s as shown in Fig. 2.17a. Assume that the initial flux is $\varphi(t_0) = 0$, hence the initial charge $q(t_0) = \hat{q}(\varphi(t_0)) = 0$. It follows that the nonvolatile memory associated with the memductance at t_0 is $\hat{q}'(\varphi(t_0)) = 0$ until an external input is applied.

Then, apply a sequence of $N + 1$ voltage pulses (cf. Fig. 2.17c) with amplitudes E_k and durations Δ_k at the instant t_k with $k = 0, 1, 2, \ldots, N$. In other words, the ideal memristor $\hat{q}(\varphi) = \frac{1}{3}\varphi^3$ is subject to a sequence of nonuniform pulses defined by a voltage source $v_s(t) = E_k$, $t \in [t_k, t_k + \Delta_k]$ and $v_s(t) = 0$, $t \in (t_k + \Delta, t_{k+1})$.

By using Eq. (2.13) in the Example 2.9, each voltage pulse increases the flux φ by $\varphi_{P_k} = E_k \Delta_k$, i.e., being $\varphi(t_0) = 0$

$$\varphi(t_k + \Delta_k) = \sum_{r=0}^{k} E_r \Delta_r, \quad (k = 0, 1, 2, \ldots, N).$$

On the other hand, the flux $\varphi(t)$ remains constant for all $t \in [t_k + \Delta_k, t_{k+1}]$ (because $v_s(t) = 0$) so that the memristor exhibits a nonvolatile memductance

$$\hat{q}'(\varphi(t_k + \Delta_k)) = \varphi^2(t_k + \Delta_k) = \varphi^2(t_{k+1}), \quad \forall t \in (t_k + \Delta_k, t_{k+1}).$$

If E_k and Δ_k are set in such a way that $\varphi_{P_k} = 1$, then four nonuniform pulses give $\varphi(t_{k+1}) = k + 1$ for $k = 0, 1, 2, 3$ (being $\varphi(t_0) = 0$); the corresponding four memductance values selected among the continuum memory states are shown by the markers in Fig. 2.18.

In conclusion, *the considered ideal memristor acts as a nonvolatile continuum memory in which a finite number of memory states can be selected by suitable*

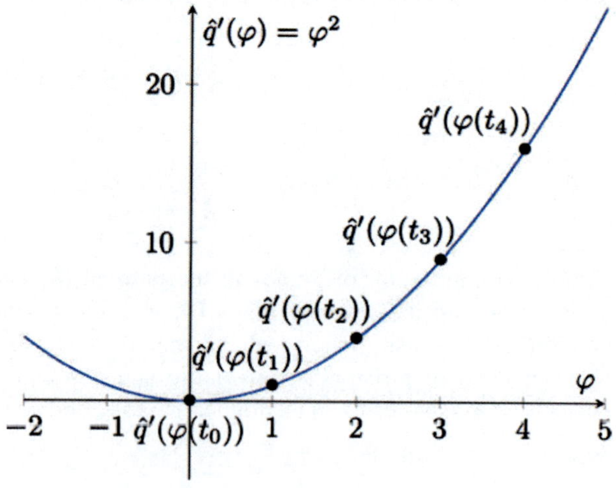

Fig. 2.18 The ideal memristor $q = \hat{q}(\varphi) = \frac{1}{3}\varphi^3$, with memductance $\hat{q}'(\varphi) = \varphi^2$, acts as a nonvolatile continuum memory in which a finite number of memory states can be selected by suitable pulses. The memductance values corresponding to the memory states are shown by the marked points when a pulse with $\varphi_{P_k} = 1$ for any $k = 0, 1, 2, 3$ is applied

pulses. A nano-scale device with resistance which can be modulated electronically would permit denser analogue circuits, multi-state memory and large-scale, solid-state artificial neural networks, where it would be crucial in implementing neuron synapses and learning [10–12].

2.2.3 Ideal Memristor as Discrete Memory Memristor

In digital computer applications, to be used as a *nonvolatile binary memory*, a flux-controlled memristor should exhibit only two sufficiently distinct memductances.

Example 2.12 Consider a memristor with the piecewise linear CR in Fig. 2.19a, thus $q = \hat{q}(\varphi)$ has a zero slope segment for $|\varphi| \leq 2.5$ Wb and two straight lines with slope 800 nS when $|\varphi| > 2.5$ Wb. Such a memristor exhibits just two memory states corresponding to the two different values 0 nS and 800 nS of memconductance. Let us perform an experiment where the initial memristor flux $\varphi(0) = 0$ and we apply a square voltage waveform as in the upper part of Fig. 2.19b. The result is a triangular waveform for φ (middle part of Fig. 2.19b) while the memductance $W(t)$ displays a square waveform switching between the two levels 0 nS and 800 nS (lower part of Fig. 2.19b).

If a flux-controlled or charge-controlled ideal memristor is defined by a piecewise linear CR with multiple slopes, then it acts as a *discrete memory memristor*, that is there exist multiple memory states corresponding to the memristance/memductance values specified by the slopes. Each memory state can be programmed by suitable input pulses.

To summarize, ***an ideal flux-controlled (resp., charge-controlled) memristor has*** *a continuum of equilibrium states*, i.e., it can memorize in a nonvolatile way infinitely many values of the flux (resp., of the charge) as stable EPs. The corresponding memorized conductances (resp., memristances) may assume a continuum of values or a discrete set of values.

At any EP we have $v = 0$ and $i = 0$, i.e., the memristor current, voltage, and absorbed power are zero. It is important to observe that if one opens or short-circuits a memristor having a given memductance (resp., memristance) at t_0, the memristor does not lose the value of memductance (resp., memristance) because both voltage and current become zero at the instant when the power is switched off, but rather holds the value unchanged thereafter! Because memristors remember their state even when the power is turned off, it is (ideally) possible to store data indefinitely, using energy only when the state of a memristor is toggled or read. We stress that is basically different from capacitors in conventional dynamic RAM (DRAM), which in practice will quickly lose their stored charge, if the power to the chip is turned off, due to leakage currents.

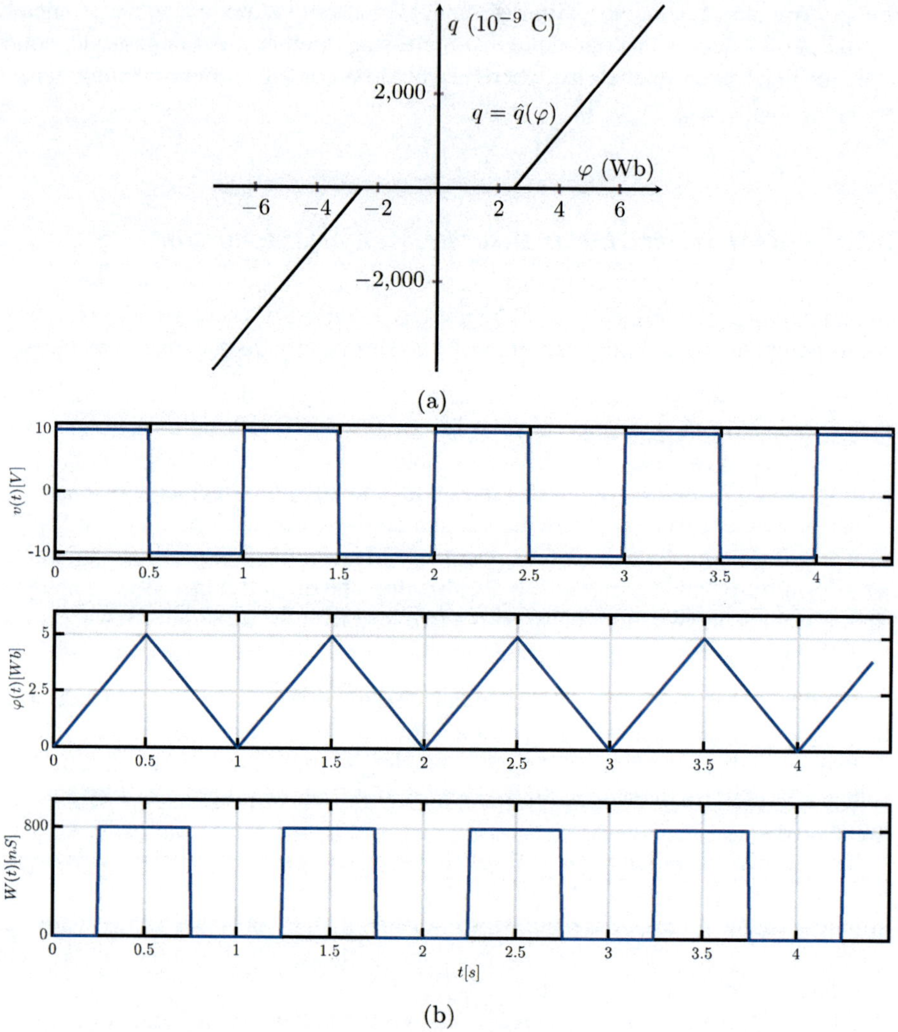

Fig. 2.19 (**a**) Memristor with a piecewise linear characteristic useful for implementing a digital memory. (**b**) Switchings between two different values of memductance when the memristor is subject to a voltage square wave. The example is adapted from [4]

2.3 Memristive Devices and Systems

Some main theoretic properties of an ideal memristor are preserved by a broader class of nonlinear dynamical systems, named *memristive devices* or in general *memristive systems*, introduced by L. O. Chua and S. Kang in the seminal article [1].

In the current-controlled case a memristive system obeys the state-dependent Ohm's law

$$v = R(\mathbf{x}, i)i \tag{2.14}$$

$$\frac{d\mathbf{x}(t)}{dt} = \mathbf{f}(\mathbf{x}(t), i(t)) \tag{2.15}$$

where $\mathbf{x} = (x_1, x_2, \ldots, x_n)^T$ denotes a vector of n internal states variables.

Note that the key property of simultaneous zero crossing $i(t) = 0 \Rightarrow v(t) = 0$ is enjoyed by a memristive device and such feature differentiates it from a general dynamical system. As stated in [1]: *"This zero-crossing property manifests itself vividly in the form of a Lissajous figure which always passes through the origin"* when a memristive device is subject to a sinusoidal current. As for an ideal memristor, Lissajous figures (i.e., pinched hysteresis loops) depend upon the excitation frequency. While at very low frequencies, memristive systems are indistinguishable from nonlinear resistors, at extremely high frequencies they reduce to linear/nonlinear resistors.

Similar considerations hold in the case of a voltage-controlled memristive device defined by the state-dependent Ohm's law

$$i = G(\mathbf{x}, v)v \tag{2.16}$$

$$\frac{d\mathbf{x}(t)}{dt} = \mathbf{g}(\mathbf{x}(t), v(t))$$

and $\mathbf{x} = (x_1, x_2, \ldots, x_n)^T$ is still a vector of n internal states variables.

Remark 2.2 It is worth to note that the original definition of memristive devices and systems given by L. O. Chua and S. Kang in [1] is in terms of Differential Algebraic Equations (DAEs) where electrical variables (i.e., voltage and current) play the role of input and output of the memristive (dynamical) system whereas $\mathbf{x}$ is the (physical) internal state vector.

2.3.1 Analogies and Differences Between Ideal Memristors and Memristive Devices

The chief signatures and properties of an ideal memristor presented in Sects. 2.1 and 2.2 can be summarized as follows:

- passivity and no energy storage
- zero-crossing property and pinched hysteresis loop
- continuum memory memristor: there exists a continuum of nonvolatile memory states, with an associated continuum of memristance/memductance values, that can be selected via suitable pulses
- discrete memory memristor: there exists a continuum of nonvolatile memory states, with an associated discrete set of memristance/memductance values, that can be selected via suitable pulses.

As shown in the previous section, memristive devices and systems exhibit the zero-crossing property by definition. The concept of passivity for a memristive system is more intricate with respect to the passivity of an ideal memristor because we cannot exploit the CR in the flux-charge domain to evaluate the memristance $\hat{\varphi}'(q)$ or the memductance $\hat{q}'(\varphi)$ (see the proof of Property 2.1). Hence, the passivity of memristive devices and systems defined in terms of DAEs results to be an inherent input–output property which may be roughly specified in relation to energy dissipation and transformation.[6] In the following, the passivity of memristor devices and systems is intended in the common meaning of electrical circuit theory, that is, the absence of any source of energy (e.g., a battery) inside the device package [7]. As a consequence, *one cannot pull more energy out of a passive (dynamic) circuit element than what was fed into it*. Clearly, a sufficient condition for passivity of a current-controlled (resp., voltage-controlled) memristive device is that we have $R(\mathbf{x}, i) \geq 0$ (resp., $G(\mathbf{x}, i) \geq 0$) for any i (resp., for any v) and any $\mathbf{x} \in \mathbb{R}^n$. Indeed, under such condition, the instantaneous power entering the one-port is always nonnegative, i.e., $p(t) = R(\mathbf{x}, i)i^2 \geq 0$ (resp., $G(\mathbf{x}, v)v^2 \geq 0$) [1]. Under such condition, energy discharge is not possible, namely, it is not possible to extract energy from the device by connecting it to a resistor (cf. Sect. 2.1.2).

A fundamental property of memristive devices, differentiating them from ideal memristors, is that *they may be either nonvolatile or volatile*. In the latter case, a memristive device has a unique state and corresponding memductance (memristance) when the input voltage (or current) is turned off. In the next section, we will see that the concept of passivity is important to figure out the use of a memristive device as a volatile or a nonvolatile memory element.

It is extremely important to observe that a volatile memory cannot be described by an ideal memristor because an ideal memristor always features a continuum of nonvolatile memory states and at least two different nonvolatile memductances. The volatility feature is then peculiar to some classes of memristor devices and systems.

Remark 2.3 Let us consider a flux-controlled ideal memristor $q = \hat{q}(\varphi)$ and suppose it can store only one memductance value, for any flux φ, when power is turned off. In other words, the slope $\hat{q}'(\varphi)$ of the CR $q = \hat{q}(\varphi)$, i.e., the memductance, is a constant G at any point φ. Then, the ideal memristor is linear ($q = G\varphi$) and it acts as a linear resistor. In conclusion, *a volatile ideal memristor cannot be distinguished from a linear resistor*.

[6] The field of passivity for nonlinear dynamical systems constitutes an active research direction. Such study is beyond the scope of this book and the reader is invited to refer to classic works [13–15].

2.3.2 Memory Memristive Devices and Power-Off-Plot (POP)

Let us discuss in more detail memristor devices behaving as a volatile or a nonvolatile memory memristor. Consider a voltage-controlled memristive device

$$i = G(x, v)v$$

where

$$\frac{dx(t)}{dt} = g(x(t), v(t))$$

and suppose for simplicity there is a single internal state variable $x \in \mathbb{R}$.

A memristive device acts as a:

- *nonvolatile memory* if and only if it exhibits at least two stable states when power is switched off
- *volatile memory* if and only if there is only one stable state when power is switched off.

Namely, set $v = 0$ and consider the scalar differential equation

$$\frac{dx}{dt} = g(x, 0). \tag{2.17}$$

The memristor is nonvolatile if and only if Eq. (2.17) has at least two different stable EPs x_A and x_B, i.e., $g(x_A, 0) = 0$ and $g(x_B, 0) = 0$, with $x_A \neq x_B$, whereas is volatile if there is only one stable EP (e.g., x_A).

A graphical approach based on the dynamic route (cf. Chap. 4) permits to understand if the memristive device is volatile or nonvolatile. The dynamic route corresponds to the plot of $g(x, 0)$ in (2.17) in the plane $(x, \dot{x})$; such dynamic route goes under the name *Power-Off-Plot* (POP). The POP is an extremely useful concept to understand the qualitative behavior of the scalar state variable x and to locate EPs.

Let us show the concept of POP in an ideal memristor.

Example 2.13 (POP in Ideal Memristor) **The POP of an ideal flux-controlled** memristor $q = \hat{q}(\varphi)$, i.e.,

$$\begin{aligned} i &= \hat{q}'(\varphi)v \\ \dot{\varphi} &= v \end{aligned} \tag{2.18}$$

corresponds to the equation $\dot{\varphi} = 0$, that is, the φ-axis in the plane $(\varphi, \dot{\varphi})$. Hence, the POP makes clear the existence of a continuum of stable EPs (i.e., any value of φ).

The next example illustrates how to use the POP to characterize volatile and nonvolatile memristive devices.

Example 2.14 (POP in Memristive Devices) Let us consider a voltage-controlled memristive device (2.16) with a scalar state variable x defined as

$$i = G(x, v)v = x^2 v$$

where

$$\frac{dx}{dt} = g(x, v) = -x + \alpha\,(|x+1| - |x-1|) + v. \tag{2.19}$$

The POP is given by

$$\frac{dx}{dt} = g(x, 0) = -x + \alpha\,(|x+1| - |x-1|)\,. \tag{2.20}$$

Parameter α allows us to change the memory and volatility properties of such memristive device. In particular, the following cases are considered:

- $\alpha = 1$ gives the POP in Fig. 2.20a. Note that there are two points $x_A \neq x_B$ where the POP intersects the x-axis with negative slope and one point x_0 where the intersection has positive slope. It follows that there exist *two (isolated) asymptotically stable EPs x_A and x_B*, and an unstable EP x_0. Arrows on the dynamic route denote that x increases (decreases) when $\dot{x} > 0$ ($\dot{x} < 0$). It follows that when $t \to \infty$ then $x(t) \to x_A$ ($x(t) \to x_B$) for any initial condition $x(0) > 0$ ($x(0) < 0$) and so the memristive device can encode two memory states into the memductance values $G(x_A, 0) = x_A^2$ and $G(x_B, 0) = x_B^2$. Hence the memristive device is nonvolatile and it acts as a *discrete memory memristor*. It is worth to note that such discrete (nonvolatile) memory states occur under zero input voltage, thus this is an inherent property of the memristive device. On the contrary, an ideal memristor can present discrete memory states only if they are induced by suitable voltage pulses (cf. Example 2.11).
- $\alpha = 0.5$ gives the POP in Fig. 2.20b, that has an interval $[x_B, x_A]$ with zero slope. Then, we have a continuum of (non-isolated) stable EPs corresponding to any value of x in the interval $[x_B, x_A]$. In such a case the memristive device operates as a *continuum memory memristor*. The memory states are all the memductance values $G(x, 0) = x^2$ with $x \in [x_B, x_A]$.
- $\alpha = 0.25$ gives the POP in Fig. 2.20c that intersects the x-axis with negative slope only in $x_A = 0$, thus there exists a unique asymptotically stable EP. In this case the memristive device serves as *volatile memory memristor*, because when the input is turned off the state variable $x(t)$ converges to x_A for $t \to \infty$ and the memristive device has a unique memductance value $G(x_A, 0) = x_A^2 = 0$ as memory state.

If we exclude situations where the POP has intervals with zero slope (i.e., there are non-isolated EPs for the memristor), we can summarize some of the previous considerations with the following result.

Fig. 2.20 POP in a voltage-controlled memristive device. (**a**) The nonvolatile memristive device is a discrete memory memristor. (**b**) The nonvolatile memristive device is a continuum memory memristor. (**c**) The memristive device acts as a volatile memory

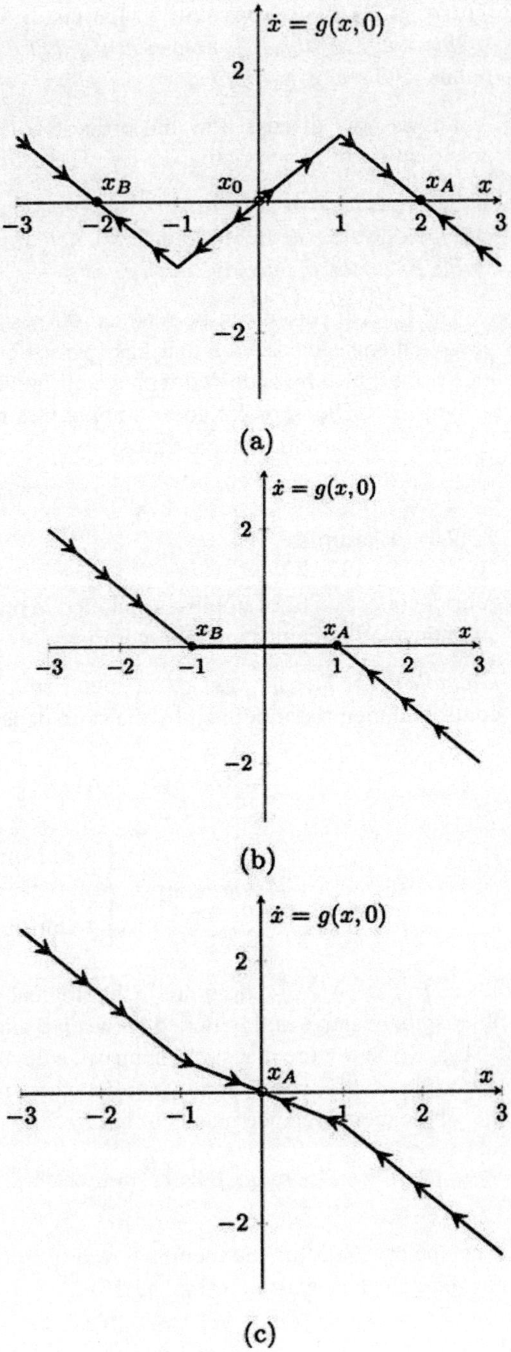

Theorem 2.1 (The Nonvolatile Memristor Theorem) *A memristive device with a scalar state variable x is nonvolatile if its POP intersects the x-axis at two or more points with a negative slope.*

Let us now discuss how the property of passivity influences the volatility or nonvolatility of a memristive device. The following can be shown to hold [6].

Theorem 2.2 (Two Is Infinity) *Suppose a passive memristive device is nonvolatile and let $x_A < x_B$ be two stable EPs. Then, any $\bar{x}$ satisfying $x_A \le \bar{x} \le x_B$ is also a stable EP of the memristive device.*

The latter theorem can be read as follows: if one can measure experimentally a passive memristive device that has two different small-signal conductance levels, then it must also be endowed with a continuum of memory states (i.e., stable EPs).

Note that Theorem 2.2 doesn't apply to a memristive device with a POP like in Fig. 2.20a because in that case passivity fails.

2.3.2.1 Examples

A brief collection of examples is included in order to illustrate the concept of volatile and nonvolatile memory in real memristive devices.

Example 2.15 The `Pt/TaOx/Ta` memristive device in [16] is a passive voltage-controlled memristor defined by the state-dependent Ohm's law

$$i = G(x, v)v = \alpha(1 - e^{-\beta v}) + \gamma x \sinh(\delta v)$$

$$\frac{dx}{dt} = g(x, v) = \begin{cases} \lambda \sinh(\eta v)(\frac{1-x}{\tau_0}), & v \ge 0 \\ \lambda \sinh(\eta v)(\frac{x}{\tau_0}), & v < 0 \end{cases}$$

where $\alpha, \beta, \gamma, \delta, \lambda, \mu, \tau_0, \eta$ are technological parameters and the scalar state variable is nonnegative and bounded between 0 and 1, i.e., $x \in [0, 1]$.

It is worth to note that such memristive device can be seen as a nonlinear resistor (i.e., $\alpha(1-e^{-\beta v})$) connected in parallel with a memristive device (i.e., $\gamma x \sinh(\delta v)$); the whole memductance results to be

$$G(x, v) = \left[\frac{\alpha}{v}(1 - e^{-\beta v}) + x\frac{\gamma}{v} \sinh(\delta v)\right].$$

It can be checked that the memductance is bounded in a neighborhood of $(x, 0)$ for any x because (see also Example 2.26)

$$G(x, 0) = \lim_{v \to 0} G(x, v) = \alpha\beta + x\delta\gamma.$$

When power is off, i.e., letting $v = 0$, we have

$$\frac{dx}{dt} = g(x, 0) = 0$$

thus the POP is zero over the domain [0, 1] and so any $\bar{x} \in [0, 1]$ is a stable EP. We conclude that the Pt/TaOx/Ta memristive device is a nonvolatile memory memristor. Since the memristive device is passive, this result is in accordance with Theorem 2.2.

Remark 2.4 In addition to the memristive device in Example 2.15, Theorem 2.2 is confirmed experimentally by several physical passive devices. Experimental observations in a memristor device based on aluminum-oxide thin-film reveal a range of nonvolatile resistive states that can be changed continuously by applying suitable voltage pulses (Fig. 12 in Hickmott's work [17]). Similar switching phenomena are also displayed by memristive device and systems based on different materials (nickel oxide, silicon monoxide, various molecular compounds as polyconjugated compounds, phthalocyanines, and charge transfer complexes and $SrZrO_3$) [18–20].

Remark 2.5 It is worth remarking that Theorem 2.2 is valid also in the special case of an ideal memristor (cf. Sect. 2.2.1).

For what concern volatile memristive devices, the chief examples are the thermistor [1, 21], the discharge tube [22], and the potassium and sodium ion channels of the Hodgkin–Huxley neuron model [23].

Example 2.16 (Thermistor) A ***Positive Temperature Coefficient (PTC) thermistor*** is a memristive device defined by

$$i = G(T)v$$

and

$$\dot{T} = \frac{\delta_P}{H_{CP}}(T_{0P} - T) + \frac{G(T)v^2}{H_{CP}}$$

where

$$G(T) = \left[R_{0P} \exp\{\beta_P (T - T_{0P})\}\right]^{-1}.$$

The state variable T is given by the PTC thermistor body temperature and $\delta_P, H_{CP}, T_{0P}, R_{0P}, \beta_P$ are positive parameters.

By switching off power, i.e., letting $v = 0$, we obtain

$$\dot{T} = \frac{\delta_P}{H_{CP}}(T_{0P} - T).$$

By exploiting the POP it is easily seen that there exists a unique EP $\bar{T} = T_{0P}$ which attracts all solutions. This means that, for any initial condition $T(0)$, the corresponding solution tends to T_{0P} as $t \to +\infty$, i.e.

$$G(T(t)) \to G(T_{0P}) = \frac{1}{R_{0P}}$$

as $t \to +\infty$.

The power-off PTC thermistor is then equivalent to a linear passive resistor with resistance equal to R_{0P} at ambient temperature 300 K. Such a memristive system is said to be volatile because the initial condition $T(0)$ (which accounts for the past history of $T(t)$) has no effect on the value of the small-signal resistance $(\delta v(t)/\delta i(t))$ measured from this device by applying an infinitesimal voltage $\delta v(t)$ across the device and measuring the corresponding current response $\delta i(t)$. Said another way, the effect of the past input signals that this memristive device has been subjected to is forgotten once a sufficient time has elapsed.

Similar considerations hold for the *Negative Temperature Coefficient (NTC) thermistor* defined by

$$i = W(T)v$$

and

$$\dot{T} = \frac{\delta_N}{H_{CN}}(T_{0N} - T) + \frac{W(T)v^2}{H_{CN}}$$

where

$$W(T) = \left[R_{0N} \exp\left\{ \beta_N \left(\frac{1}{T} - \frac{1}{T_{0N}} \right) \right\} \right]^{-1}$$

and δ_N, H_{CN}, T_{0N}, R_{0N}, β_N are positive parameters.

Example 2.17 (Discharge Tube) A *discharge tube* can be described as a current-controlled memristive device via the state-dependent Ohm's law

$$v = \left(\frac{F}{n} \right) i$$

$$\frac{dn}{dt} = -\beta n + \alpha \left(\frac{F}{n} \right) i^2$$

where α, β, and F are device constants depending on the dimensions of the tube and gas filling, and n is the internal scalar state variable. The POP gives the unique EP $\bar{n} = 0$ that attracts all solutions, thus the discharge tube operates as an open circuit at the steady state.

Example 2.18 (Potassium and Sodium Ions Channels in the Hodgkin–Huxley Model) The renowned *Hodgkin–Huxley neuron model* (HH model) provides a description of neurophysiology and action potential propagation in the squid giant axon membrane. The squid was chosen by Hodgkin and Huxley because it has axons of almost one millimeter in diameter and the *Hodgkin–Huxley equation* was derived strictly by empirical methods.

The electric behavior of the Hodgkin–Huxley axon membrane is described by the net ion flow through a great number of ion channels. The Hodgkin–Huxley axon circuit model (see [23]) takes into account the electric current flowing through the cell membrane during activation due to four current components: (1) current $i_{Na}(t)$ carried by sodium ions; (2) current $i_K(t)$ carried by potassium ions; (3) current $i_L(t)$ carried by other ions (designated *leakage current*, mainly due to chloride ions); (4) capacitive current. As discussed in [1], the Hodgkin–Huxley axon circuit model may be replaced by the memristor-based Hodgkin–Huxley axon circuit model because:

- the potassium ion-channel, described by a potassium time-varying conductance $g_K[n]$, is a first-order memristor defined by (denoting $v = v_K + E_K$):

$$\frac{dn}{dt} = f_K(v_K, n) = \frac{0.01(v+10)}{\exp\left(\frac{v+10}{10}\right) - 1}(1-n) \\ -0.125\exp\left(\frac{v}{80}\right) n \tag{2.21}$$

$$i_K = g_K[n]v_K = \left(\bar{g}_K\, n^4\right) v_K \tag{2.22}$$

- the sodium ion-channel, described by a sodium time-varying conductance, $g_{Na}[m, h]$ is a second-order memristor defined by (denoting $v_{Na} = v + E_{Na}$)

$$\frac{dm}{dt} = \frac{0.1(v+25)}{\exp\left(\frac{v+25}{10}\right) - 1}(1-m) \\ -4\exp\left(\frac{v}{18}\right) m \tag{2.23}$$

$$\frac{dh}{dt} = 0.07\exp\left(\frac{v}{20}\right)(1-h) \\ -\frac{1}{\exp\left(\frac{v+30}{10}\right) + 1} h \tag{2.24}$$

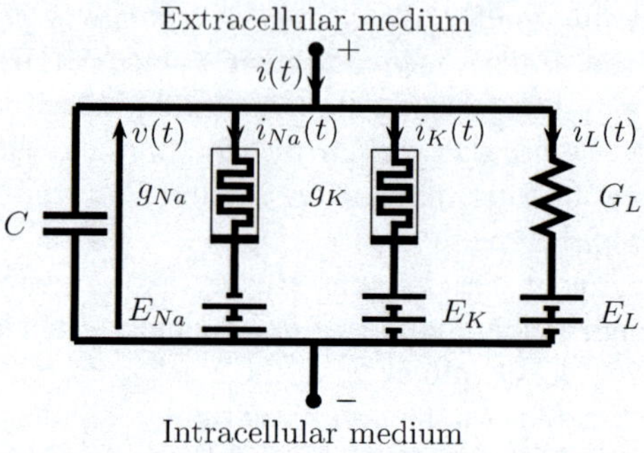

Fig. 2.21 Memristor-based Hodgkin–Huxley axon membrane circuit model derived in [1]. Circuit parameters are: $C = 1\,\mu\text{F/cm}^2$, $G_L = 0.3\,\text{mS/cm}^2$, $E_{Na} = 115\,\text{mV}$, $E_K = 12\,\text{mV}$, $E_L = -10.613\,\text{mV}$, $g_K[n(t)] = \bar{g}_K\, n^4(t)$, and $g_{Na}[m(t), h(t)] = \bar{g}_{Na}\, h(t)\, m^3(t)$, where $\bar{g}_K = 36\,\text{mS/cm}^2$, $\bar{g}_{Na} = 120\,\text{mS/cm}^2$ whereas $n(t)$, $m(t)$, and $h(t)$ are governed by Eqs. (2.21), (2.23), and (2.24), respectively

$$i_{Na} = g_{Na}[m, h]v_{Na} = \left(\bar{g}_{Na}\, h\, m^3\right)\, v_{Na}$$

The memristor-based Hodgkin–Huxley axon circuit model, with the time-varying potassium and sodium conductances respectively replaced by a first- and second-order memristor, is shown in Fig. 2.21, where the values of all the circuit parameters are also reported. The POP associated with the potassium ion-channel memristor shows that it is volatile. The sodium ion-channel memristor cannot be studied by means of the POP because it has two internal state variables (m, h).

2.4 Genealogy of Memristor Devices

For practical and didactic purposes, it is useful to introduce a hierarchical classification of memristive devices and systems [4, 24]. The following categorization and nomenclature are commonly used to distinguish the different complexity in the state-dependent Ohm's law describing a memristive device:

- *ideal* memristor
- *ideal generic* memristor
- *generic* memristor
- *extended* memristor

Hereinafter, we refer to the nomenclature reported above.

2.4.1 Ideal Memristor

The *ideal memristor* has been defined in Chap. 1 and its properties are extensively discussed in Sects. 2.1 and 2.2. For the sake of completeness of this section, here we repeat the CR in the flux-charge domain of flux-controlled (resp., charge-controlled) ideal memristor $q = \hat{q}(\varphi)$ (resp., $\varphi = \hat{\varphi}(q)$). Hence, the corresponding state-dependent Ohm's law (DAE) in the voltage-current domain is:

- $i = \hat{q}'(\varphi)v$ and $\dot{\varphi} = v$ for a flux-controlled ideal memristor
- $v = \hat{\varphi}'(q)i$ and $\dot{q} = i$ for a charge-controlled ideal memristor.

No passive solid-state device with a memristive behavior was known when L. O. Chua theoretically envisioned the existence of memristor [2]. To support the concept of memristor, L. O. Chua proposed an ideal memristor emulator based on a two-port network named *mutator*, as illustrated next.

Example 2.19 (Memristor Implementation via a Mutator) The schematic of a type-1 memristor-resistor (MR) mutator is shown in Fig. 2.22. This is an active dynamic two-port network whose basic implementation requires the use of two controlled sources. The CRs of a type-1 MR mutator are

$$v_1 = \frac{d}{dt}v_2$$

$$i_1 = -\frac{d}{dt}i_2.$$

Suppose to connect to port 2 of the MR mutator a nonlinear resistor with CR $f(v_R, i_R) = 0$ (Fig. 2.23). Then, we obtain

$$v_R(t) = v_2(t) = \int_{-\infty}^{t} v_1(\tau)d\tau = \varphi_1(t)$$

and

$$i_R(t) = -i_2(t) = \int_{-\infty}^{t} i_1(\tau)d\tau = q_1(t).$$

The electrical variables at port 1 thus satisfy

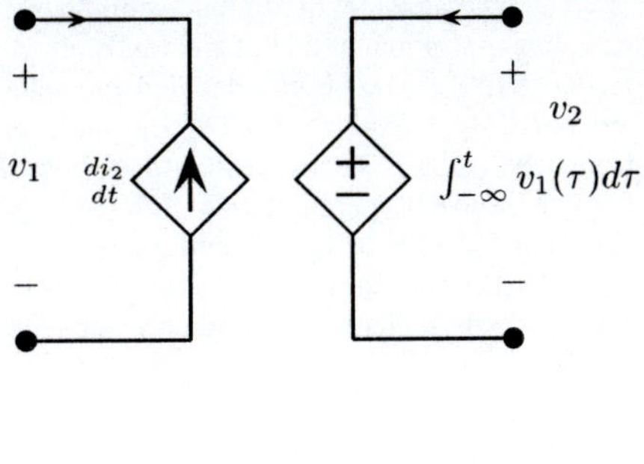

Fig. 2.22 Schematic of an active dynamic two-port network implementing a type-1 memristor-resistor mutator

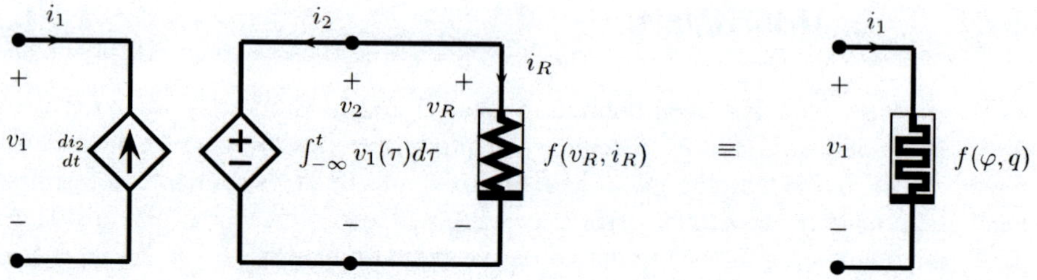

Fig. 2.23 A type-1 memristor-resistor mutator transforms a nonlinear resistor with CR $f(v_R, i_R) = 0$ into a memristor with CR $f(\varphi, q) = 0$

$$f(\varphi_1, q_1) = 0$$

hence the type-1 MR mutator transforms the nonlinear resistor $f(v_R, i_R) = 0$ in a memristor with CR $f(\varphi, q) = 0$.

Overall there are six types of different mutators that enable to transform a nonlinear resistor R, inductor L, or capacitor C into a memristor [2].

L. O. Chua also proposed a mutator realization using active elements. Such an implementation is quite complex since it needs a relatively large number of transistors and operational amplifiers.[7]

2.4.2 *Ideal Generic Memristor and the HP Memristor*

A memristive device is called an *ideal generic memristor* if it's defined by a state-dependent Ohm's law that assumes the form:

- current-controlled ideal generic memristor:

$$v = R(\mathbf{x})i \tag{2.25}$$

$$\frac{d\mathbf{x}}{dt} = \mathbf{f}(\mathbf{x})i \tag{2.26}$$

[7]More recently, other mutator implementations with a simpler structure have been devised [25]. Several other analog or digital techniques have been proposed to implement devices whose behavior approximates that of an ideal memristor without resorting to using a mutator. We refer the reader to [25] and [26] for a detailed account of the current state of the art. Such main techniques are based on using various current-mode building blocks such as second-generation Current Conveyor (CCII), electronically tunable CCII, operational transconductance amplifier (OTA), current feedback operational amplifier (CFOA), differential difference current conveyor, current conveyor transconductance amplifier (CCTA), and differential voltage CCTA. Also, relatively simple structures based on MOS transistors have been proposed (see [26] and references therein). Wave-digital emulators of memristors have been devised in [27].

- voltage-controlled ideal generic memristor:

$$i = G(\mathbf{x})v \tag{2.27}$$

$$\frac{d\mathbf{x}}{dt} = \mathbf{g}(\mathbf{x})v \tag{2.28}$$

where $\mathbf{x} = (x_1, x_2, \ldots, x_n)^T \in \mathbb{R}^n$ is the vector of state variables. If $n = 1$ and $f(x) = 1$ ($g(x) = 1$), then $x = q$ ($x = \varphi$), and so an ideal generic memristor reduces to an ideal memristor, namely:

- charge-controlled ideal memristor:

$$v = R(q)i = \hat{\varphi}'(q)i \tag{2.29}$$

$$\frac{dq}{dt} = i \tag{2.30}$$

- flux-controlled ideal memristor:

$$i = G(\varphi)v = \hat{q}'(\varphi)v \tag{2.31}$$

$$\frac{d\varphi}{dt} = v. \tag{2.32}$$

It is noted that, in Chap. 1, we used the notation $M(q)$ (resp., $W(\varphi)$) for the memristance (resp., memductance) of an ideal memristor.

Given an ideal memristor, it is possible to create an infinite number of ideal generic memristor siblings by choosing a one-to-one differentiable function [4]. On the other hand, the one-to-one correspondence can be also used to recover an ideal memristor from its ideal generic memristor siblings.

The next examples illustrate such procedure for charge-controlled memristors having a single scalar state variable x, but a similar approach can be used, *mutatis mutandis*, for flux-controlled memristors.

Example 2.20 (Create an Ideal Generic Memristor Sibling from an Ideal Memristor) Let us consider a charge-controlled ideal memristor with CR

$$\varphi = \hat{\varphi}(q) = q + \frac{1}{3}q^3$$

and its corresponding state-dependent Ohm's law in the (v, i)-domain

$$v = \hat{\varphi}'(q)i = (1 + q^2)i \tag{2.33}$$

$$\dot{q} = i. \tag{2.34}$$

By introducing a cubic function $q = x^{\frac{1}{3}}$, i.e., a differentiable one-to-one mapping $q = h(x)$ between the charge q and the internal scalar state variable x, then the CR becomes

$$\varphi = \hat{\varphi}(h(x)) = x^{\frac{1}{3}} + \frac{1}{3}x$$

and so the state-dependent Ohm's law of an ideal generic memristor sibling is obtained

$$v = (1 + x^{\frac{2}{3}})i = R(x)i$$

$$\dot{x} = (3x^{\frac{2}{3}})i = f(x)i.$$

A different choice of the function $q = h(x)$ gives a different state-dependent Ohm's law for the ideal generic memristor sibling, but all siblings are equivalent to the original ideal memristor.

It is easily derived that $\dot{x} = f(x)i = f(x)\dot{q}$ allows us to derive

$$q = h(x) = \int \frac{1}{f(x)} dx.$$

Example 2.21 (Recover an Ideal Memristor from Its Ideal Generic Memristor Sibling) Let us consider a current-controlled ideal generic memristor (2.25) and (2.26). The following steps permit to recover the ideal memristor from its siblings:

1. Calculate

$$q = h(x) = \int \frac{1}{f(x)} dx$$

2. Calculate the inverse function

$$x = h^{-1}(q)$$

3. Substitute $h^{-1}(q)$ for x in $R(x)$, that is

$$R(x)_{|x=h^{-1}(q)} = \hat{\varphi}'(q)$$

4. Write the state-dependent Ohm's law for the charge-controlled ideal memristor

$$v = \hat{\varphi}'(q)i$$

$$\dot{q} = i.$$

By using $f(x) = 3x^{\frac{2}{3}}$ and $R(x) = 1 + x^{\frac{2}{3}}$ in the Example 2.20, Step 1 provides $q = x^{\frac{1}{3}} = h(x)$, Step 2 gives $x = q^3$, and then $\hat{\varphi}'(q) = 1 + q^2$ is derived in Step 3. Finally, the ideal memristor in (2.33) and (2.34) is obtained.

The above examples make clear that:

- *every ideal memristor spawns an infinite family of equivalent ideal generic memristor siblings*;
- *two ideal generic memristors whose state variables are related by a one-to-one differentiable transformation have identical signatures and properties* (e.g., pinched hysteresis loops). This property is named *memristor state variable transform property*.

For example, the HP memristor [28] is a sibling of an ideal memristor and this is true also for memristor devices described by a mathematical models obtained from the HP memristor and based on window functions introduced to bound the state variables domain [29–32].

2.4.2.1 The HP Memristor

The basic structure of the HP memristor [28] is shown in Fig. 2.24. The HP nano-scale device is a thin film of Titanium Dioxide ($\mathtt{TiO_2}$) sandwiched between two Platinum ($\mathtt{Pt}$) contacts. The film consists of a conductive layer of Oxygen-deficient Titanium Dioxide ($\mathtt{TiO}_{2-x}$, where $\mathtt{x} = 5\,\%$) and of an insulating layer of stoichiometric Titanium Dioxide ($\mathtt{TiO_2}$). The film conductivity may be suitably adjusted by applying an external input to the nano-device. The linear model, originally proposed in [28, 33], assumes constant ionic drift rate throughout the film length under application of an external input.

Denote by D the thickness of the semiconductor film, and by w, where $0 \leq w \leq D$, the input-controlled length of conductive layer (i.e., the thickness of the doped part having low resistance). Furthermore, $v(t)$ and $i(t)$ respectively denote voltage across and current through the nano-structure.

Fig. 2.24 Structure of the HP memristor device

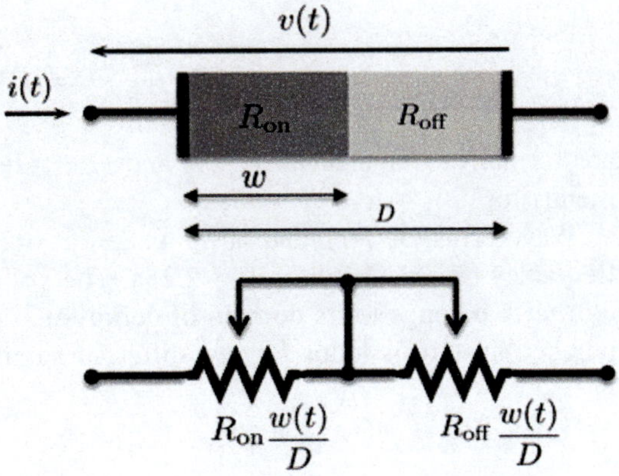

The state-dependent Ohm's law of HP memristor can be written as[8]

$$v(t) = \left(R_{\text{on}} \frac{w(t)}{D} + R_{\text{off}} \left(1 - \frac{w(t)}{D} \right) \right) i(t)$$

where R_{off} is the resistance when $w = 0$ and the full device is undoped, while R_{on} is the resistance when $w = D$, i.e., it is fully doped.

Vacancies (positive ions) can move when subject to a vector field. For simplicity, a uniform electric field is assumed through the device and it is also supposed a linear relationship between drift-diffusion velocity and the electric field, that is ions have a constant average mobility μ_V. Such physical assumptions allow to write the differential equation governing the (scalar) state variable $w(t)$

$$\frac{dw(t)}{dt} = \mu_V \frac{R_{\text{on}}}{D} i(t). \tag{2.35}$$

By integrating we have

$$w(t) = \mu_V \frac{R_{\text{on}}}{D} q(t) \tag{2.36}$$

and, assuming $R_{\text{on}} \ll R_{\text{off}}$, it turns out that

$$v(t) = R(q(t))i(t) \tag{2.37}$$

where we have let

$$R(q(t)) = R_{\text{off}} \left(1 - \frac{\mu_V R_{\text{on}}}{D^2} q(t) \right). \tag{2.38}$$

Equation (2.37), together with

$$\frac{dq(t)}{dt} = i(t)$$

are formally equivalent to (2.29) and (2.30) describing a charge-controlled ideal memristor.

Waveforms of $i(t)$ and $w(t)$ under a sinusoidal voltage $v(t)$ with angular frequency ω_0 are shown in Fig. 2.25a. The voltage amplitude is set in such a way that $w(t)$ belongs to its domain of definition, i.e., $w \in [0, D]$. The corresponding pinched hysteresis loops for two different angular frequencies ω_0 and $10\omega_0$ are in

[8]The dimensionless state variable $x(t) = \frac{w(t)}{D}$ is commonly used in several publications. In such a case x is bounded within the domain [0, 1].

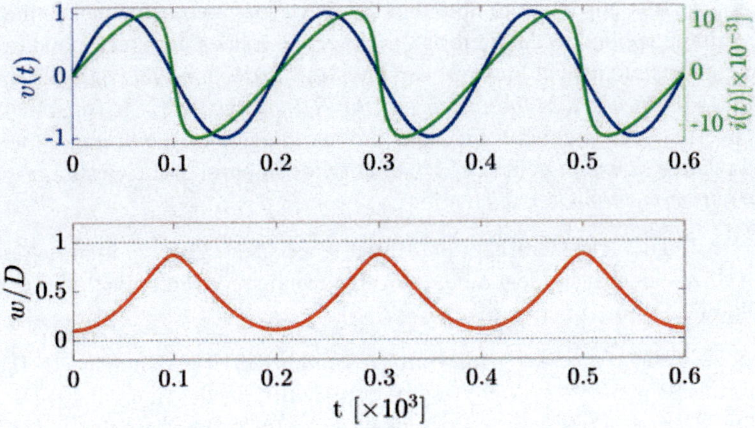

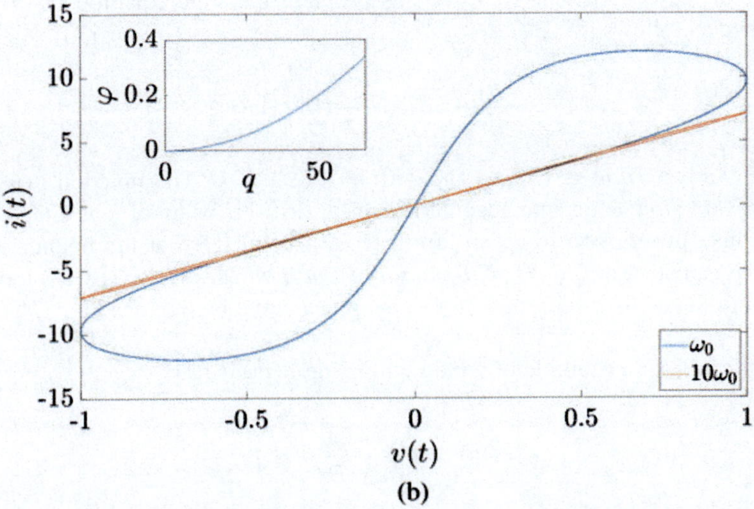

Fig. 2.25 (**a**) Time-domain behavior of the current $i(t)$ and of $w(t)$ when the HP memristor is subject to a sinusoidal voltage $v(t)$ with angular frequency ω_0. (**b**) Hysteresis loops displayed in response to sinusoidal voltages with the same amplitude and different angular frequencies ω_0 and $10\omega_0$. The inset represents the monotone increasing flux-charge characteristic of the memristor

Fig. 2.25b. According to the signatures of ideal memristor, the pinched hysteresis loop tends to a straight line for large frequencies (see the red line in Fig. 2.25b).

In addition, the passive HP memristor has a ***monotone flux-charge characteristic*** obtained by integrating $R(q)$, i.e.,

$$\varphi = \hat{\varphi}(q) = R_{\text{off}}q - \frac{1}{2}\frac{\mu_V R_{\text{on}} R_{\text{off}}}{D^2}q^2 \tag{2.39}$$

for not too large values of $|q|$, as shown in the inset of Fig. 2.25b.

Remark 2.6 It is important to note that the flux φ has to be intended as the integral of the voltage applied to the memristor. However, it does not correspond to the flux of a magnetic field as that known from physics.[9] Indeed, *no flux of a magnetic field appears to be involved in the physics of the HP memristor* [28]. To make clear the difference between magnetic flux and flux as integral of the voltage, the latter is named *voltage momentum* in [24]. By duality the integral of the current results to be the *current momentum*, a.k.a. charge [24].

Remark 2.7 The HP memristor is actually an ideal generic memristor and its sibling correspond to an ideal memristor that can be recovered via the differentiable globally one-to-one function given in (2.36).

Remark 2.8 (HP Memristor with Window Functions) The length w of the doped part of an HP memristor is physically constrained in the interval $[0, D]$, but the equation (2.35) describing the dynamics of w does not guarantee that this constraint is automatically satisfied. To solve this problem the state equation (2.35) can be modified as follows [29, 30]:

$$\frac{dw(t)}{dt} = \mu_V \frac{R_{\text{on}}}{D} i(t)\zeta(w(t))$$

where $\zeta(w) > 0$, $w \in (0, D)$, and $\zeta(0) = \zeta(D) = 0$. The function $\zeta(w)$, a.k.a. *window function*, takes into account nonlinear drift-diffusion of ions and vacancies and it also prevents w to escape from $[0, D]$, being zero at the boundaries, i.e., when w can stick at 0 or D. A *Boundary Condition Memristor* (BCM) model has been introduced to specify complex phenomena at the boundaries [31, 32].

Example 2.22 An example of window function $\zeta(w)$ is [29]

$$\zeta(w) = 1 - \left(\frac{2w}{D} - 1\right)^{2p}$$

where p is a positive integer. Appendix 2 reports further details on window functions that are used to develop mathematical models of memristor devices.

2.4.3 *Generic Memristor*

A memristor is called a *generic memristor* if it's defined by the following state-dependent Ohm's law:

---

[9] It is worth to observe that there exists no magnetic field in a synthetic inductor obtained via a gyrator [7], thus in such a case the flux is just the integral of the voltage!

- current-controlled generic memristor

$$v = R(\mathbf{x})i \tag{2.40}$$

$$\frac{d\mathbf{x}}{dt} = \mathbf{f}(\mathbf{x}, i) \tag{2.41}$$

- voltage-controlled generic memristor

$$i = G(\mathbf{x})v \tag{2.42}$$

$$\frac{d\mathbf{x}}{dt} = \mathbf{g}(\mathbf{x}, v) \tag{2.43}$$

where $\mathbf{x} = (x_1, x_2, \ldots, x_n)^T$ is once more a vector of internal state variables.

Note the ideal generic memristor differs from a generic memristor only in the form of the differential equation for the state variables; in the Eq. (2.26) (resp., (2.28)) the r.h.s. is factorized so that $\mathbf{f}(\cdot)$ (resp., $\mathbf{g}(\cdot)$) depends only on the state vector $\mathbf{x}$, whereas in the r.h.s. of (2.41) (resp., (2.43)) $\mathbf{f}(\cdot)$ (resp., $\mathbf{g}(\cdot)$) depends on both $\mathbf{x}$ and i (resp., v).

The PTC and NTC thermistors in Example 2.16, the discharge tube in Example 2.17, and the potassium and sodium ion channels of the Hodgkin–Huxley neuron model in Example 2.18 belong to the class of generic memristors. The following example presents a generic memristor as well.

Example 2.23 The ThrEshold Adaptive Memristor (TEAM) model proposed in [34] is a passive Generic Memristor satisfying

$$v = R(x)i$$

where

$$\frac{dx(t)}{dt} = \begin{cases} k_{\text{off}} \left(\frac{i(t)}{i_{\text{off}}} - 1 \right)^{\alpha_{\text{off}}} f_{\text{off}}(x(t)), \; 0 < i_{\text{off}} < i \\ 0, \qquad\qquad\qquad\qquad\qquad i_{\text{off}} < i < i_{\text{on}} \\ k_{\text{on}} \left(\frac{i(t)}{i_{\text{on}}} - 1 \right)^{\alpha_{\text{on}}} f_{\text{on}}(x(t)), \; i < i_{\text{on}} < 0. \end{cases}$$

In these equations, $k_{\text{off}}, k_{\text{on}}, \alpha_{\text{off}}, \alpha_{\text{on}}$ are parameters ($k_{\text{off}} > 0$, $k_{\text{on}} < 0$) while i_{off} and i_{on} ($i_{\text{on}} < 0 < i_{\text{off}}$) represent current thresholds. Functions $f_{\text{off}}(x) > 0$ and $f_{\text{on}}(x) > 0$, which are not necessarily equal, have the role of window functions and they serve to constrain the state variable x in the interval $[x_{\text{on}}, x_{\text{off}}]$.

By turning off power, i.e., letting $i = 0$, we simply obtain

$$\frac{dx(t)}{dt} = 0$$

which means that any $x \in [x_{\text{on}}, x_{\text{off}}]$ is a stable EP. The TEAM model is then a nonvolatile generic memristor. This result is in accordance with Theorem 2.2.

2.4.4 Extended Memristor

A memristor is called an *extended memristor* if it's defined by a state-dependent Ohm's law in which the memristance $R(\mathbf{x}, i)$ (resp. memductance $G(\mathbf{x}, v)$) has the following properties:

(a) it is a function of not only the state variables $\mathbf{x}$, but also of the input current i (resp., voltage v)
(b) it assumes only finite values when $i = 0$ (resp. $v = 0$), thus the memristance $R(\mathbf{x}, 0)$ (resp., the memductance $G(\mathbf{x}, 0)$) is a bounded and differentiable function in a neighborhood of $(\mathbf{x}, 0)$ for any $\mathbf{x} \in \mathbb{R}^n$.

Hence, an extended memristor is described by the following DAEs in the (v, i)-domain:

- current-controlled extended memristor

$$v = R(\mathbf{x}, i)i \tag{2.44}$$

$$\frac{d\mathbf{x}}{dt} = \mathbf{f}(\mathbf{x}, i) \tag{2.45}$$

  where $R(\mathbf{x}, i)$ is bounded in a neighborhood of $(\mathbf{x}, 0)$ for any $\mathbf{x}$
- voltage-controlled extended memristor

$$i = G(\mathbf{x}, v)v \tag{2.46}$$

$$\frac{d\mathbf{x}}{dt} = \mathbf{g}(\mathbf{x}, v) \tag{2.47}$$

  where $G(\mathbf{x}, v)$ is bounded and differentiable in a neighborhood of $(\mathbf{x}, 0)$ for any $\mathbf{x} \in \mathbb{R}^n$.

The $\mathtt{Pt/TaO_x/Ta}$-memristor in Example 2.15 is an extended memristor. A further example is as follows.

Example 2.24 The $\mathtt{NbO_2}$–Mott memristor device in [35] is a passive voltage-controlled extended memristor defined by the state-dependent Ohm's law

$$i = G(x, v)v$$
$$= \left[\sigma_0 e^{-\frac{0.301}{2k_B T}} A \left\{ \left(\frac{k_B T}{\omega}\right)^2 \left(1 + \left(\frac{\omega\sqrt{v/d}}{k_B T} - 1\right) e^{\frac{\omega\sqrt{v/d}}{k_B T}}\right) + \frac{1}{2d} \right\}\right] v$$

where $\sigma_0, A, k_B, \omega, d$ are parameters and k_B is the Boltzmann constant. The state variable is given by the temperature T which obeys Newton's law of cooling

$$\frac{dT}{dt} = \frac{iv}{C_{\text{th}}} - \frac{T - T_{\text{amb}}}{C_{\text{th}} R_{\text{th}}(T)}$$

where $T_{\text{amb}} = 300$ K is the ambient temperature, C_{th} is the thermal capacitance, and $R_{\text{th}}(T)$ is the temperature-dependent effective thermal resistance of the device modeled by

$$R_{\text{th}}(T) = \begin{cases} 1.4 \times 10^6, & T \le T_C \\ 2.0 \times 10^6, & T > T_C \end{cases}$$

where $T_C = 1,070$ K is the Mott metal-insulator transition temperature.

By turning off power, i.e., letting $v = 0$, we have

$$\frac{dT}{dt} = -\frac{T - T_{\text{amb}}}{C_{\text{th}} R_{\text{th}}(T)}$$

which has a unique EP $T = T_{\text{amb}}$ attracting all solutions. We conclude that, as it happens for PTC and NTC thermistors in Example 2.16, the NbO_2-Mott memristor device is volatile.

Example 2.25 (The First Example of Extended Memristor Made of Passive Electronic Components) **The first and unique example of extended memristor made of** passive electronic components is reported in [36]. Consider the circuit of Fig. 2.26. A voltage source v_g **is applied across a full-wave rectifier cascaded with a RLC series filter. The two-terminal RLC filter connected to the output terminal of the** diode bridge can be substituted by any linear dynamical bipole [37]. The relation between the voltage across and the current through diode D_k, named as v_k and i_k respectively ($k = \{1, 2, 3, 4\}$), is modeled as

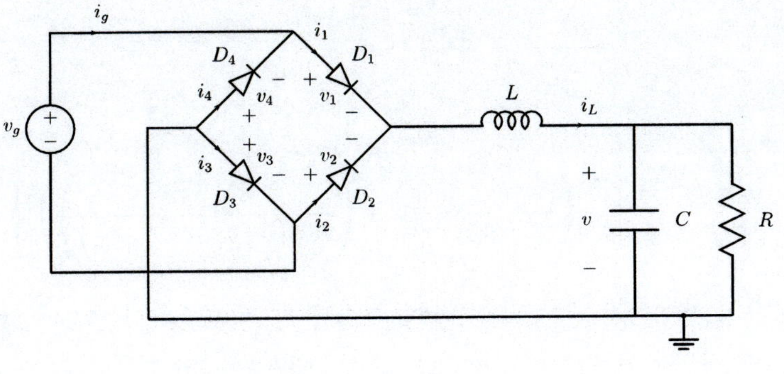

Fig. 2.26 Extended memristor based on a four diode-bridge and standard passive components

$$i_k = I_S\left(\exp\left(\frac{v_k}{nV_T}\right) - 1\right)$$

where I_S symbolizes the reverse saturation current, n is the emission coefficient, and $V_T = KTq^{-1}$ stands for the thermal voltage, where $K = 1.38 \cdot 10^{-23}$ J K^{-1} is the Boltzmann's constant, T represents the absolute temperature, and $q = 1.6 \cdot 10^{-19}$ C refers to the elementary electronic charge.

The application of Kirchhoff's laws to the diode bridge permits to show that at each time instant the voltages across diodes satisfy the following constrains:

$$v_1 = v_3$$

and

$$v_2 = v_4$$

where $v_2 = v_1 - v_g$ and v_1 is found to be given by

$$v_1 = nV_T \ln\left(\frac{i_L + 2I_S}{2I_S \exp\left(-\frac{v_g}{2nV_T}\right)\cosh\left(\frac{v_g}{2nV_T}\right)}\right). \tag{2.48}$$

Using the CRs of the dynamic elements (i.e., C and L), some simple algebraic manipulations permit to derive a closed-form expression for i_g and then the following state-dependent Ohm's law is obtained

$$i_g = (i_L + 2I_S)\tanh\left(\frac{v_g}{2nV_T}\right) \tag{2.49}$$

where the state variables $x_1 = v/V_T$ and $x_2 = i_L/I_S$ are governed by the differential equations

$$\frac{d\mathbf{x}}{d\tau} = \mathbf{g}(\mathbf{x}, v_g) \tag{2.50}$$

with

$$\mathbf{g}(\mathbf{x}, v_g) = \begin{bmatrix} \beta(x_2 - \alpha x_1) \\ \gamma\left(\frac{v_g}{2nV_T} - x_1 - 2\ln\left(\frac{x_2+2}{2\exp\left(-\frac{v_g}{2nV_T}\right)\cosh\left(\frac{v_g}{2nV_T}\right)}\right)\right) \end{bmatrix}$$

where $\tau = \frac{t}{t_0}$, $\alpha = \frac{V_T}{RI_S}$, $\beta = \frac{I_S t_0}{CV_T}$, and $\gamma = \frac{V_T t_0}{LI_S}$ are dimensionless parameters and $t_0 = 2\pi/\omega_0$ stands for the time normalization factor and $\omega_0 = [(LC)^{-1} - (RC)]^{\frac{1}{2}}$ denotes the resonant frequency of the second-order low-pass. Equations (2.49) and (2.50) are nothing but the defining Eqs. (2.46) and (2.47) for the extended memristor

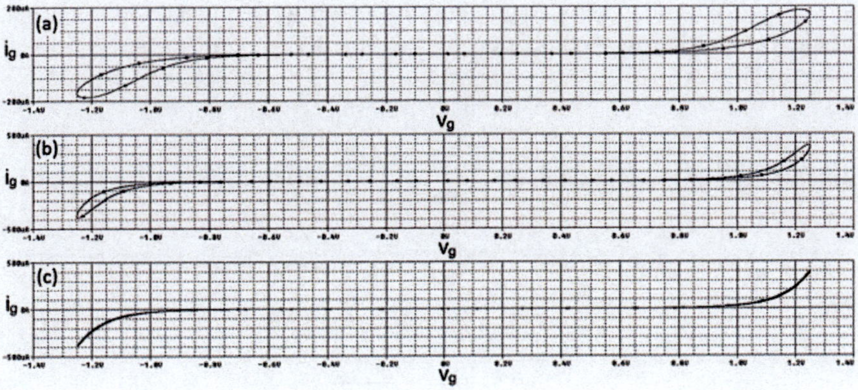

Fig. 2.27 Current-voltage characteristics derived from PSpice simulations of the circuit of Fig. 2.26 under sine-wave input with frequency equal to 10 (plot (**a**)), 100 (plot (**b**)), and 1000 Hz (plot (**c**))

of the proposed circuit. In particular, the memductance $G(\mathbf{x}, v)$ in (2.46) can be derived from the r.h.s. of (2.49) as follows:

$$G(\mathbf{x}, v_g) = \frac{i_L + 2I_S}{v_g} \tanh\left(\frac{v_g}{2nV_T}\right)$$

that is bounded in a neighbor of $(\mathbf{x}, 0)$ (see also the next Example 2.26). This proves that the elementary circuit of Fig. 2.26 is a *second-order extended memristor*. Note that the key mechanisms at the origin of its memristive behavior are the voltage constraints involving each pair of parallel diodes, i.e., $v_1 = v_3$ and $v_2 = v_4$. Figure 2.27 shows the pinched hysteresis loops derived from PSpice simulations of the circuit in Fig. 2.26 subject to sine-wave inputs with different frequencies.

It is worth to note that if we drop the condition (b) on the boundedness of $R(\mathbf{x}, i)$ or $G(\mathbf{x}, i)$ then in general (2.44) and (2.45) or (2.46) and (2.47) do not define a memristor, since the zero-crossing property might fail ([4, p. 339]). The next Example 2.26 shows how to exploit condition (b) to write the state-dependent Ohm's law in the form (2.46) whereas Example 2.27 makes clear that if such a property (b) is allowed to fail then we may easily arrive at a nonsense where a two-terminal circuit element without memristive behavior might fall into the class of extended memristors.

Example 2.26 (State-Dependent Ohm's Law in the Form (2.46)) Consider a voltage-controlled extended memristor defined by a general DAE

$$i = h(x, v)$$
$$\dot{x} = g(x, v)$$

where $x \in \mathbb{R}$. We aim to show that $h(x, v)$ can be factorized in the form of a state-dependent Ohm's law (2.44) under the assumption that, for any $x \in \mathbb{R}$

$$\lim_{v \to 0} h(x, v) = 0.$$

In other words, we want to write

$$i = h(x, v) = G(x, v)v$$

where

$$G(x, v) = \frac{h(x, v)}{v}$$

is the memductance. Note that $G(x, 0) = 0/0$ is indeterminate, however, the memductance $G(x, 0)$ is bounded in a neighbor of $(x, 0)$ for any x. By de l'Hôpital rule we obtain

$$G(x, 0) = \lim_{v \to 0} \frac{h(x, v)}{v} = \lim_{v \to 0} \frac{\frac{\partial h(x,v)}{\partial v}}{\frac{\partial (v)}{\partial v}} = \lim_{v \to 0} \frac{\partial h(x, v)}{\partial v} \doteq \frac{\partial h(x, 0)}{\partial v}.$$

As an application, reconsider the $\mathtt{Pt/TaO_x/Ta}$ memristor device in the Example 2.15, which is defined as

$$i = h(x, v) = \alpha(1 - e^{-\beta v}) + \gamma x \sinh(v\delta) = G(x, v)v$$

for any $x \in [0, 1]$. According to de l'Hôpital rule, we obtain

$$G(x, 0) = \frac{\partial h(x, 0)}{\partial v} = \alpha\beta + \gamma\delta x.$$

Example 2.27 (Extended Memristor is Not for Everything!) Consider a two-terminal circuit element defined by a DAE

$$v = R(x, i)i \tag{2.51}$$

where

$$R(x, i) = \frac{x}{i}$$

and the scalar state variable $x \in \mathbb{R}$ satisfies the equation

$$\frac{dx}{dt} = f(x, i) = i.$$

It is easily seen that such two-terminal device, defined by equations similar to (2.44) and (2.45) without the constrain (b) on the boundedness of $R(x, i)$, would be classified as an extended memristor.

To check the zero-crossing property and then pinched hysteresis loops, let us apply a current source $i = \sin(t)$ and assume the initial state $x(0) = -1$. This yields

$$x(t) = x(0) + \int_0^t \sin(\tau)d\tau = -\cos(t).$$

Substituting for x in (2.51), we obtain

$$v(t) = \left[\frac{x(t)}{i(t)}\right] i(t).$$

The corresponding Lissajoux figure in the (v, i) plane is the *unit circle* shown in Fig. 2.28a, which is *not pinched* at the origin $(v, i) = (0, 0)$!

This is a *shocking result* suggesting that one should accept the notion that not all memristors exhibit the ***pinched hysteresis loop fingerprint***.

Resolving this anomaly leads to a *logical inconsistency* which we will now demonstrate. Following the systematic approach shown in the Examples 2.4 and 2.5, it is easily understood that the problem is the plot of the Lissajoux figure in the wrong plane. Considering the plot of $v(t)$ as function $x(t)$, i.e., the Lissajoux figure in the $v - x$ plane, a single-valued curve without any hysteresis is obtained (Fig. 2.28b). Such curve represents the correct characteristic of the considered two-terminal device and, from the previous equations, it is easily derived that

$$v(t) = x(t) = \int_{-\infty}^t i(\tau)d\tau \doteq q(t)$$

which is just the definition of a 1 *Farad capacitor*! It would certainly be **nonsense to refer to a capacitor as (extended) memristor!**

Remark 2.9 Although an extensive discussion of zero-crossing property and pinched hysteresis loops is reported in Sect. 2.1.3, the Examples 2.4, 2.5, and 2.27 allow us to summarize the following chief statements:

- **the pinched hysteresis loop is the hallmark of all memristors**, and it can be used for an experimental definition of memristors, whether they are ideal or not. The *pinched hysteresis loop* signature of memristors *must* hold for *all* periodic voltage or current inputs with zero time-average. It follows that the measurement of a non-pinched hysteresis loop for even just one such testing signal would disqualify a device from being classified as a memristor

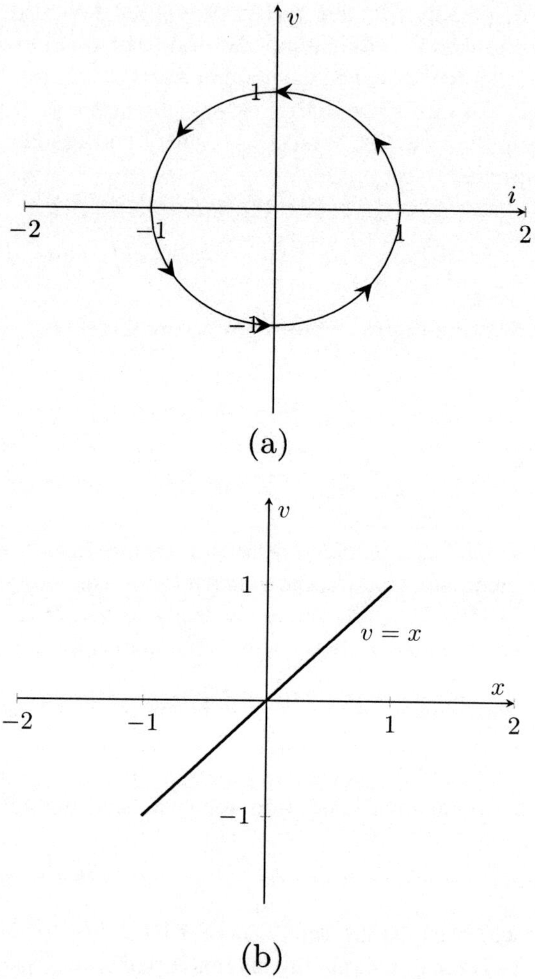

Fig. 2.28 (**a**) Anomalous circular and hence *non-pinched* hysteresis loop. (**b**) The same loop becomes a straight line when plotted in the (v, x) plane (here we have $x = q$)

- **pinched hysteresis loops don't have predicting ability** because they represent just the device response to a specific (sinusoidal or periodic) input.

In conclusion, zero-crossing property is a key necessary condition to be satisfied for any memristor device

If It's Not Pinched, It's Not a Memristor!

but

Pinched hysteresis loops are not models (i.e., constitutive relations).

2.5 Memcapacitors and Meminductors: Properties and Signatures

Memcapacitors and meminductors have been originally introduced by L. O. Chua in [3] as an extension of the memristor concept to lossless elements.[10] As discussed in Chap. 1, these circuit elements can be also axiomatically introduced via a black-box approach based on the periodic table of circuit elements. The approach used in Sect. 2.4 to construct a hierarchical classification of memristor devices from ideal to generic and then to extended memristors permits to generalize the definitions of memcapacitor and meminductor (Chap. 1), as follows:

- voltage-controlled memcapacitor
  - ideal memcapacitor

$$q = \hat{\sigma}'(\varphi)v = C(\varphi)v$$
$$\dot{q} = i$$

  - generic memcapacitor

$$q = C(\mathbf{x})v$$
$$\dot{\mathbf{x}} = \mathbf{g}(\mathbf{x}, v)$$
$$\dot{q} = i$$

    where $\mathbf{x} = (x_1, x_2, \ldots, x_n)^T \in \mathbb{R}^n$ is a vector of state variables
  - extended memcapacitor

$$q = C(\mathbf{x}, v)v$$
$$\dot{\mathbf{x}} = \mathbf{g}(\mathbf{x}, v)$$
$$\dot{q} = i$$

    where $C(\mathbf{x}, 0)$ is a bounded differentiable function in a neighbor of $(\mathbf{x}, 0)$ for any $\mathbf{x}$

- current-controlled meminductor
  - ideal meminductor

$$\varphi = \hat{\rho}'(q)v = L(q)i$$
$$\dot{\varphi} = v$$

[10]The first name and equations of a memcapacitor and meminductor were introduced by L. Chua in the last part of the keynote lecture given at the 1st Symposium on Memristor and Memristive Systems, University of California, Berkeley, November 2008—available online at https://www.youtube.com/watch?v=QFdDPzcZwbs.

– generic meminductor

$$\varphi = L(\mathbf{x})i$$
$$\dot{\mathbf{x}} = \mathbf{f}(\mathbf{x}, i)$$
$$\dot{\varphi} = v$$

where once more $\mathbf{x} = (x_1, x_2, \ldots, x_n)^T \in \mathbb{R}^n$ is a vector of state variables

– extended meminductor

$$\varphi = L(\mathbf{x}, i)i$$
$$\dot{\mathbf{x}} = \mathbf{f}(\mathbf{x}, i)$$
$$\dot{\varphi} = v$$

where $L(\mathbf{x}, 0)$ is a bounded differentiable function in a neighbor of $(\mathbf{x}, 0)$ for any $\mathbf{x}$. By duality, both charge-controlled memcapacitors and flux-controlled meminductors can be classified as ideal, generic, and extended.[11] Recall that the ideal memcapacitor (resp., ideal meminductor) corresponds to the $(-1, -2)$-element (resp., $(-2, -1)$-element) in the Periodic Table of Circuit Elements (Fig. 1.12 in Chap. 1).

Next, we briefly discuss some main properties and signatures that characterize memcapacitors and meminductors.

2.5.1 *Passivity and Losslessness*

Addressing passivity and losslessness of memcapacitors and meminductors is a nontrivial task that is still the subject of investigation. A specific condition that guarantees passivity and losslessness in ideal memcapacitors and ideal meminductors can be derived when they are subject to a sinusoidal input [38, Sect. 2]. We remark that the losslessness property is of special practical interest since it means that, differently from a memristor, in principle memcapacitors and meminductors could store data without dissipating energy.

[11] Ideal generic memcapacitors and meminductors are not introduced because they can be obtained by the corresponding ideal elements via a one-to-one transformation as shown for memristors in the Example 2.20.

2.5.2 *Volatile and Nonvolatile Memory Properties*

Volatile and nonvolatile memory properties of memcapacitors and meminductors can be derived, *mutatis mutandis*, as shown for memristors. In particular, the following attributes are easily obtained:

- ideal memcapacitors (resp., meminductors) present a continuum of nonvolatile memory states associated with the memcapacitance (resp., meminductance) values.

  For a flux-controlled ideal memcapacitor (resp., charge-controlled ideal meminductor) a nonvolatile memory state can be programmed by suitable voltage (resp., current) pulses (similar to Example 2.11 for memristors)
- a generic and extended memcapacitor (resp., meminductor) can act as:
  - nonvolatile memory memcapacitor (resp., meminductor) having both a discrete number or a continuum of memory states
  - volatile memory memcapacitor (resp., meminductor) having a unique memory state.

  Volatile and nonvolatile features can be investigated by means of the POP in voltage-controlled memcapacitors (resp., current-controlled meminductors) with just a scalar state variable x (similar to the Example 2.14 for memristors).

2.5.3 *Zero-Crossing Property and Pinched Hysteresis Loops*

The comparison between the hierarchical definition of ideal, generic, and extended (voltage-controlled) memcapacitor given at the beginning of this section and the corresponding ideal, generic, and extended memristors (respectively in (2.31)–(2.32), (2.42)–(2.43), and (2.46)–(2.47)) makes it clear that the pair (v, q) in memcapacitors plays the role of (v, i) in memristors. Hence, it is apparent that the zero-crossing property between voltage v and current i in memristors turns into the zero-crossing property between voltage v and charge q in memcapacitors.

Thus, the charge $q = 0$ at any instant in which the voltage $v = 0$. However, since $i = 0$ does not imply $v = 0$, a memcapacitor can store energy and deliver the previously absorbed energy back to a circuit.

By duality, zero-crossing property holds for the flux φ and the current i in meminductors as well.

A direct consequence of the coincidence between the zeros of v (resp., i) and those of q (resp., φ) is that the Lissajoux figure of a memcapacitor (meminductor) in the (v, q) plane (resp., (i, φ) plane) gives a *pinched hysteresis loop*.

In conclusion, we can summarize the signatures of memcapacitors and meminductors as done at the end of the previous section:

- **the pinched hysteresis loop in the (v, q) (resp., (i, φ)) plane is the hallmark of all memcapacitors (resp., meminductors)**, and it can be used for an experimental definition of memcapacitors (resp., meminductors), whether they are ideal or not. The *pinched hysteresis loop* signature of memcapacitors (resp., meminductors) *must* hold for *all* periodic voltage or current inputs with zero time-average. It follows that the measurement of a non-pinched hysteresis loop for even just one such testing signal would disqualify a device from being classified as a memcapacitor (resp., meminductor)
- **pinched hysteresis loops of memcapacitors (resp., meminductors) don't have predicting ability** because they represent just the device response to a specific (sinusoidal or periodic) input.

Lastly, zero-crossing property is a key necessary condition to be satisfied for any mem-element device

If It's Not Pinched, It's Not a Mem-Element!

but

Pinched Hysteresis Loops of Mem-Elements Are Not Models!

Remark 2.10 According to the considerations reported above, the example of memcapacitor in [39] exhibiting a *non-pinched hysteresis loop* in the (v, q) plane is misleading. Actually, such an error can be traced back to the associated capacitance tending to infinity at the origin [40]. The paradox is easily solved as shown in the Example 2.27.

Another two-terminal device that supposedly behaves as a memcapacitor and displays a non-pinched hysteresis loop in the (v, q) plane is reported in [41]. As remarked in [40], that device needs to be modeled by the connections of different circuit element types including standard linear capacitors and resistors. Then, on one hand it cannot be classified as just one single memcapacitor. On the other hand, the observed non-pinched hysteresis loop can be traced back to the presence of both standard capacitive and resistive elements, as it can be verified via a simple circuit analysis [40].

Remark 2.11 The article [42] examined a ferromagnetic inductor (FML) to see what new insights might be gained from a nonlinear circuit analysis of this familiar element. The FML is realized using a nanocrystalline ferromagnetic toroidal core of composition $Fe_{73.5}Si_{13.5}B_9Nb_3Cu_1$ and a wire winding of known resistance. The experiments in that paper show that the device can be modeled in a first approximation by a nonlinear inductor, i.e., a $(-1, 0)$-element in series with the winding resistance. It is worth remarking that, contrary to what is claimed elsewhere [39], the element does not correspond to a meminductor, i.e., a $(-2, -1)$ element of the periodic table.

In the next, an example of a generic memcapacitor is presented.

Example 2.28 (Memory Capacitance in Biomimetic Membranes) The work [43] reports the first example of a volatile, voltage-controlled memcapacitor where capacitive memory arises from reversible and hysteretic geometrical changes in a lipid bilayer that mimics the composition and structure of biomembranes.

The system consists of an elliptical, planar lipid bilayer that forms at the interface between two lipid-coated aqueous droplets (about 200 nL each; L denotes Liter) in oil.

The charge is given by

$$q = C(R, W)v$$

where $R(t)$ is the membrane radius, $W(t)$ is the hydrophobic thickness, v is the applied voltage, and C is the memcapacitance. For a parallel-plate capacitor with planar ellipticity a, and equivalent dielectric constant ε (ε_0 is the vacuum dielectric constant), we have

$$C(R, W) = \frac{\varepsilon\varepsilon_0(a\pi R^2(t))}{W(t)}.$$

The state variables $R(t)$ and $W(t)$ are governed by

$$\frac{dR(t)}{dt} = \frac{1}{\xi_{ew}}\left(\frac{a\varepsilon\varepsilon_0}{2W(t)}v^2(t) - k_{ew}(R(t) - R_0)\right)$$

and

$$\frac{dW(t)}{dt} = \frac{1}{\xi_{ec}}\left(\frac{-a\varepsilon\varepsilon_0\pi R^2(t)}{2W^2(t)}v^2(t) - k_{ec}(W(t) - W_0)\right)$$

where $\xi_{ew}, k_{ew}, \xi_{ec}, k_{ec}$ are positive physical parameters (defined in [43]).

Such biomimetic membrane actually falls into the class of generic memcapacitors and it exhibits volatile memory features. In fact, when power is off, i.e., we set $v = 0$, then the state variables $R(t)$ and $W(t)$ tend to a unique steady state (R_0, W_0).

The biomimetic membrane shows a pinched hysteresis loop in the (v, q) plane at low and intermediate frequencies. Hysteretic and reversible changes in the memcapacitance $C(R, W)$ yield complex nonlinear dynamical behaviors, including capacitive short-term facilitation and depression. The biomimetic membrane is then capable of adaptive signal processing and learning via synapse-like, short-term capacitive plasticity. This is potentially useful to develop low-energy, biomolecular neuromorphic mem-elements, which, in turn, could also serve as models to study capacitive memory and signal processing in neuronal membranes. In addition to their possible implementation as memcapacitive synapses for spike recurrent neural networks supporting online learning and computation, biomimetic membranes are expected to have impact in two major areas: (1) modular, low-power materials that could interface with cells and tissues for biosensing and processing of biological signals due to their soft and biocompatible nature; and (2) as a model system to study capacitive excitability in neuronal membranes.

Appendix 1: Memristive Devices: Materials and Complex Physics

The axiomatic approach introduced in Chap. 1 suggests that a realistic circuit model of any molecular/nanodevice will in general require an appropriate interconnection of building blocks chosen from the periodic table of ideal circuit elements. In this context, the "concept of ideal memristor" as introduced by L. O. Chua in 1971 [2] is useful as long as it permits to capture with a suitable degree of accuracy and within a broad range of operation the electrical behavior at the external terminals of real memristor devices. Thus, *the intent of ideal memristor is to be useful in the design of a circuit model of real devices* (e.g., the Josephson junction in Example 1.9 of Chap. 1). Such concepts can be also used to conceive a systematic modus operandi for tuning chemical/physical properties of materials to approximate under suitable conditions an ideal memristor and to empower real memristor devices with novel computation and information processing potentials.

Although the main focus of the book is on nonlinear dynamical circuits and systems including ideal mem-elements, this section aims to provide a basic summary of inorganic/organic materials, nanoscale/molecular structures, and physical phenomena giving rise to an electrical behavior ascribed to memristor. The complexity of technological realizations and of chemical/physical processes results in physical/behavioral models that keep the structure of the original idealized memristor [2] and encapsulate the distinctive property of tuning the resistive state according to the history of the applied input (voltage or current). Along this line, in 2008, researchers at HP headed by S. Williams were the first to show, by *"using a simple analytical example, that memristance arises naturally in nanoscale systems in which solid-state electronic and ionic transport are coupled under an external bias voltage"* [28]. The resistive state of the HP memristor is modulated by the bias voltage applied across two `Pt` electrodes of a metal-insulator-metal structure because the electric field is capable of shifting the interface between $\mathtt{TiO}_2$ and $\mathtt{TiO}_{2-x}$ due to the migration of oxygen vacancies (see Sect. 2.5.3 for more details).

Since 2008, memristor conceptualization has attracted the attention of many researchers in different fields ranging from material science, storage memory devices, nonlinear circuits and complex systems, and unconventional/neuromorphic computing architectures. During the last decade several review works and books have been published in material science and neuromorphic systems. The remaining part of this section gives a compendious list of material structures and physical factors that influence the motion of electron/ion charges for: (a) drift phenomena (electric potential gradient); (b) electromigration (electron kinetic energy); (c) Fick diffusion (concentration gradient); (d) thermophoresis (temperature gradient). Any of the four factors, or a combination of them, may occur to produce different idealized type of resistance modulation or switching (field/thermal dominated unipolar/bipolar) in real-world memristor (switching) devices.

- Material structures
  - **Oxide based memristor**: Similar to the HP memristor, almost all memristors exploit a capacitor like structure with metal electrodes and a switching oxide layer. Switching layers are usually made up of dielectric materials. In particular, binary oxides such as titanium oxide TiO_2, tantalum oxide Ta_2O_5, silicon oxide SiO_2, copper oxide CuO, nickel oxide NiO, zinc oxide ZnO, hafnium oxide HfO_2, and aluminum oxide Al_2O_3 are mainly used. Memristors based on these binary oxide present a good resistive switching performance. The resistive switching properties are typically either a sharp resistance change between two resistive states R_{on} and R_{off} (suitable for binary storage memories) or a progressive resistance adaptation (useful in analog memories for neuromorphic computing). The operation of resistance change is usually referred to as "write operation" and it is performed by applying voltage/current pulses with appropriate amplitude, duration, and polarity on the memristor electrodes. Input pulses with low amplitude are used in "read operation" in order to sense the resistive state of the memristor without inducing any resistance change. Voltage pulses trigger Joule-heating effects that play a crucial role in the formation and destruction of localized conducting paths in the oxide. Although many switching oxide layers are plausible for realizing memristors, a few of them result to be CMOS compatible and lead to superior device performance. For instance, Ta_2O_5, HfO_2, SiO_2, and Al_2O_3 are used in crossbar configurations to implement arrays with cross-point devices embedding a selector device and a memristor (1T1R). Oxide based memristors are also referred to as Resistive Random Access Memories (RRAMs) or Resistive Switching Devices.
  - **Phase Change Memory** (PCM): Memristor based on phase change materials are mainly assembled by means of a chalcogenide. $Ge_2Sb_2Te_5$ (GST) material is one of the most used chalcogenide to build PCM devices. Voltage/current pulses are applied to change the phase of the chalcogenide between crystalline and amorphous due to Joule heating. Read-from and write-to PCM devices are operations that involve quite different current and voltage levels. Although PCM is one of the most promising technology for memristor devices, there are still many technological issues.
  - **Other memristors**: Many studies explore other switching materials besides binary oxides. Such materials include: (a) heterogeneous materials (e.g., cerium oxide CeO_2 and strontium titanate $SrTiO_3$); (b) organic materials [44, 45]; (c) electrolyte materials such as copper sulfide (CuS). Unconventional fabrication methods, CMOS compatibility, and unreliable physical

characteristics limit the use of these materials for memristor devices. Looking beyond the electrical domain, it is not difficult to find physical systems whose characteristics may be conveniently represented by a memristor. Especially interesting is the discussion of mechanical [46, 47], hydraulic [48], and thermal [49] physical arrangements acting as a memristor.

The physical mechanisms underlying the change between different resistive states of switching materials can be summarized as follows:

- Physical mechanisms [50, Figure 1], [51]
  - Schottky emission: thermally activated electrons injected over the barrier into the conduction band
  - Fowler–Nordheim tunneling: electrons tunnel from the cathode into the conduction band; usually occurs at high electric field
  - Direct tunneling: electrons tunnel from cathode to anode directly; only when the oxide is thin enough. When the insulator has localized states (traps) caused by disorder, off-stoichiometry or impurities, trap-assisted transport contributes to additional conduction, including:

    (a) tunneling from cathode to traps;
    (b) emission from traps to the conduction band (Poole-Frenkel emission);
    (c) tunneling from trap to conduction band;
    (d) trap-to-trap hopping or tunneling, ranging from Mott hopping between localized states to metallic conduction through extended states;
    (e) tunneling from traps to anode.

The lists above are far from being complete and are intended to provide a minimum basis of materials and physics for memristors.

Manufacturing processes compatible with CMOS technology, sustainable scaling, and superior computer efficiency are the key factors for the use of memristor devices in information technology markets. The material challenges and integration strategies for memristor-based computing systems include the accurate control of device-to-device and cycle-to-cycle variability, electrical programming (e.g., analog/digital resistive memory, $R_{\text{on}}/R_{\text{off}}$ ratio, etc.), and switching properties (e.g., endurance, retention, speed, etc.). Reliability is among the most important concerns for RRAM devices. Frequent migration of ions/atoms under a high local field, high current density, high power dissipation, and high temperature can degrade the electrodes and the active material. Another high priority is cycling endurance, especially for memory applications where the memory is accessed multiple times from the CPU for in-memory applications. As a nonvolatile memory, memristors must demonstrate data retention at both room temperature and elevated temperature, which is mandatory for meeting specifications for embedded memory and/ or automotive applications. The breakthrough for achieving a complete control of the peculiar memristor properties are complete understanding of the complex physics in the devices and a suitable process to model the electrical characteristics.

Resistive Switching Devices and Memristor

In this section, the historical links between the research on memristor and that on resistive switching devices are briefly discussed.

The first instances of resistive switching were identified by Hickmott in 1962 [17], where he used several different materials for the insulator layer, as $\mathtt{SiO}$, $\mathtt{Al_2O_3}$, $\mathtt{Ta_2O_5}$, $\mathtt{ZrO_2}$, and $\mathtt{TiO_2}$, showing that all of them exhibit reversible resistive switching. It is argued in [17] that although the electric fields present across the thin-films of oxide materials are large enough to induce dielectric breakdown (the process of sudden irreversible increase in conductance of an insulator under a large electric field), the observed negative resistance (decreasing conductance with constant or increasing voltage) shows that this is not actually the case. The interest in resistive switching increased following Hickmott's observations [52, 53], and led to the practical idea to exploit these effects as nonvolatile memory devices [19]. However, research into resistive switching devices drastically and quickly declined due mainly to the difficulties in understanding the underlying physical mechanisms.

Almost in the same period, and apparently in an unrelated fashion, Chua published his seminal 1971 paper [2] on the theoretic foundation of the memristor. Quite curiously, the paper [2] does not mention the results of Hickmott and others on resistive switching, although the qualitative similarity can be clearly observed. The quoted paper [2] postulates the existence and the necessity to introduce a fourth fundamental circuit element, in addition to resistor, capacitor, and inductor, acting as a *nonlinear resistor capable of memorizing its resistive state*. The memristor exhibits the distinctive property of tuning its resistive state according to the history of the applied input (voltage or current). In addition, when the input is off, the final resistive state is (ideally) kept forever.

In 1971 there was no known solid-state electronic device displaying a memristive behavior. Then, L. O. Chua proposed an *emulator* of a passive memristor built using active circuits, namely, a two-port network named mutator connected to a nonlinear resistor (Example 2.19). One main practical problem of such mutator-based emulator is that the implementation of the mutator needs a large number of active elements as transistors and operational amplifiers.

However, the possibility to come across physical devices embedding memristor features is reported in [2] by the following prophetic statement: *"Although no physical memristor has yet been discovered in the form of a physical device without internal power supply, ... a monotonically increasing φ–q curve could be invented, if not discovered accidentally. It is perhaps not unreasonable to suppose that such a device might already have been fabricated as a laboratory curiosity but was improperly identified! After all, a memristor with a simple φ–q curve will give rise to a rather peculiar—if not complicated hysteretic—v-i curve when erroneously traced in the current-versus-voltage plane. Perhaps, our perennial habit of tracing the v–i curve of any new two-terminal device has already misled some of our device-oriented colleagues and prevented them from discovering the true essence of some new device, which could very well be the missing memristor."*

The prophecy has come true 37 years later when, in 2008, the team at HP laboratories led by S. Williams published a paper entitled "The missing memristor found" [28] where, in their words, they present *"the logical and scientific basis for the existence of a new two-terminal circuit element called the memristor (contraction for memory resistor) which has every right to be as basic as the three classical circuit elements already in existence, namely, the resistor, inductor, and capacitor."* In that paper it is also shown, *"using a simple analytical example, that memristance arises naturally in nanoscale systems in which solid-state electronic and ionic transport are coupled under an external bias voltage."*

The HP memristor [28] is a capacitor-like Metal-Insulator-Metal (MIM) structure consisting of a thin titanium oxide TiO_2 film (50 nm) sandwiched between two metal (platinum) contacts (Fig. 2.24). Actually, TiO_2 is split into two chemically different layers. Stoichiometric TiO_2 (the ratio of oxygen to titanium was perfect, exactly 2 to 1) and closer to the top platinum electrode, the titanium dioxide is missing a tiny amount of its oxygen, between 2 and 3 %, called oxygen-deficient titanium dioxide TiO_{2-x}, where x is about 0.05 %. The TiO_2 is electrically insulating (actually a semiconductor), but the TiO_{2-x} is conductive, because its oxygen vacancies are donors of electrons, which makes the vacancies themselves positively charged. Then, the semiconductor film has a region with a low resistance R_{on}, while the remaining part has a much higher resistance R_{off}. The total resistance of the device is thus determined by two variable resistors connected in series (Fig. 2.24). The resistive state of the HP memristor is modulated by the bias voltage applied across the electrodes because the electric field is capable of shifting the interface between TiO_2 and TiO_{2-x} due to the migration of oxygen vacancies.

As remarked in [28], *"vacancies ... can be pushed up and down at will in the titanium dioxide material because they are electrically charged."* In fact, if a positive voltage is applied to the top electrode of the device, it will repel the (also positive) oxygen vacancies in the TiO_{2-x} layer down into the pure TiO_2 layer. That turns the TiO_2 layer into TiO_{2-x} and makes it more conductive. A negative voltage has the opposite effect: the vacancies are attracted upward and back out of the TiO_2, and thus the thickness of the TiO_2 layer increases and the device becomes less conductive. In [28] it is also stressed that the *"resistance of these devices stayed constant whether we turned off the voltage or just read their states (interrogating them with a voltage so small it left the resistance unchanged). The oxygen vacancies didn't roam around; they remained absolutely immobile until we again applied a positive or negative voltage. That's memristance: the devices remembered their current history. We had coaxed Chua's mythical memristor off the page and into being."*

Both fields, resistive switching and memristor theory, share a similar goal, namely, describing and controlling the process of storing information in the resistive state of an electronic device. Initiated around the same time and in parallel, both have been largely ignored because of the lack of an immediate practical application and because CMOS technology was already performing similar tasks at reduced cost.

However, the slowdown of Moore's Law [54] has shifted interest back to unconventional devices that can potentially outperform CMOS established technologies. Indeed, the situation has drastically changed after the announcement of HP in 2008 that has made clear how the *memristor approach permits to formulate an accurate circuit model of resistive switching devices*. The integration of the two fields, i.e., memristor theory and resistive switching devices, is also fundamental due to the emerging multidisciplinary research in bio-inspired computing architectures, which fosters a broader understanding of complex physical phenomena in material science and biological systems. Furthermore, advancements in measurement instrumentation and analytical tools allowed for observing and quantifying the resistive switching mechanism at fine volume resolution, also promoting the idea that memristive effects are more prominent at the nanoscale. In this context, the memristor theory developed by L. O. Chua offers a full framework which HP has adapted to include the complex physical processes responsible for resistive switching in the form of a wrap-around macromodel [55]. As a result, the coalition of the two fields has sparked an explosion of interest in the research community and the paradigm in the resistive switching community has somewhat changed. The majority of measured resistive switching effects are now described from a macroscopic point of view as memristive effects, by extending memristor theory to memristive device and systems (see Sect. 2.3). Following this line of reasoning, L. O. Chua has promoted the idea that all resistive switching memories are memristors [5]: *"All 2-terminal non-volatile memory devices based on resistance switching are memristors, regardless of the device material and physical operating mechanisms. They all exhibit a distinctive fingerprint characterized by a pinched hysteresis loop confined to the first and the third quadrants of the* $v - i$ *plane whose contour shape in general changes with both the amplitude and frequency of any periodic ssinewave-like input voltage source, or current source."* Based on this fundamental observation, many solid-state and/or nano-resistive switching devices can now be regarded as memristors. These include several types of Random Access Memories (RAMs) as ferroelectric random access memory (FeRAM), magnetic RAM (MRAM), Phase-Change RAM (PCRAM) and Resistive RAM (ReRAM), and Atomic Switch [12, 55–58].

Appendix 2: Memristor Modeling and Simulation

Much progress has been achieved over the past few years to develop accurate models for the dynamics of resistance switching memories, to establish solid foundations on memristor theory and to embed the mathematical descriptions of these devices into commercially available software packages, but a comprehensive body of knowledge allowing a thorough exploration of the full potential of memristors in future electronics has not been developed yet. The fact that there exist numerous physical memristors, composed of both organic and inorganic materials, and that researchers continue to investigate new materials or fabrication process methodologies to improve the performance of these devices in their diverse applications, clearly explains why there is still a lot of work to be carried out in memristor modeling, theory, and simulation. Of course some memristor technologies are more mature than others, and in some cases the physics underlying the dynamics of resistance switching memories are well understood, allowing the development of accurate models, their extensive simulation, and their analysis to establish solid theoretic foundations on circuits and systems based upon them.

In the following, we summarize various memristor mathematical/circuit models proposed in the literature. Some models assume that the control waveform is in current form (the voltage v-current i relationship is expressed by a state-dependent Ohm's law), views the memristance as the series between two variable resistances, associated with the insulating and conductive layers of the nano-film, and sets the width w of the conductive layer, normalized with respect to the entire length D of the device, as the state $x = \frac{w}{D} \in [0, 1]$ of the system. An instance of this kind is the linear drift model from Williams [28] (cf. Sect. 2.4.2.1), where the time derivative of the state is proportional to the input waveform in current form. Such model is valid under the assumption of low electric field, since it does not take into account the boundary behavior.

In the nonlinear drift models from [29, 30] and [59] the rate of change of the state is proportional to the product between the input waveform in current form and a window function accounting for nonlinear dynamical behavior and imposing suitable boundary conditions.

In Joglekar's model [29] the window function is defined as $\zeta_J(x) = 1-(2x-1)^{2p}$ (p is a positive integer). Such window describes the suppression of dopant drift close to the extremities, but is not vertically scalable (i.e., its maximum value may not be up- or down-shifted) and introduces the so-called "terminal-state problem" [59], since if the state is at either of its two bounds it may not leave it for any subsequent time instant. Note that for $p = 1$ Joglekar's window is a scaled (by a factor of 4) version of yet another window previously derived by Strukov in [28], i.e., $\zeta_S(x) = x(1 - x)$. Benderli [60] presented a circuit realization of Strukov's model [28], where the use of comparators and logic gates allowed the emulation of the state clipping at or release from either bound.

In Biolek's model [30] the window function depends on both the state x and input current i, being defined as $\zeta_B(x, i) = 1 - (x - \text{stp}(-i))^{2p}$, where $\text{stp}(x) = 1$ for $x \geq 0$ and $\text{stp}(x) = 0$ otherwise (p is a positive integer). Such window resolves the "terminal-state problem," but has limited scalability (in particular, its maximum value may not exceed $+1$ [59]). PSpice implementations of Joglekar's and Biolek's models are reported in [30].

In the versatile model proposed by Prodomakis [59] the window function $\zeta_P(x) = j(1 - ((x - 0.5)^2 - 0.75)^p)$ has two positive real control parameters j and p and is vertically scalable, i.e., $0 \leq \max\{\zeta_P(x)\}$ and $\max\{\zeta_P(x)\} \gtreqless 1$. A PSpice version of such model may be easily derived by modifying the PSpice code file available in [30].

Another model endowed with a PSpice circuit implementation was developed by Cserey [61]. In this model the state evolution function in Strukov's model [28] was augmented with an additive state-dependent linear term to resolve the "terminal-state problem."

One of the finest circuit emulators of memristor behavior is credited to Shin and Kang [62], which proposed a general model where the control waveform may be in either current or voltage form and the state is defined as the memristance. Their model, from which the charge-flux relationship of the memristor under modeling may be easily extracted, may be suitably tuned through the introduction of a window function depending on the memristor charge.

Kavehei [63] proposed a memristor model based on the specification of a piecewise-linear charge q-flux φ relationship. In such model the state and output equations are not specified. Its PSpice implementation is based on Chua's [2] first circuit realization of a memristor through a type-1 memristor-resistor mutator.

An interesting model was presented in [64] to explain the memristor behavior of nanoparticle assemblies.

The nonlinear dependence of the time derivative of the state on the input signal is taken into account in Lehtonen's model [65], inspired by the experimental work from [66], where the current is related to the voltage by means of a rectifying exponential function in the off state (as in a diode) and of a sinh function in the on state (typical of electron tunneling). This model, where the control waveform is in voltage form, was implemented in PSpice to describe the neighborhood connections among cells in cellular neural networks (CNNs) [67, 68].

An even more highly nonlinear function of the input governs the state equation in the voltage-controlled model from Poikonen [69], which studied the transition between non-programming and programming phases in memristor devices.

In the memristor emulator circuit from [70], used as basic building block of a 4-memristor bridge synapse for neuromorphic applications, the memristance, modeled by the input impedance of an active circuit, is made proportional to the time integral of the memristor current by constraining the voltage at one of the input terminals of an operational amplifier to be the analogue multiplication between the voltage across a resistor, proportional to the memristor current, and the voltage across a capacitor, proportional to the time integral of the memristor current.

In [71] Strukov and Williams demonstrated the exponential relationship between drift velocity and local electric field. Since this discovery a number of models have been introduced to support threshold-activated state dynamics. Among them, one which merits mention is the physics-based Pickett's model from [72], in which the dependency of the rate of change of the state on the current-form input is strongly nonlinear. In such model the memristor is seen as the series between a low resistance associated with the conductive layer of the nano-film and Simmons' electron tunneling barrier [73], whose width is chosen as the system state. A PSpice version of the latter was presented in [74].

More recently Kvatinski developed a simplified version of the Pickett's model [72] and named it as ThrEshold Adaptive Memristor (TEAM) model [34]. In such model for input current magnitude below a certain adaptable threshold no state change occurs, otherwise the state evolution rule may be tuned to the memristor element under modeling through specification of an appropriate set of control parameters and of suitable window and memristance functions. The PSpice architecture of the TEAM model is similar to the one originally presented in [62].

Another activation-type state model, where the state variable expresses the memristance and the control signal is in voltage form, embedded in the PSpice software program [75], enabled to capture the adaptive behavior of a unicellular organism named amoeba through a simple memristor-based oscillator [76].

A further interesting model with threshold-activated state dynamics was proposed in [77] to explain Spike-Timing-Dependent-Plasticity (STDP) in neural synapses.

An additional insightful discussion on the models available in the literature was recently published in [78], where a novel model inspired from Simmons' electron tunneling theory [73], endowed with programming threshold capability and PSpice circuit implementation, was also proposed.

The Boundary Condition Memristor (BCM) model is a simple yet accurate boundary condition-based mathematical model for memristor nano-structures made up of two layers with different conductivity levels, whose longitudinal extensions depend on the time history of the input. In comparison with the classical BCM [31], the generalized version [32] is augmented with programming threshold capability [78], i.e., with tunable nonvolatile behavior.

Recently, in [79], assuming Pickett's model [72] as reference for comparison, various memristor models, including Biolek's, the TEAM, and the BCM models, were first compared on the basis of the ability to reproduce (after an optimization process) the dynamics of the reference model in a particular simulation scenario, and second employed in a couple of memristor-based circuits to investigate the variance in the nonlinear dynamical behaviors they give rise to. The latter study revealed the model-dependency of the dynamics of the memristor-based circuits, and thus raised a warning against a blind faith in the memristor models and pointed out the necessity to develop a universal mathematical model for exploring the full potential of the memristor and unfolding its unique properties.

Appendix 3: Memristor Circuits and Systems

Due to its nature and properties (e.g., CMOS process compatibility, lower cost, zero standby power, nanosecond switching speed, great scalability, and high density), the memristor devices are enabling the exploration of alternative computing architectures and algorithms. Neuromorphic computing, computation-in-memory, and accelerate architectures embedding memristors will be able to address emerging applications, which are extremely demanding and/or have surpassed the capabilities of today's computation architectures and technologies. These new architectures will have to realize at least partially the following: (a) eliminate the communication and memory bottleneck, (b) support massive parallelism to increase the overall performance, (c) drastically enhance energy efficiency (both static and dynamic) to improve the computation efficiency, (d) be cheaper to manufacture, etc. Reliable, mainstream memristors-CMOS circuits can be used in applications that require adaptability and nonvolatility. Memristors offer a compact solution for the future nonvolatile memory which could be used in Internet-of-Things (IoT), robotics, medical and memory applications. In particular, highly predictable memristors with high yield and long retention would support an immense range of possible applications exploiting highly compact analog memories, such as precision circuits with low overhead trimming, precise crossbar computing applications (e.g., very fast matrix multiplications, with high impact for deep neural networks and machine learning at large), and adaptive neural (and non-neural) like circuits. As memristor technology is constantly developing, more and more practical application concepts based on memristor devices are being demonstrated in various research papers. Great interest has been raised in the field of neuromorphic applications as various memristive technologies/devices (such as oxide based memristors) can have their resistance shifted in small steps which renders them ideal in emulating the operation of physical synapses. Other novel applications, for example the memristor-based humanoid robot control system presented in [80, 81], exploit the ability of memristive devices to store and process data in the same physical location. In all cases, memristors are utilized to replace conventional designs (CMOS-based designs with Von Neumann architectures) for reducing total power consumption and increasing energy efficiency.

There is also a crucial need for accurate and reliable models for the nonlinear dynamics of resistance switching memories, especially those composed of novel unexplored materials, where the physical mechanisms underlying the inherently complex dynamics emerging in the nanodevices are still under study. Provided a reliable mathematical description of a memristor is available, a thorough analysis of the model through the application of methodologies from nonlinear dynamic system theory [82, 83] may allow to predict the set of input/initial condition combinations under which it is safe to operate the nano-device. Furthermore, the adoption of concepts from stability theory and nonlinear dynamics may provide insights into the biasing circuit arrangement necessary to induce specific dynamical phenomena in memristors, e.g., locally active behaviors, bifurcations, and complex attractors [84, 85]. All in all, the development of accurate, reliable, and numerically stable memristor circuit models, but also the analysis through concepts and methodologies from the theory of dynamical systems, are essential tools to provide circuit designers with a comprehensive picture of the nonlinear response of the nanostructures to input/initial condition combinations expected in the application of interest, helping them to take more conscious decisions on the most suitable circuit topology to meet prescribed specifications.

Appendix 4: Memristor Applications

Though not currently offered as part of standard semiconductor processes, a number of companies are working towards integrating memristors with standard CMOS processes. Memristor devices have been so far proven useful to multiple application areas ranging from neuromorphic computing as artificial synapses to biosensors. This section provides an overview of memristor applications.

Artificial Synapses

Brain-inspired computing is an emerging field, which can extend the capabilities of information technology beyond the Von Neumann paradigm. Biologically inspired systems, aiming to emulate the nervous system of living beings, are the best solution to solve ill-posed problems, such as real time interaction with the external environment or pattern recognition. In biological systems, neurons are interconnected and they interact among them through synapses, which can change their strength in order to inhibit (synaptic depression) or facilitate (synaptic potentiation) the connection between two neurons. This ability is called synaptic plasticity and it is a key mechanism used by the brain in learning processes. In neuromorphic architectures, whilst CMOS technology is typically used to design artificial neurons, the implementation of artificial synapses poses a serious challenge. In fact, synapses outnumber neurons by 3–4 orders of magnitude, therefore calling for high-density and low-power devices. Moreover, they should be CMOS-compatible in order to be easily integrated with CMOS-based neurons [86]. In this scenario, Resistive Switching (RS) devices, or memristors, are considered good candidates to be used as artificial synapses due to their high scalability and low power consumption. Memristors are two-terminal devices able to change their conductance from a High Conductance State (HCS) to a Low Conductance State (LCS), and vice versa. The change is induced upon application of proper electrical stimuli. This behavior is used to emulate synaptic plasticity. More specifically, the transition from HCS to LCS emulates a depression operation whereas the reverse transition emulates a potentiation operation [86]. Several other applications of memristive devices as artificial synapses include: circuits for encryption/decryption of medical data [87], computer arithmetic systems [88], spiking networks implementing unsupervised learning [89], real-time encoding and compression of neuronal spikes [90], attractor networks [91], oscillatory networks for unconventional computing [92], unbiased generation of random numbers [93], emulation of short-term synaptic dynamics [94], modeling of biochemical reactions [95], maze-solving computations [96], and circular buffer for real-time signal processing [97]. Furthermore, memristive devices can have other functions in the emulation of biological computation, e.g., the implementation of neural dynamics.

Memory Applications

The use of emerging memristor materials for advanced electrical devices such as multi-valued logic is expected to outperform today's binary logic digital technologies. While conventional memory cells can store only 1 bit, memristor-based multi-bit cells can store more information within single device thus increasing the information storage density. Such devices can potentially utilize the nonlinear resistance of memristor bio-inspired materials for efficient information storage [98]. Unlike other passive, two-terminal devices such as resistors or capacitors, the hysteresis of a memristive device can be used for information storage (e.g., resistive memory RRAM), with low resistance and reconfigurable signal routing (opening and closing the connections between the nanowire electrodes). The memristor-based crossbar network structure [99], being used as memory, can offer the following advantages: (1) it allows ultra-high density memory storage with relatively small number of control electrodes and cross points can be accessed by n-rows and n-columns in the crossbar; (2) it offers large connectivity between devices; and (3) it facilitates reconfigurable circuits by changing the conductance of the memristive devices at selected cross points [51].

In-memory Computing

Conventional computers are based on Von Neumann architecture, where processing and storing of the data are performed by different units (namely, CPU and memory). Over the last few decades, the performance of processors has improved in a much higher pace than that of memories, which has led to today several orders of magnitude performance gap between processor and memory. This gap causes a bottleneck for transferring data between memory and processor, which is usually called the memory wall. One of today's critical challenge is data-intensive and big-data problems in data storage and analysis. The increase of the data size has already surpassed the capabilities of today's computation architectures, which suffer from the limited bandwidth, programmability overhead, energy inefficiency, and limited scalability. Memristor-based architectures for data-intensive applications have been reported with the potential to solve data-intensive problems by increasing the computation efficiency, solving the communication bottleneck and reducing the leakage currents. Attempts for reducing the memory wall problem by getting the memory closer to the processing unit have been tried before. One of the leading trials is processing near memory (PNM) architecture. In PNM, processing units, on which part of the program is executed, are added to the off-chip DRAM memory and a part of the program is executed in the off-chip memory by a co-processor. One well-known implementation of PNM is Berkeley's Intelligent RAM (IRAM) project. The IRAM project had not become commercially successful, mainly due to the bad integration between DRAM and logic technologies. Processing within memory using the same cells for both memory and logic is desired. Such integration can be performed using novel emerging nonvolatile memory technologies. These emerging technologies include RRAM, PCM, STT MRAM, and others. Due to their speed, low power, scalability, and high endurance, memristors that store data as resistance values are considered as attractive candidates to replace conventional memory technologies (e.g., DRAM and Flash). Furthermore, memristive technologies have also been explored for additional applications such as logic circuits, whereas some of the proposed logic may be computed within a memristive memory structure, allowing both storage and processing within the same cells and without changing the topology of the memristive memory array [100].

Biosensors

Nanowire-based field-effect transistors (FETs) are currently attracting a strong attention due to their potential to deliver promising miniaturized diagnostic tools with a high and label-free sensitivity. Recently, their memristive property was demonstrated as biosensing principle, showing that the binding of charged biomolecules brings a violation of its zero crossing signature by opening a voltage gap in the current minima when the source-to-drain voltage was swept [101, 102]. A new application gives proof that the coupling of DNA-aptamers and silicon nanowire-arrays with memristive electrical response leads to a high-performance biosensors for the detection of disease biomarkers as well as monitoring of therapeutic compounds [102, 103]. Charged residues up-taken on the surface of these special nanodevices play a pivotal role on the resulting electrical response. These devices are used to implement novel efficient and accurate biosensors based on the change of hysteretic properties before and after the bio-modification due to biological processes.

Tunable Electronic Components

Memristors have proven to be an attractive feature for memory, logic-in-memory, and neuromorphic computing. Recently, radio-frequency memristive switches (RFMSs) have exhibited promising high-frequency performance, opening the possibility of their use in radio-frequency integrated circuit applications. Novel topologies of Tunable Inductors using memristors have been reported using a switched tunable inductor and the multi-layer stacked inductor switched by an RFMS single-pole double-throw. The two-inductor topologies are fully passive and are tuned by electrochemical metallization memristors. Memristive devices improve the performance of tunable inductors, as they provide low area overhead, low-energy switching, and nonvolatility, resulting in more compact and energy-efficient devices [104]. Memristive elements, indeed, by virtue of the temperature-

and electric-field driven ionic migration at the base of their operation, can generate complex nonlinear dynamical evolution of their resistance that can be exploited to build nanoscale oscillators potentially described by low-order mathematical/circuit models. Networks of interconnected and interacting oscillators can develop cooperative and collective dynamics, e.g., phase synchronization and other self-organizing spatiotemporal phenomena for alternative computing schemes that overpass the limits of conventional digital and Boolean computation. Various classes of memristor-based relaxation oscillators displaying a tunable range of periodic and chaotic self-oscillations have been implemented in recent years. This book presents in Chap. 5 a new method, named Flux-Charge Analysis Method (FCAM), to study a wide class of nonlinear circuits containing ideal memristors in the flux-charge domain. FCAM gives a clear picture of the global dynamics and the main peculiar dynamic aspects, such as the presence of invariant manifolds, the coexistence of different dynamics for the same set of (fixed) circuit parameters, and the new interesting phenomenon of bifurcations without parameters, i.e., bifurcations due to changing the initial conditions for the state variables for a fixed set of circuit parameters.

References

1. L.O. Chua, S.M. Kang, Memristive devices and systems. Proc. IEEE **64**(2), 209–223 (1976)
2. L.O. Chua, Memristor-The missing circuit element. IEEE Trans. Circuit Theory **18**(5), 507–519 (1971)
3. L. Chua, Introduction to memristors. IEEE, New York (2009). https://ieeexplore.ieee.org/xpl/articleDetails.jsp?arnumber=EDP091&contentType=Education+%26+Learning
4. L. Chua, Everything you wish to know about memristors but are afraid to ask. Radioengineering **24**(2), 319–368 (2015)
5. L. Chua, Resistance switching memories are memristors. Appl. Phys. A **102**(4), 765–783 (2011)
6. L. Chua, Five non-volatile memristor enigmas solved. Appl. Phys. A **124**(8), 563 (2018)
7. L.O. Chua, C.A. Desoer, E.S. Kuh, Linear and Nonlinear Circuits (McGraw-Hill, New York, 1987)
8. L.O. Chua, Nonlinear circuit foundations for nanodevices. I. The four-element torus. Proc. IEEE **91**(11), 1830–1859 (2003)
9. D. Biolek, Z. Biolek, V. Biolkova, Pinched hysteretic loops of ideal memristors, memcapacitors and meminductors must be 'self-crossing'. Electron. Lett. **47**(25), 1385–1387 (2011)
10. H. Kim, M. Sah, C. Yang, T. Roska, L.O. Chua, Memristor bridge synapses. Proc. IEEE **100**(6), 2061–2070 (2012)
11. M. Sah, H. Kim, L.O. Chua, Brains are made of memristors. IEEE Circuits Syst. Mag. **14**(1), 12–36 (2014)
12. D. Ielmini, Brain-inspired computing with resistive switching memory (RRAM): devices, synapses and neural networks. Microelectron. Eng. **190**, 44–53 (2018)
13. M. Vidyasagar, *Nonlinear Systems Analysis*, vol. 42 (SIAM, Philadelphia, 2002)
14. J.C. Willems, Dissipative dynamical systems part I: general theory. Arch. Ration. Mech. Anal. **45**(5), 321–351 (1972)
15. J.C. Willems, Dissipative dynamical systems part II: linear systems with quadratic supply rates. Arch. Ration. Mech. Anal. **45**(5), 352–393 (1972)

16. T. Zhang, M. Yin, X. Lu, Y. Cai, Y. Yang, R. Huang, Tolerance of intrinsic device variation in fuzzy restricted Boltzmann machine network based on memristive nano-synapses. Nano Futur. **1**(1), 015003 (2017)
17. T.W. Hickmott, Low-frequency negative resistance in thin anodic oxide films. J. Appl. Phys. **33**(9), 2669–2682 (1962)
18. J.F. Gibbons, W.E. Beadle, Switching properties of thin NiO films. Solid-State Electron. **7**(11), 785–790 (1964)
19. J.G. Simmons, R.R. Verderber, New conduction and reversible memory phenomena in thin insulating films. Proc. R. Soc. Lond. A Math. Phys. Sci. **301**(1464), 77–102 (1967)
20. Y.G. Kriger, N.F. Yudanov, I.K. Igumenov, S.B. Vashchenko, Study of test structures of a molecular memory element. J. Struct. Chem. **34**(6), 966–970 (1993)
21. M. Sapoff, R.M. Oppenheim, Theory and application of self-heated thermistors. Proc. IEEE **51**(10), 1292–1305 (1963)
22. V.J. Francis, *Fundamentals of Discharge Tube Circuits* (Methuen, London, 1948)
23. A.L. Hodgkin, A.F. Huxley, A quantitative description of membrane current and its application to conduction and excitation in nerve. J. Physiol. **117**(4), 500–544 (1952)
24. F. Corinto, P.P. Civalleri, L.O. Chua, A theoretical approach to memristor devices. IEEE J. Emerg. Sel. Top. Circuits Syst. **5**(2), 123–132 (2015)
25. L. Chua, G. Sirakoulis, A. Adamatzky, (Eds.), *Handbook of Memristor Networks. Vol. 1 and 2* (Springer, New York, 2019)
26. J. Vista, A. Ranjan, A simple floating MOS-memristor for high-frequency applications. IEEE Trans. Very Large Scale Integr. Syst. **27**(5), 1186–1195 (2019)
27. E. Solan, K. Ochs, Wave digital emulation of general memristors. Int. J. Circuit Theory Appl. **46**(11), 2011–2027 (2018)
28. D.B. Strukov, G.S. Snider, D.R. Stewart, R.S. Williams, The missing memristor found. Nature **453**(7191), 80–83 (2008)
29. Y.N. Joglekar, S.J. Wolf, The elusive memristor: properties of basic electrical circuits. Eur. J. Phys. **30**(4), 661 (2009)
30. Z. Biolek, D. Biolek, V. Biolkova, SPICE model of memristor with nonlinear dopant drift. Radioengineering **18**(2), 210–214 (2009)
31. F. Corinto, A. Ascoli, A boundary condition-based approach to the modeling of memristor nanostructures. IEEE Trans. Circuits Syst. I Regul. Pap. **59**(11), 2713–2726 (2012)
32. A. Ascoli, F. Corinto, R. Tetzlaff, Generalized boundary condition memristor model. Int. J. Circuit Theory Appl. **44**(1), 60–84 (2016)
33. R. Stanley Williams, How we found the missing memristor. IEEE Spectr. **45**(12), 28–35 (2008)
34. S. Kvatinsky, E.G. Friedman, A. Kolodny, U.C. Weiser, TEAM: threshold adaptive memristor model. IEEE Trans. Circuits Syst. I Regul. Pap. **60**(1), 211–221 (2013)
35. G.A. Gibson, S. Musunuru, J. Zhang, K. Vandenberghe, J. Lee, C.-C. Hsieh, W. Jackson, Y. Jeon, D. Henze, Z. Li, et al., An accurate locally active memristor model for S-type negative differential resistance in nbox. Appl. Phys. Lett. **108**(2), 023505 (2016)
36. F. Corinto, A. Ascoli, Memristive diode bridge with LCR filter. Electron. Lett. **48**(14), 824–825 (2012)
37. A. Ascoli, F. Corinto, R. Tetzlaff, A class of versatile circuits, made up of standard electrical components, are memristors. Int. J. Circuit Theory Appl. **44**(1), 127–146 (2016)
38. M. Itoh, L.O. Chua, Parasitic effects on memristor dynamics. Int. J. Bifurcat. Chaos **26**(6), 1630014 (2016)
39. Y.V. Pershin, M. Di Ventra, Memory effects in complex materials and nanoscale systems. Adv. Phys. **60**(2), 145–227 (2011)
40. L.O. Chua, Reply to comment on 'If it's pinched it's a memristor'. Semicond. Sci. Technol. **34**(9), 098002 (2019)
41. J. Martinez-Rincon, M. Di Ventra, Y.V. Pershin, Solid-state memcapacitive system with negative and diverging capacitance. Phys. Rev. B **81**(19), 195430 (2010)

42. S. Kumar, R. Stanley Williams, Tutorial: experimental nonlinear dynamical circuit analysis of a ferromagnetic inductor. IEEE Circuits Syst. Mag. **18**(2), 28–34 (2018)
43. J.S. Najem, M.S. Hasan, R. Stanley Williams, R.J. Weiss, G.S. Rose, G.J. Taylor, S.A. Sarles, C. Patrick Collier, Dynamical nonlinear memory capacitance in biomimetic membranes. Nat. Commun. **10**(1), 3239 (2019)
44. S. Battistoni, V. Erokhin, S. Iannotta, Organic memristive devices for perceptron applications. J. Phys. D Appl. Phys. **51**(28), 284002 (2018)
45. I. Valov, M. Kozicki, Non-volatile memories: organic memristors come of age. Nat. Mater. **16**(12), 1170 (2017)
46. G. Oster, A note on memristors. IEEE Trans. Circuits Syst. **21**(1), 152–152 (1974)
47. J.-S. Pei, J.P. Wright, M.D. Todd, S.F. Masri, F. Gay-Balmaz, Understanding memristors and memcapacitors in engineering mechanics applications. Nonlinear Dyn. **80**(1–2), 457–489 (2015)
48. D. Biolek, Z. Biolek, V. Biolkova, Z. Kolka, Reliable modeling of ideal generic memristors via state-space transformation. Radioengineering **24**(2), 393–407 (2015)
49. F. Yang, M.P. Gordon, J.J. Urban, Theoretical framework of the thermal memristor via a solid-state phase change material. J. Appl. Phys. **125**(2), 025109 (2019)
50. J. Joshua Yang, R. Stanley Williams, Memristive devices in computing system: promises and challenges. ACM J. Emerg. Technol. Comput. Syst. **9**(2), 11 (2013)
51. Q. Xia, J. Joshua Yang, Memristive crossbar arrays for brain-inspired computing. Nat. Mater. **18**(4), 309–323 (2019)
52. F. Argall, Switching phenomena in titanium oxide thin films. Solid-State Electron. **11**(5), 535–541 (1968)
53. D.P. Oxley, Electroforming, switching and memory effects in oxide thin films. Act. Passive Electron. Compon. **3**(4), 217–224 (1977)
54. M.A. Zidan, J.P. Strachan, W.D. Lu, The future of electronics based on memristive systems. Nat. Electron. **1**(1), 22 (2018)
55. J. Joshua Yang, M.D. Pickett, X. Li, D. A.A. Ohlberg, D.R. Stewart, R. Stanley Williams, Memristive switching mechanism for metal/oxide/metal nanodevices. Nat. Nanotechnol. **3**(7), 429 (2008)
56. J.M. Slaughter, Materials for magnetoresistive random access memory. Annu. Rev. Mater. Res. **39**, 277–296 (2009)
57. K.M. Kim, D.S. Jeong, C.S. Hwang, Nanofilamentary resistive switching in binary oxide system; a review on the present status and outlook. Nanotechnology **22**(25), 254002 (2011)
58. K. Terabe, T. Hasegawa, T. Nakayama, M. Aono, Quantized conductance atomic switch. Nature **433**(7021), 47 (2005)
59. T. Prodromakis, B.P. Peh, C. Papavassiliou, C. Toumazou, A versatile memristor model with nonlinear dopant kinetics. IEEE Trans. Electron. Dev. **58**(9), 3099–3105 (2011)
60. S. Benderli, T.A. Wey, On SPICE macromodelling of TiO2 memristors. Electron. Lett. **45**(7), 377–379 (2009)
61. Á. Rák, G. Cserey, Macromodeling of the memristor in SPICE. IEEE Trans. Comput.-Aided Des. Integr. Circuits Syst. **29**(4), 632–636 (2010)
62. S. Shin, K. Kim, S.-M. Kang, Compact models for memristors based on charge-flux constitutive relationships. IEEE Trans. Comput.-Aided Des. Integr. Circuits Syst. **29**(4), 590–598 (2010)
63. O. Kavehei, A. Iqbal, Y.S. Kim, K. Eshraghian, S.F. Al-Sarawi, D. Abbott, The fourth element: characteristics, modelling and electromagnetic theory of the memristor. Proc. Roy. Soc. A Math. Phys. Eng. Sci. **466**(2120), 2175–2202 (2010)
64. T.H. Kim, E.Y. Jang, N.J. Lee, D.J. Choi, K.-J. Lee, J.-T. Jang, J.-S. Choi, S.H. Moon, J. Cheon, Nanoparticle assemblies as memristors. Nano Lett. **9**(6), 2229–2233 (2009)
65. E. Lehtonen, M. Laiho, CNN using memristors for neighborhood connections, in *2010 12th International Workshop on Cellular Nanoscale Networks and their Applications (CNNA 2010)* (IEEE, Piscataway, 2010), pp. 1–4

66. J. Joshua Yang, M.D. Pickett, X. Li, D.A.A. Ohlberg, D.R. Stewart, R.S. Williams, Memristive switching mechanism for metal/oxide/metal nanodevices. Nat. Nanotechnol. **3**(7), 429 (2008)
67. L.O. Chua, L. Yang, Cellular neural networks: theory. IEEE Trans. Circuits Syst. **35**(10), 1257–1272 (1988)
68. T. Roska, L.O. Chua, The CNN universal machine: an analogic array computer. IEEE Trans. Circuits Syst. II **40**, 163–173 (1993)
69. E. Lehtonen, J. Poikonen, M. Laiho, W. Lu, Time-dependency of the threshold voltage in memristive devices, in 2011 IEEE Int. Symp. Circuits Syst. (IEEE, Piscataway, 2011), pp. 2245–2248
70. M.P. Sah, C. Yang, H. Kim, L. Chua, A voltage mode memristor bridge synaptic circuit with memristor emulators. Sensors **12**(3), 3587–3604 (2012)
71. D.B. Strukov, R. Stanley Williams, Exponential ionic drift: fast switching and low volatility of thin-film memristors. Appl. Phys. A **94**(3), 515–519 (2009)
72. M.D. Pickett, D.B. Strukov, J.L. Borghetti, J.J. Yang, G.S. Snider, D.R. Stewart, R.S. Williams, Switching dynamics in titanium dioxide memristive devices. J. Appl. Phys. **106**(7), 074508 (2009)
73. J.G. Simmons, Generalized formula for the electric tunnel effect between similar electrodes separated by a thin insulating film. J. Appl. Phys. **34**(6), 1793–1803 (1963)
74. H. Abdalla, M.D. Pickett, SPICE modeling of memristors, in *2011 IEEE International Symposium on Circuits and Systems* (IEEE, Piscataway, 2011), pp. 1832–1835
75. Y.V. Pershin, M. Di Ventra, SPICE model of memristive devices with threshold (2012). arXiv:1204.2600
76. Y.V. Pershin, S.L. Fontaine, M. Di Ventra, Memristive model of amoeba learning. Phys. Rev. E **80**(2), 021926 (2009)
77. B. Linares-Barranco, T. Serrano-Gotarredona, Memristance can explain spike-time-dependent-plasticity in neural synapses. Nat. Prec. 1 (2009)
78. K. Eshraghian, O. Kavehei, K.-R. Cho, J.M. Chappell, A. Iqbal, S.F. Al-Sarawi, D. Abbott, Memristive device fundamentals and modeling: applications to circuits and systems simulation. Proc. IEEE **100**(6), 1991–2007 (2012)
79. A. Ascoli, F. Corinto, V. Senger, R. Tetzlaff, Memristor model comparison. IEEE Circuits Syst. Mag. **13**(2), 89–105 (2013)
80. A. Ascoli, D. Baumann, R. Tetzlaff, L.O. Chua, M. Hild, Memristor-enhanced humanoid robot control system–part I: theory behind the novel memcomputing paradigm. Int. J. Circuit Theory Appl. **46**(1), 155–183 (2018)
81. D. Baumann, A. Ascoli, R. Tetzlaff, L.O. Chua, M. Hild, Memristor-enhanced humanoid robot control system–part II: circuit theoretic model and performance analysis. Int. J. Circuit Theory Appl. **46**(1), 184–220 (2018)
82. D. Kaplan, L. Glass, *Understanding Nonlinear Dynamics* (Springer, Berlin, 2012)
83. L.P. Shil'nikov, *Methods of Qualitative Theory in Nonlinear Dynamics*, vol. 5 (World Scientific, Singapore, 2001)
84. R. Devaney, *An Introduction to Chaotic Dynamical Systems* (CRC Press, Boca Raton, 2018)
85. S. Sastry, *Nonlinear Systems: Analysis, Stability, and Control*, vol. 10 (Springer, Berlin 2013)
86. E. Covi, S. Brivio, A. Serb, T. Prodromakis, M. Fanciulli, S. Spiga, HfO2-based memristors for neuromorphic applications, in *2016 IEEE International Symposium on Circuits and Systems* (IEEE, Piscataway, 2016), pp. 393–396
87. A. Ascoli, V. Senger, R. Tetzlaff, N. Du, O.G. Schmidt, H. Schmidt, BiFeO3 memristor-based encryption of medical data, in *2016 IEEE International Symposium on Circuits and Systems* (IEEE, Piscataway, 2016), pp. 1602–1605
88. G. Papandroulidakis, I. Vourkas, G.C. Sirakoulis, S.G. Stavrinides, S. Nikolaidis, Multi-state memristive nanocrossbar for high-radix computer arithmetic systems, in *2015 IEEE International Conference on Nanotechnology* (IEEE, Piscataway, 2015), pp. 625–628

89. E. Covi, S. Brivio, A. Serb, T. Prodromakis, M. Fanciulli, S. Spiga, Analog memristive synapse in spiking networks implementing unsupervised learning. Front. Neurosci. **10**, 482 (2016)
90. I. Gupta, A. Serb, A. Khiat, R. Zeitler, S. Vassanelli, T. Prodromakis, Real-time encoding and compression of neuronal spikes by metal-oxide memristors. Nat. Commun. **7**, 12805 (2016)
91. V. Milo, D. Ielmini, E. Chicca, Attractor networks and associative memories with STDP learning in RRAM synapses, in *2017 IEEE International Electron Devices Meeting* (IEEE, Piscataway, 2017), pp. 11.2.1–11.2.4
92. F. Corinto, M. Forti, Complex dynamics in arrays of memristor oscillators via the flux–charge method. IEEE Trans. Circuits Syst. I Regul. Pap. **65**(3), 1040–1050 (2017)
93. S. Balatti, S. Ambrogio, R. Carboni, V. Milo, Z. Wang, A. Calderoni, N. Ramaswamy, D. Ielmini, Physical unbiased generation of random numbers with coupled resistive switching devices. IEEE Trans. Electron Dev. **63**(5), 2029–2035 (2016)
94. R. Berdan, E. Vasilaki, A. Khiat, G. Indiveri, A. Serb, T. Prodromakis, Emulating short-term synaptic dynamics with memristive devices. Sci. Rep. **6**, 18639 (2016)
95. H.A. Hanna, L. Danial, S. Kvatinsky, R. Daniel, Modeling biochemical reactions and gene networks with memristors, in *2017 IEEE Biomedical Circuits and Systems Conference* (IEEE, Piscataway, 2017), pp. 1–4
96. V. Ntinas, I. Vourkas, G.C. Sirakoulis, A.I. Adamatzky, Oscillation-based slime mould electronic circuit model for maze-solving computations. IEEE Trans. Circuits Syst. I Regul. Pap. **64**(6), 1552–1563 (2017)
97. C. Sichonidis, I. Vourkas, N. Mitianoudis, G.C. Sirakoulis, A memristive circular buffer for real-time signal processing, in *2016 5th International Conference on Modern Circuits and Systems Technologies* (IEEE, Piscataway, 2016), pp. 1–4
98. Z. Lv, Y. Zhou, S.-T. Han, V. Roy, From biomaterial-based data storage to bio-inspired artificial synapse. Mater. Today **21**(5), 537–552 (2018)
99. S. Pi, C. Li, H. Jiang, W. Xia, H. Xin, J.J. Yang, Q. Xia, Memristor crossbar arrays with 6-nm half-pitch and 2-nm critical dimension. Nat. Nanotechnol. **14**(1), 35 (2019)
100. Y. Li, Z. Wang, R. Midya, Q. Xia, J. Yang, Review of memristor devices in neuromorphic computing: materials sciences and device challenges. J. Phys. D Appl. Phys. **51**(50), 503002 (2018)
101. S. Carrara, D. Sacchetto, M.-A. Doucey, C. Baj-Rossi, G. De Micheli, Y. Leblebici, Memristive-biosensors: a new detection method by using nanofabricated memristors. Sensors Actuators B Chem. **171**, 449–457 (2012)
102. I. Tzouvadaki, N. Aliakbarinodehi, G. De Micheli, S. Carrara, The memristive effect as a novelty in drug monitoring. Nanoscale **9**(27), 9676–9684 (2017)
103. I. Tzouvadaki, P. Jolly, X. Lu, S. Ingebrandt, G. De Micheli, P. Estrela, S. Carrara, Label-free ultrasensitive memristive aptasensor. Nano Lett. **16**(7), 4472–4476 (2016)
104. N. Wainstein, S. Kvatinsky, TIME-tunable inductors using memristors. IEEE Trans. Circuits Syst. I Regul. Pap. **65**(5), 1505–1515 (2017)

Chapter 3
RLC Networks Equations and Analysis Methods

Let us consider a circuit[1] N made of an arbitrary interconnection of (two-terminal) circuit elements. In general, analysis and design of a circuit require to solve a network problem where numerous mathematical (circuit) variables are established and then a set of equations is generated describing the behavior of the composite network. Fundamental principles of *Circuit Theory* permit to show that circuit equations depend on (a) the structure of the network and (b) the properties of the devices which it contains. If the network contains nonlinear elements, then the equations are usually more difficult to solve.

A *graph* is an abstraction of a physically implemented network, i.e., it represents the structure or skeleton of the network. *Graph Theory* is therefore expected to be valuable in the network analysis. In particular, the general results of graph theory provide the theoretical basis to produce circuit equations in the simplest form. Since the choice of the method should be based on mathematical convenience, a method yielding equations in the simplest (normal) form is indeed highly desirable. In the following, the laws of Kirchhoff with some fundamental concepts of graph theory are summarized in order to show how many electrical variables and how many circuit equations are needed to fully describe a network N. Several different methods for deriving the set of circuit equations are available in the literature. It is well known that methods based on classical dynamics (e.g., Lagrange and Hamiltonian principles) yield equations similar to those derived by Kirchhoff laws.

The chapter starts reviewing Kirchhoff current and voltage laws and then, after discussing the basics of graph theory, arrives at introducing the main forms of tableau equations to describe dynamic circuits, i.e., circuits containing inductors and capacitors in addition to resistive elements. Tableau equations include in general linear algebraic equations corresponding to Kirchhoff laws and dynamic equations (given in integral or differential forms) representing the constitutive relations of

[1] In many instances the term *network* is also used in literature. Hereinafter, "circuit" and "network" are used interchangeably.

F. Corinto et al., *Nonlinear Circuits and Systems with Memristors*,
https://doi.org/10.1007/978-3-030-55651-8_3

capacitors and inductors. The chapter then addresses the relevant issue of how to write a state equation (SE) representation for a given dynamic network, pointing out that networks that do not admit an SE representation might be ill-defined from a mathematical and a physical viewpoint. Conditions ensuring the existence of the SEs are provided. Since they are couched in topological form, they can often be easily checked by inspection. The discussion and application examples refer to networks containing two-terminal elements as (possibly) nonlinear resistors, inductors, and capacitors, in addition to independent voltage or current sources. Such networks are referred to in the following as RLC networks. All the results on graphs and on SEs may be extended also to networks containing multiterminal or multiport resistors (e.g., ideal operational amplifiers or resistive controlled sources), capacitors, and inductors.

3.1 Introduction to Kirchhoff Laws

The laws governing the interactions among the circuit elements in a network N are the two Kirchhoff laws, which are briefly introduced by means of the following simple topological concepts:

- a *node* is the connecting point of at least two terminals of distinct circuit elements
- a *branch* is associated with each two-terminal element (a.k.a. one-port) connected between two nodes
- a *loop* is any closed path starting from any node, passing over different branches and nodes and ending at the same node, where just only two branches are incident with each node.

It follows that a branch is represented by a line with two endpoints and a node includes at least two endpoints.

The two Kirchhoff laws can be stated as follows:

Definition 3.1 (Kirchhoff Current Law (KCL)) The algebraic sum of all currents entering (leaving) any node of any network N is zero at all times t.

Definition 3.2 (Kirchhoff Voltage Law (KVL)) The algebraic sum of all voltages around any loop of a network N is zero at all times t.

Although Kirchhoff laws can be derived from Maxwell's equations in a more general circuit model based on the electromagnetic field theory, KCL and KVL are given as postulates or experimental laws in the spirit of this book. The equations obtained by applying Kirchhoff laws are independent of the nature of the single elements, but they depend only on the way in which elements are interconnected. Their extension to a circuit also including elements with more than two terminals is straightforward if one is referring to the circuit graph.

For the sake of clarity, let us assume that a network consists of b branches and n nodes. By using basic concept of graph theory, it is easy to show that KCL and KVL provide $(n-1)$ *node equations* and $b-(n-1)$ *loop equations* that are independent of each other. Thus, overall Kirchhoff laws allow to write b independent linear algebraic equations in terms of the $2b$ unknowns given by the branch variables, i.e., the branch currents $i_1, i_2, \ldots, i_b$ and the branch voltages $v_1, v_2, \ldots, v_b$.

By joining the b equations derived from the constitutive relations (CRs) associated with the b two-terminal circuit elements in the network and the independent b equations obtained applying Kirchhoff laws we obtain a system of $2b$ equations in $2b$ unknowns. It is apparent that such a system is linear/nonlinear algebraic/dynamic according to the mathematical properties of the CRs. The circuit analysis methods derived from this approach are briefly discussed in the next Sect. 3.2. Before presenting circuit analysis methods it is convenient to introduce a compact matrix formulation of KCLs and KVLs via basic notions of graph theory.

3.1.1 Basics of Graph Theory

The interconnection properties of a network are fully described by its *graph*, according to the next definition.

Definition 3.3 (Graph of a Network) A graph $\mathcal{G}$ is specified by the set of n nodes $\{①, ②, \ldots, ⓝ\}$ together with a set of b branches $\{\beta_1, \beta_2, \ldots, \beta_b\}$. If each branch is given an orientation, then $\mathcal{G}$ is a directed graph (or digraph).

Hereinafter the *associated reference direction* is considered for each branch, i.e., the branch current is assumed to go from the terminal "+" to the terminal "−" associated with the branch voltage.[2] In addition, each branch in $\mathcal{G}$ is oriented according to the branch current. The direction of a branch is indicated by an arrow in the same direction of the positive branch current. When we are not interested with the reference directions of branch voltages and currents, all the arrows in $\mathcal{G}$ may be removed. Such simpler graph is called *undirected graph* associated with the network N.

Example 3.1 (Directed Graph $\mathcal{G}$) Figure 3.1 shows a network N and its associated directed graph $\mathcal{G}$ with $n = 4$ nodes and $b = 6$ branches.

We assume that $\mathcal{G}$ is a (directed or undirected) *connected graph*, that is, there exists a path between any two nodes of the graph. The path between two nodes ⓘ and ⓙ is defined as a set of p branches $\{\beta_1, \beta_2, \ldots, \beta_p\}$ such that: (a) consecutive branches β_i and β_{i+1} always have a common end point; (b) no node is the endpoint of more than two branches in the set; (c) ⓘ is the endpoint of exactly one branch in the set, and so is ⓙ.

[2] Sometimes, for drawing convenience, in the subsequent chapters the reference direction for the branch voltage is also denoted by an arrow, where the arrow-head corresponds to the "+" terminal and the arrow-end to the "−" terminal.

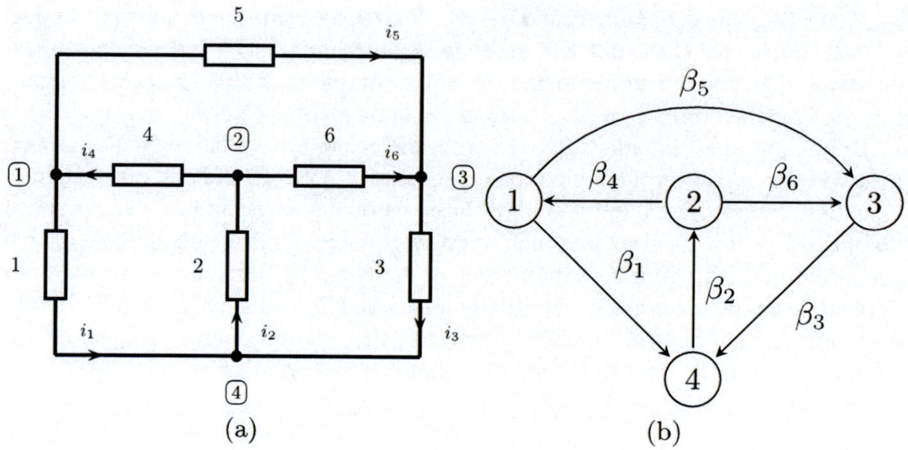

Fig. 3.1 (**a**) Network N of two-terminal circuit elements and (**b**) corresponding digraph $\mathcal{G}$. The associated reference direction is assumed for each element

Two of the most important concepts in graph theory are the cut-set and the tree of a graph.

Definition 3.4 (Cut-Set of a Graph) Given a connected digraph $\mathcal{G}$, a set of branches $\mathcal{C}$ of $\mathcal{G}$ is called a cut-set if and only if:

- the removal of all the branches of the cut-set reduces $\mathcal{G}$ to an unconnected digraph
- the removal of all but any one branch of $\mathcal{C}$ leaves $\mathcal{G}$ connected.

Definition 3.5 (Tree of a Graph) A tree $\mathcal{T}$ of a connected digraph $\mathcal{G}$ is a subgraph such that:

- $\mathcal{T}$ is connected
- $\mathcal{T}$ contains all nodes of $\mathcal{G}$
- $\mathcal{T}$ has no loops.

A graph $\mathcal{G}$ may have many trees. It can be shown that for a complete graph there exist n^{n-2} distinct trees. The branches that belong to a tree $\mathcal{T}$ are called *tree branches* (or, in short, *twigs*), and those which do not belong to a tree are called *links* (the terms *chords* is also used by some authors). All the links of a given tree $\mathcal{T}$ form what is called *cotree* with respect to the tree $\mathcal{T}$.

Example 3.2 (Trees and Cut-Sets of a Graph $\mathcal{G}$) Figure 3.2 shows four distinct trees of the digraph in Fig. 3.1b. The cut-sets

$$\mathcal{C}_1 = \{\beta_1, \beta_2, \beta_3\} \tag{3.1}$$

$$\mathcal{C}_2 = \{\beta_1, \beta_3, \beta_4, \beta_6\} \tag{3.2}$$

$$\mathcal{C}_3 = \{\beta_3, \beta_5, \beta_6\} \tag{3.3}$$

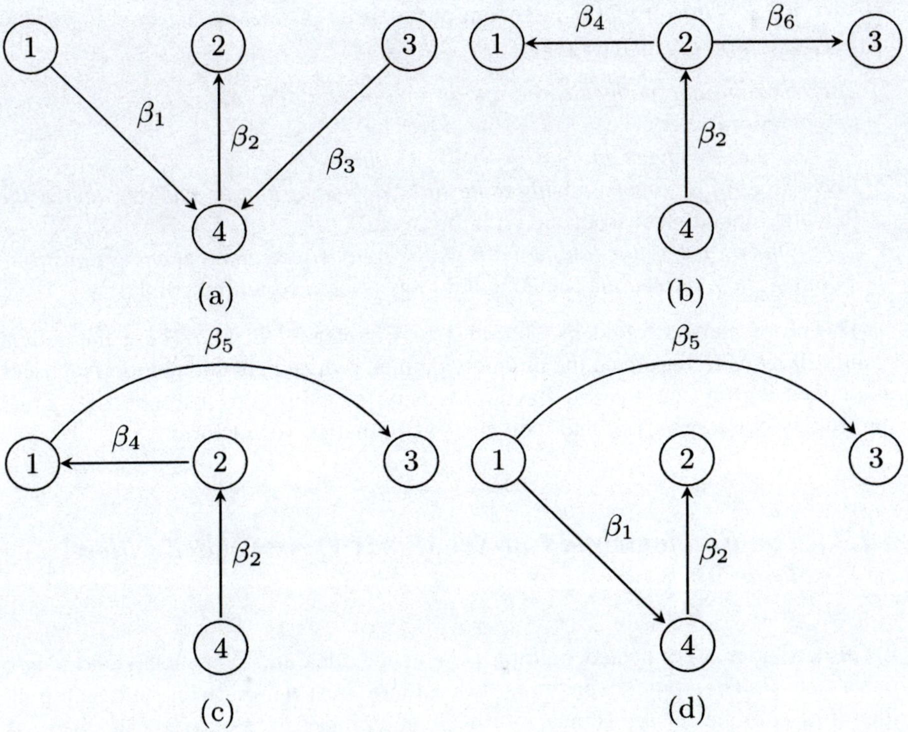

Fig. 3.2 Trees associated with the digraph $\mathcal{G}$ in Fig. 3.1b. (**a**) Tree $\mathcal{T}_1$. (**b**) Tree $\mathcal{T}_2$. (**c**) Tree $\mathcal{T}_3$. (**d**) Tree $\mathcal{T}_4$

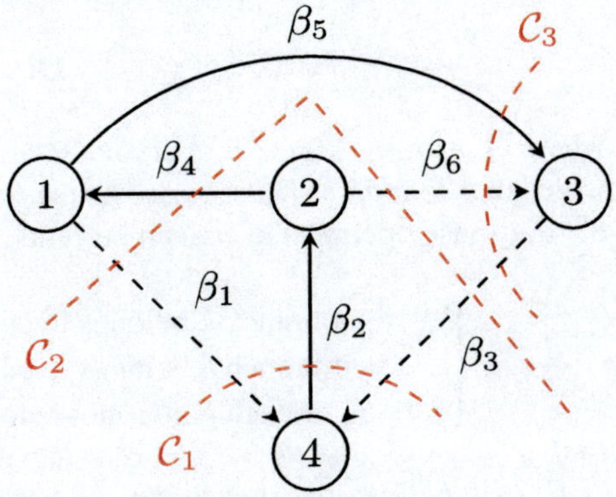

Fig. 3.3 Cut-sets $\mathcal{C}_1$, $\mathcal{C}_2$ and $\mathcal{C}_3$ associated with the tree $\mathcal{T}_3$ of $\mathcal{G}$ in Fig. 3.1b. The corresponding cotree is made of the links $\{\beta_1, \beta_3, \beta_6\}$ shown by dashed lines. Each cut-set is specified by one of the twigs β_2, β_4, and β_5

associated with the tree $\mathcal{T}_3 = \{\beta_2, \beta_4, \beta_5\}$ are shown in Fig. 3.3. It is worth noting that each cut-set includes just one twig of the tree, i.e., $\mathcal{C}_1$, $\mathcal{C}_2$ and $\mathcal{C}_3$ are associated with β_2, β_4, and β_5, respectively.

Theorem 3.1 (Fundamental Theorem of Graphs) *Given a connected digraph $\mathcal{G}$ with n nodes and b branches and a tree $\mathcal{T}$ of $\mathcal{G}$:*

1. *There is a unique path (dismiss the branch orientation) along the tree between any pairs of nodes.*
2. *There are $n - 1$ twigs and $l = b - (n - 1)$ links.*
3. *Every twig of $\mathcal{T}$ together with some links defines a unique cut-set, named the* fundamental cut-set associated with the twig.
4. *Every link of $\mathcal{T}$ and the unique path on the tree between its two nodes constitute a unique loop, named the* fundamental loop associated with the link.

The proof can be found in Section 3 of Chapter 12 in [1]. These theoretical results allow us to introduce the uniquely defined reduced cut-set and loop matrices associated with a tree, thereby paving the way for expressing Kirchhoff laws and developing cut-set analysis and loop analysis in matrix-vector form.

3.1.2 The Fundamental Cut-Set Matrix Q Associated with a Tree

Let us consider a connected digraph $\mathcal{G}$ with n nodes and b branches and single out a tree $\mathcal{T}$. Theorem 3.1 guarantees that there exist $n - 1$ twigs and each twig identifies a unique cut-set. Hence, $n-1$ KCL equations can be derived from the $n-1$ fundamental cut-sets. Certainly, these $n-1$ KCL equations are linearly independent because each equation has one and only one twig current. In matrix form the KCL equations based on the fundamental cut-sets of a tree can be written as:

$$\mathbf{Qi} = \mathbf{0} \tag{3.4}$$

where $\mathbf{Q}$ is an $(n - 1) \times b$ matrix called the *fundamental cut-set matrix* associated with a tree $\mathcal{T}$ and $\mathbf{i} = (i_1, \ldots, i_b)^T$ is the $b \times 1$ column vector (the symbol T denotes the transpose operator) of branch currents. The jk-th entry of $\mathbf{Q}$ is defined as:

$$q_{jk} = \begin{cases} +1 & \text{if branch } k \text{ belongs to cut-set } j \text{ and has the same direction} \\ -1 & \text{if branch } k \text{ belongs to cut-set } j \text{ and has the opposite direction} \\ 0 & \text{if branch } k \text{ does not belong to cut-set } j. \end{cases} \tag{3.5}$$

Here, the direction of each cut-set j is defined as the direction of its associated twig. By labeling the b branches of the graph $\mathcal{G}$ in such a way that first there are the twigs, i.e., $1, 2, \ldots, n - 1$, and then the links, i.e., $n, n + 1, \ldots, b$, the fundamental cut-set matrix associated with the corresponding tree $\mathcal{T}$ can be written in a convenient form as follows:

$$\mathbf{Q} = [\mathbf{I}_{n-1}, \mathbf{Q}_l] \tag{3.6}$$

where $\mathbf{I}_{n-1}$ is the $(n-1) \times (n-1)$ identity matrix and $\mathbf{Q}_l$ is a submatrix of $n-1$ rows and l columns with entries 0, 1, and -1. Clearly, the vector of the branch currents $\mathbf{i}$ can be decomposed in the corresponding $\mathbf{i}_t = (i_1, i_2, \dots, i_{n-1})^T$ twig currents and $\mathbf{i}_l = (i_n, i_{n+1}, \dots, i_b)^T$ link currents. As a consequence, the $n-1$ KCL equations can be written as follows:

$$\mathbf{i}_t = -\mathbf{Q}_l \mathbf{i}_l. \tag{3.7}$$

The corresponding decomposition of the branch voltage vector $\mathbf{v} = (v_1, v_2, \dots, v_b)^T$ into the twig voltage vector $\mathbf{v}_t = (v_1, v_2, \dots, v_{n-1})^T$ and link voltage vector $\mathbf{v}_l = (v_n, v_{n+1}, \dots, v_b)^T$ permits to write the KVL equations in the form

$$\mathbf{v} = \mathbf{Q}^T \mathbf{v}_t \tag{3.8}$$

from which it is easily derived that

$$\mathbf{v}_l = \mathbf{Q}_l^T \mathbf{v}_t. \tag{3.9}$$

Equations (3.4) and (3.8) provide the matrix form of Kirchhoff equations based on the fundamental cut-set matrix $\mathbf{Q}$.

3.1.3 *The Incidence Matrix* A

Any cut-set separates the set of nodes {①, ②, ..., ⓝ} of $\mathcal{G}$ into two subsets. Writing the KCL at each node in such subsets, and adding the results, we obtain the cut-set equations. In general, Kirchhoff equations (3.4) and (3.8) represent a generalization of KCL and KVL based on the nodes and the incidence matrix. It can be shown that for many digraphs, a tree $\mathcal{T}$ can be picked in such a way that $\mathbf{Q}$ is identical to the reduced incident matrix for a particular datum node.

Suppose that for the connected digraph $\mathcal{G}$, we write the n KCL equations for each node. Then we choose a datum node (e.g., node ⓝ) and we throw away the corresponding KCL equation. The remaining $n-1$ KCL equations at nodes ①, ②, ..., (n-1) are linearly independent and they can be written in matrix form by means of the *reduced incident matrix* $\mathbf{A}$ which is of dimension $(n-1) \times b$. The corresponding $n-1$ KCLs read

$$\mathbf{Ai} = \mathbf{0} \tag{3.10}$$

where the elements of $\mathbf{A}$ are specified as follows:

$$a_{jk} = \begin{cases} +1 & \text{if branch } k \textit{ leaves} \text{ node } \boxed{j} \\ -1 & \text{if branch } k \textit{ enters} \text{ node } \boxed{j} \\ 0 & \text{if branch } k \text{ does not touch node } \boxed{j} \end{cases} \tag{3.11}$$

A set of complete linearly independent KVL equations can be derived in terms of the *node-to-datum voltage vector* $\mathbf{v}_e = (v_{e,1}, v_{e,2}, \ldots, v_{e,n-1})^T$ as follows:

$$\mathbf{v} = \mathbf{A}^T \mathbf{v}_e. \tag{3.12}$$

Comparing (3.8) with (3.12), it is apparent that $\mathbf{v}_e$ coincides with $\mathbf{v}_t$ and $\mathbf{A} = \mathbf{Q}$.

Example 3.3 (Fundamental Cut-Set Matrix Associated with the Tree $\mathcal{T}_3$ in Fig. 3.3) With regard to the cut-sets $\mathcal{C}_1$, $\mathcal{C}_2$, and $\mathcal{C}_3$ in Fig. 3.3, the fundamental cut-set matrix associated with tree $\mathcal{T}_3 = \{\beta_2, \beta_4, \beta_5\}$ reads

$$\mathbf{Q} = \left(\begin{array}{ccc|ccc} +1 & 0 & 0 & -1 & -1 & 0 \\ 0 & +1 & 0 & -1 & -1 & +1 \\ 0 & 0 & +1 & 0 & -1 & +1 \end{array} \right) \tag{3.13}$$

where twig currents and voltages are $\mathbf{i}_t = (i_2, i_4, i_5)^T$ and $\mathbf{v}_t = (v_2, v_4, v_5)^T$, respectively, whereas link currents and voltages are $\mathbf{i}_l = (i_1, i_3, i_6)^T$ and $\mathbf{v}_l = (v_1, v_3, v_6)^T$, respectively. It follows that the KCL equations can be written as

$$i_2 - i_1 - i_3 = 0 \tag{3.14}$$

$$i_4 - i_1 - i_3 + i_6 = 0 \tag{3.15}$$

$$i_5 - i_3 + i_6 = 0 \tag{3.16}$$

whereas the KVL equations are

$$v_1 = -v_2 - v_4 \tag{3.17}$$

$$v_3 = -v_2 - v_4 - v_5 \tag{3.18}$$

$$v_6 = v_4 + v_5. \tag{3.19}$$

3.1.4 The Fundamental Loop Matrix B Associated with a Tree

A dual approach with respect to that based on the fundamental cut-sets permits to express Kirchhoff equations by means of link currents instead of twig voltages. Theorem 3.1 guarantees that there exist $l = b - (n-1)$ links and each link identifies a unique loop. Hence, KVL equations can be derived from the l fundamental loops. Such KVL equations are linearly independent because each loop equation has one and only one link voltage. In matrix form the KVL equations based on the fundamental loops of a tree can be written as

$$\mathbf{B}\mathbf{v} = \mathbf{0} \tag{3.20}$$

where $\mathbf{B}$ is an $l \times b$ matrix called the *fundamental loop matrix* associated with a tree $\mathcal{T}$ and $\mathbf{v} = (i_1, \ldots, i_b)^T$ is the $b \times 1$ column vector of branch voltages. The jk-th entry of $\mathbf{B}$ is defined as:

$$b_{jk} = \begin{cases} +1 & \text{if branch } k \text{ is in loop } j \text{ and their reference directions are the same} \\ -1 & \text{if branch } k \text{ is in loop } j \text{ and their reference directions are opposite} \\ 0 & \text{if branch } k \text{ is not in loop } j. \end{cases} \tag{3.21}$$

Here, the reference direction of each loop j is defined by the direction of its associated link. Using the same labeling scheme as before, i.e., the twigs are marked from 1 to $n-1$, and the links from n to b, the fundamental loop matrix $\mathbf{B}$ can be decomposed as follows:

$$\mathbf{B} = [\mathbf{B}_{n-1}, \mathbf{I}_l] \tag{3.22}$$

and KCL equations using link currents are of the form

$$\mathbf{i} = \mathbf{B}^T \mathbf{i}_l \tag{3.23}$$

from which it is derived that

$$\mathbf{i}_t = \mathbf{B}_{n-1}^T \mathbf{i}_l. \tag{3.24}$$

3.1.5 Link Between Q and B: Tellegen's Theorem

The following fundamental result expresses the relationship between the two matrices $\mathbf{Q}$ and $\mathbf{B}$ [1].

Theorem 3.2 *Let* $\mathbf{Q}$ *and* $\mathbf{B}$ *be the fundamental cut-set matrix and the fundamental loop matrix, respectively, of a connected digraph* $\mathcal{G}$ *for a specified tree* $\mathcal{T}$*. Then, we have*

$$\mathbf{B}\mathbf{Q}^T = \mathbf{0}. \tag{3.25}$$

The transpose of (3.25) gives $\mathbf{Q}\mathbf{B}^T = \mathbf{0}$. Premultiplying such equation by $\mathbf{v}_t^T$ and then postmultiplying by $\mathbf{i}_l$ we obtain

$$\mathbf{v}_t^T \mathbf{Q}\mathbf{B}^T \mathbf{i}_l = 0 \tag{3.26}$$

which reduces to Tellegen's theorem [1]

$$\mathbf{v}^T\mathbf{i} = 0 \tag{3.27}$$

by using (3.8) and (3.23).

Remark 3.1 Tellegen's theorem can also be proved by means of the reduced incidence matrix $\mathbf{A}$. In fact, deriving $\mathbf{v}^T = \mathbf{v}_e^T\mathbf{A}$ from (3.12), and multiplying by $\mathbf{i}$, it follows that

$$\mathbf{v}^T\mathbf{i} = \mathbf{v}_e^T\mathbf{A}\mathbf{i} = 0 \tag{3.28}$$

due to KCL equations $\mathbf{A}\mathbf{i} = \mathbf{0}$ in (3.10).

3.2 Tableau Analysis

The tableau analysis is based on the network equations obtained by combining KCL and KVL equations with the branch characteristics expressing the CRs of circuits elements composing the network. The following *nonlinear integro-differential-algebraic equation* describes the k-th (α_k, β_k)-element

$$f_k(v_k^{(\alpha_k)}, i_k^{(\beta_k)}) = 0 \tag{3.29}$$

where $v_k^{(\alpha_k)}$ and $i_k^{(\beta_k)}$ $(k = 1, 2, \ldots, b)$ are the higher-order integral or derivative of the branch voltage and current, respectively, as defined in Chap. 1. Combining CRs and Kirchhoff equations we have the following three forms of the tableau equations:

- *cut-set tableau equations*: basic variables are the twig voltages $\mathbf{v}_t$

$$\begin{cases} \mathbf{Q}\mathbf{i} = \mathbf{0} \\ \mathbf{v} = \mathbf{Q}^T\mathbf{v}_t \\ \mathbf{f}(\mathbf{v}^{(\alpha)}, \mathbf{i}^{(\beta)}) = \mathbf{0} \end{cases} \tag{3.30}$$

- *node tableau equations*: basic variables are the node-to-datum voltages $\mathbf{v}_e$

$$\begin{cases} \mathbf{A}\mathbf{i} = \mathbf{0} \\ \mathbf{v} = \mathbf{A}^T\mathbf{v}_e \\ \mathbf{f}(\mathbf{v}^{(\alpha)}, \mathbf{i}^{(\beta)}) = \mathbf{0} \end{cases} \tag{3.31}$$

- *loop tableau equations*: basic variables are the link currents $\mathbf{i}_l$

$$\begin{cases} \mathbf{B}\mathbf{v} = \mathbf{0} \\ \mathbf{i} = \mathbf{B}^T \mathbf{i}_l \\ \mathbf{f}(\mathbf{v}^{(\alpha)}, \mathbf{i}^{(\beta)}) = \mathbf{0} \end{cases} \tag{3.32}$$

where $\mathbf{v}^{(\alpha)} = (v_1^{(\alpha_1)}, v_2^{(\alpha_2)}, \ldots, v_b^{(\alpha_b)})^T$, $\mathbf{i}^{(\beta)} = (i_1^{(\beta_1)}, i_2^{(\beta_2)}, \ldots, i_b^{(\beta_b)})^T$, and $\mathbf{f} = (f_1, f_2, \ldots, f_b)^T : \mathbb{R}^{2b} \to \mathbb{R}^b$.

Remark 3.2 Whenever possible, it is convenient from a mathematical viewpoint to avoid the use of integral equations, i.e., to write the tableau equations avoiding $\alpha_k, \beta_k < 0$. This yields via the tableau a system of differential algebraic equations (DAEs).

Example 3.4 (Cut-Set Tableau Analysis) Let us consider a class of networks $\mathcal{N}$ with topology as in Fig. 3.1a and described by the digraph in Fig. 3.1b. Different classes of networks can be specified according to the properties of the two-terminal circuit elements in $\mathcal{N}$. In this example the cut-set tableau analysis is presented for the following classes:

(1) the class $\mathcal{N}_1$ of circuits in $\mathcal{N}$ made of just resistive linear elements. An example of resistive linear circuit in $\mathcal{N}_1$ is shown in Fig. 3.4;
(2) the class $\mathcal{N}_2$ of circuits in $\mathcal{N}$ made of linear/nonlinear resistive elements. An example of resistive nonlinear circuit in $\mathcal{N}_2$ including just one nonlinear resistive element is shown in Fig. 3.5;
(3) the class $\mathcal{N}_3$ of circuits in $\mathcal{N}$ made of linear/nonlinear resistive/dynamic elements. An example of dynamic nonlinear circuit in $\mathcal{N}_3$ including just one nonlinear resistive element and one capacitor is shown in Fig. 3.6.

The cut-set tableau equations (3.30) for the three classes $\mathcal{N}_1$, $\mathcal{N}_2$ and $\mathcal{N}_3$ can be written by means of the fundamental cut-set matrix given in (3.13). In the following, these three cases are presented in detail. Similar conclusions can be obtained by

Fig. 3.4 A resistive linear circuit in $\mathcal{N}_1$

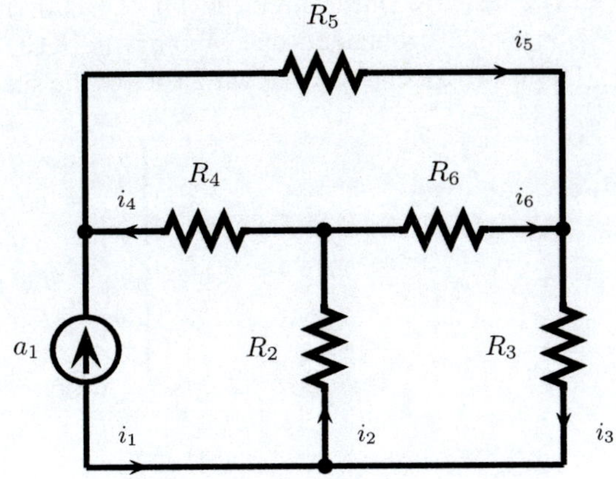

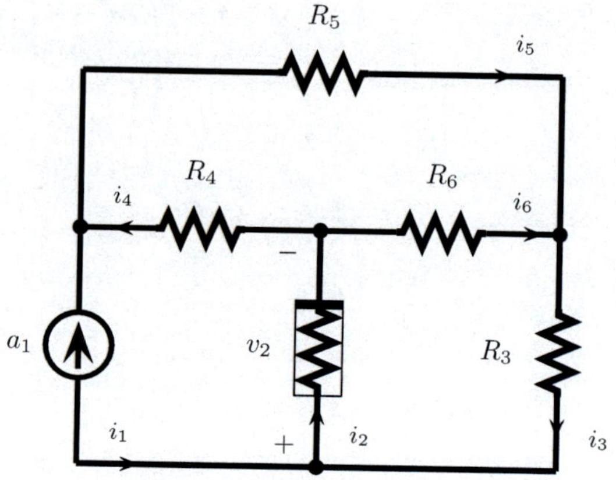

Fig. 3.5 A resistive nonlinear circuit in $\mathcal{N}_2$

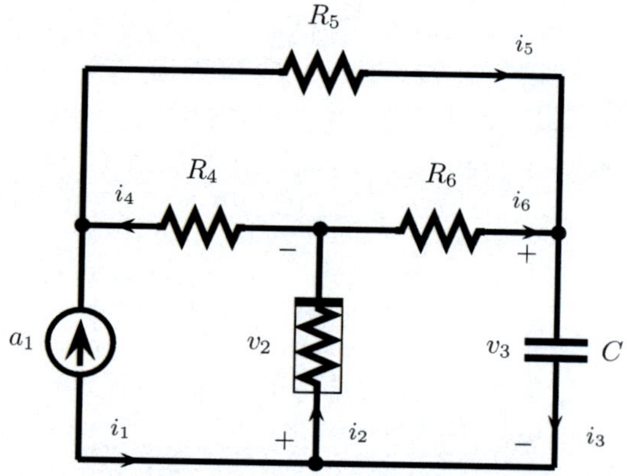

Fig. 3.6 A dynamic nonlinear circuit in $\mathcal{N}_3$

exploiting the node tableau analysis and the loop tableau analysis based on the matrices **A** and **B**, respectively.

- The resistive linear circuit in Fig. 3.4 is described by the cut-set tableau equations (3.30) obtained by combing the KCL equations (3.14), the KVL equations (3.17), and the following CRs of the six circuit elements

$$\begin{cases} i_1 = -a_1 \\ v_2 = R_2 i_2 \\ v_3 = R_3 i_3 \\ v_4 = R_4 i_4 \\ v_5 = R_5 i_5 \\ v_6 = R_6 i_6. \end{cases} \tag{3.33}$$

Thus, the set of 12 *linear algebraic equations* (3.14), (3.17), and (3.33) in the 12 unknowns v_k and i_k ($k = 1, 2, \dots, 6$) permit to describe and solve the circuit in Fig. 3.4.

Remark 3.3 Introducing the vectors of branch currents $\mathbf{i} = (\mathbf{i}_t, \mathbf{i}_l)^T$ and voltages $\mathbf{v} = (\mathbf{v}_t, \mathbf{v}_l)^T$ organized in terms of twig and link variables of the tree $\mathcal{T}_3$ in Fig. 3.2c, i.e.,

$$\mathbf{i}_t = (i_2, i_4, i_5)^T$$
$$\mathbf{v}_t = (v_2, v_4, v_5)^T$$
$$\mathbf{i}_l = (i_1, i_3, i_6)^T$$
$$\mathbf{v}_l = (v_1, v_3, v_6)^T$$

the cut-set tableau equations can be written in the compact form

$$\mathbf{Qi} = \mathbf{0} \tag{3.34}$$

$$\mathbf{v} = \mathbf{Q}^T \mathbf{v}_t \tag{3.35}$$

$$\mathbf{i} = \mathbf{Gv} - \mathbf{a} \tag{3.36}$$

where (3.36) represents the matrix form of (3.33). Matrix $\mathbf{G}$ is a 6×6 diagonal matrix including the conductances of resistors, i.e., $\mathbf{G} = \mathrm{diag}(G_2, G_4, G_5, 0, G_3, G_6)$ and $\mathbf{a} = (0, 0, 0, a_1, 0, 0)^T$. Simple algebraic manipulations permit to reduce the cut-set tableau equations to

$$\mathbf{QGQ}^T \mathbf{v}_t = \mathbf{Q}^T \mathbf{a} \tag{3.37}$$

which includes just three equations. It is important to observe that if a tree $\mathcal{T}$ is picked in such a way that $\mathbf{Q}$ coincides with $\mathbf{A}$, then (3.37) describes the *nodal analysis method*. Further details on the nodal analysis (and the dual mesh analysis) can be found in [1].

- The resistive nonlinear circuit in Fig. 3.5 is obtained by the circuit in Fig. 3.4 by replacing R_2 with a voltage-controlled nonlinear resistor with CR $i_2 = \hat{i}(v_2)$. It follows that the CRs of the six circuit elements are

$$\begin{cases} i_1 = -a_1 \\ i_2 = \hat{i}(v_2) \\ v_3 = R_3 i_3 \\ v_4 = R_4 i_4 \\ v_5 = R_5 i_5 \\ v_6 = R_6 i_6. \end{cases} \tag{3.38}$$

Hence, the resistive nonlinear circuit in Fig. 3.5 is described by a set of *nonlinear algebraic equations* given by (3.14), (3.17), and (3.38). A comprehensive discussion of algorithms and numerical methods for solving nonlinear algebraic equations is available in [2].

- When the resistor R_3 of the resistive nonlinear circuit in Fig. 3.5 is substituted by a linear capacitor C_3, the dynamic nonlinear circuit in Fig. 3.6 is attained and the set of CRs result in

$$\begin{cases} i_1 = -a_1 \\ i_2 = \hat{\imath}(v_2) \\ i_3 = C_3 \frac{d\,v_3}{dt} \\ v_4 = R_4 i_4 \\ v_5 = R_5 i_5 \\ v_6 = R_6 i_6. \end{cases} \tag{3.39}$$

The cut-set tableau equations for the dynamic nonlinear circuit in Fig. 3.6 is consequently a set of nonlinear DAEs resulting from (3.14), (3.17), and (3.39).

The following simplified form of DAEs can be derived by using the wye-delta transformation of the resistors R_4, R_5, and R_6

$$C_3 \frac{d\,v_3}{dt} = \hat{\imath}(v_2) + a_1 \tag{3.40}$$

$$(\hat{R}_2 + \hat{R}_3)\hat{\imath}(v_2) + v_2 + v_3 + \hat{R}_3 a_1 = 0 \tag{3.41}$$

where $\hat{R}_1 = (R_4 R_6)/R$, $\hat{R}_2 = (R_4 R_5)/R$, $\hat{R}_3 = (R_5 R_6)/R$, $R = R_4 + R_5 + R_6$ and we suppose $R \neq 0$.

An interesting problem is *whether the obtained DAEs can be put or not into the form of an ordinary differential equation (ODE).*

Assume the voltage-controlled nonlinear resistor described by $\hat{\imath}(\cdot)$, and the circuit parameters, permit to derive v_2 in terms of v_3 from (3.41). Namely, function $(\hat{R}_2 + \hat{R}_3)\hat{\imath}(v_2) + v_2$ is globally invertible and then it is possible to solve (3.41) in the form $v_2 = h(v_3)$, where $h(\cdot)$ is a nonlinear function. Then, (3.40) and (3.41) reduce to the ODE

$$C_3 \frac{d\,v_3}{dt} = \hat{\imath}(h(v_3)) + a_1 \tag{3.42}$$

that is expressed in terms of the (state) variable v_3 (the voltage across the capacitor C_3). The investigation of nonlinear dynamics in the circuit in Fig. 3.6 via the differential equation (3.42) (a.k.a. state equation) results to be more advantageous with respect to the approach based on directly studying the DAEs (3.40)–(3.41). We also stress that when the stated invertibility assumption is not satisfied, then it is *not possible* to recast (3.40) and (3.41) in the form of an ODE (SE).

Starting from this observation, the next section focuses on the description of dynamic nonlinear networks via the *state equations* perspective.

The tableau equations (3.30), (3.31), and (3.32) are frequently referred to as *sparse tableau equations* because there always exist a large number of zeros in the matrices involved in these equations. For the class of linear dynamic networks, a conventional method for solving *linear integro-differential equations* (3.30), (3.31), and (3.32) exploits the Laplace transform and efficient algorithms for sparse systems.

The next sections discuss the state equation approach for *RLC networks*. Readers interested in a deep understanding of circuit analysis methodologies are also invited to refer to the specialized books [1] and [2].

3.3 State Equations

The *state equations* (SEs) of any dynamic lumped network are written in normal form

$$\dot{\mathbf{x}} = \mathbf{f}(\mathbf{x}), \qquad t \geq t_0 \tag{3.43}$$

where $\mathbf{x} = (x_1, x_2, \ldots, x_n)^T \in \mathbb{R}^n$ is the vector of state variables, t_0 is a given finite initial instant,

$$\mathbf{x}(t_0) \in \mathbb{R}^n \tag{3.44}$$

is the initial state (initial condition), and $\mathbf{f} : \mathbb{R}^n \to \mathbb{R}^n$ is the vector field defining the SEs. The *order* of the SEs (3.43) is the dimension n of the vector $\mathbf{x}$ of state variables.

Equations (3.43), together with the initial condition (3.44), define an *initial value problem* (IVP) (or Cauchy problem) associated with the SEs. If $\mathbf{x}(t)$, $t \in I$, where $I \subseteq \mathbb{R}$ is an interval such that $t_0 \in I$, is a solution of the IVP, then the locus $\{\mathbf{x}(t) \in \mathbb{R}^n, t \in I\}$ is a curve in $\mathbb{R}^n$ passing through $\mathbf{x}_0$ that is named *trajectory* or *orbit* through $\mathbf{x}_0$ of (3.43) and $\mathbb{R}^n$ is named *state space* or *phase space*.

An equilibrium point (EP) is a constant (i.e., a stationary) solution of (3.43). Note that $\bar{\mathbf{x}}$ is an EP if and only if we have $\mathbf{f}(\bar{\mathbf{x}}) = \mathbf{0}$. A (nontrivial) periodic solution is a nonequilibrium solution of (3.43) such that $\mathbf{x}(t+T) = \mathbf{x}(t)$ for some $T > 0$ and any t. While the image in the state space of an EP is a singleton, the image of a periodic solution is a closed trajectory usually called a periodic orbit or a closed orbit. An isolated periodic orbit is called a limit cycle. The reader is referred to [3] for the concept of local stability, asymptotic stability, and instability of an EP or a periodic solution.

Equation (3.43) is usually called an *autonomous* SE in the literature and includes dynamic circuits containing only linear time-invariant elements and dc sources. In order to include *non-autonomous* circuits with time-varying elements or sources (e.g., a sinusoidal source) we need to consider the class of non-autonomous SEs in normal form

$$\dot{\mathbf{x}} = \mathbf{f}(\mathbf{x}, t), \qquad t \geq t_0$$

with the initial condition

$$\mathbf{x}(t_0) \in \mathbb{R}^n$$

where now the vector field $\mathbf{f} : \mathbb{R}^{n+1} \to \mathbb{R}^n$ has an explicit dependence on time t.

There are some important reasons for writing the SEs of a dynamic circuit. Since it is usually impossible to solve the dynamic equations of a nonlinear circuit explicitly, the interest is in the study of the *qualitative properties* of its solutions (existence, uniqueness and boundedness of solutions, equilibrium points, stability, oscillations, and complex dynamics). It is well known that there are effective tools and a well-developed mathematical theory for studying the qualitative properties of SEs in the normal form (3.43). It is remarkable for example that it is usually possible to provide simple and easily checkable conditions on the vector field to guarantee that, given any initial state, there exists a unique solution of the IVP (3.43) and (3.44) on some time interval. More physically, this means that the knowledge of the initial condition, the structure of the circuit, and the applied sources guarantees a unique evolution of the state for $t > t_0$. Another important reason is that any properly modeled circuits have a well-defined SE. In fact, as we will discuss in more detail in Chap. 4, circuits that do not admit an SE representation[3] may be bad modeled from a physical viewpoint due to the presence of singular points, named impasse points, where the solutions cannot be prolonged forward or backward in time. Finally, it is noted that most numerical methods for solving nonlinear differential equations are formulated in terms of the standard form (3.43).

Several techniques are available to write the SEs of a dynamic circuit. For simple low-dimensional circuits it is often possible to proceed by inspection [1]. Otherwise, one can first write the DAEs describing the tableau, the node, or the loop equations and then try to cast them by substitution into an SE (cf. Example 3.4). Next, we discuss a systematic and effective technique where use is made of the decomposition of the network into a dynamic part and a resistive part. The hybrid representation of a multiport resistive network obtained via this decomposition is then used to write the SEs. For simplicity, in the remainder of this chapter we suppose that all elements, except possibly the voltage and current sources, are time-invariant. The extension to the time-varying case is straightforward. For didactic reasons, we start with the discussion of linear *RLC* networks and then address the case of nonlinear *RLC* networks.

[3] We have already encountered one such circuit in Example 3.4.

3.3.1 State Equations of Linear RLC Circuits

The SEs of linear circuits containing ideal (linear) resistors, capacitors, inductors, and independent voltage and current sources can be written by means of an effective procedure based on considering the storage elements connected to a resistive multiport network and using the hybrid representation of the multiport resistive network. The procedure is illustrated in the case where there is one or two storage elements, but can be easily generalized to any number of storage elements. From classical circuit theory the state variables are chosen as the capacitor voltages and inductor currents.

3.3.1.1 First-Order Linear Circuits

Consider a circuit with one capacitor connected to a two-terminal element N containing linear resistors and independent voltage or current sources, as shown in Fig. 3.7a. Assume N is voltage-controlled and consider the circuit obtained by replacing N with its Norton equivalent [1], where G_{eq} is the equivalent conductance and $i_{\mathrm{sc}}(t)$ is the short-circuit current (Fig. 3.7b).

Accounting for the reference directions of currents and voltages, and the CR of the linear capacitor and N, we have $i = G_{\mathrm{eq}}v + i_{\mathrm{sc}}(t) = G_{\mathrm{eq}}v_C + i_{\mathrm{sc}}(t) = -i_C = -C dv_C/dt$, hence the circuit satisfies the first-order linear SE in normal form

$$\frac{dv_C}{dt} = -\frac{G_{\mathrm{eq}}}{C}v_C - \frac{i_{\mathrm{sc}}(t)}{C}.$$

This should be solved using the initial condition $v_C(t_0) = v_{C_0}$.

The dual circuit in Fig. 3.8 where the linear capacitor is replaced with a linear inductor can be analyzed similarly. If N is current-controlled, by replacing it with its Thevenin equivalent [1], we obtain $v = R_{\mathrm{eq}}i + v_{\mathrm{oc}}(t) = R_{\mathrm{eq}}i_L + v_{\mathrm{oc}}(t) = -v_L = -L di_L/dt$, where R_{eq} is the equivalent resistance and $v_{\mathrm{oc}}(t)$ is the open circuit voltage of N. Hence, we obtain the first-order linear SE in normal form

$$\frac{di_L}{dt} = -\frac{R_{\mathrm{eq}}}{L}i_L - \frac{v_{\mathrm{oc}}(t)}{L}.$$

This should be solved with the initial condition $i_L(t_0) = i_{L_0}$.

3.3.1.2 Second-Order Linear Circuits

Consider first a two-capacitor configuration. By extracting the capacitors we can always redraw this configuration as in Fig. 3.9a, where the two-port network N contains the linear resistors and the independent voltage and current sources.

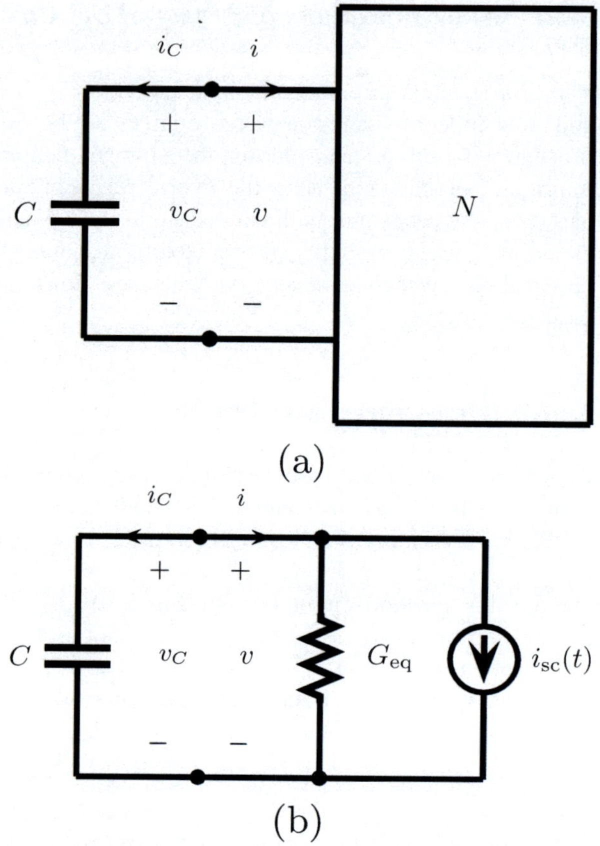

Fig. 3.7 (**a**) First-order linear circuit with a capacitor and (**b**) circuit obtained by replacing N with its Norton equivalent

Accounting for the reference directions of currents and voltages, and the CRs of the linear capacitors, we have

$$i_{C_j} = C_j \frac{dv_{C_j}}{dt} = -i_j, \quad v_{C_j} = v_j, \, j = 1, 2.$$

To write the SEs, we have to express i_{C_j}, $j = 1, 2$, as a function of the state variables v_{C_j}, $j = 1, 2$. To this end, suppose to replace the capacitors by voltage sources v_j, $j = 1, 2$, as shown in Fig. 3.9b. If the resistive circuit thus obtained is uniquely solvable for any v_j, $j = 1, 2$, and any value of the voltages and currents impressed by the independent sources, a standard two-port representation theorem yields the hybrid representation of N [1]

$$i_1(t) = g_{11}v_1(t) + g_{12}v_2(t) + i_{s_1}(t) \tag{3.45}$$

$$i_2(t) = g_{21}v_1(t) + g_{22}v_2(t) + i_{s_2}(t) \tag{3.46}$$

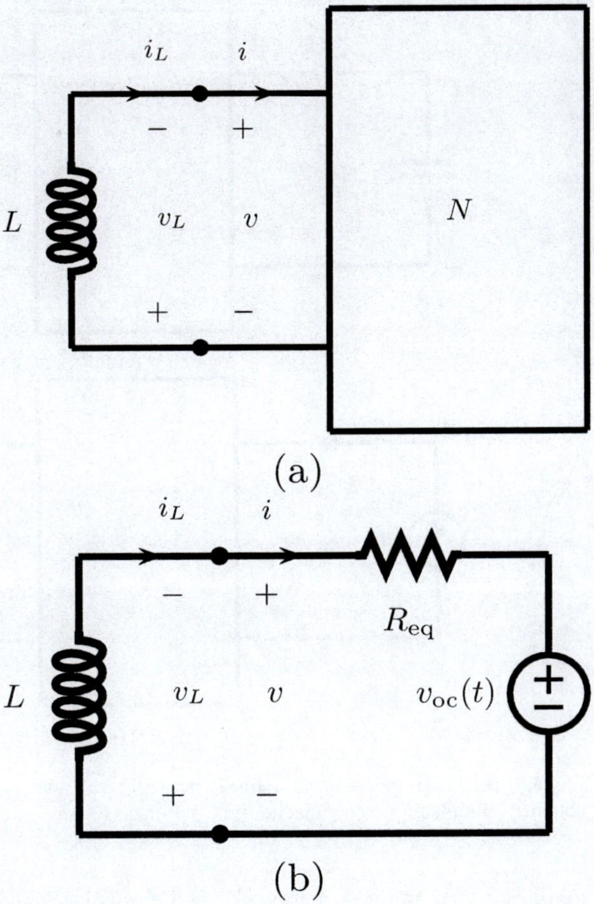

Fig. 3.8 (**a**) First-order linear circuit with an inductor and (**b**) circuit obtained by replacing N with its Thevenin equivalent

where $g_{11}, g_{12}, g_{21}, g_{22}$ are constants and $i_{s_1}(t), i_{s_2}(t)$ are time functions that depend on the independent sources within N.

By substitution, we obtain the following SEs in normal form for the two-capacitor configuration

$$\begin{pmatrix} \frac{dv_{C_1}(t)}{dt} \\ \frac{dv_{C_2}(t)}{dt} \end{pmatrix} = -\begin{pmatrix} \frac{g_{11}}{C_1} & \frac{g_{12}}{C_1} \\ \frac{g_{21}}{C_2} & \frac{g_{22}}{C_2} \end{pmatrix} \begin{pmatrix} v_{C_1}(t) \\ v_{C_2}(t) \end{pmatrix} - \begin{pmatrix} \frac{i_{s_1}(t)}{C_1} \\ \frac{i_{s_2}(t)}{C_2} \end{pmatrix}.$$

The initial conditions are $v_{C_1}(t_0) = v_{C_{10}}$ and $v_{C_2}(t_0) = v_{C_{20}}$.

Remark 3.4 The unique solvability assumption for the network in Fig. 3.9b is satisfied if and only if v_1 and v_2 are independent variables. A necessary condition is that there is no loop formed exclusively by C_1, C_2 and independent voltage sources [1]. It can be shown that such condition is also sufficient if all resistors within N are positive.

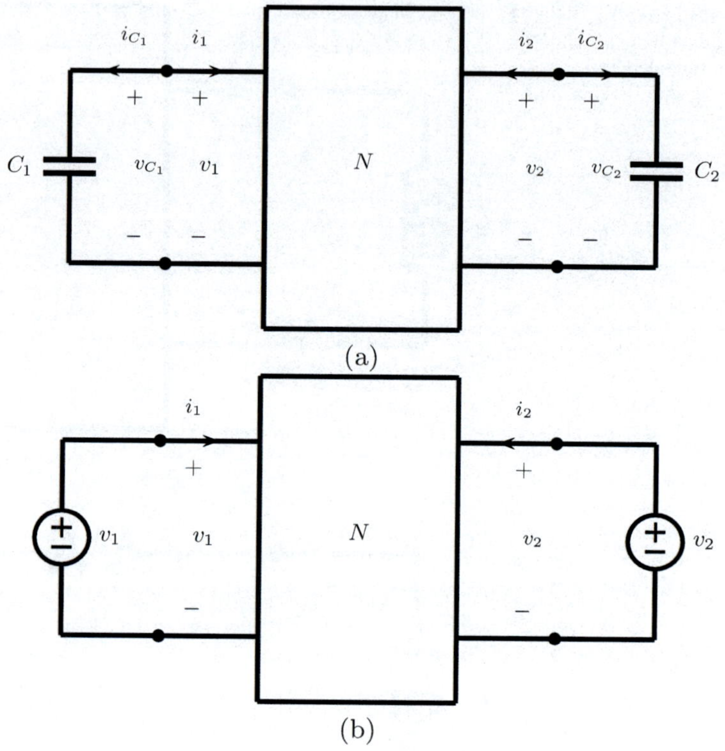

Fig. 3.9 (**a**) Decomposition of a linear network with two capacitors and (**b**) resistive circuit obtained by replacing each capacitor with a voltage source

Remark 3.5 If there is a loop formed exclusively by capacitors and independent voltage sources, then the capacitor voltages are dependent variables and the SE representation thus described does not exist. Yet, it may be in general still possible to write an SE using a reduced number of state variables. Systematic methods to write the SEs when there are such loops (and, dually, there are cut-sets made by inductors and current sources only) are discussed in [2]. Such methods, which can be applied also to nonlinear *RLC* networks, usually involve, however, a huge amount of algebra. An alternative and more effective technique is to use suitable circuit transformations that enable to eliminate such undesired capacitor loops (and inductor cut-sets) [4, Th. 2].

Consider now the configuration with a capacitor and an inductor as in Fig. 3.10a, where the two-port network N contains the linear resistors and the independent voltage and current sources. We have

$$i_{C_1} = C_1 \frac{dv_{C_1}}{dt} = -i_1, \quad v_{C_1} = v_1$$

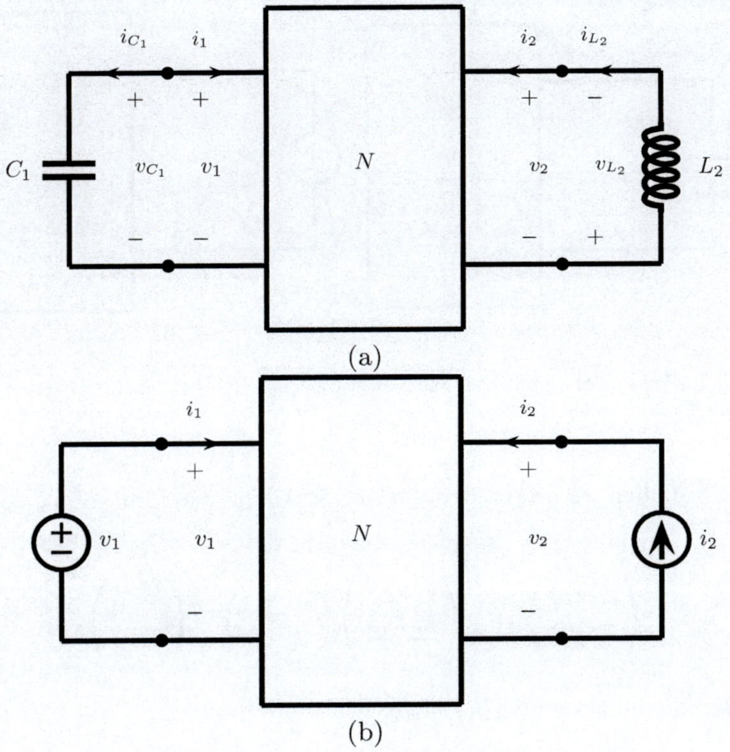

Fig. 3.10 (**a**) Decomposition of a linear network with a capacitor and an inductor and (**b**) resistive circuit obtained by replacing the capacitor with a voltage source and the inductor with a current source

and

$$v_{L_2} = L_2 \frac{di_{L_2}}{dt} = -v_2, \quad i_{L_2} = i_2.$$

Suppose to replace the capacitor by a voltage sources v_1 and the inductor by a current source i_2, as shown in Fig. 3.10b. If the resistive circuit thus obtained is uniquely solvable for any v_1 and i_2, and for any value of the currents and voltages impressed by the independent sources, i.e., v_1 and i_2 are independent variables, a standard two-port representation theorem for N yields the hybrid representation [1]

$$i_1(t) = h_{11}v_1(t) + h_{12}i_2(t) + i_{s_1}(t) \tag{3.47}$$

$$v_2(t) = h_{21}v_1(t) + h_{22}i_2(t) + v_{s_2}(t) \tag{3.48}$$

where $h_{11}, h_{12}, h_{21}, h_{22}$ are constants and $i_{s_1}(t), v_{s_2}(t)$ are time functions that depend on the independent sources within N.

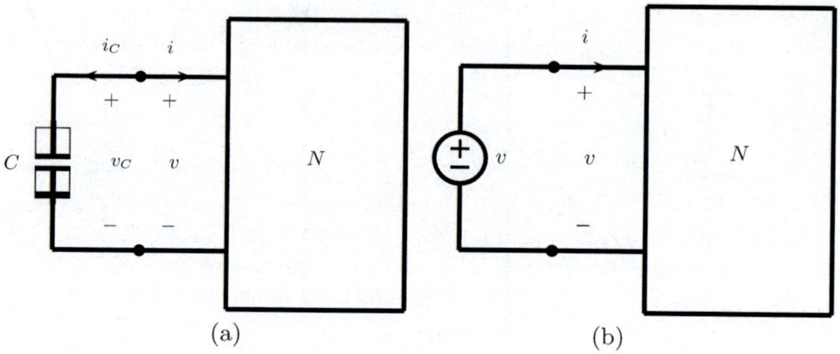

Fig. 3.11 (**a**) First-order circuit with a nonlinear capacitor and (**b**) circuit obtained by replacing the capacitor with a voltage source

By substitution we obtain the following SEs in normal form for the considered configuration

$$\begin{pmatrix} \frac{dv_{C_1}(t)}{dt} \\ \frac{di_{L_2}(t)}{dt} \end{pmatrix} = -\begin{pmatrix} \frac{h_{11}}{C_1} & \frac{h_{12}}{C_1} \\ \frac{h_{21}}{L_2} & \frac{h_{22}}{L_2} \end{pmatrix} \begin{pmatrix} v_{C_1}(t) \\ i_{L_2}(t) \end{pmatrix} - \begin{pmatrix} \frac{i_{s_1}(t)}{C_1} \\ \frac{v_{s_2}(t)}{L_2} \end{pmatrix}.$$

The initial conditions are $v_{C_1}(t_0) = v_{C_{1_0}}$ and $i_{L_2}(t_0) = i_{L_{2_0}}$.

3.3.2 *State Equations of Nonlinear RLC Circuits*

The previous technique can be generalized to RLC circuits containing nonlinear resistors, inductors, and capacitors, in addition to independent voltage and current sources. We first exemplify the technique in two cases with one and two storage elements, respectively. Then, in the next section, the technique is extended to any number of nonlinear capacitors and inductors.

3.3.2.1 First-Order Nonlinear Circuits

Consider a circuit with a nonlinear capacitor connected to a resistive two-terminal element N containing nonlinear resistors and independent voltage or current sources as in Fig. 3.11a. Replace C by a voltage source and assume N is voltage-controlled, i.e., we can write

$$i = \hat{i}(v, t).$$

If C is voltage controlled, i.e., $q_C = \hat{q}_C(v_C)$, we obtain

$$C(v_C)\frac{dv_C}{dt} = i_C = -i = -\hat{\imath}(v, t) = -\hat{\imath}(v_C, t)$$

where the small-signal capacitance is given by $C(v_C) = \hat{q}'_C(v_C)$ (cf. Chap. 1). Assuming $C(v_C) \neq 0$, for any v_C, we obtain that the circuit is described by the first-order SE in normal form

$$\frac{dv_C}{dt} = -\frac{1}{C(v_C)}\hat{\imath}(v_C, t).$$

The state variable is v_C and the initial condition is $v_C(t_0) = v_{C0}$.

Remark 3.6 It is worth to stress that if $C(v_C) = 0$ for some v_C, then it is not possible to obtain an SE in normal form using v_C. We refer the reader to [4] for further considerations and examples discussing this singular case.

If C is charge-controlled, i.e., $v_C = \hat{v}(q_C)$, then from $i_C = -\hat{\imath}(v_C, t)$ we obtain the first-order SE in normal form

$$\frac{dq_C}{dt} = -\hat{\imath}(\hat{v}_C(q_C), t).$$

In this case the state variable is q_C and the initial condition $q_C(t_0) = q_{C0}$.

A dual treatment holds in the case the capacitor is replaced by an inductor and N is assumed to be current-controlled. The details are left to the reader.

3.3.2.2 Second-Order Nonlinear Circuits

Consider first the case where there is a voltage-controlled capacitor

$$q_{C_1} = \hat{q}_{C_1}(v_{C_1})$$

hence

$$i_{C_1} = C_1(v_{C_1})\frac{dv_{C_1}}{dt} = -i_1, \quad v_{C_1} = v_1$$

where $C_1(v_{C_1}) = \hat{q}'_{C_1}(v_{C_1})$ and a current-controlled inductor

$$\varphi_{L_2} = \hat{\varphi}_{L_2}(i_{L_2})$$

hence

$$v_{L_2} = L_2(i_{L_2})\frac{di_{L_2}}{dt} = -v_2, \quad i_{L_2} = i_2$$

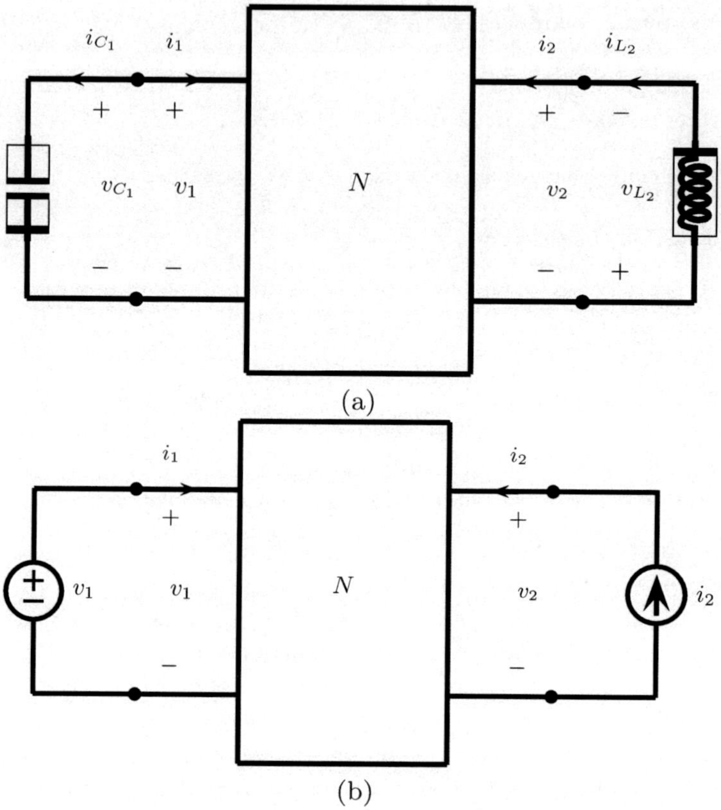

Fig. 3.12 (**a**) Second-order circuit with a nonlinear capacitor and a nonlinear inductor and (**b**) resistive circuit obtained by replacing the capacitor with a voltage source and the inductor with a current source

where $L_2(i_{L_2}) = \hat{\varphi}'_{L_2}(i_{L_2})$, as shown in Fig. 3.12a.

Let us use v_{C_1} and i_{L_2} as state variables. Suppose to replace the capacitor by a voltage sources v_1 and the inductor by a current source i_2 as shown in Fig. 3.12b. If the resistive circuit in this figure is uniquely solvable for any v_1 and i_2, and any value of currents and voltages impressed by the independent sources, then we can consider the hybrid representation of N

$$\begin{aligned} i_1 &= \hat{i}_1(v_1, i_2, t) \\ v_2 &= \hat{v}_2(v_1, i_2, t). \end{aligned} \tag{3.49}$$

Assuming the incremental capacitance $C_1(v_{C_1}) \neq 0$ for any v_{C_1}, and the incremental inductance $L_2(i_{L_2}) \neq 0$ for any i_{L_2}, we obtain the SEs in normal form

$$\frac{dv_{C_1}}{dt} = -\frac{1}{C_1(v_{C_1})}\hat{i}_1(v_{C_1}, i_{L_2}, t) \tag{3.50}$$

$$\frac{di_{L_2}}{dt} = -\frac{1}{L_2(i_{L_2})}\hat{v}_2(v_{C_1}, i_{L_2}, t) \tag{3.51}$$

where once more the explicit dependence on t accounts for time-varying sources. The initial conditions are $v_{C_1}(t_0) = v_{C_{1_0}}$ and $i_{L_2}(t_0) = i_{L_{2_0}}$.

Suppose now that there is a charge-controlled capacitor

$$v_{C_1} = \hat{v}_{C_1}(q_{C_1}) = v_1$$

and a flux-controlled inductor

$$i_{L_2} = \hat{i}_{L_2}(\varphi_{L_2}) = i_2.$$

Let us use q_{C_1} and φ_{L_2} as state variables. Suppose to replace the capacitor by a voltage source v_1 and the inductor by a current source i_2 and consider under the unique solvability assumption the same representation (3.49) as before for N. Then, noting that $i_{C_1} = dq_{C_1}/dt = -i_1$ and $v_{L_2} = d\varphi_{L_2}/dt = -v_2$ we obtain the SEs in normal form

$$\frac{dq_{C_1}}{dt} = -\hat{i}_1(\hat{v}_{C_1}(q_{C_1}), \hat{i}_{L_2}(\varphi_{L_2}), t) \tag{3.52}$$

$$\frac{d\varphi_{L_2}}{dt} = -\hat{v}_2(\hat{v}_{C_1}(q_{C_1}), \hat{i}_{L_2}(\varphi_{L_2}), t). \tag{3.53}$$

The initial conditions are $q_{C_1}(t_0) = q_{C_{1_0}}$ and $\varphi_{L_2}(t_0) = \varphi_{L_{2_0}}$.

Example 3.5 Consider the network in Fig. 3.13a with a linear capacitor C_1, a linear inductor L_2, and two nonlinear resistors $i_{R_1} = e^{v_{R_1}}$, $v_{R_2} = i_{R_2}^4$. By replacing C_1 with a voltage source and L_2 with a current source, we obtain the network in Fig. 3.13b.

By the KCL at the cut-set $\mathcal{C}$ we obtain

$$i_1 = \hat{i}_1(v_1, i_2) = v_1 + i_{R_1} - i_2 - 1 = v_1 + e^{v_1} - i_2 - 1.$$

Applying KVL at the loop formed by L_2, C_1 and R_2 we obtain

$$v_2 = \hat{v}_2(v_1, i_2) = v_{R_2} + v_1 = (1 + i_2)^4 + v_1.$$

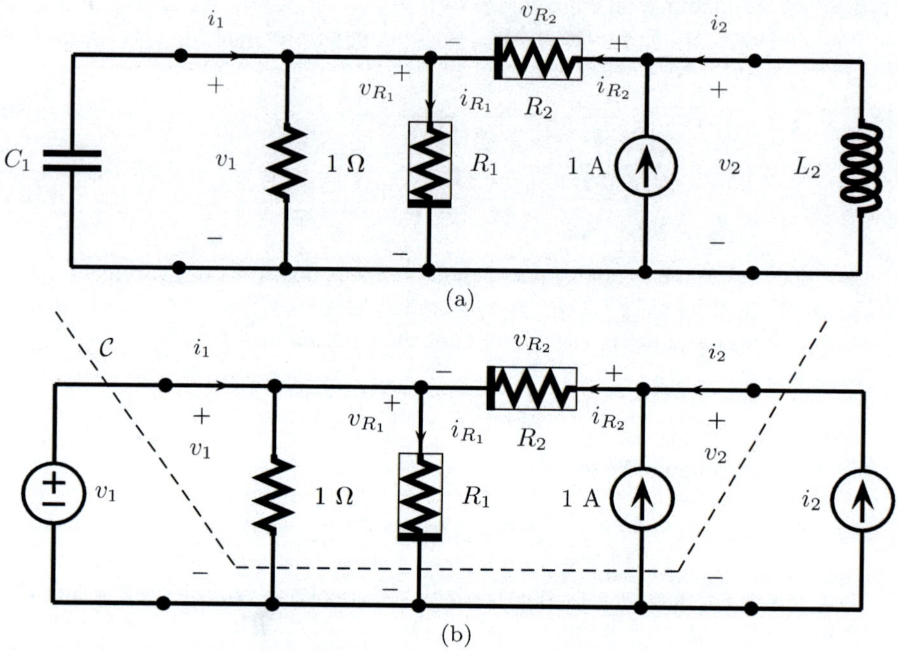

Fig. 3.13 (**a**) A network with two linear storage elements and nonlinear resistors and (**b**) network obtained by replacing the capacitor with a voltage source and the inductor by a current source

Then, considering that $v_{C_1} = v_1$ and $i_{L_2} = i_2$, we find that the circuit obeys the second-order SE in normal form

$$\frac{dv_{C_1}}{dt} = \frac{1}{C_1}(v_{C_1} + e^{v_{C_1}} - i_{L_2} - 1)$$

$$\frac{di_{L_2}}{dt} = -\frac{1}{L_2}((1 + i_{L_2})^4 + v_{C_1}).$$

Example 3.6 Consider the network in Fig. 3.14a with a nonlinear charge-controlled capacitor $v_{C_1} = q_{C_1}^2 + 1$ and a nonlinear flux-controlled inductor $i_{L_2} = \sin\varphi_{L_2}$ that are connected to a linear resistive network with a sinusoidal current source. By replacing C_1 with a voltage source and L_2 with a current source, we obtain the network in Fig. 3.14b.

Analyzing the linear memoryless network in Fig. 3.14b we easily obtain

$$i_1 = \hat{i}_1(v_1, i_2, t) = v_1 - \cos(2t) + \frac{v_1}{2} - \frac{i_2}{2} = \frac{3}{2}v_1 - \frac{1}{2}i_2 - \cos(2t)$$

and

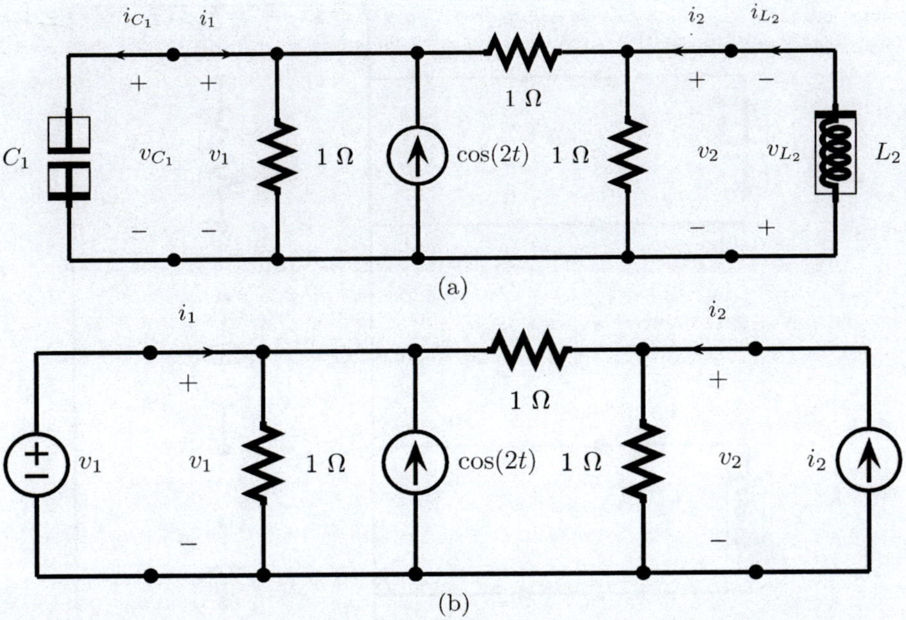

Fig. 3.14 (**a**) A network with two nonlinear storage elements and linear resistors and (**b**) network obtained by replacing the capacitor with a voltage source and the inductor by a current source

$$v_2 = \hat{v}_2(v_1, i_2, t) = \frac{1}{2}v_1 + \frac{1}{2}i_2.$$

By using the CR of the capacitor and inductor, we obtain that the circuit satisfies the second-order non-autonomous SEs in normal form

$$\frac{dq_{C_1}}{dt} = -\frac{3}{2}(q_{C_1}^2 + 1) + \frac{1}{2}\sin\varphi_{L_2} + \cos(2t)$$
$$\frac{d\varphi_{L_2}}{dt} = -\frac{1}{2}(q_{C_1}^2 + 1) - \frac{1}{2}\sin\varphi_{L_2}.$$

3.3.2.3 General Nonlinear *RLC* Circuits

Consider an *RLC* circuit containing an arbitrary number of (possibly) nonlinear resistors, inductors, capacitors, and independent voltage and current sources. Let n_C be the number of capacitors, n_L that of inductors and also let $n = n_C + n_L$. By extracting the storage elements, the circuit can be redrawn as in Fig. 3.15, where the resistive n-port network N contains only (linear and nonlinear) resistors and independent voltage or current sources.

Suppose there exists the hybrid representation of the n-port N

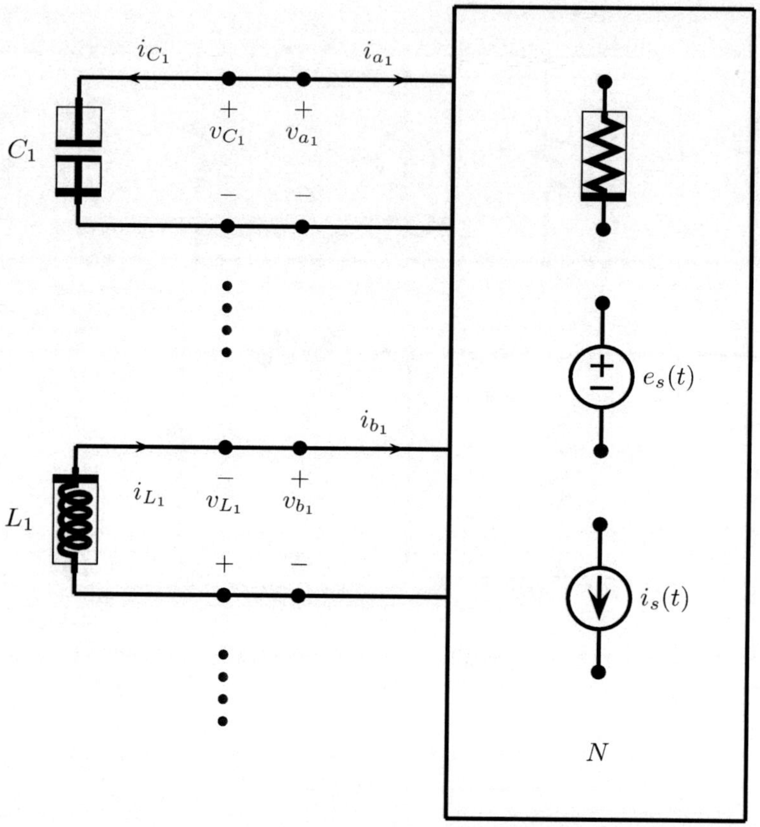

Fig. 3.15 Decomposition of a nonlinear *RLC* network

$$
\begin{aligned}
\mathbf{i}_a &= \mathbf{h}_a(\mathbf{v}_a, \mathbf{i}_b, \mathbf{u}_s(t)) \\
\mathbf{v}_b &= \mathbf{h}_b(\mathbf{v}_a, \mathbf{i}_b, \mathbf{u}_s(t))
\end{aligned}
\tag{3.54}
$$

where

$$\mathbf{u}_s(t) = (\mathbf{e}_s(t), \mathbf{i}_s(t))$$

and $\mathbf{e}_s(t)$, $\mathbf{i}_s(t)$ are the vectors of independent voltage and current sources, respectively.[4]

We consider next two basic SE formulations. In the first one, we assume the capacitors are voltage-controlled, i.e., they satisfy the CRs in vector form

[4]Note that these formulas explicitly highlight the dependence on the vectors of independent sources.

$$\mathbf{q}_C = \hat{\mathbf{q}}_C(\mathbf{v}_C)$$

and the inductors are current-controlled, i.e., they satisfy the CRs

$$\boldsymbol{\varphi}_L = \hat{\boldsymbol{\varphi}}_L(\mathbf{i}_L).$$

Let us choose $\mathbf{v}_C$ and $\mathbf{i}_L$ as state variables. The total number of state variables is $n = n_C + n_L$. Assume a non-singular incremental small-capacitance matrix $\mathbf{C}(\mathbf{v}_C) = \hat{\mathbf{q}}'_C(\mathbf{v}_C) = \text{diag}(C_1(v_{C_1}), C_2(v_{C_2}), \ldots, C_{nC}(v_{C_{nC}}))$ for any $\mathbf{v}_C$ and a non-singular small-signal inductance matrix $\mathbf{L}(\mathbf{i}_L) = \hat{\boldsymbol{\varphi}}'_L(\mathbf{i}_L) = \text{diag}(L_1(i_{L_1}), L_2(i_{L_2}), \ldots, L_{n_L}(i_{L_{n_L}}))$ for any $\mathbf{i}_L$. Since $\mathbf{i}_C = \dot{\mathbf{q}}_C = \mathbf{C}(\mathbf{v}_C)\dot{\mathbf{v}}_C = -\mathbf{i}_a$ and $\mathbf{v}_L = \dot{\boldsymbol{\varphi}}_L = \mathbf{L}(\mathbf{i}_L)(d\mathbf{i}_L/dt) = -\mathbf{v}_b$, substituting into the hybrid representation (3.54) of N, we obtain the SEs in normal form in the state variables $\mathbf{v}_C$ and $\mathbf{i}_L$

$$\dot{\mathbf{v}}_C = -\mathbf{C}^{-1}(\mathbf{v}_C)\mathbf{h}_a(\mathbf{v}_C, \mathbf{i}_L, \mathbf{u}_s(t)) \tag{3.55}$$

$$\frac{d\mathbf{i}_L}{dt} = -\mathbf{L}^{-1}(\mathbf{i}_L)\mathbf{h}_b(\mathbf{v}_C, \mathbf{i}_L, \mathbf{u}_s(t)). \tag{3.56}$$

The initial conditions are $\mathbf{v}_C(t_0) = \mathbf{v}_{C_0}$ and $\mathbf{i}_L(t_0) = \mathbf{i}_{L_0}$.

In the second formulation we assume the capacitors are charge-controlled, i.e., they satisfy the CRs

$$\mathbf{v}_C = \hat{\mathbf{v}}_C(\mathbf{q}_C)$$

and the inductors are flux-controlled, i.e., they satisfy the CRs

$$\mathbf{v}_L = \hat{\mathbf{i}}_L(\boldsymbol{\varphi}_L).$$

Now, we choose $\mathbf{q}_C$ and $\boldsymbol{\varphi}_L$ as state variables. The number of state variables is again $n = n_C + n_L$. Since $\mathbf{i}_C = \dot{\mathbf{q}}_C = -\mathbf{i}_a$ and $\mathbf{v}_L = \dot{\boldsymbol{\varphi}}_L = -\mathbf{v}_b$, substituting into the hybrid representation (3.54) of N, we obtain the SEs in normal form in the state variables $\mathbf{q}_C$ and $\boldsymbol{\varphi}_L$

$$\dot{\mathbf{q}}_C = -\mathbf{h}_a(\hat{\mathbf{v}}_C(\mathbf{q}_C), \hat{\mathbf{i}}_L(\boldsymbol{\varphi}_L), \mathbf{u}_s(t)) \tag{3.57}$$

$$\dot{\boldsymbol{\varphi}}_L = -\mathbf{h}_b(\hat{\mathbf{v}}_C(\mathbf{q}_C), \hat{\mathbf{i}}_L(\boldsymbol{\varphi}_L), \mathbf{u}_s(t)). \tag{3.58}$$

The initial conditions are $\mathbf{q}_C(t_0) = \mathbf{q}_{C_0}$ and $\boldsymbol{\varphi}_L(t_0) = \boldsymbol{\varphi}_{L_0}$.

To apply this technique, it is needed that there exists the hybrid representation (3.54) of the n-port N, or, equivalently, that the resistive network obtained by replacing each capacitor with a voltage source and each inductor with a current source is uniquely solvable for any currents and voltages impressed by the independent sources.

The next result can be proved (cf. Theorem 2 in [4]).

Property 3.1 A set of sufficient conditions for the existence of the hybrid representation (3.54) of the considered RLC circuit is as follows.

1. There is no loop formed exclusively by capacitors, inductors, and/or independent voltage sources. Furthermore, there is no cut-set formed exclusively by capacitors, inductors, and/or independent current sources.
2. Each voltage-controlled (but not current-controlled) resistor is in parallel with a capacitor. Each current-controlled (but not voltage-controlled) resistor is in series with an inductor.
3. Each remaining two-terminal resistor is strongly passive or else it is either in parallel with a capacitor or in series with an inductor.

A nonlinear resistor $v = \hat{v}_R(i)$ is said to be strongly passive if there exist constants $0 < \gamma' < \gamma''$ such that $\gamma' \leq (\hat{v}_R(i_1) - \hat{v}_R(i_2))/(i_1 - i_2) \leq \gamma''$ for any $i_1 \neq i_2$.

Note that the conditions in Property 3.1 are couched in topological terms and as such can be usually checked by inspection on a given circuit. When these conditions fail, the SE representation of a nonlinear RLC circuit is not guaranteed to exist globally due for example to the presence of impasse points (see Chap. 4).

Remark 3.7 In the treatment we assumed that capacitors, inductors, and resistors are uncoupled. It would be not difficult to extend the treatment to multiterminal or multiport capacitors, inductors, or resistors. The interested readers can find a detailed discussion in [4].

Example 3.7 Consider the network in Fig. 3.16a with a linear capacitor C_1, a nonlinear flux-controlled inductor L_1 defined by $i_{L_1} = \varphi_{L_1}^3$, and a linear inductor L_2, that are connected to a resistive networks with a nonlinear resistor $i_{R_2} = v_{R_2}^2$.

It can be immediately checked that the conditions for the existence of the SE representation are satisfied. By replacing C_1 with a voltage source and L_1, L_2 with current sources, we obtain the network in Fig. 3.16b. By analyzing the linear memoryless network in Fig. 3.16b we easily obtain the following.

KCL at cut-set C yields

$$i_{a_1} = v_{a_1}^2 + \frac{v_{a_1}}{R_3} - i_{b_1} - i_{b_2} - \frac{\sin(2t)}{R_3}.$$

We can also write the two KVLs

$$v_{b_1} = v_{a_1} + R_1 i_{b_1}$$

and

$$v_{b_2} = v_{a_1} - \sin(2t).$$

We have $v_{C_1} = v_{a_1}, i_{L_1} = i_{b_1} = \varphi_{L_1}^3$, and $i_{L_2} = i_{b_2}$, while $i_{C_1} = -i_{a_1}, v_{L_1} = -v_{b_1}$, and $v_{L_2} = -v_{b_2}$. By substitution, we obtain the third-order non-autonomous SE in

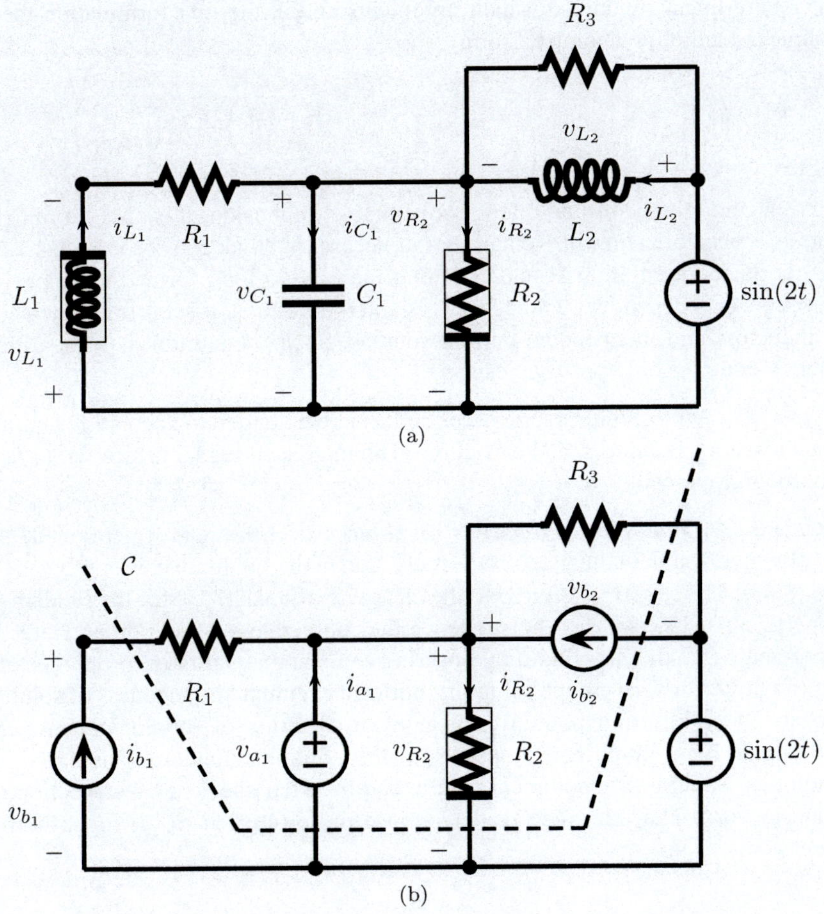

Fig. 3.16 (**a**) Third-order nonlinear network and (**b**) resistive network for finding the hybrid representation

normal form

$$\frac{dv_{C_1}}{dt} = -\frac{1}{C_1}(v_{C_1}^2 + \frac{v_{C_1}}{R_3} - \varphi_{L_1}^3 - i_{L_2} - \frac{\sin(2t)}{R_3})$$

$$\frac{d\varphi_{L_1}}{dt} = -v_{C_1} - R_1\varphi_{L_1}^3$$

$$\frac{di_{L_2}}{dt} = -\frac{1}{L_2}(-v_{C_1} + \sin(2t)).$$

Remark 3.8 Let us consider the special case where the multi-port N in Fig. 3.15 contains only linear resistors and independent voltage or current sources. Then, $\mathbf{h}_a$

and $\mathbf{h}_b$ are linear functions of their arguments and in the first formulation the SEs assume the simplified compact form

$$\begin{pmatrix} \frac{d\mathbf{v}_C}{dt} \\ \frac{d\mathbf{i}_L}{dt} \end{pmatrix} = -\text{diag}(\mathbf{C}^{-1}(\mathbf{v}_C), \mathbf{L}^{-1}(\mathbf{i}_L))\mathbf{H}\begin{pmatrix} \mathbf{v}_C \\ \mathbf{i}_L \end{pmatrix} + \hat{\mathbf{H}}\begin{pmatrix} \mathbf{e}_s(t) \\ \mathbf{i}_s(t) \end{pmatrix}$$

where $\mathbf{H}$ and $\hat{\mathbf{H}}$ are constant matrices of suitable dimension. It is also worth to note that, as a generalization of Remark 3.4, a necessary condition for the existence of the hybrid representation $\mathbf{H}$ and $\hat{\mathbf{H}}$ is that there is no loop formed exclusively by capacitors or independent voltage sources and there is no cut-set formed exclusively by inductors and independent current sources. Such a condition is also sufficient when N contains only positive resistors.

Remark 3.9 All techniques here described can be extended to account for multiterminal resistors, inductors, and capacitors. The interested reader is referred to [4] for a thorough treatment.

Remark 3.10 (Qualitative Properties of Solutions) As already mentioned, there are effective mathematical tools to study the main qualitative properties of the SEs (3.43). The reader is referred to the classic textbooks [3, 5] for the fundamental properties of SEs, such as the existence and uniqueness of solutions. Here, it is important to remark that conditions ensuring some main qualitative properties of the SEs (no finite-forward escape solutions, uniform eventual boundedness of solutions, stability of equilibrium points) can be given in terms of topological properties of the underlying graph and properties of the nonlinear functions involved. Such conditions, being of a topological nature, can be often checked by inspection on a given network. The interested reader is once more referred to [4] for a thorough discussion.

References

1. L.O. Chua, C.A. Desoer, E.S. Kuh, *Linear and Nonlinear Circuits* (McGraw-Hill, New York, 1987)
2. L.O. Chua, P.-M. Lin, *Computer Aided Analysis of Electronic Circuits, Algorithms and Computational Techniques* (Prentice-Hall, Englewood Cliffs, 1975)
3. H.K. Khalil, *Nonlinear Systems* (Prentice Hall, Englewood Cliffs, 2002)
4. L.O. Chua, Dynamic nonlinear networks: state-of-the-art. IEEE Trans. Circuits Syst. **27**(11), 1059–1087 (1980)
5. M.W. Hirsch, R.L. Devaney, S. Smale, *Differential Equations, Dynamical Systems, and Linear Algebra*, vol. 60 (Academic, Cambridge, 1974)

Chapter 4
Nonlinear Dynamics and Bifurcations in Autonomous RLC Circuits

In this brief chapter we discuss some fundamental dynamic phenomena that can be observed in nonlinear circuits containing time-invariant resistors, inductors, capacitors, and dc sources (autonomous RLC circuits). In Chap. 6 we will study analogous dynamic phenomena for nonlinear circuits containing also memristors. While in first-order autonomous circuits any bounded solution converges to an equilibrium point (EP), second-order circuits with locally active nonlinear resistors can display nonvanishing oscillations, as negative resistance oscillators belonging to the class of Van der Pol oscillators. More complex dynamics, as chaotic dynamics, can be observed in third-order autonomous circuits, the most famous example being Chua's oscillator [1].

We then consider bifurcations due to changing parameters in autonomous circuits briefly recalling some basic types of local bifurcations of EPs, as the saddle-node and Hopf bifurcation, and local bifurcations of cycles, as period-doubling bifurcations, leading to the birth of complex attractors.

The theme of bifurcations is especially of interest. In fact, later in the book, we will discuss a different type of bifurcations that can be observed for structural reasons in circuits containing memristors. Those bifurcations will be named bifurcations without parameters since they are caused by changing the initial conditions and are observable even in a circuit where parameters are held fixed (see Chap. 6).

The treatment on bifurcations in this chapter is basically descriptive. The interested reader may further investigate these aspects in the classical books [2–4].

4.1 First-Order Circuits

Consider an autonomous first-order circuit made of one linear capacitor (or one linear inductor) connected to a nonlinear resistor. The resistor may be replaced,

F. Corinto et al., *Nonlinear Circuits and Systems with Memristors*,
https://doi.org/10.1007/978-3-030-55651-8_4

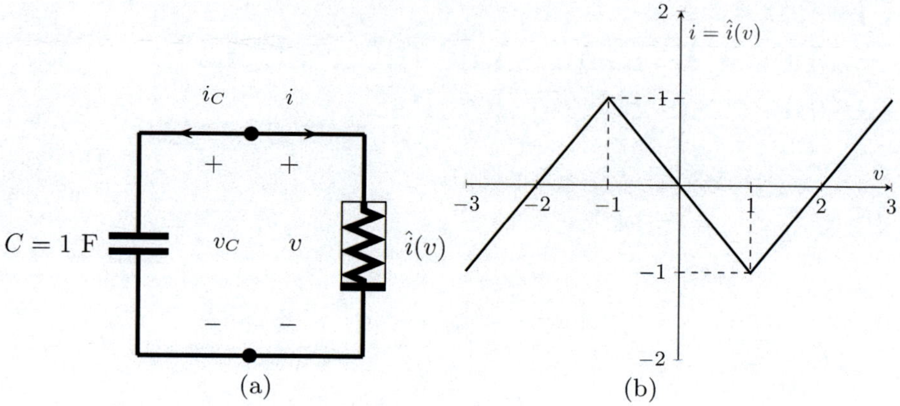

Fig. 4.1 (**a**) First-order circuit with a capacitor and a voltage-controlled nonlinear resistor and (**b**) nonlinear characteristic of the resistor

more generally, by any nonlinear resistive two-terminal element obtained by interconnecting nonlinear resistors and dc sources.

4.1.1 Dynamic Route

A useful tool to analyze first-order nonlinear circuits is the *dynamic route*. This is illustrated via a simple example.

Example 4.1 Consider for $t \geq 0$ the first-order circuit in Fig. 4.1 with a capacitor $C = 1$ F and a nonlinear resistor with characteristic

$$i = \hat{\imath}(v) = v - (|v + 1| - |v - 1|).$$

Note that the resistor is voltage-controlled but not current-controlled.

It can be easily checked that the circuit satisfies the hypotheses for the existence of the SE representation given in Property 3.1 of Chap. 3, since the voltage-controlled resistor is in parallel with a capacitor. Since

$$i_C = \frac{dv_C}{dt} = -i = -\hat{\imath}(v) = -\hat{\imath}(v_C)$$

the SE is given by

$$\frac{dv_C}{dt} = -\hat{\imath}(v_C)$$

with initial condition $v_C(0) = v_{C_0}$.

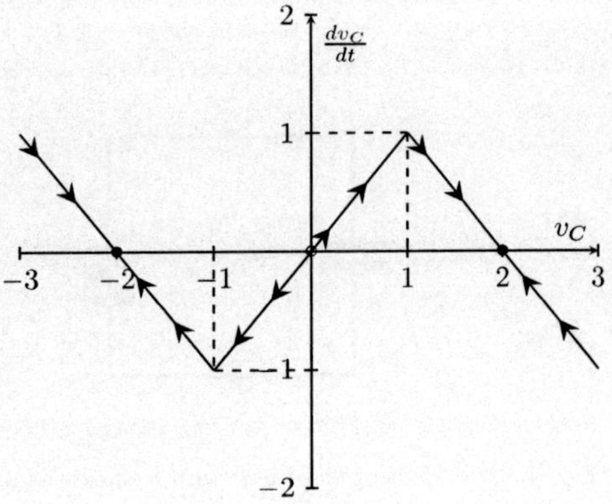

Fig. 4.2 Dynamic route. The symbol "filled circle" (resp., "open circle") denotes an asymptotically stable (resp., an unstable) EP of the circuit

An EP is a stationary solution of the circuit. The EPs can be found by letting $dv_C/dt = 0$ and are given by the solutions of the nonlinear algebraic equation

$$\hat{\imath}(v_C) = 0.$$

It can be easily checked that there are three distinct EPs given by $\bar{v}_{C_1} = -2$, $\bar{v}_{C_2} = 0$ and $\bar{v}_{C_3} = 2$.

To further investigate the global dynamics of the circuit, and the stability of EPs, let us draw a diagram reporting $dv_C/dt = -\hat{\imath}(v_C)$ as a function of v_C (Fig. 4.2). This is simply obtained by flipping the resistor characteristic with respect to the v_C axis. On this basis we can immediately find the "route" and "direction," i.e., the *dynamic route* where the motion takes place. Since $(v_C, dv_C/dt) = (v, -\hat{\imath}(v_C))$, it is clear that a point must move along the route given by the characteristic $-\hat{\imath}(v_C)$. To find the direction of motion, note that for any point above the v_C axis we have $dv_C/dt > 0$, hence the point should move toward the right along the characteristic. Conversely, any point below the v_C axis is such that $dv_C/dt < 0$ and must move toward the left. We can then attach arrowheads along the characteristic as shown in Fig. 4.2 by simply following the rule of thumb:

"If North go East" and "If South go West."

Analysis by inspection of the dynamic route thus obtained permits to conclude that any solution is bounded, moreover any solution not starting at the EP 0 must necessarily converge to the EP 2 or to the EP −2, depending on the initial condition v_{C0}. It is also seen that the EP 0 is unstable, since it is repelling solutions starting nearby, while the EPs 2 and −2 are asymptotically stable (the reader is referred to [5] for the definitions of an asymptotically stable and an unstable EP). This yields a simple and clear global portrait of the dynamics of the first-order circuit.

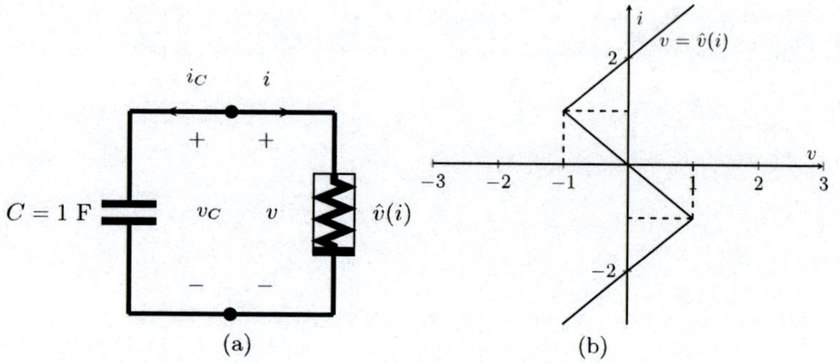

Fig. 4.3 (**a**) First-order circuit with a capacitor and a current-controlled nonlinear resistor and (**b**) nonlinear characteristic of the resistor

4.1.2 Impasse Points

Next we discuss a variant of the previous example where the voltage-controlled nonlinear resistor is replaced by a current-controlled one. In this case it is shown that it is not possible to write an SE and there are also quite unexpected dynamic complications due to the presence of impasse points.

Example 4.2 Consider a circuit composed by a linear capacitor $C = 1$ F and a nonlinear resistor as shown in Fig. 4.3. Note that the resistor is current-controlled, i.e., the voltage $v = \hat{v}(i)$ is a single-valued function of the current, but not voltage controlled. The resistor can be implemented using an operational amplifier and a resistive circuit, namely, a negative impedance converter, as discussed in [6, p. 192].

The description in terms of a differential algebraic equation (DAE) of the circuit is obtained as

$$\begin{aligned} \frac{dv_C}{dt} &= i_C \\ v &= \hat{v}(i) \\ v_C &= v \\ i_C &= -i \end{aligned}$$

and so we have

$$\begin{aligned} \frac{dv_C}{dt} &= -i \\ v_C &= \hat{v}(i). \end{aligned} \tag{4.1}$$

This yields the dynamic route shown with arrowheads in Fig. 4.4. A point on the characteristic must move toward the right in the upper plane and toward the left in the lower plane.

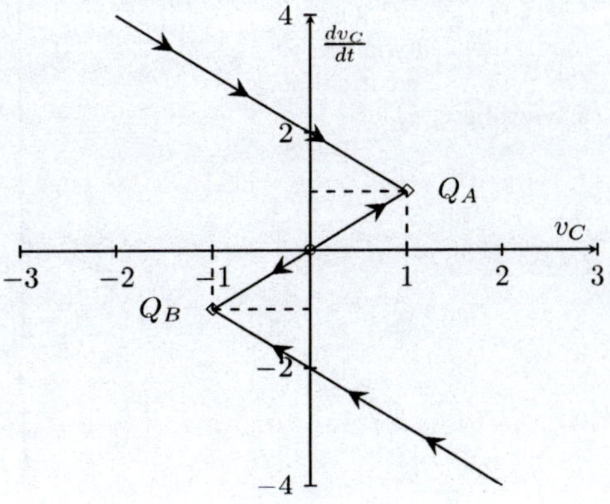

Fig. 4.4 Dynamic route with forward impasse points Q_A and Q_B. The symbol "open diamond" denotes an impasse point of the circuit

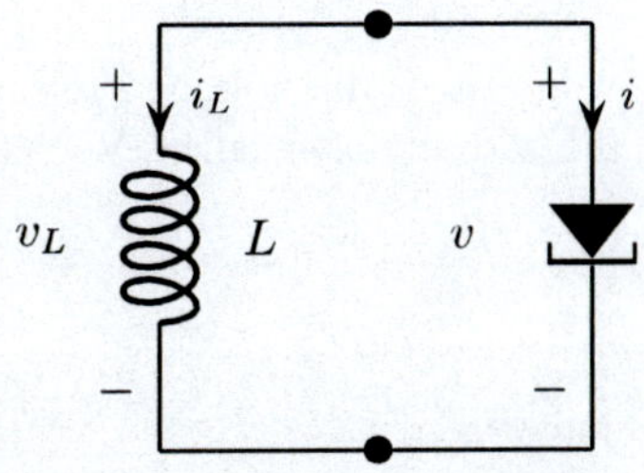

Fig. 4.5 Circuit with inductor and tunnel diode

The EPs are obtained by letting $dv_C/dt = 0$. There is only one EP such that $i_C = -i = 0$ and hence $v_C = v = \hat{v}(0) = 0$.

It can be easily checked that in finite time any solution not starting at the EP reaches one of the breakpoints Q_A or Q_B of the characteristic. However, these points are not EPs since the capacitor current is not 0. Note that the dynamic route cannot be continued forward in time starting from Q_A or Q_B. We conclude that there is no way to further prolong in time the solution after it has reached one of these points. For these reasons such points are called (forward) *impasse points*. The circuit thus considered results to be bad modeled from a physical viewpoint since solutions are not defined up to $+\infty$.

Let us discuss in more detail this result in relation to the techniques for writing the SEs in Chap. 3. Consider again the DAE (4.1). To write a global SE using v_C as a state variable starting from the DAE we would need to express $i = (\hat{v})^{-1}(v_C)$ and substitute in the first equation in (4.1). However, this is not possible since $\hat{v}(\cdot)$ is not invertible. Then, we cannot obtain an SE description for the circuit. This is in agreement with results in Chap. 3, in fact the circuit does not satisfy the condition for the existence of the SE in Property 3.1 in Chap. 3, since the current-controlled (but not voltage-controlled) nonlinear resistor is not in series with an inductor.

Example 4.3 Consider the circuit with an inductor and a tunnel diode in Fig. 4.5. The nonmonotone characteristic $i = g(v)$ of the tunnel diode[1] is shown in Fig. 4.6

[1] Note that, for convenience, we have used the simplified notation $i = g(v)$, instead of the notation $i = \hat{i}(v)$ introduced in Chap. 1, to denote a voltage-controlled resistor.

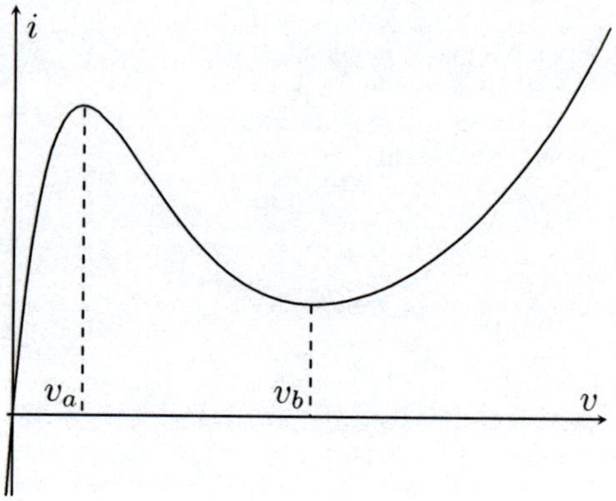

Fig. 4.6 Non-monotone tunnel diode characteristic $i = g(v)$ with a current peak and a valley

where v_a is the voltage such that the current reaches a peak while v_b is that where the current has a valley. We have

$$L\frac{di_L}{dt} = v_L = v$$

and

$$i = -i_L$$

yielding the DAE description

$$L\frac{di_L}{dt} = v$$
$$i_L = -g(v).$$

As in Example 4.2, in order to write a global SE we would need to express v as a function of i_L; however, this is not possible since function $g(\cdot)$ is not invertible.

Suppose for simplicity $L = 1$. Since $di/dt = -di_L/dt = -v$, it follows that

$$\frac{di}{dt} < 0, \quad v > 0$$

while

$$\frac{di}{dt} > 0, \quad v < 0.$$

Thus we can draw the dynamic route in Fig. 4.7. It is easily seen that the origin is the unique EP and it is asymptotically stable since it attracts solutions starting nearby. Instead, solutions starting with voltages larger than v_a are seen to converge in finite time to point Q_B. This point however is not an EP since the inductor current does

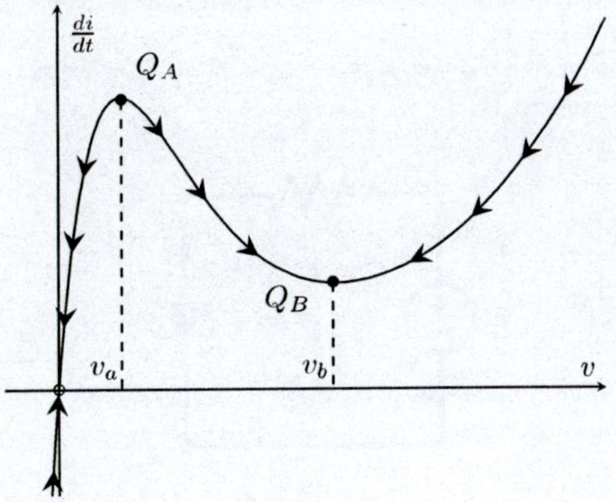

Fig. 4.7 Dynamic route of circuit with inductor and tunnel diode

not vanish at Q_B, hence, Q_B is a forward impasse points and solutions of the circuit cannot be continued forward in time after they reach Q_B. Similarly, it can be seen that solutions staring with voltages less than v_B, reach in finite backward time point Q_A and they cannot be further prolonged backward in time. Since Q_A is not an EP, such a point represents a backward impasse point.

Example 4.4 (Multiple Operating Point Paradox) Consider the resistive circuit with a dc voltage source E and a tunnel diode in Fig. 4.8a (cf. Example 2.8 in Ch. 2). The dc solutions of the circuit, which are named dc operating points, are obtained by intersecting the load line $E = ri + v$ with the diode characteristic $i = g(v)$. Suppose E and R are chosen so that there are three distinct operating points $\bar{v}_1$, $\bar{v}_2$, and $\bar{v}_3$ (Fig. 4.8b).

The existence of multiple solutions leads to a paradoxical situation since in any actual single laboratory experiment on the corresponding physical circuit we would observe only one operating point! The considered circuit is clearly not a realistic model of the physical circuit. In fact, in physical circuits, parasitic inductances and capacitances always exist. When the circuit has a unique solution, parasitics can often be neglected without significant errors. Instead, in circuits exhibiting multiple operating points, one or more of these parasitic elements *cannot be neglected*, no matter how small their values are.

To solve the previous paradox, suppose to insert in parallel to the tunnel diode a small parasitic inductance C_p modeling the electric field between the diode terminals as shown in Fig. 4.8c. We then obtain a dynamic first-order nonlinear circuit obeying the SE

$$\frac{dv_C}{dt} = -g(v_C) + \frac{E - v_C}{R}$$

where v_C is the capacitor voltage and we let $C = 1$. It is easily seen that there are three EPs $\bar{v}_{C_1} = \bar{v}_1$, $\bar{v}_{C_2} = \bar{v}_2$, and $\bar{v}_{C_3} = \bar{v}_3$ that coincide with the operating

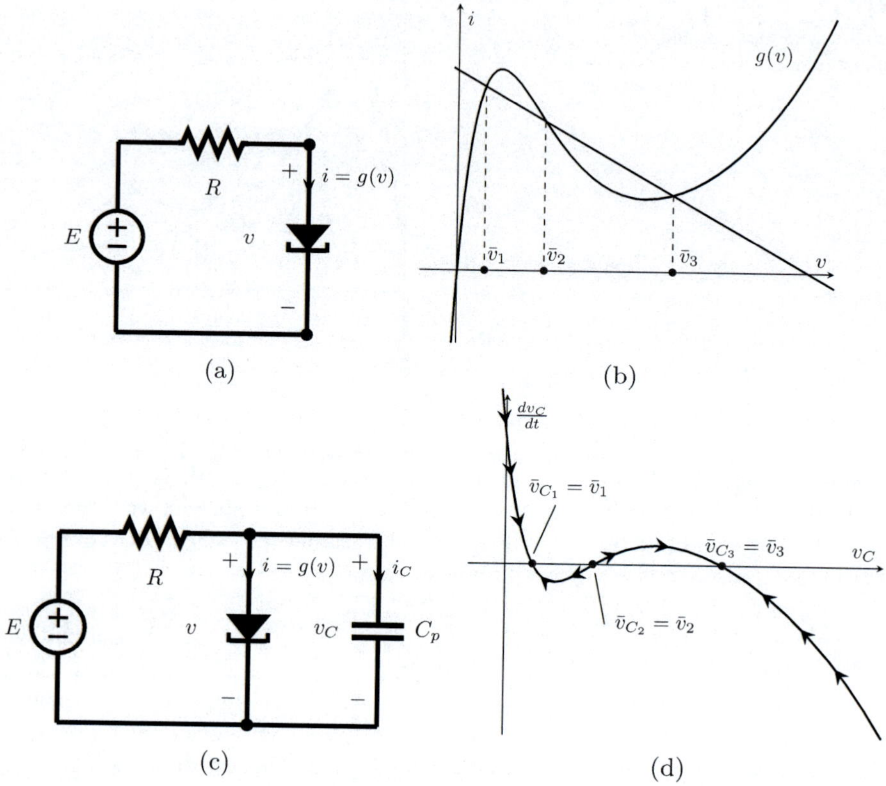

Fig. 4.8 (**a**) Circuit with tunnel diode and (**b**) three different operating points of the circuit. (**c**) Augmented circuit with a parasitic capacitance and (**d**) its dynamic route

points of the resistive circuit. This is not surprising since the EPs are simply found by letting $dv_C/dt = 0$, i.e., by open circuiting the capacitor.

For a dynamic circuit we can speak about stability of each EP. From the dynamic route (Fig. 4.8d) it follows that $\bar{v}_{C_1}$ and $\bar{v}_{C_3}$ are asymptotically stable EPs, while $\bar{v}_{C_2} = \bar{v}_2$ is an unstable EP. In a practical experiment on the physical circuit we may observe, depending on the initial condition $v_C(0)$, only $\bar{v}_{C_1}$ or $\bar{v}_{C_3}$. Instead, the unstable EP $\bar{v}_{C_2}$ is not observable in practice due to electronic noise.

In conclusion, the three operating points in the resistive circuit can be interpreted as EPs in the remodeled dynamic circuit with an added parasitic. Which of these three operating points is actually measured depends on the stability of the corresponding EP as well as on the initial condition.

4.2 Second-Order Circuits

The dynamics of a first-order autonomous nonlinear circuit are quite limited. In fact, it can be proved that any solution is either unbounded or otherwise it converges monotonically to an EP (see, e.g., [3]). We encountered a case in point in Example 4.1. The situation is more interesting for second-order nonlinear autonomous circuits, since such circuits can in general display persistent oscillations. Next, we discuss a basic class of second-order oscillators, named *negative resistance oscillators*, which includes the famous Van der Pol oscillator as a special case. We also discuss how to break impasse points in the first-order circuit in Example 4.2 by adding a suitable parasitic element and thus obtaining a special case of negative-resistance oscillators named relaxation oscillators.

4.2.1 Negative Resistance Oscillators

The basic structure of a negative resistance oscillator is depicted in Fig. 4.9a. The inductor and capacitor are linear and passive, while the resistor is nonlinear and locally active. Assume more precisely that the nonlinear resistor is current-controlled and its characteristic $v = \hat{v}(i)$ satisfies

$$\hat{v}(0) = 0, \quad \hat{v}'(0) < 0 \tag{4.2}$$

and

$$\lim_{i\to+\infty} \hat{v}(i) = +\infty, \quad \lim_{i\to-\infty} \hat{v}(i) = -\infty. \tag{4.3}$$

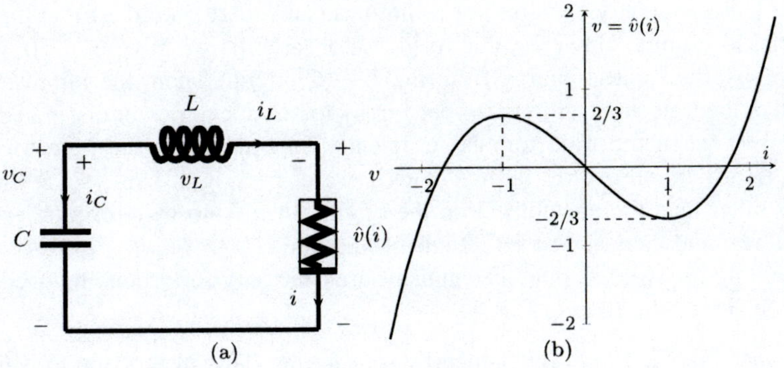

Fig. 4.9 (**a**) Second-order negative resistance oscillator and (**b**) nonlinear characteristic of the current-controlled resistor

Note that the resistor is locally active, i.e., $\hat{v}'(i) < 0$ for small $|i|$, while it is eventually passive, i.e., $vi = i\hat{v}(i) > 0$ for large $|i|$.

The SEs are easily derived as

$$\frac{dv_C}{dt} = -\frac{i_L}{C}$$
$$\frac{di_L}{dt} = \frac{v_C - \hat{v}(i_L)}{L}.$$

It is seen that the circuit has a unique EP at the origin, i.e., $(\bar{v}_C, \bar{i}_L) = (0, 0)$, and that the Jacobian of the vector field defining the SEs at the EP is given by

$$\mathbf{J} = \begin{pmatrix} 0 & -\frac{1}{C} \\ \frac{1}{L} & -\frac{\hat{v}'(0)}{L} \end{pmatrix}.$$

We have

$$T = \text{tr}\mathbf{J} = -\frac{\hat{v}'(0)}{L}$$

and

$$\Delta = \det \mathbf{J} = \frac{1}{LC}.$$

Because $T > 0$ and $\Delta > 0$, it follows that the eigenvalues of $\mathbf{J}$ are real positive, or they are complex conjugate with positive real part. In any case the unique EP is unstable and it repels nearby solutions [5].

Consider now the nonlinear resistor characteristic. Since $\hat{v}(0) = 0$ and $\hat{v}'(0) < 0$, for small $|i|$ we have $vi < 0$ so that the resistor supplies electric energy to the $L - C$ circuit. This physically explains why solutions can depart from the EP at the origin and head to infinity. However, due to (4.3), for large $|i|$ we have $vi < 0$, i.e., the resistor absorbs electric energy from the $L - C$ circuit. Then, the initial outward motion of the trajectory will be damped out by losses due to power dissipated inside the resistor when the trajectory is sufficiently far out. Soon, the trajectory must "grind to a halt" and start "falling" back toward the origin. Now, since there are no EPs other than the unstable EP at the origin, there is no way for a trajectory to come at rest and then it must set into an oscillation.

Summing up, two physical mechanisms are necessary for producing oscillations using the circuit structure in Fig. 4.9:

1. An unstable EP which repels nearby trajectories. This in turn requires that the nonlinear resistor must be active at least in a small neighborhood around the EP.
2. A dissipative mechanism which restrains the trajectories from running away to infinity. This in turn requires that the nonlinear resistor is eventually passive.

To fix ideas, suppose the nonlinear characteristic of the resistor is the cubic function

$$v = \hat{v}(i) = -i + \frac{1}{3}i^3$$

as shown in Fig. 4.9b. The oscillator with a cubic nonlinearity satisfies the SEs

$$\begin{aligned} \frac{dv_C}{dt} &= -\frac{i_L}{C} \\ \frac{di_L}{dt} &= \frac{v_C + i_L - \frac{1}{3}i_L^3}{L} \end{aligned} \tag{4.4}$$

and is known as *Van der Pol oscillator*.

It is possible to reduce the number of parameters by writing the SEs of the Van der Pol oscillator in *adimensional form*. To this end, define the dimensionless time

$$\tau = \frac{t}{\sqrt{LC}}$$

so that

$$\frac{dv_C}{dt} = \frac{1}{\sqrt{LC}}\frac{dv_C}{d\tau}$$

and

$$\frac{di_L}{dt} = \frac{1}{\sqrt{LC}}\frac{di_L}{d\tau}.$$

Substituting in (4.4) we obtain the following equivalent SE in terms of the dimensionless time τ

$$\begin{aligned} \frac{dv_C}{d\tau} &= -\frac{1}{\epsilon}i_L \\ \frac{di_L}{d\tau} &= \epsilon\left[v_C + i_L - \frac{1}{3}i_L^3\right] \end{aligned} \tag{4.5}$$

where we let

$$\epsilon = \sqrt{\frac{C}{L}}.$$

Now, we have only one parameter ϵ, so the geometric features of the family of trajectories in the phase plane (phase portrait) of the Van der Pol oscillator can be effectively analyzed by varying only ϵ. Three such phase portraits corresponding to a small, medium, and large value of ϵ are depicted in Fig. 4.10. It is worth to stress that, in each case, all trajectories tend to a unique periodic solution (a stable limit cycle). This can be rigorously shown using Poincaré–Bendixson theorem [5]. For small ϵ (say $\epsilon < 0.2$), the limit cycle is approximately a smooth ellipse, and the waveforms of $v_C(t)$ and $i_L(t)$ are approximately sinusoidal (see Fig. 4.10 for $\epsilon = 0.1$). For medium values of ϵ (say $0.3 < \epsilon < 4$), the limit cycle becomes distorted as shown in the same figure for $\epsilon = 3$. In this case, the waveforms $v_C(t)$ and $i_L(t)$ are no longer sinusoidal and do not admit of even an approximate closed form expression. For large ϵ (say $\epsilon > 20$), the limit cycle is seen to cycling closely to the curve $v_C = -i_L + \frac{1}{3}i_L^3$, except at the corners, where it becomes nearly vertical (cf. Fig. 4.10 for $\epsilon = 30$). The corresponding waveforms for $v_C(t)$ and $i_L(t)$ consist of nearly instantaneous transition from the lower branch to the upper branch, and vice versa, a peculiar feature also named *jump phenomenon*. The oscillation waveforms of $v_C(t)$ and $i_L(t)$ are far from being sinusoidal. Such oscillators are named *relaxation oscillators*.

4.2.2 How to Break Impasse Points

The impasse points in the circuit in Fig. 4.3 of Example 4.2 can be removed when taking into account suitable parasitic elements. Indeed, suppose to insert in the circuit in Fig. 4.3, in series with C (recall that we have chosen a normalized value $C = 1$), a *small inductance* L_p modeling for instance the parasitic inductance of connecting wires, thus obtaining the circuit in Fig. 4.11. Note that this is in the form of a negative resistance oscillator.

The conditions for the existence of a global SE are now satisfied and the SEs have been already obtained as

$$\begin{aligned} \frac{dv_C}{dt} &= -i_{L_p} \\ \frac{di_{L_p}}{dt} &= \frac{v_C - \hat{v}(i_{L_p})}{L_p}. \end{aligned} \tag{4.6}$$

Since $\hat{v}(\cdot)$ is piecewise-linear, the vector field defining the SEs (4.6) satisfies a Lipschitz condition, hence local existence and uniqueness of the solution for any initial condition $(v_C(0), i_{L_p}(0))^T$ are guaranteed [5, Theorem 3.1]. Furthermore, the norm of the same vector field increases at most linearly with the norm of the state vector $(v_C, i_{L_p})^T$, hence any solution is defined in the whole interval $t \in [0, +\infty)$ [5, Theorem 3.2]. This rules out in particular the presence of impasse points.

Clearly, any value of the inductance L_p would rule out the presence of impasse points. Namely, impasse can be broken also by deliberating inserting an inductance in series as in Fig. 4.11 without the need to account for parasitic inductances.

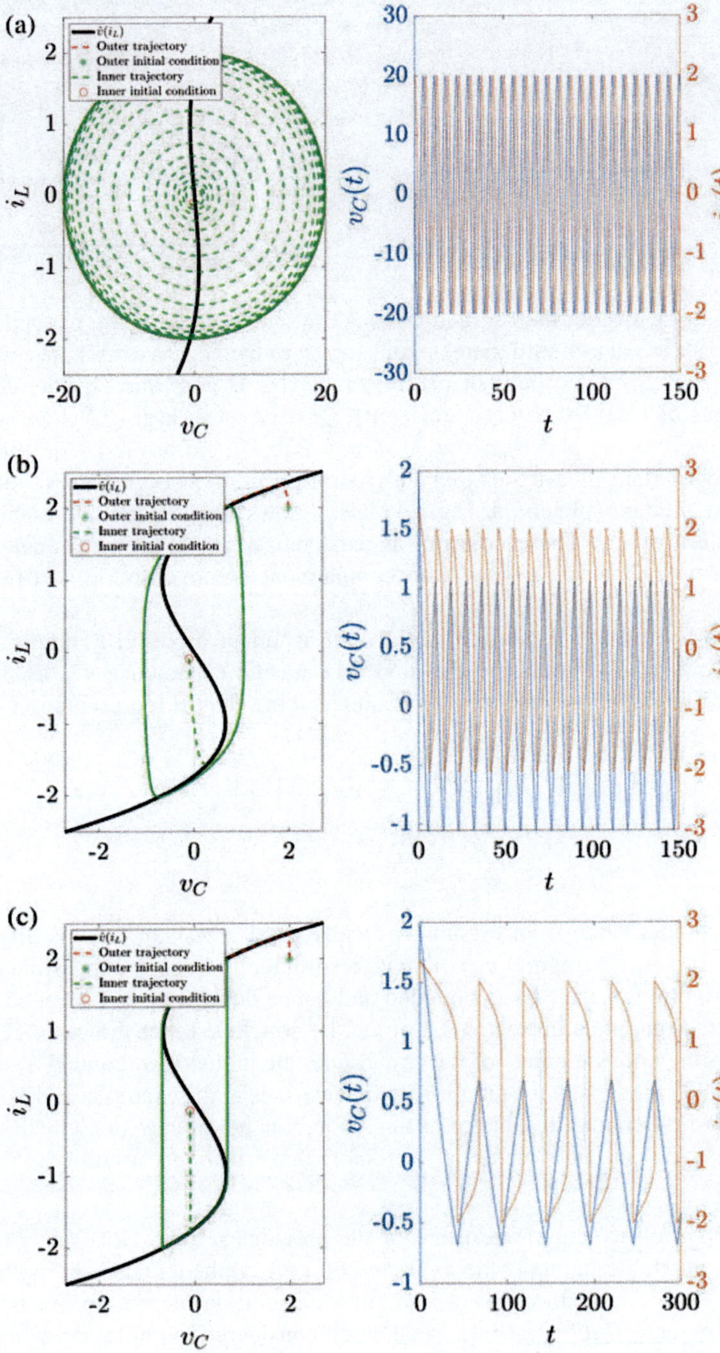

Fig. 4.10 Trajectories in the state space and time-domain evolution of state variables of the Van der Pol oscillator for a small ($\epsilon = 0.1$, upper plot), a medium ($\epsilon = 3$, middle plot), and a large ($\epsilon = 30$, lower plot) value of the parameter

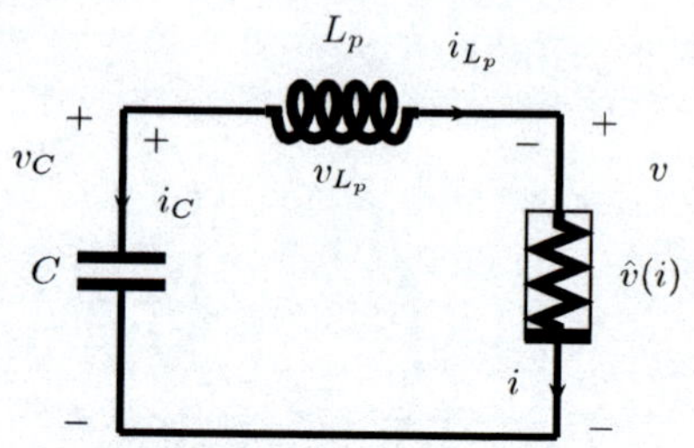

Fig. 4.11 Relaxation oscillator obtained by inserting a small parasitic inductance L_p in the circuit with impasse points of Fig. 4.3

Example 4.5 Consider the oscillator (4.6). For small values of the parasitic inductance L_p, we are in a condition analogous to that of the negative resistance oscillator in Sect. 4.2.1 for high values of parameter ϵ. This is confirmed by the computer simulations of (4.6) in the case $L_p = 10^{-3}$ as shown in Fig. 4.12 (compare with Fig. 4.10). From the phase portrait it is noted that there are quick jumps of the inductor current, i.e., we are dealing with a strongly nonlinear oscillator (a relaxation oscillator). For completeness, Fig. 4.12 also shows the simulations obtained for larger values of L_p. These confirm, as expected, a global scenario analogous to that of a negative resistance oscillator (compare once more with Fig. 4.10).

Example 4.6 Consider again the circuit with a tunnel diode in Example 4.3 and suppose to insert, in parallel to the diode, a parasitic capacitance C_p as shown in Fig. 4.13. The second-order circuit thus obtained has the SE representation

$$\frac{dv_C}{dt} = -\frac{1}{C_p}(i_L + g(v_C))$$

$$\frac{di_L}{dt} = \frac{v_C}{L}.$$

Since the tunnel diode is an eventually locally passive resistor, i.e., its differential resistance is positive for any $|v| > \bar{v}$, with $\bar{v}$ sufficiently large, it is possible to show that any solution of the SEs is bounded and hence defined up to $t = +\infty$ (cf. [7, Sect. VI]). Arguing as in Sect. 4.2.2, it can be concluded that the insertion of C_p eliminates the impasse points of the circuit with the inductor and tunnel diode.

By slightly modifying the circuit with tunnel diode in the example, and exploiting the negative differential resistance of the diode, it is possible to implement second-order negative resistance relaxation oscillators that are widely used in the technical applications.

Remark 4.1 Let us summarize some of the previous results. Both analytical and experimental studies support the existence of a jump phenomenon in the resistor-capacitor ($R - C$) circuit in Fig. 4.3 whenever a solution reaches an impasse point such as Q_A or Q_B (cf. Fig. 4.4). Analogous considerations hold for a first-order $R - L$ circuit. This allows us to state the following property.

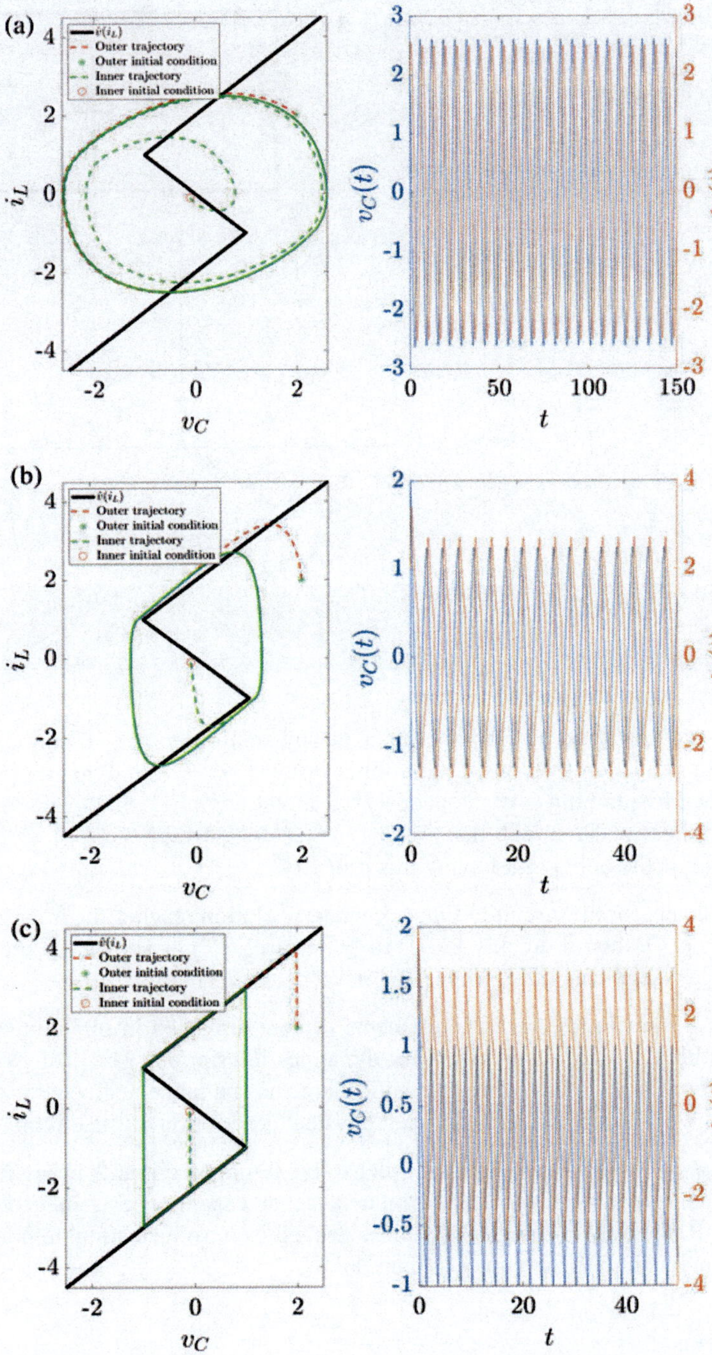

Fig. 4.12 Phase portrait and time evolution of state variables for a relaxation oscillator obtained by inserting a parasitic inductance L_p in the circuit with impasse points of Fig. 4.3. Simulations are shown for three different values of L_p, namely, $L_p = 1$ (upper plot), $L_p = 0.1$ (middle plot), and $L_p = 10^{-3}$ (lower plot)

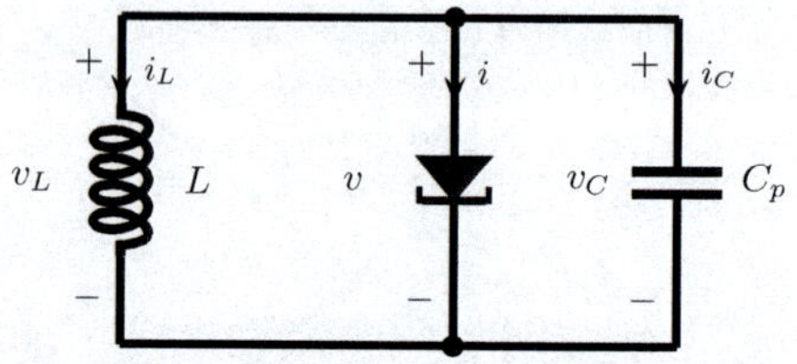

Fig. 4.13 Circuit with inductor, tunnel diode, and parasitic capacitance inserted for breaking impasse points

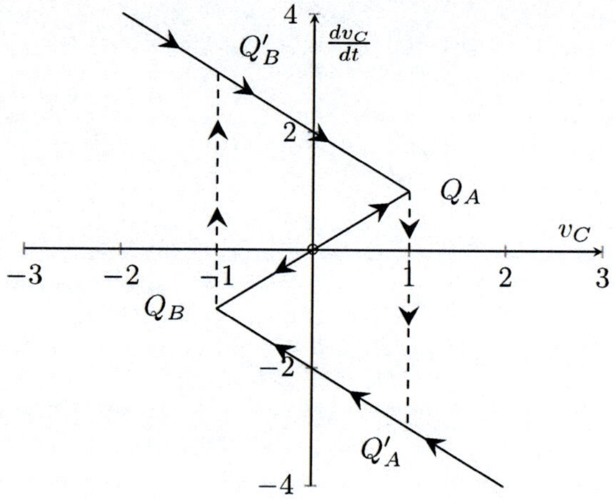

Fig. 4.14 Dynamic route with jumps (vertical dashed segments) in a relaxation oscillator obtained by accounting for a small parasitic inductance for breaking impasse points

Property 4.1 Let Q be an impasse point of any first-order $R - C$ circuit (resp., $R - L$ circuit). Upon reaching Q at a finite instant $t = T$, the dynamic route can be continued by jumping (instantaneously) to another point Q' on the characteristic of the nonlinear resistor such that $v_C(T^+) = v_C(T^-)$ [resp., $i_L(T^+) = i_L(T^-)$] provided Q' is the only point having this property.

For instance, for the circuit in Fig. 4.3, when a solution reaches the impasse point Q_A (resp., Q_B), then it quickly jumps to Q'_A (resp., Q'_B) as shown via the arrows on the dashed segments in Fig. 4.14.

Remark 4.2 Sometimes, parasitic elements can be neglected in modeling a given circuit without losing accuracy in its dynamic description. The last examples, however, introduce circuits where parasitic elements cannot be neglected if we wish to obtain a well-posed mathematical and physical description of the circuit.

Remark 4.3 It can be shown that, in order to break impasse points in higher-order circuits, the insertion of a parasitic inductance or capacitance is in general not sufficient. Rather, as discussed in [8], it is needed to introduce suitable high order (α, β)-elements into the circuit (cf. Chap. 1).

4.3 Third-Order Circuits

The possible dynamic behaviors of a second-order autonomous circuit, although of practical interest, continue to be quite limited. In fact, it can be shown that in a generic second-order autonomous system of differential equations defined by a smooth vector field the limit set of each trajectory[2] is either an EP or a cycle (periodic attractor) [10]. On the other hand, third-order autonomous dynamical circuits can display much more complicated dynamics, named chaotic dynamics, that are characterized by an erratic, non-periodic, behavior of solutions and limit sets with a complicated fractal structure. One of the most famous examples of this kind is Chua's oscillator [1], which is briefly discussed in the next section.

4.3.1 *Chua's Oscillator*

Chua's oscillator is a third-order autonomous nonlinear circuit with three passive reactive elements (two linear capacitors C_1, C_2 and a linear inductor L), two passive linear resistors (r and R) and a locally active voltage-controlled nonlinear resistor N_R (Fig. 4.15).

In the piecewise-linear case N_R is an active nonlinear resistor with characteristic

$$i = \hat{\imath}(v) = G_b v + \frac{1}{2}(G_a - G_b)[|v + E| - |v - E|]$$

where $G_a, G_b < 0$, and $E > 0$ (Fig. 4.16a). The nonlinear resistor may be also modeled by the smooth differentiable cubic nonlinearity

$$i = \hat{\imath}(v) = -av + bv^3 \tag{4.7}$$

where $a, b > 0$, which corresponds to a locally active, eventually passive, nonlinear resistor (Fig. 4.16b).

Fig. 4.15 Chua's oscillator

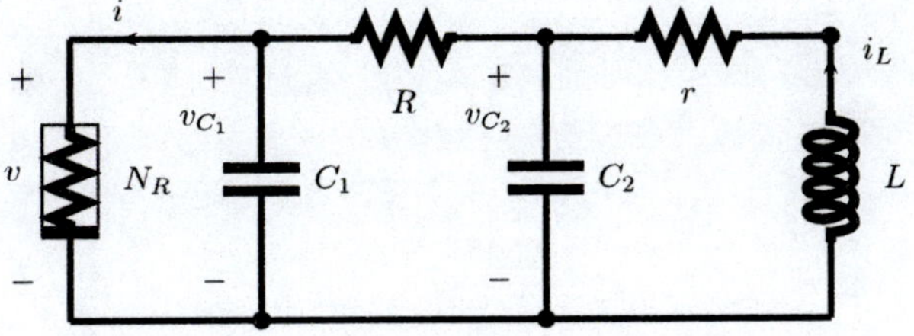

[2]The (positive) limit set of a trajectory is the set of points that are approached by the trajectory as $t \to +\infty$ [9].

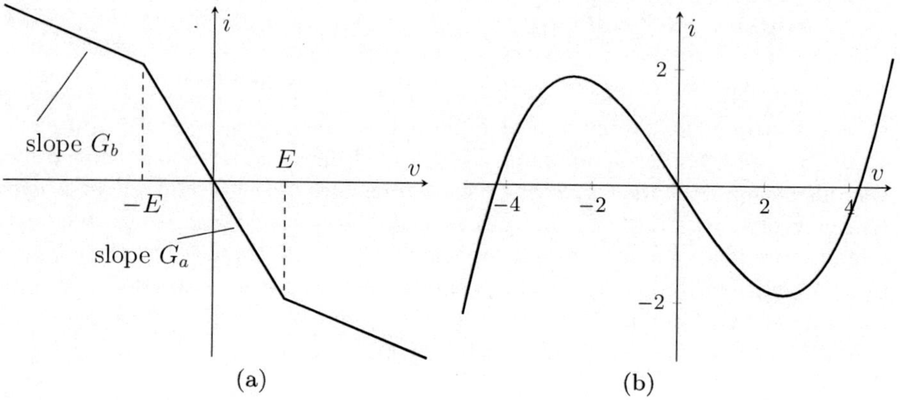

Fig. 4.16 (**a**) Piecewise-linear and (**b**) cubic characteristic ($a = 8/7, b = 4/63$) of the nonlinear resistor in Chua's oscillator

The dynamics of Chua's oscillator is described by the third-order SEs (Appendix)

$$\frac{dv_{C_1}}{dt} = \frac{1}{C_1}\left[\frac{1}{R}(v_{C_2} - v_{C_1}) - \hat{i}(v_{C_1})\right]$$
$$\frac{dv_{C_2}}{dt} = \frac{1}{C_2}\left[\frac{1}{R}(v_{C_1} - v_{C_2}) + i_L\right]$$
$$\frac{di_L}{dt} = \frac{1}{L}[-v_{C_2} - ri_L].$$

The change of variables

$$x = v_{C_1}, y = v_{C_2}, z = Ri_L$$

yields the following adimensional form of SEs

$$\frac{dx}{d\tau} = \alpha[y - x - n(x)]$$
$$\frac{dy}{d\tau} = x - y + z$$
$$\frac{dz}{d\tau} = -\beta y - \gamma z$$

where

$$\alpha = \frac{C_2}{C_1}, \beta = \frac{R^2 C_2}{L}, \gamma = \frac{RrC_2}{L}, \tau = \frac{t}{RC_2}$$

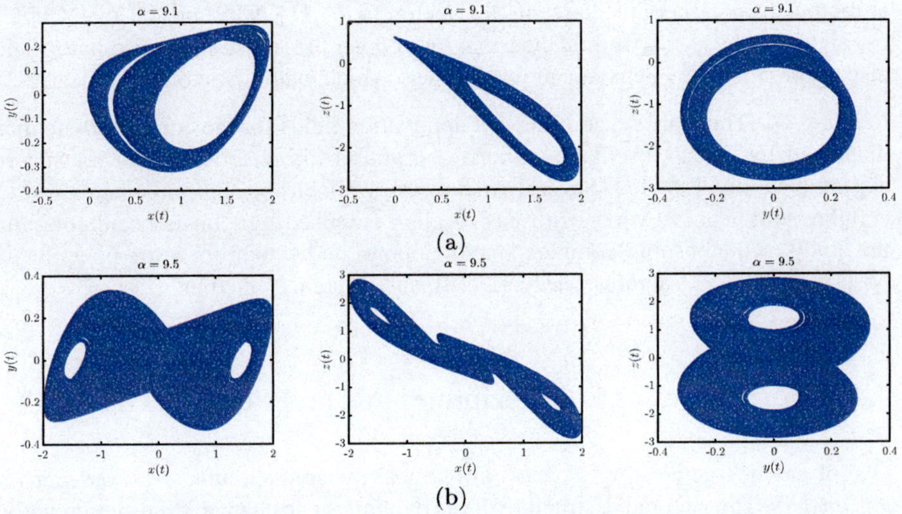

Fig. 4.17 Some attractors displayed by Chua's oscillator for different values of parameter α. The initial conditions are $(x(0), y(0), z(0)) = (0.1, 0.1, 0.1)$. (**a**) Single scroll attractors for $\alpha = 9.1$ (projection onto the $x - y$, $x - z$, and $y - z$ plane). (**b**) Double-scroll attractor for $\alpha = 9.5$

and

$$n(x) = R\hat{i}(x).$$

Chua's oscillator [1] is the simplest electronic circuit exhibiting chaotic behavior and for which a rigorous mathematical proof of chaos is available [11]. A discussion on the genesis and a chronological bibliography on the main achievements on Chua's oscillator are available in [12] and [1], respectively. Chua's oscillator is known to display an immense variety and shapes of complex attractors. The circuit is indeed considered as a paradigm for complex dynamic phenomena observable in nonlinear circuits [13].

Example 4.7 Consider Chua's oscillator with a cubic nonlinearity (4.7) with $a = 8/7$ and $b = 4/63$. Choose $r = 0$, hence $\gamma = 0$ and $\beta = 15$. Figure 4.17 shows some attractors displayed by Chua's oscillator for different values of parameter α. When $\alpha = 9.1$, the solution starting at $(x(0), y(0), z(0)) = (0.1, 0.1, 0.1)$ tends to a single-scroll chaotic attractor, while for $\alpha = 9.5$ we can observe convergence to a double-scroll attractor. In Sect. 4.4.3 we will see how a complex attractor originates from a cascade of period doubling bifurcations. We refer the reader to [1, 11, 14, 15] for a zoo of attractors for other sets of parameters or different choices of nonlinearities in Chua's oscillator.

Remark 4.4 Chua's oscillator is the most general but structurally the simplest system capable of reproducing all possible dynamical phenomena and complex attractors from a certain class of 3D vector fields [1]. This means that Chua's oscillator can be used to mimic the behavior of other piecewise-linear oscillators and also approximate the behavior of many others which exhibit smooth nonlinearities.

Remark 4.5 There are several relevant application fields for the complex dynamics displayed by Chua's oscillator as data encryption via chaotic signals for secure communications [16, 17]. Several studies are available on the implementation of cellular nonlinear networks with basic cells composed by Chua's oscillators and the study of the ensuing complex spatiotemporal and synchronization phenomena observed in these structures, see, e.g., [18], and references therein.

4.4 Bifurcations of Equilibrium Points and Periodic Orbits

One of the most interesting aspects of nonlinear dynamic circuits is *dependence on parameters*. The qualitative structure of the dynamics can change, even significantly, when parameters are varied. For example, there may be the birth or disappearance of EPs, or the change of stability of an EP, when a parameter reaches a certain value. Something analogous may happen for periodic orbits (cycles). Such qualitative changes in the dynamics are called *bifurcations* and the parameter values at which they occur are named bifurcation points or critical parameter values. In this chapter we discuss at an elementary level some basic bifurcations of EPs and periodic orbits that are of interest for the topics dealt with in the book. The reader is referred to the fundamental textbooks [2, 3] for a thorough treatment.

4.4.1 *Saddle-Node Bifurcations*

The saddle-node bifurcation is the basic mechanism by which EPs are created or destroyed. As a parameter is varied, two EPs of a nonlinear circuit move toward each other, collide, and mutually annihilate. We refer the reader to [3] for other types of bifurcations of an EP, as the transcritical and pitchfork bifurcation.

Example 4.8 Consider the first-order autonomous nonlinear circuit with a tunnel diode and a dc current source in Fig. 4.18. The diode has a nonmonotone charac-

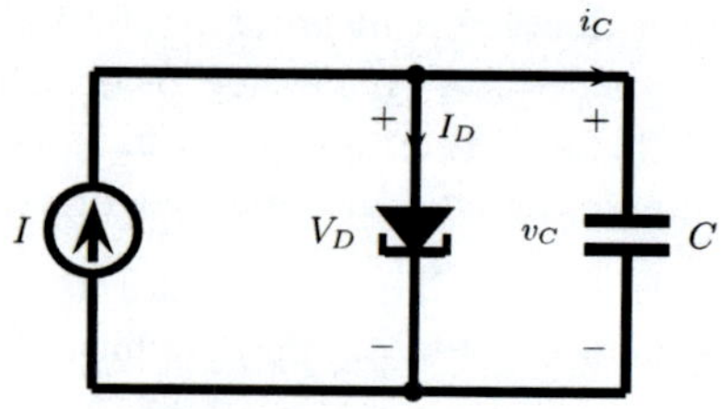

Fig. 4.18 Circuit with a tunnel diode for studying saddle-node bifurcations of EPs

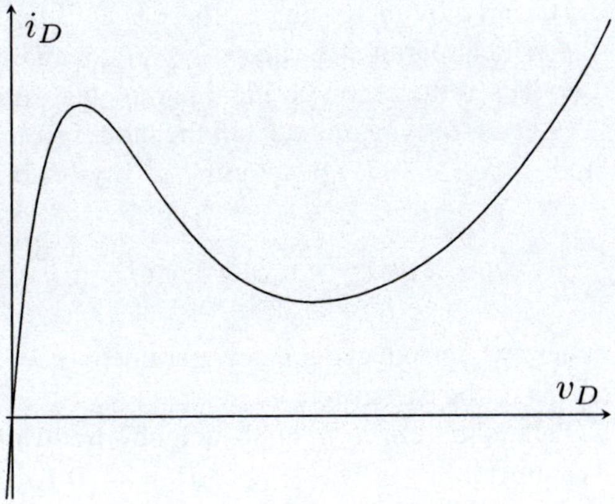

Fig. 4.19 Nonlinear nonmonotone characteristic of a tunnel diode

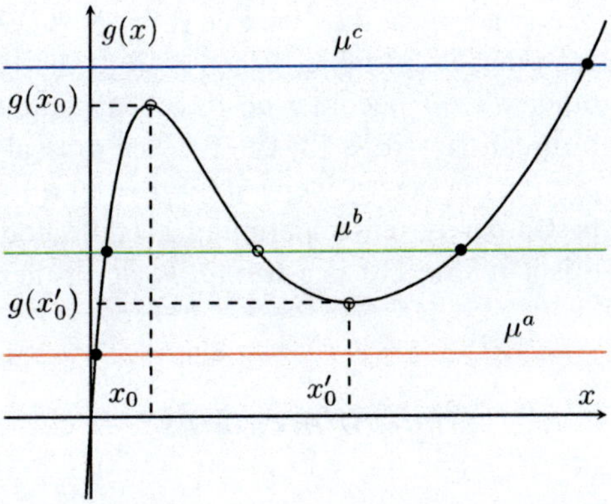

Fig. 4.20 EPs of the circuit with tunnel diode as a function of parameter μ. For $\mu < g(x'_0)$ (for example, μ_a) and $\mu > g(x_0)$ (for example, μ_c) there is a unique EP, while there are three EPs for $g(x'_0) < \mu < g(x_0)$ (for example, μ_b)

teristic $I_D = g(V_D)$ as shown in Fig. 4.19. The SE describing the circuit dynamics is

$$\frac{dv_C}{dt} = \frac{I}{C} - \frac{1}{C}g(v_C).$$

If we suppose for simplicity $C = 1$ F, and let $x = v_C$, we obtain

$$\frac{dx}{dt} = \mu - g(x)$$

where the source current I is chosen as a parameter μ.

The EPs are the solutions of

$$g(x) = \mu$$

and can be found graphically as shown in Fig. 4.20. Define x_0, x'_0, $g(x_0)$, and $g(x'_0)$ as in the figure. It is seen that when $\mu > g(x_0)$ or $\mu < g(x'_0)$ the circuit has a unique EP, while there are three distinct EPs when $g(x'_0) < \mu < g(x_0)$.

Clearly, by increasing μ, there is a bifurcation with the appearance of two EPs at the critical parameter value $\mu = g(x_0')$ and a bifurcation with the disappearance of two EPs at the second critical parameter value $\mu = g(x_0)$.

Let us study in more detail the latter bifurcation. Note that $g(x_0) > 0$, $g'(x_0) = 0$, and $g''(x_0) < 0$. By developing in Taylor series g in a neighborhood of x_0 we have

$$\frac{dx}{dt} = \mu - g(x_0) - g'(x_0)(x - x_0) - \frac{1}{2}g''(x_0)(x - x_0)^2 = r - \frac{1}{2}g''(x_0)(x - x_0)^2 \tag{4.8}$$

where we introduced the new parameter $r = \mu - g(x_0)$ and we omitted higher-order terms of the expansion.

The situation in a small neighborhood of x_0, and for three different values of r, namely, $r < 0$ ($\mu < g(x_0)$), $r = 0$ ($\mu = g(x_0)$), and $r > 0$ ($\mu > g(x_0)$) is represented in Fig. 4.21. When $r < 0$ there are two distinct EPs in a neighborhood of x_0 and, from the dynamic route, it is seen that one is asymptotically stable while the other is unstable. When $r = 0$ the two EPs collide and then annihilate each other when r becomes positive. This situation corresponds to a typical saddle-node bifurcation of EPs [3, Ch. 3]. The critical parameter value at which a saddle-node bifurcation occurs is given by $\mu = g(x_0)$ ($r = 0$). The form in (4.8) is typical for a first-order system undergoing a saddle-node bifurcation. In the bifurcation literature this is referred to as a *normal form* for the saddle-node bifurcation.

4.4.2 Hopf Bifurcations

In this case we are dealing with an autonomous nonlinear system depending on a parameter μ whose variation causes an EP to change its local stability properties. The main issue is whether the change in stability of the EP can be associated with the appearance of a periodic solution (a limit cycle). Roughly speaking, the Hopf bifurcation theory says that if a pair of complex conjugate eigenvalues of the linearization about the EP cross the imaginary axis as μ varies through certain critical values, then for near-critical values of μ there are limit cycles close to the EP.

Example 4.9 Consider the second-order nonlinear autonomous circuit with a tunnel diode shown in Fig. 4.22. The nonmonotone diode characteristic $I_D = g(V_D)$ is shown in Fig. 4.23. The SEs of the circuit are easily written as

$$\begin{aligned}\frac{di_L}{dt} &= \frac{v_C}{L}\\ \frac{dv_C}{dt} &= -\frac{i_L}{C} + \frac{1}{C}g(E - v_C).\end{aligned}$$

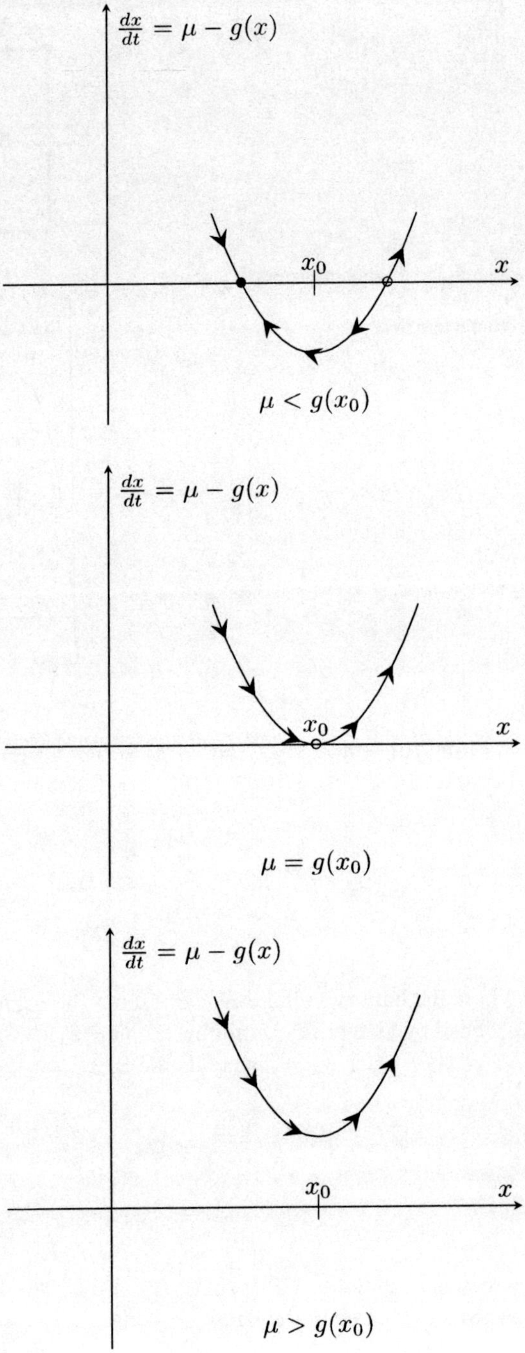

Fig. 4.21 Illustration of a saddle-node bifurcation of EPs. The figure depicts the behavior of $dx/dt = \mu - g(x)$ in a neighborhood of x_0 for three values of μ. A small black (resp., white) circle denotes an asymptotically stable (resp., unstable) EP

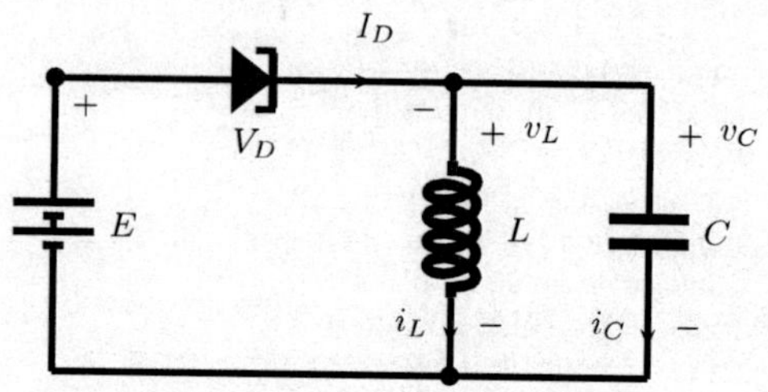

Fig. 4.22 Circuit with tunnel diode exhibiting a Hopf bifurcation

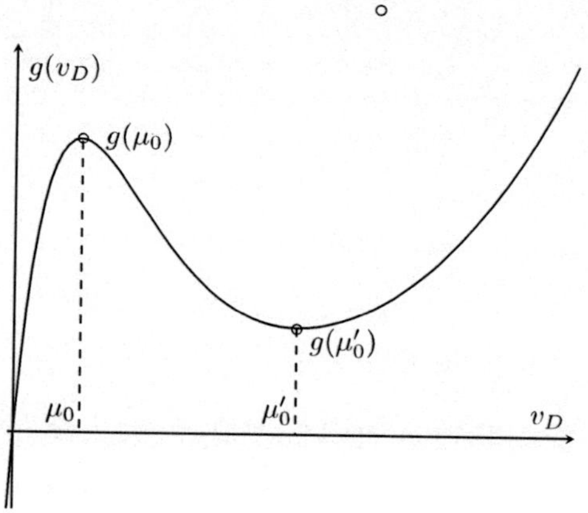

Fig. 4.23 Main parameters of a tunnel diode characteristic

Assume for simplicity $C = 1$, $L = 1$, and let $x_1 = i_L$, $x_2 = v_C$. Then the SEs become

$$\frac{dx_1}{dt} = x_2$$
$$\frac{dx_2}{dt} = -x_1 + g(\mu - x_2)$$

where the battery voltage E is chosen as a parameter μ.

For any μ there is a unique EP $(\bar{x}_1, \bar{x}_2) = (g(\mu), 0)$. The change of variables $y_1 = x_1 - g(\mu)$, $y_2 = x_2$ yields the SEs

$$\frac{dy_1}{dt} = y_2 \tag{4.9}$$
$$\frac{dy_2}{dt} = -y_1 - g(\mu) + g(\mu - y_2) \tag{4.10}$$

which have a unique EP at (0, 0) for any μ. The Jacobian of the vector field defining the SEs evaluated at the EP is

$$\mathbf{J} = \begin{pmatrix} 0 & 1 \\ -1 & -g'(\mu) \end{pmatrix}$$

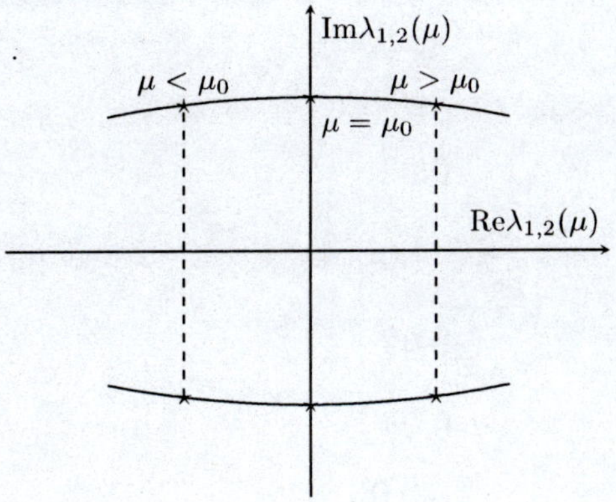

Fig. 4.24 Eigenvalues of the linearization at the EP crossing the imaginary axis as parameter μ varies

and its eigenvalues are given by

$$\lambda_{1,2}(\mu) = -\frac{g'(\mu)}{2} \pm j\sqrt{1 - \left(\frac{g'(\mu)}{2}\right)^2}$$

for sufficiently small $|g'(\mu)|$.

Now, let $\mu_0 > 0$ be as in Fig. 4.23. Note that $g(\mu_0) > 0$, $g'(\mu_0) = 0$, and $g''(\mu_0) < 0$. Then, if μ belongs to a neighborhood of μ_0, and μ increases, it is seen that $\lambda_{1,2}(\mu)$ transversely cross the imaginary axis from left to right as shown in Fig. 4.24, moreover, $\lambda_{1,2}(\mu_0) = \pm j$. Then, there is an asymptotically stable EP (a stable focus) when $\mu < \mu_0$, while for $\mu > \mu_0$ the EP becomes unstable (more precisely, an unstable focus) [5]. Such a loss of stability of the EP is seen to originate the birth of a stable limit cycle surrounding the EP. The qualitative portrait of this bifurcation, which is named Hopf bifurcation, is represented in Fig. 4.25 in the space $(y_1, y_2) \times \mu$ in relation to a specific example (cf. Example 4.10). A precise technical statement concerning the Hopf bifurcation, and a rigorous mathematical proof, can be found in [19]. Actually, the observed bifurcation is called a supercritical Hopf bifurcation, to distinguish it from a subcritical Hopf bifurcation where, as a parameter varies, an unstable limit cycle collides with a stable EP and transfers its instability to the EP (see [3] for details).

Example 4.10 Consider system (4.10) with a tunnel diode characteristic g as in Fig. 4.19. By varying μ, i.e., the battery voltage, we obtain the scenario depicted in Fig. 4.25. In this example we have $\mu_0 = 0.344$ and $\mu_0' = 1.5$. Figure 4.25 shows the birth of limit cycle when μ reaches the value μ_0 according to the supercritical Hopf bifurcation mechanism. For μ slightly greater than μ_0 the cycle has a small size, but the size quickly increases by increasing μ. From simulations it is seen that for any $\mu_0 < \mu < \mu_0'$ there is a stable limit cycle for the considered circuit and that the cycle disappears via a reverse supercritical Hopf bifurcation when μ reaches the value μ_0'.

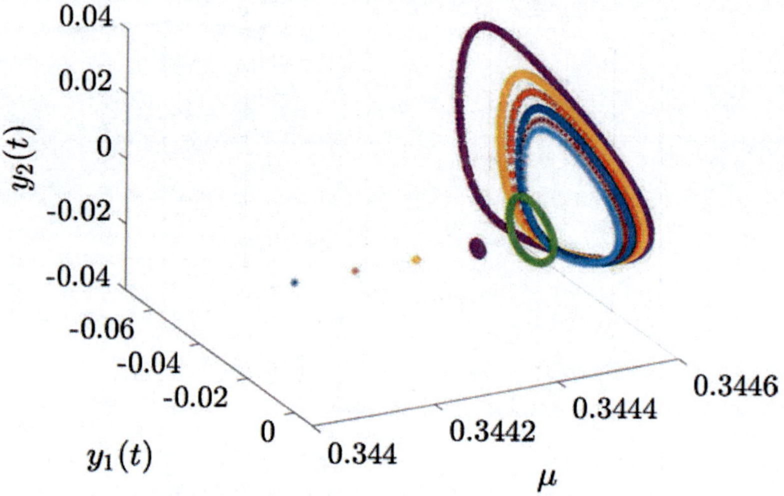

Fig. 4.25 Supercritical Hopf bifurcation for the system considered in Example 4.10 when μ reaches the critical value $\mu_0 = 0.344$

4.4.3 Period-Doubling Bifurcations

In a nonlinear circuit it may happen that a cycle loses its stability when a parameter reaches a critical value and at the same time a stable limit cycle with almost twice the period appears close to the original cycle. In quite a general situation, by varying the parameter beyond the critical value, such a doubling process repeats infinitely many times and leads to the onset of complicated oscillations and an erratic behavior of solutions, a behavior usually referred to as chaos. Such a cascade of period-doubling bifurcations represents a typical route to chaos in nonlinear systems [2]. It can be observed for instance in periodically forced second-order oscillators, where the parameter may be the amplitude or the frequency of the forcing signal. It can also be observed in third-order autonomous nonlinear circuits depending upon one parameter, the most famous example being Chua's oscillator [1].

Example 4.11 Let us consider again Chua's oscillator with the same cubic nonlinearity and parameters $\gamma = 0$ and $\beta = 15$ as in Example 4.7. By varying parameter α the circuit undergoes to a cascade of period-doubling bifurcations leading to the birth of a double-scroll attractor as shown in Fig. 4.26. More precisely, for $\alpha = 7$ the solution starting at $(x(0), y(0), z(0)) = (0.1, 0.1, 0.1)$ converges to a small-size attracting limit cycle, while for $\alpha = 8$ we have convergence to a cycle with larger size. For $\alpha = 9$ the solution converges to a cycle with period 2 and for $\alpha = 9.03$ we observe a cycle with period 4. By further increasing α the period-doubling process continues until a complex double-scroll attractor is observed.

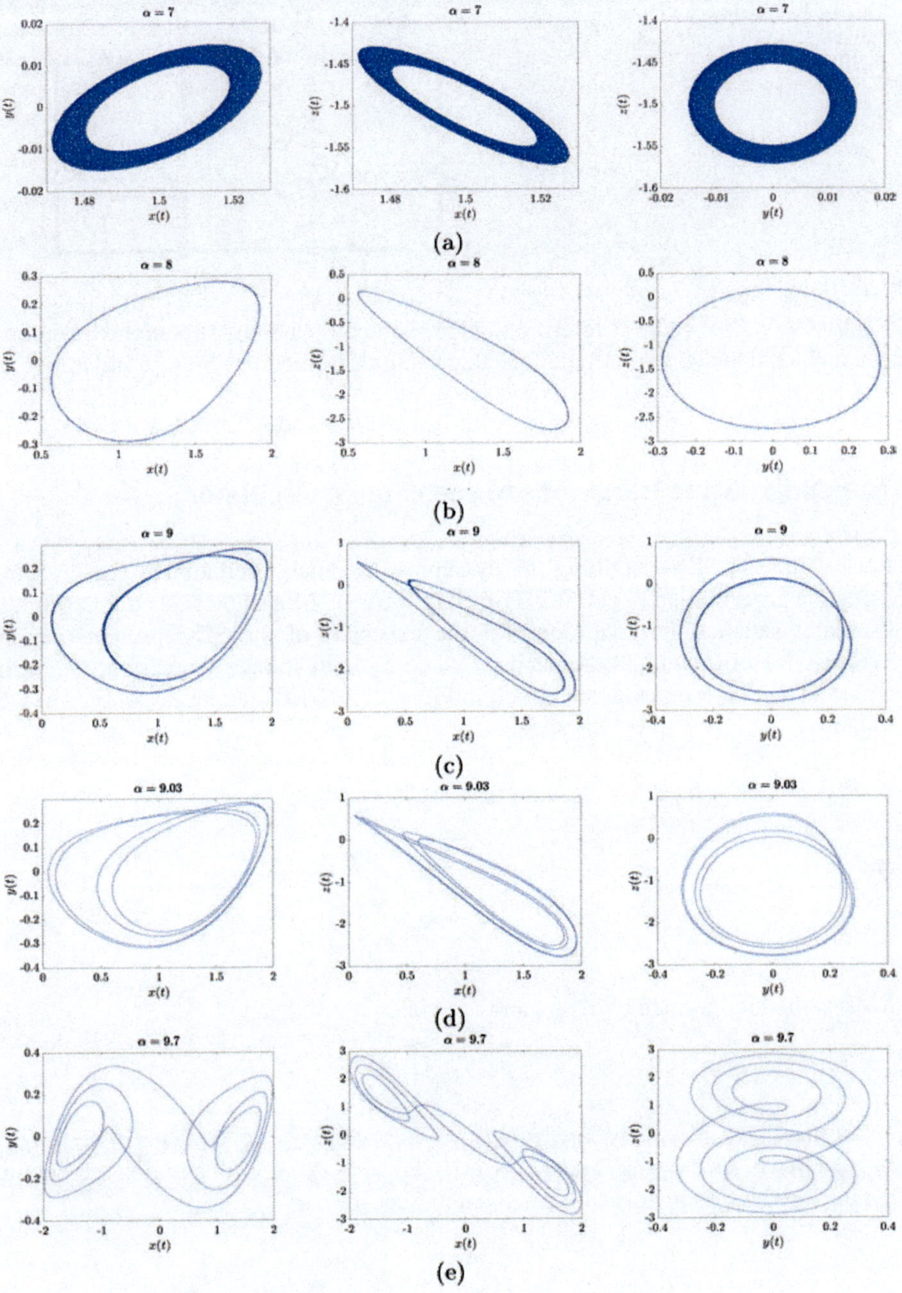

Fig. 4.26 Period-doubling bifurcations displayed by Chua's oscillator when varying parameter α. (**a**) Small size limit cycle for $\alpha = 7$ (projection onto the $x - y$, $x - z$, and $y - z$ plane). (**b**) Larger size limit cycle for $\alpha = 8$. (**c**) Cycle with period 2 for $\alpha = 9$. (**d**) Cycle with period 4 for $\alpha = 9.03$. (**e**) Cycle with period 3 for $\alpha = 9.7$

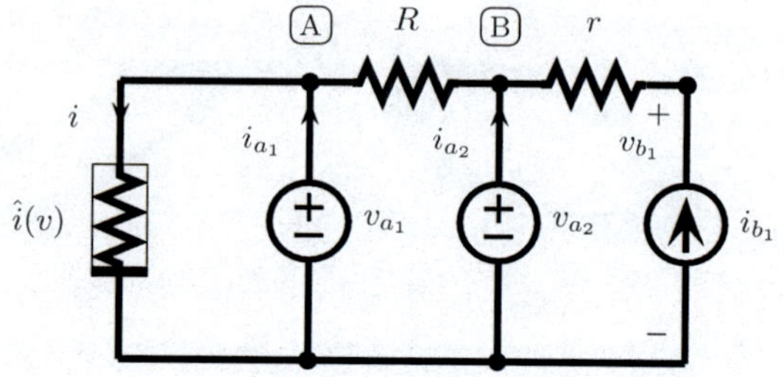

Fig. 4.27 Resistive nonlinear network derived from Chua's oscillator for finding the SEs

Remark 4.6 The reader is referred to [15] for a compendium of complex phenomena and local and global bifurcations that can be observed in Chua's oscillator.

Appendix: State Equations of the Chua's Oscillator

Let us find the SEs describing the dynamics of Chua's oscillator in Fig. 4.15 by using the procedure in Sect. 3.3.2.3 of Chap. 3. It can be easily checked that Chua's oscillator satisfies the conditions for the existence of the SE representation in Property 3.1 of Chap. 3. By replacing C_1 and C_2 with voltage sources and L_1 with a current source, we obtain the circuit in Fig. 4.27. By KCL at nodes (A) and (B) we obtain

$$i_{a_1} = \hat{\imath}(v_{a_1}) + \frac{v_{a_1} - v_{a_2}}{R}$$

and

$$i_{a_2} = \frac{v_{a_2} - v_{a_1}}{R} - i_{b_1}.$$

KVL at the loop formed by i_{b_1}, r, and v_{a_2} yields

$$v_{b_1} = r i_{b_1} + v_{a_2}.$$

We have $v_{C_1} = v_{a_1}$, $v_{C_2} = v_{a_2}$, $i_L = i_{b_1}$, while $i_{C_1} = C_1 dv_{C_1}/dt = -i_{a_1}$, $i_{C_2} = C_2 dv_{C_2}/dt = -i_{a_2}$, and $v_L = L di_L/dt = -v_{b_2}$. By substitution we obtain the third-order SEs in normal form given in Sect. 4.3.1.

References

1. L.O. Chua, Global unfolding of Chua's circuit. IEICE Trans. Fundam **E76-A**(5), 948–962 (1993)
2. J. Guckenheimer, P. Holmes, *Nonlinear Oscillations, Dynamical Systems, and Bifurcations of Vector Fields* (Springer, New York, 1983)
3. S.H. Strogatz, *Nonlinear Dynamics and Chaos with Student Solutions Manual: With Applications to Physics, Biology, Chemistry, and Engineering* (CRC Press, Boca Raton, 2018)
4. Y.A. Kuznetsov, *Elements of Applied Bifurcation Theory*, vol.112 (Springer, Berlin, 2013)
5. H.K. Khalil, *Nonlinear Systems* (Prentice Hall, Englewood Cliffs, 2002)
6. L.O. Chua, C.A. Desoer, E.S. Kuh, *Linear and Nonlinear Circuits* (McGraw-Hill, New York, 1987)
7. L.O. Chua, Dynamic nonlinear networks: state-of-the-art. IEEE Trans. Circuits Syst. **27**(11), 1059–1087 (1980)
8. L. Chua, Device modeling via basic nonlinear circuit elements. IEEE Trans. Circuits Syst. **27**(11), 1014–1044 (1980)
9. J.K. Hale, *Ordinary Differential Equations* (Wiley, New York, 1969)
10. M.M. Peixoto, Structural stability on two-dimensional manifolds Topology **1**(2), 101–120 (1962)
11. L. Chua, M. Komuro, T. Matsumoto, The double scroll family. IEEE Trans. Circuits Syst. **33**(11), 1072–1118 (1986)
12. L.O. Chua, The genesis of Chua's circuit. Archiv Elektronik Ubertragungstechnik **46**(4), 250–257 (1992)
13. R. Madan, Special issue on Chua's circuit: a paradigm for chaos, part I introduction and applications. J. Circuit Syst. Comput. **3**, (1993)
14. E. Bilotta, P. Pantano, *A Gallery of Chua Attractors*, vol. 61 (World Scientific, Singapore, 2008)
15. L. Pivka, C.W. Wu, A. Huang, Chua's oscillator: a compendium of chaotic phenomena. J. Franklin Inst. **331**(6), 705–741 (1994)
16. T. Yang, A survey of chaotic secure communication systems. Int. J. Comput. Cognit. **2**(2), 81–130 (2004)
17. L.O. Chua, R.N. Madan, T. Matsumoto, (Eds.), *Special Section On: Chaotic Systems*. Proceedings of IEEE (1987)
18. L.O. Chua (Ed.), Special issue on nonlinear waves, patterns and spatio-temporal chaos in dynamic arrays. IEEE Trans. Circuits Syst. I Fundam. Theory Appl. **42**(10), 557–823 (1995)
19. A. Mees, L. Chua, The Hopf bifurcation theorem and its applications to nonlinear oscillations in circuits and systems. IEEE Trans. Circuits Syst. **26**(4), 235–254 (1979)

Part II
Flux-Charge Analysis Method (FCAM)

Chapter 5
Flux-Charge Analysis Method of Memristor Circuits

Let us consider a relevant class of nonlinear networks, denoted by $\mathcal{LM}$, containing at least one ideal memristor in addition to ideal (linear) resistors, inductors, capacitors, and independent voltage or currents sources. Thus, $\mathcal{LM}$ describes *nonlinear dynamic networks including ideal memristors*.

This chapter provides an effective systematic methodology to write the dynamic equations of any circuit in $\mathcal{LM}$ with the aim of exploring qualitative and quantitative properties of nonlinear dynamic behavior. Memristor networks in the class $\mathcal{LM}$ could be analyzed in the traditional voltage-current (v, i)-domain, but the CR of a memristor is given in terms of flux and charge variables (see Chap. 2 and the next Sect. 5.1 for a brief recap). Hence, due to the presence of the memristor, it is logical to seek an answer to the following question:

> Which is the domain, if any, where qualitative and quantitative nonlinear dynamic properties of a memristor circuit in $\mathcal{LM}$ are described and can be studied in the most effective way?

Hereinafter, the ultimate goal of the book is to demonstrate that ***the flux-charge*** (φ, q)*-domain is the most effective framework for memristor circuits in the class* $\mathcal{LM}$. The chief pillar of the systematic method for the analysis of memristor circuits in $\mathcal{LM}$ is the writing of Kirchhoff laws and CRs of circuit elements in the (φ, q)-domain, i.e., using flux $\varphi = v^{(-1)}$ and charge $q = i^{(-1)}$ as port variables of each two-terminal element. This new technique is named *Flux-Charge Analysis Method* (or FCAM, for short).

The application of FCAM to the class $\mathcal{LM}$ makes clear that the term "most effective way" results into multiple practical and theoretical aspects. A concise summary of the main advantages of FCAM with respect to traditional methods in the (v, i)-domain is reported below:

- the analysis in the (φ, q)-domain via FCAM allows one to unfold the nonlinear dynamics of memristor circuits by *unveiling some new and peculiar features* that are not typically observable in standard RLC circuits. These include the existence invariants of motion and invariant manifolds, the coexistence of

F. Corinto et al., *Nonlinear Circuits and Systems with Memristors*,
https://doi.org/10.1007/978-3-030-55651-8_5

different dynamics and attractors for the same set of circuit parameters and the existence of special bifurcations phenomena named bifurcations without parameters (Chap. 6).

- FCAM permits to obtain a *reduction of order* when dynamic circuit equations of a memristor circuit in $\mathcal{LM}$ are derived in the (φ, q)-domain with respect to their derivation in the (v, i)-domain. From a mathematical viewpoint, the reduction of order enables a simpler dynamical analysis by exploiting standard theory of low-order dynamical systems.
- FCAM in the (φ, q)-domain paves the way to programme different dynamics and attractors in a systematic way by means of pulse voltage or current sources (Chap. 7). Memristor programming is a crucial issue in several applications including neuromorphic systems.

It is also worth to note that all the strong points reported above result from the application of the FCAM to memristor circuits in $\mathcal{LM}$ in the (φ, q)-domain, because it is more intricate to get a thorough understanding of nonlinear dynamics by means of traditional circuit methodologies in the (v, i)-domain.

FCAM has been originally developed mainly in a series of works [1–3]. This chapter is based on [1], but the treatment, presented in a didactic form, is also extended along several directions, including multiterminal and multiport elements, time-varying elements, etc. Finally, several examples are presented to facilitate the acquisition of basic theoretic aspects of FCAM.

5.1 The Importance of Choosing the Correct Pair of Variables

One of the chief concepts presented in Sect. 2.1.3 of Chap. 2 is that pinched hysteresis loops displayed in the $v-i$ plane by a memristor in response to any (zero-mean) periodic signals don't permit to derive a CR, thus pinched hysteresis loops are not models! Changing the amplitude and frequency of the excitation would give rise to a complicated scenario where the whole $v-i$ plane is filled with differently shaped hysteresis loops. The hysteresis loops are only the response to specific excitations but they don't have predicting ability since they do not permit to obtain the response to other types of excitations. The only CR of a memristor is instead obtained when using the correct pair of electric variables, i.e., flux and charge. The CR is indeed given by a nonlinear characteristic relating flux and charge. For any possible signal applied to the memristor, charge and flux lie on such characteristic, thus implying that the CR in the flux-charge domain has predicting ability.

Hence, *the selection of the correct pair of electric variables is of paramount importance for characterizing and modeling any two-terminal circuit element*. Flux and charge are the proper variable for memristor devices. Next, we discuss three additional examples, the first one in the electrical domain and the others in the mechanical field.

Example 5.1 (Time-Varying Capacitor) Consider the CR

$$i = C\frac{dv}{dt} \tag{5.1}$$

of a linear time-invariant capacitor C in the (v, i)-domain. Then, consider a time-varying capacitor $C(t)$ obtained by varying the distance between plates via the application of a mechanical force. We can ask whether there holds the obvious generalization

$$i = C(t)\frac{dv}{dt}.$$

Unfortunately, laboratory experiments would show that this generalized CR would not give correct results for current i for any applied signal v. The problem is that we have chosen the incorrect pair of variables for the CR of a time-varying capacitor.

Suppose instead to consider the pair of variables $(v, q) = (v, i^{(-1)})$ to describe C, i.e., to use the CR

$$q = Cv$$

which is obtained by integrating (5.1) in time. The correct CR of the time-varying capacitor $C(t)$ is simply given as

$$q = C(t)v$$

from which it is possible to derive the right CR in the (v, i)-domain, that is

$$i = C(t)\frac{dv}{dt} + \frac{dC(t)}{dt}v.$$

Note the extra term $(dC(t)/dt)v$ that is needed to predict the correct values of the time-varying capacitor current.

Example 5.2 (Rocket Launching) An example similar to the time-varying capacitor can be found in mechanics concerning a rocket launching.

The rocket mass $m(t)$ changes with time due to fuel consumption. We can ask whether the following generalization of Newton's Law of Motion holds

$$f = m(t)\frac{dv}{dt}$$

where f is the force and $v = dx/dt$ is the velocity. Again, such formula turns out to fail to predict the experimental results, i.e., it has no predicting ability.

Integrating in time Newton's Law of Motion $f = mdv/dt$ for a constant mass m yields

$$p = mv$$

where the momentum of f is defined as

$$p(t) = \int_{-\infty}^{t} f(\tau)d\tau.$$

The correct Newton's Law of Motion for a time-varying mass reads as follows:

$$p = m(t)v$$

where

$$\frac{dp}{dt} = f.$$

This yields

$$f = m(t)\frac{dv}{dt} + \frac{dm(t)}{dt}v(t).$$

The correct pair of variables for a time-varying mass is then given by $(p, v) = (f^{(-1)}, x^{(1)})$.

Example 5.3 (Laws of Motion) Section II-E in [4] reports a thorough discussion on the concept of predicting the motion (i.e., dynamical behaviors) of a physical object (e.g., stones, material bodies, planets, stars, etc.) by a law relating two observable (i.e., measurable) attributes (i.e., physical variables).

The first attempts date back at least 2300 years ago, when Greek philosopher Aristotle proposed the law $f = mv$. We now realize that Aristotle's Law of Motion is not valid because he had chosen an incorrect pair of physical variables (force and velocity) to characterize a body in motion.

We had to wait almost 2000 years to replace Aristotle's Law with the Newton's Law of Motion

$$f = m\frac{dv}{dt}$$

which is correct at least for mundane velocities $v \ll c$, where c is the velocity of light. It is interesting to note that, actually, Newton's Law was originally expressed in the equivalent form using the momentum $p(t) = \int_{-\infty}^{t} f(\tau)d\tau$ of f as

$$f = \frac{dp}{dt}; \quad p(t) = mv(t).$$

Unfortunately, as is the case in all models, even the celebrated Newton's Second Law is an approximation of reality, and it loses its predictive ability (in the sense that it gives answers that do not agree with measurements) when the velocity v of a body approaches the velocity of light c. Thanks to Einstein's Special Theory of Relativity [5], it can be shown that Newton's Second Law must be replaced by its relativistic version

$$f = \frac{m_0 a}{\sqrt{\left(1 - \frac{v^2}{c^2}\right)^3}} \tag{5.2}$$

where $a = dv/dt = v^{(1)}$ and m_0 is a constant called the rest mass. Note that (5.2) involves three variables, namely, f, v, and $v^{(1)}$. Fortunately, it turns out that (5.2) is an exact differential in the sense that if we integrate both sides of the Eq. (5.2), we would obtain

$$p(t) = \int_{-\infty}^{t} f(\tau)d\tau = \frac{m_0 v}{\sqrt{1 - \frac{v^2}{c^2}}}$$

a law involving only the pair of variables p and v. In conclusion, the following law of motion has so far predicted all measurement outcomes correctly

$$f = \frac{dp}{dt}; \quad p = \frac{m_0 v}{\sqrt{1 - \frac{v^2}{c^2}}}$$

and the variable pair $(p, v) = (f^{(-1)}, x^{(1)})$ is once more the correct choice for characterizing the motion of a body traveling close to the velocity of light.

The three examples reported above and all the theoretical concepts of the previous chapters bear witness to flux and charge as the proper electrical variables to model memristor devices via a CR with predictive ability (see also the memristor as basic algebraic element in Fig. 1.13).[1] The next two sections make clear the importance of writing Kirchhoff laws by means of the flux and charge variables in developing FCAM when at least one memristor is included in a nonlinear RLC circuits.

5.2 Forms of Kirchhoff Laws in the Flux-Charge Domain

Before presenting FCAM, we discuss a number of examples illustrating with simple circuits some fundamental issues that need to be taken into account when writing Kirchhoff laws in the (φ, q)-domain and some basic ideas underlying FCAM.

[1] Additional physical variables can be included in the CR of a type-M dynamic elements defined in (1.21) (refer to [6] and Fig. 1.13).

In the standard (v, i)-domain, KCL states that the algebraic sum of all currents in a cut-set vanishes for all times t (Sect. 3.1.2 in Chap. 3). This can be written mathematically as

$$\sum_k i_k(t) = 0$$

for any cut-set and any t. In the same domain KVL states that the algebraic sum of all voltages around a loop vanishes for all t (Chap. 3), which can be written as

$$\sum_k v_k(t) = 0$$

around any loop and for any t.

By integrating in time KCL and KVL we next establish three different forms that can be given to Kirchhoff laws in the (φ, q)-domain.

1. *First Form of Kirchhoff Laws in the* (φ, q)*-domain.*

   If we integrate KCL and KVL with respect to time in the interval $(-\infty, t)$, we would obtain

$$\sum_k q_k(t) = 0 \tag{5.3}$$

for any cut-set and any t and

$$\sum_k \varphi_k(t) = 0 \tag{5.4}$$

around any loop and for any t, where

$$q_k(t) = \int_{-\infty}^{t} i_k(\tau)d\tau$$

is the charge and

$$\varphi_k(t) = \int_{-\infty}^{t} v_k(\tau)d\tau$$

is the flux of the k-th two-terminal element. In practice, $-\infty$ is the instant where each two-terminal element is manufactured. Clearly, $q_k(-\infty) = 0$ and $\varphi_k(-\infty) = 0$.

   Equations (5.3) and (5.4) express the *principles of conservation of charge and flux* [7]: Charge and flux can neither be created nor destroyed, i.e., the quantity of charge and flux is always conserved.

2. *Second Form of Kirchhoff Laws in the* (φ, q)*-domain.*
   Suppose that $-\infty < t_0 < +\infty$ is a *finite initial instant* and we are interested in applying Kirchhoff laws to analyze a circuit for $t \geq t_0$. If we integrate KCL and KVL in the interval $[t_0, t]$, where $t \geq t_0$, we obtain

$$\sum_k q_k(t) = \sum_k q_k(t_0) \tag{5.5}$$

   for any cut-set and any $t \geq t_0$ and

$$\sum_k \varphi_k(t) = \sum_k \varphi_k(t_0) \tag{5.6}$$

   around any loop and for any $t \geq t_0$, where

$$q_k(t_0) = \int_{-\infty}^{t_0} i_k(\tau)d\tau$$

   and

$$\varphi_k(t_0) = \int_{-\infty}^{t_0} v_k(\tau)d\tau$$

   are the *initial charge and initial flux* at t_0, respectively.
3. *Third Form of Kirchhoff Laws in the* (φ, q)*-domain.*
   Suppose again that t_0 is a finite initial instant and define for any $t \geq t_0$ the *incremental charge and incremental flux* of a two-terminal element as follows:

$$q_k(t; t_0) = q_k(t) - q_k(t_0) = \int_{t_0}^{t} i_k(\tau)d\tau$$

   and

$$\varphi_k(t; t_0) = \varphi_k(t) - \varphi_k(t_0) = \int_{t_0}^{t} v_k(\tau)d\tau$$

   respectively. Then, we can write

$$\sum_k q_k(t; t_0) = 0 \tag{5.7}$$

   for any cut-set and any $t \geq t_0$ and

$$\sum_k \varphi_k(t; t_0) = 0 \tag{5.8}$$

   around any loop and for any $t \geq t_0$, respectively.

Equations (5.7) and (5.8) express the *principles of conservation of incremental charge and flux*: Incremental charge and flux can neither be created nor destroyed, i.e., the quantity of incremental charge and flux is always conserved for $t \geq t_0$.

5.3 Fundamental Examples on Kirchhoff Laws

The examples in this section and the next one illustrate possible difficulties to use the first two forms of Kirchhoff laws in the (φ, q)-domain. The same examples show that these difficulties can be completely overcome by using the third form of Kirchhoff laws in the (φ, q)-domain, that is based on incremental values of charge and flux of each two-terminal element. We start with examples concerning linear circuits and then we consider in the next section a crucial basic example of memristor circuit.

5.3.1 *Linear Circuits*

Example 5.4 Consider the circuit with two ideal switches S_1 and S_2 in Fig. 5.1 and suppose S_1 is closed since a long time $(-\infty)$ and it opens at $t_0 = 0$, while S_2 is open since a long time $(-\infty)$ and it closes at $t_0 = 0$. Assume C_1, C_2 are discharged at $-\infty$. The circuit changes topology at $t_0 = 0$ but has a fixed topology for $t > t_0 = 0$.

In the interval $(-\infty, 0)$ the battery creates the initial conditions by charging C_1 at a voltage $v_{C_1}(0^-)$ almost equal to E (henceforth, we assume for simplicity $v_{C_1}(0^-) = E$). We also have $v_{C_2}(0^-) = 0$. The state variables v_{C_1} and v_{C_2} are continuous when the switches commutate, hence $v_{C_1}(0^+) = v_{C_1}(0) = v_{C_1}(0^-) = E$ and $v_{C_2}(0^+) = v_{C_2}(0) = v_{C_2}(0^-) = 0$ are the initial conditions at $t = 0$ needed to study the transient for $t \geq 0$.

We now wish to analyze the transient for $t \geq 0$ first in the (v, i)-domain and then in the (φ, q)-domain.

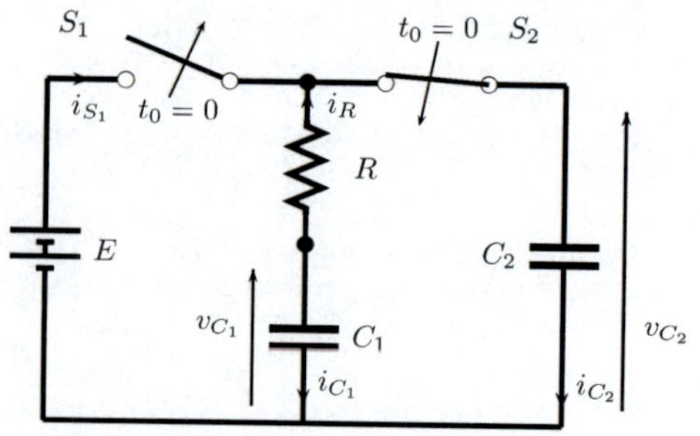

Fig. 5.1 Linear circuit with two capacitors changing topology at $t_0 = 0$

Analysis in the (v, i)*-Domain*

Let us start with the analysis in the (v, i)-domain. For $t \geq 0$, C_1 discharges on C_2 through the resistor R with a known transient. From KCLs we have for $t \geq 0$

$$i_{C_1} = -i_R = -i_{C_2}$$

while KVL at the loop formed by C_1, R and C_2 yields

$$v_{C_1} = Ri_R + v_{C_2}$$

where $i_{C_1} = C_1 dv_{C_1}/dt$ and $i_{C_2} = C_2 dv_{C_2}/dt$. By substitution, we easily arrive at the following second-order SE for $t \geq 0$

$$\frac{dv_{C_1}}{dt} = -\frac{v_{C_1} - v_{C_2}}{RC_1}$$
$$\frac{dv_{C_2}}{dt} = \frac{v_{C_1} - v_{C_2}}{RC_2}$$

with initial conditions $v_{C_1}(0) = E$ and $v_{C_2}(0) = 0$. The solution is known from textbooks in circuit theory and is given by

$$v_{C_1}(t) = Ee^{-\frac{t}{\tau}} + \frac{EC_2}{C_1 + C_2}(1 - e^{-\frac{t}{\tau}})$$

and

$$v_{C_2}(t) = \frac{EC_1}{C_1 + C_2}(1 - e^{-\frac{t}{\tau}})$$

where

$$\tau = R\frac{C_1 C_2}{C_1 + C_2}.$$

Note that to solve the circuit for $t \geq 0$ it is enough to know the initial conditions $v_{C_1}(0)$ and $v_{C_2}(0)$, while we do not need to know in detail how these initial conditions are created for $t \leq 0$. This is consistent with the dynamical approach based on the SEs described in Chap. 3.

Analysis in the (φ, q)*-Domain*

Suppose now we wish to analyze the dynamics for $t \geq t_0 = 0$ in the (φ, q)-domain. Let us start by writing the conservation of charge for $t \geq t_0 = 0$. One may think that it is possible to split the circuit in two parts and consider for $t \geq 0$ the subcircuit C_1-R-C_2 as *isolated* from the remaining part of the circuit for $t \geq 0$, as shown in Fig. 5.2. This would yield, according to the first form (5.3) of Kirchhoff law in the (φ, q)-domain, applied to cut-set C', the conservation of charge

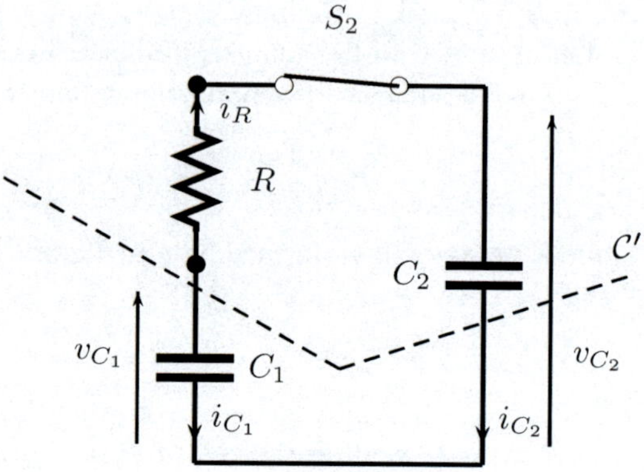

Fig. 5.2 Sub-circuit for $t \geq 0$ given by $C_1 - R - C_2$

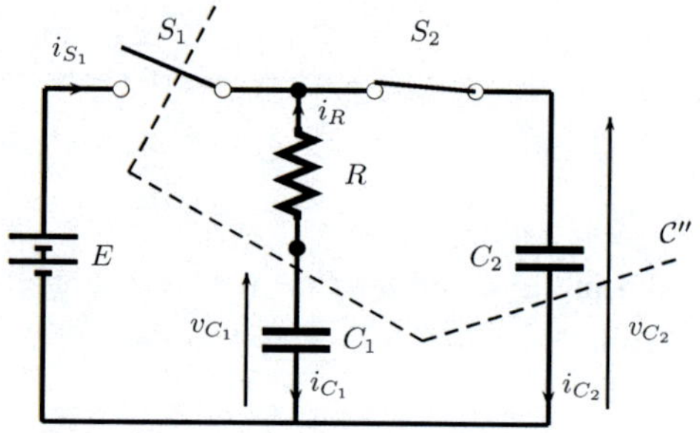

Fig. 5.3 Circuit as in Fig. 5.1 for $t \geq 0$

$$q_{C_1}(t) + q_{C_2}(t) = 0 \tag{5.9}$$

for any $t \geq 0$. However, this is *incorrect*. In fact, at $t = 0$ we have $q_{C_1}(0) = C_1 E$ and $q_{C_2}(0) = 0$, hence $q_{C_1}(0) + q_{C_2}(0) = C_1 E \neq 0$.

The problem is that $q_{C_1}(t)$ and $q_{C_2}(t)$ are *quantities involving the whole history of currents in* C_1 *and* C_2 *from* $-\infty$ *to the current instant* t. Actually, in order to write a correct conservation equation for the charge we should consider for instance the cut-set C'' as shown in Fig. 5.3, or the cut-set C_1–C_2 and the battery. In the first case we have

$$q_{C_1}(t) + q_{C_2}(t) + q_{S_1}(t) = 0$$

for $t \geq 0$, where $q_{S_1(t)} = \int_{-\infty}^{t} i_{S_1}(\tau)d\tau$ is the charge through the switch. In general, for a complicated circuit, it may be difficult to deal with the charge $q_{S_1}(t)$ through a switch, since this involves knowing the history of the current in the switch for $t \leq t_0$. There are similar problems for dealing for instance with the charge in a battery or flux in a current source.

Consider now the third form of conservation of incremental charge as in (5.7). Note that to evaluate the incremental charge of elements we need to *consider only currents for* $t \geq 0$, hence we can correctly split the circuit in two parts and, for the cut-set C_1-C_2, we can write

$$q_{C_1}(t;0) + q_{C_2}(t;0) = 0$$

for any $t \geq 0$. This yields

$$q_{C_1}(t;0) + q_{C_2}(t;0) = C_1 v_{C_1}(t) - C_1 v_{C_1}(0) + C_2 v_{C_2}(t) - C_2 v_{C_2}(0) = 0$$

hence

$$q_{C_1}(t) + q_{C_2}(t) = C_1 v_{C_1}(0) = C_1 E$$

for $t \geq 0$, which is indeed the correct equation describing the conservation of charge for $t \geq 0$. Note that here we used only the information on the initial conditions $v_{C_1}(0) = E$ and $v_{C_2}(0) = 0$.

For the moment we stop here with the analysis in the (φ, q)-domain of this circuit, since the current goal is to illustrate the difficulties in using the first form (5.3) of conservation of charge. Later in the chapter, after developing the flux-charge analysis method, we will come back to this circuit and complete its analysis in the (φ, q)-domain (see Example 5.20).

Example 5.5 Consider again the sub-circuit C_1-R-C_2 as in Fig. 5.2 of Example 5.4 for $t \geq t_0 = 0$. Let $v_{C_1}(0) \neq 0$, $v_{C_2}(0) = 0$. Suppose such initial conditions are available from measurements at $t = 0$ but we do not know the history of the circuit for $t \leq 0$, i.e., we do not know how the initial conditions are created for $t \leq 0$. As already discussed, the charge conservation law in the form $q_{C_1}(t) + q_{C_2}(t) = 0$ for $t \geq 0$ is not valid. Since the information on the history of the circuit for $t \leq 0$ is not available, we are forced to use the third form of conservation of charge, i.e., the incremental form

$$q_{C_1}(t;0) + q_{C_2}(t;0) = 0$$

which involves only the behavior of the circuit for $t \geq 0$. Note that this is all we need to arrive at the correct result $q_{C_1}(t) + q_{C_2}(t) = C_1 v_{C_1}(0)$ for $t \geq 0$.

Example 5.6 This example aims at highlighting difficulties in using the second form given in (5.5) and (5.6) of Kirchhoff laws in the (φ, q)-domain.

Consider a simple C-R circuit as in Fig. 5.4 for $t \geq t_0 = 0$ and suppose the initial condition $v_C(0) = v_{C0} \neq 0$ is available from a measurement. Since we are not aware of the history of the circuit for $t \leq 0$, the values of $\varphi_C(0)$, $\varphi_R(0)$, and $q_R(0)$ are actually not known. It is not difficult to see that a given v_{C0} may correspond to

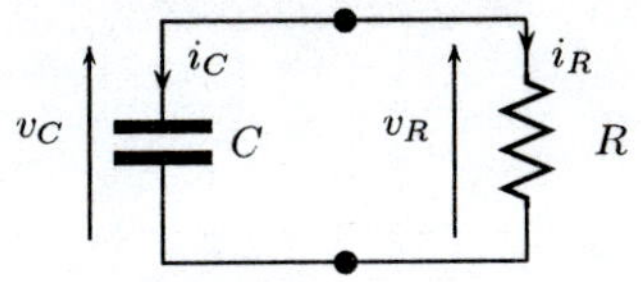

Fig. 5.4 A simple linear $R - C$ circuit

any possible value of $\varphi_C(0)$. To see this, suppose to disconnect C from R for $t \leq 0$ and use a current source connected to C such that $i_a(t) = I_0$, $-T \leq t \leq 0$, and $i_a(t) = 0$ for $t < -T$. If we choose $I_0 = Cv_{C_0}/T$, then $v_C(0^-) = v_{C_0}$. However, we have $\varphi_C(0) = v_{C_0}T/2$ and, clearly, by varying T, $\varphi_C(0)$ may assume any value. The same holds for $\varphi_R(0)$ and $q_R(0)$. Instead, $q_C(0) = v_{C_0}/C$ is known, since it is proportional to a state variable in the (v, i)-domain.

In general, it can be seen that *working with the initial values $q_k(0)$ (resp., $\varphi_k(0)$) for elements such that $q_k(t)$ (resp., $\varphi_k(t)$) is not a state variable in the (v, i)-domain may be problematic.*

This discussion shows that there are problems also in using the second form of Kirchhoff laws in the (φ, q)-domain, since these laws involve the initial quantities $q_k(t_0)$ and $\varphi_k(t_0)$ that are not necessarily known. Again, the problem can be overcome via the use of the third form of Kirchhoff laws (5.7) and (5.8) in the (φ, q)-domain since this form with incremental charge and flux no longer involves the initial quantities $q_k(t_0)$ and $\varphi_k(t_0)$.

Let us now solve this simple circuit in the (φ, q)-domain. We can write the conservation of incremental charge as

$$q_C(t; 0) + q_R(t; 0) = 0$$

and the conservation of incremental flux as

$$\varphi_C(t; 0) = \varphi_R(t; 0)$$

for $t \geq 0$. For C we have $q_C(t) = Cv_C(t)$, hence

$$q_C(t; 0) = Cv(t) - Cv_{C_0} = C\frac{d\varphi_C(t)}{dt} - Cv_{C_0} = C\frac{d\varphi_C(t; 0)}{dt} - Cv_{C_0}$$

while for R we have $v_R(t) = Ri_R(t)$ and hence

$$q_R(t; 0) = \frac{\varphi_R(t; 0)}{R}.$$

These yield

$$\begin{aligned} C\frac{d\varphi_C(t; 0)}{dt} &= q_C(t; 0) + Cv_{C_0} \\ &= -q_R(t; 0) + Cv_{C_0} \\ &= -\frac{\varphi_R(t; 0)}{R} + Cv_{C_0} \\ &= -\frac{\varphi_C(t; 0)}{R} + Cv_{C_0}. \end{aligned}$$

For $t > 0$ the C-R circuit is thus described in the (φ, q)-domain by the first-order SE

$$\frac{d}{dt}\varphi_C(t; 0) = -\frac{\varphi_C(t; 0)}{RC} + v_{C_0}$$

in the state variable $\varphi_C(t; 0)$. Note that the initial condition vanishes by definition of incremental flux, i.e.,

$$\varphi_C(0; 0) = 0.$$

From the theory of linear differential equations the solution is given by

$$\varphi_C(t; 0) = RCv_{C_0}(1 - e^{-\frac{t}{RC}})$$

for $t \geq 0$. By differentiation this yields the well-known transient (discharge of C on R)

$$v_C(t) = \frac{d}{dt}\varphi_C(t; 0) = v_{C_0}e^{-\frac{t}{RC}}$$

for $t \geq 0$.

5.4 Kirchhoff Laws for Memristor Circuits

For circuits without memristors like those considered in the previous examples, as it is intuitively reasonable, the analysis in the (φ, q)-domain may result more complex than that in the (v, i)-domain. However, when there are memristors, it is found that it is more convenient to analyze the circuit in the (φ, q)-domain. Some advantages of the analysis in the (φ, q)-domain are illustrated via an elementary memristor circuit in the next example. The remaining part of the chapter then develops in a systematic way the flux-charge analysis approach for a broad class of memristor circuits.

Example 5.7 (Memristor-Capacitor Circuit) Consider for $t \geq t_0$, where t_0 is a finite initial instant, a simple circuit composed by one flux-controlled memristor M with CR $q_M = f(\varphi_M)$[2] and$\varphi_M(t)$ for $t < t_0$ as well (i.e., S_1 and S_3 are closed while S_2 is open). The two-terminal circuit elements $\mathcal{L}_a$ and $\mathcal{L}_b$ can be any linear networks made of resistors, capacitors, inductors, voltage, and current sources and

[2]For convenience, from now on in the book we will use the simplified notation $q_M = f(\varphi_M)$, instead of the notation $q_M = \hat{q}_M(\varphi_M)$ introduced in Chap. 1, to denote a flux-controlled memristor. Analogously, we will use $\varphi_M = h(q_M)$, instead of $\varphi_M = \hat{\varphi}_M(q_M)$, to denote a charge-controlled memristor.

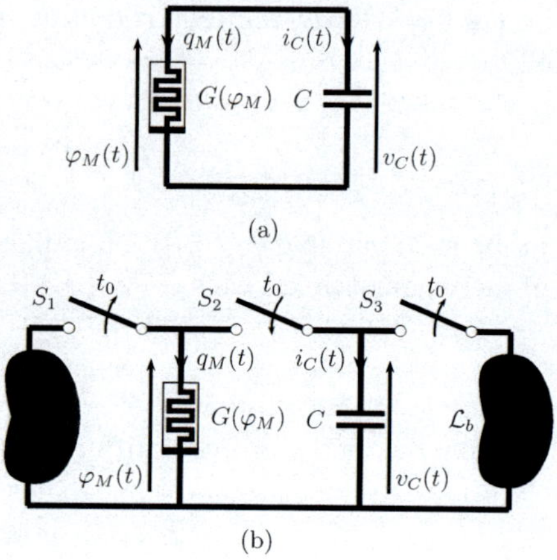

Fig. 5.5 The simplest memristor-based circuit. (**a**) Memristor-capacitor ($M - C$) circuit for $t \geq t_0$. (**b**) $M - C$ circuit including networks $\mathcal{L}_a$ and $\mathcal{L}_b$ that set independent initial conditions v_{C_0} and φ_{M_0} through the evolution of the electrical variables $v_C(t)$ and $\varphi_M(t)$ for $t < t_0$

are used to set suitable initial conditions at t_0 for the state variables $v_C(t_0) = v_{C0}$ and $\varphi_M(t_0) = \varphi_{M_0}$ in the (v, i)-domain.

It is clear that we can set *independent initial conditions* v_{C_0} *and* φ_{M_0} *for the state variables* by means of the dynamics for $t < t_0$ of the circuits $\mathcal{L}_a$–M (with S_1 closed), and $\mathcal{L}_b$–C (with S_3 closed), respectively. We suppose $v_C(-\infty) = 0$, $\varphi_M(-\infty) = 0$.

The $M - C$ circuit has a fixed topology for $t > t_0$ and we are interested in studying the transient following the closing and opening of switches at t_0. Analysis by inspection of the M–C circuit permits to derive that the state variables $v_C(t)$ and $\varphi_M(t)$ obey the following *Initial Value Problem (IVP) for a second-order SE* in the (v, i)-domain

$$C\frac{dv_C(t)}{dt} = -G(\varphi_M(t))v_C(t) \tag{5.10}$$

$$\frac{d\varphi_M(t)}{dt} = v_C(t) \tag{5.11}$$

$$v_C(t_0) = v_{C0}$$

$$\varphi_M(t_0) = \varphi_{M0}$$

for any $t \geq t_0$, where $G(\varphi_M) = f'(\varphi_M)$ is the memductance.

Now note that the right-hand-side (r.h.s.) of (5.10) can be written as

$$-G(\varphi_M(t))v_C(t) = -\frac{df(\varphi_M(t))}{d\varphi_M}\frac{d\varphi_M(t)}{dt} = -\frac{d}{dt}f(\varphi_M(t)) \tag{5.12}$$

therefore, by integrating (5.10) in t over the interval $[t_0, t]$, where $t \geq t_0$, we have

$$C\left(v_C(t) - v_C(t_0)\right) = -f(\varphi_M(t)) - f(\varphi_M(t_0))$$

$$= -\left(q_M(t) - q_M(t_0)\right). \tag{5.13}$$

From (5.13) we can draw two basic results.

(i) First, the nonlinear dynamical behavior for $t \geq t_0$ of the M–C circuit is described in the (φ, q)-domain by the following initial value problem (IVP) *for a first-order SE* (derived from (5.13) and (5.11))

$$\begin{aligned} \frac{d\varphi_M(t)}{dt} &= -\frac{f(\varphi_M(t))}{C} + \frac{f(\varphi_{M_0})}{C} + v_{C_0} \\ \varphi_M(t_0) &= \varphi_{M_0} \end{aligned} \tag{5.14}$$

where the state variable is $\varphi_M(t)$ and the initial condition is the known quantity φ_{M_0}. It is also worth remarking that the initial conditions v_{C_0} and φ_{M_0} for the state variable in the (v, i)-domain appear as constant inputs in the r.h.s. of (5.14).

We can summarize these observations as follows. *The IVP* (5.10) *and* (5.11) *for a second-order SE in the* (v, i)*-domain can be reduced to the IVP* (5.14) ***for a first-order SE*** *in the* (φ, q)*-domain where the r.h.s. depends on the initial conditions* v_{C_0} *and* φ_{M_0} *at* t_0 *for the state variables in the* (v, i)*-domain.*

This reduction is crucial in analyzing nonlinear dynamics and bifurcations in the M–C circuit. We will pursue such an analysis in detail later in the book in Example 5.15. Dual considerations hold in a circuit composed of one charge-controlled memristor connected to an inductor. In such a case the state variable in the (φ, q)-domain is the memristor charge $q_M(t)$ and the initial condition is $q_M(t_0)$.

(ii) Second, Eq. (5.13)—which yields (5.14)—is obtained from the integration of KCL over $[t_0, t]$, where $t > t_0$, and it can be formulated as "the sum of the *capacitor incremental charge*

$$q_C(t; t_0) = q_C(t) - q_C(t_0) = C(v_C(t) - v_C(t_0))$$

and the *memristor incremental charge*

$$q_M(t; t_0) = q_M(t) - q_M(t_0) = f(\varphi_M(t)) - f(\varphi_M(t_0))$$

is zero" for any $t \geq t_0$.

It turns out that the fundamental step in reducing by one the order of SEs in the IVP (5.10) and (5.11) is the *integration of the KCL in* (t_0, t), stating that "the algebraic sum of the *incremental charge* in any cut-set is zero," exactly as in the third form of Kirchhoff law in the (φ, q)-domain given in (5.7). With reference to the M–C circuit in Fig. 5.5a, we can indeed rewrite (5.13) as follows for $t \geq t_0$

$$q_C(t; t_0) + q_M(t; t_0) = 0. \tag{5.15}$$

Remark 5.1 Since $\varphi_M(t; t_0) = \varphi_C(t; t_0)$, (5.14) can also be rewritten as an IVP for the first-order SE in the state variable $\varphi_C(t; t_0)$

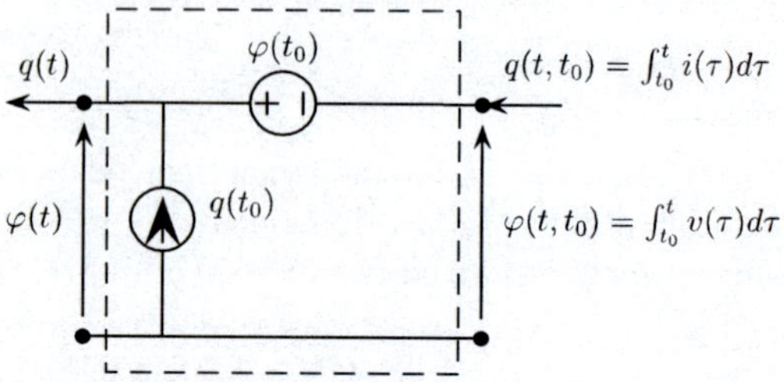

Fig. 5.6 Two-port network where the charge source $q(t_0)$ and the flux source $\varphi(t_0)$ take into account the initial conditions at a finite instant t_0

$$\frac{d\varphi_C(t; t_0)}{dt} = -\frac{f(\varphi_C(t; t_0) + \varphi_{M0})}{C} + \frac{f(\varphi_{M0})}{C} + v_{C0}$$
$$\varphi_C(t_0; t_0) = 0. \tag{5.16}$$

This shows as in Example 5.6 that a natural choice of the state variable in the (φ, q)-domain is the incremental capacitor flux $\varphi_C(t; t_0)$. Note that with this choice, by construction, the initial condition $\varphi_C(t_0; t_0) = 0$.

Remark 5.2 The key idea enabling to develop the method is the conceptual difference between charge and flux and incremental charge and flux, respectively. The following property, which follows directly from the definition of charge and flux, and their incremental counterparts, further clarifies such concept.

Property 5.1 (Incremental Charge Versus Charge) The incremental charge $q_k(t; t_0)$ and flux $\varphi_k(t; t_0)$ of any two-terminal element reduce to the charge $q_k(t)$ and flux $\varphi_k(t)$, respectively, if and only if $t_0 \to -\infty$, i.e., the circuit topology is invariant for any $t \in (-\infty, +\infty)$.

On the other hand, the past dynamics over $(-\infty, t_0)$ has to be considered in order to set independent initial conditions of a circuit switching its topology at a finite instant t_0. The two-port network in Fig. 5.6 provides a symbolic circuit representation of the result in Property 5.1 including a charge source $q(t_0)$ and a flux source $\varphi(t_0)$ to take into account initial conditions for charge and flux at t_0.

Remark 5.3 The examples discussed so far, although they refer to elementary circuits, permit to highlight the properties and draw the conclusions listed next. These properties will be shown to hold for a general class of memristor circuits via the method developed in the next sections.

- Simple dynamic circuits, with or without memristors, can be analyzed both in the (v, i)-domain, via KCL and KVL, and in the (φ, q)-domain, via Kirchhoff laws in terms of charge and flux.
- Circuits without memristors are in general easier to analyze in the (v, i)-domain. In particular, no reduction of order is obtained by analyzing them in the (φ, q)-domain.
- Circuits with memristors are described by a lower-order SE in the (φ, q)-domain with respect to the (v, i)-domain. The possibility of a simplified dynamic analysis in the (φ, q)-domain can then be envisaged.

- The SE in the (φ, q)-domain keeps track of the initial conditions for the state variables in the (v, i)-domain. Such initial conditions in fact appear as constant inputs in the r.h.s. of the SE in the (φ, q)-domain.
- The natural choice of the state variables in the (φ, q)-domain is given by the incremental flux of capacitors and incremental charge of inductors. The memristor has no associated state variable in the (φ, q)-domain since, as it will be discussed in more detail in Sect. 5.5.2, it is described by a nonlinear algebraic relation between charge and flux in the (φ, q)-domain.
- For the analysis in the (φ, q)-domain, special care is needed to use Kirchhoff laws in terms of charge and flux. The most effective form of Kirchhoff laws in the (φ, q)-domain is that based on incremental charge and incremental flux of two-terminal elements.

5.5 Flux-Charge Analysis Method (FCAM) for Class $\mathcal{LM}$ of Memristor Circuits

We consider the class $\mathcal{LM}$ of memristor circuits containing at least one ideal flux-controlled memristor (or one ideal charge-controlled memristor) in addition to ideal (linear) resistors, capacitors, inductors, and independent voltage or current sources. All elements in $\mathcal{LM}$, possibly with the exception of the independent sources, are assumed to be time-invariant. The extension to the case where all elements may be time-varying is straightforward and is briefly discussed in Sect. 5.6.2.

Consider a circuit $N \in \mathcal{LM}$ and suppose it has a fixed topology for $t \geq t_0$, where t_0 is a finite initial instant. Our goal is to show that we can develop a method, named flux-charge analysis method (FCAM), enabling to analyze the circuit dynamics for $t \geq t_0$ in the (φ, q)-domain. To this end we will address the following main issues:

- how to write Kirchhoff laws in the (φ, q)-domain;
- how to write the CR of each element in the class $\mathcal{LM}$ in the (φ, q)-domain.

We then address the issue of writing in a systematic way the DAEs and SEs describing the circuit dynamics in the (φ, q)-domain.

5.5.1 *Incremental Form of Kirchhoff Laws in the Flux-Charge Domain*

Let b be the number of two-terminal elements and n the number of nodes of a circuit $N \in \mathcal{LM}$. Then, KCL permits to write $n-1$ fundamental cut-set equations in the form (Chap. 3)

$$\mathbf{Ai}(t) = \mathbf{0}$$

where $\mathbf{A} \in \mathbb{R}^{(n-1)\times b}$ is the reduced incidence matrix and $\mathbf{i} \in \mathbb{R}^b$ is the vector of two-terminal element currents. Moreover, KVL yields $b - n + 1$ fundamental loop equations (Chap. 3)

$$\mathbf{B}\mathbf{v}(t) = \mathbf{0}$$

where $\mathbf{B} \in \mathbb{R}^{(b-n+1)\times b}$ is the fundamental loop matrix and $\mathbf{v} \in \mathbb{R}^b$ is the vector of two-terminal element voltages. By integrating between t_0 and $t \geq t_0$ we obtain

$$\mathbf{A}\mathbf{q}(t) = \mathbf{A}\mathbf{q}(t_0) \tag{5.17}$$

$$\mathbf{B}\boldsymbol{\varphi}(t) = \mathbf{B}\boldsymbol{\varphi}(t_0) \tag{5.18}$$

for any $t \geq t_0$, where $\mathbf{q}(t)$ (resp., $\boldsymbol{\varphi}(t)$) is the vector of charges (resp., fluxes) of the b two-terminal elements.

These equations are expressed in the (φ, q)-domain, however they involve all the initial conditions $\mathbf{q}(t_0)$, $\boldsymbol{\varphi}(t_0)$ of two-terminal circuit elements. As already discussed in Example 5.6, the initial conditions $q_{C_k}(t_0) = C_k v_{C_k}(t_0)$ and $\varphi_{L_k}(t_0) = L_k i_{L_k}(t_0)$ can be obtained by the measurement at the instant t_0 of voltages $v_{C_k}(t_0)$ across capacitors and currents $i_{L_k}(t_0)$ through inductors by means of a voltmeter or an ammeter. Instead, initial conditions as $q_{L_k}(t_0) = \int_{-\infty}^{t_0} i_{L_k}(t)dt$, $\varphi_{C_k}(t_0) = \int_{-\infty}^{t_0} v_{C_k}(t)dt$ and $q_{R_k}(t_0) = \int_{-\infty}^{t_0} i_{R_k}(t)dt$ cannot be obtained via measurements at t_0. Rather, to evaluate them we would need to know the past circuit history for $t < t_0$ (see for example Fig. 5.5b), an information that is usually unavailable or difficult to obtain in practice. In addition, it may happen that $\mathbf{q}_a(t_0)$ ($\boldsymbol{\varphi}_e(t_0)$) are not finite for some current (voltage) ideal sources.[3]

Kirchhoff Laws can be made independent of initial conditions by using the incremental fluxes and charges of two-terminal elements defined as

$$\varphi_k(t; t_0) \doteq \varphi_k(t) - \varphi_k(t_0) = \int_{t_0}^{t} v_k(\tau)d\tau \tag{5.19}$$

$$q_k(t; t_0) \doteq q_k(t) - q_k(t_0) = \int_{t_0}^{t} i_k(\tau)d\tau \tag{5.20}$$

for any $t \geq t_0$, where $v_k(t)$ is the voltage across and $i_k(t)$ is current through any two-terminal element in $\mathcal{LM}$, respectively. Indeed, using the vector of incremental charges $\mathbf{q}(t; t_0)$, the Kirchhoff Charge Law (KqL) takes the simpler form

$$\mathbf{A}\mathbf{q}(t; t_0) = \mathbf{0} \tag{5.21}$$

[3] It is readily derived that if $a(t) = A$ ($e(t) = E$) for all $t \in (-\infty, t_0]$, with $A \neq 0$ ($E \neq 0$) constant, then $q_a(t_0) = \int_{-\infty}^{t_0} A\,d\tau$ ($\varphi_e(t_0) = \int_{-\infty}^{t_0} E\,d\tau$) are infinite.

which no longer involves the initial charges $\mathbf{q}(t_0)$, but just expresses the constraints on the incremental charges due to the circuit topology.

Similarly, the Kirchhoff Flux Law (KφL) using the vector of incremental fluxes $\boldsymbol{\varphi}(t; t_0)$ can be written as

$$\mathbf{B}\boldsymbol{\varphi}(t; t_0) = \mathbf{0}. \tag{5.22}$$

These laws will be referred to henceforth as *conservation of incremental charge and flux, respectively*. It is known that these equations give in overall b independent topological constraints on $\mathbf{q}(t; t_0)$ and $\boldsymbol{\varphi}(t; t_0)$ in the (φ, q)-domain (Chap. 3).

Remark 5.4 It is worth remarking that the incremental fluxes and charges have a physical meaning also in that they satisfy Tellegen's Theorem. In fact, it is known from Chap. 3 that KCL can be written as

$$\mathbf{Ai}(t) = \mathbf{0}$$

while KVL can be expressed as

$$\boldsymbol{v}(t) = \mathbf{A}^T \mathbf{v}_e(t)$$

where $\mathbf{v}_e(t)$ is the vector of node voltages.

By integrating in time we obtain for $t > t_0$

$$\mathbf{Aq}(t; t_0) = \mathbf{0}$$

and

$$\boldsymbol{\varphi}(t; t_0) = \mathbf{A}^T \boldsymbol{\varphi}_e(t; t_0)$$

where $\boldsymbol{\varphi}_e(t; t_0) = \int_{t_0}^{t} \mathbf{v}_e(\tau) d\tau$ is the vector of incremental fluxes of nodes. It follows that

$$\boldsymbol{\varphi}(t; t_0)^T \mathbf{q}(t; t_0) = \boldsymbol{\varphi}_e(t; t_0)^T \left(\mathbf{A}^T\right)^T \mathbf{q}(t; t_0) = \boldsymbol{\varphi}_e(t; t_0)^T \mathbf{Aq}(t; t_0) = 0.$$

Hence

$$\boldsymbol{\varphi}(t; t_0)^T \mathbf{q}(t; t_0) = \sum_{k=1}^{b} \varphi_k(t; t_0) q_k(t; t_0) = 0.$$

Remark 5.5 We have just expressed in vector form, using the reduced incidence matrix $\mathbf{A}$, Kirchhoff laws in the (φ, q)-domain. It is useful to note that, in their simplest and most fundamental form, such laws can be stated as follows.

Consider a circuit N with a fixed topology for $t \geq t_0$, where t_0 is a finite initial instant.

Definition 5.1 (Kirchhoff (Incremental) Charge Law (KqL)) The algebraic sum of all incremental charges entering (leaving) any cut-set of N is zero at all times $t \geq t_0$.

Definition 5.2 (Kirchhoff (Incremental) Flux Law (KφL)) The algebraic sum of all incremental fluxes around any loop of N is zero at all times $t \geq t_0$.

5.5.2 *Constitutive Relations in the Flux-Charge Domain*

The discussion in Sect. 5.5.1 shows that it is convenient to write the KφL and KqL equations in the (φ, q)-domain by using the incremental charge $q_k(t; t_0)$ and incremental flux $\varphi_k(t; t_0)$ of each two-terminal element in $\mathcal{LM}$. Then, we are led in what follows to describe each two-terminal element by using its incremental charge $q_k(t; t_0)$ and incremental flux $\varphi_k(t; t_0)$ as port variables, i.e., to express the CR in terms of $q_k(t; t_0)$ and $\varphi_k(t; t_0)$. Once each circuit element is described in the (φ, q)-domain by incremental flux and charge at its terminals, *any circuit in $\mathcal{LM}$ is represented in the (φ, q)-domain by connecting circuit elements using such terminals*. As a consequence, KφL and KqL result to be independent of initial conditions and provide the "usual" circuit topology constraints.

Next we discuss in detail how to obtain CRs and equivalent circuits in the (φ, q)-domain for any two-terminal element in the class $\mathcal{LM}$ (we drop the index k to simplify the notation).

5.5.2.1 Ideal Capacitor

Let us consider an ideal capacitor C described by

$$q_C(t) = Cv_C(t) = C\frac{d\varphi_C(t)}{dt}.$$

Let $q_C(t_0) = q_{C_0} = Cv_C(t_0)$ be the initial charge at t_0. We obtain for $t \geq t_0$

$$q_C(t; t_0) = q_C(t) - q_C(t_0) = C\frac{d\varphi_C(t)}{dt} - Cv_C(t_0)$$

that is

$$q_C(t; t_0) = C\frac{d\varphi_C(t)}{dt} - q_{C_0}. \tag{5.23}$$

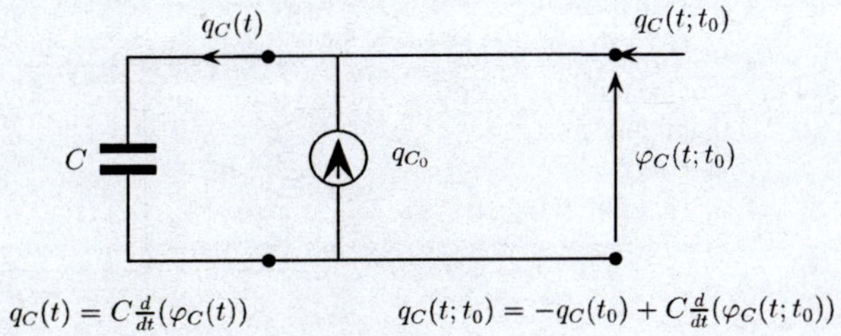

Fig. 5.7 Equivalent circuit for an ideal capacitor in terms of the incremental charge $q_C(t; t_0)$ and flux $\varphi_C(t; t_0)$

By noting that

$$C\frac{d\varphi_C(t)}{dt} = C\frac{d\varphi_C(t; t_0)}{dt}$$

the *CR of the ideal capacitor C in terms of the incremental variables* $q_C(t; t_0)$ *and* $\varphi_C(t; t_0)$ is obtained as

$$q_C(t; t_0) = C\frac{d\varphi_C(t; t_0)}{dt} - q_{C_0}. \tag{5.24}$$

In the (φ, q)-domain the ideal capacitor C described by (5.24) has the equivalent circuit representation in Fig. 5.7.[4] It is important to remark that the equivalent circuit in the (φ, q)-domain has an independent charge source proportional to the initial condition $v_C(t_0) = v_{C_0} = q_{C_0}/C$ for the state variable $v_C(t)$ in the (v, i)-domain. Moreover, according to (5.24), a natural choice for the *state variable* of a capacitor in the (φ, q)-domain is the incremental flux $\varphi_C(t; t_0)$.

5.5.2.2 Ideal Inductor

Let us consider an ideal inductor L described by

$$\varphi_L(t) = Li_L(t) = L\frac{dq_L(t)}{dt}.$$

Following the same approach used for the ideal capacitor, we can introduce the initial flux at t_0, i.e., $\varphi_L(t_0) = \varphi_{L_0} = Li_L(t_0)$, so that

$$\varphi_L(t; t_0) = \varphi_L(t) - \varphi_L(t_0) = L\frac{dq_L(t)}{dt} - Li_L(t_0)$$

that is

[4]The circuit in Fig. 5.7 represents, in the (φ, q)-domain, the dual of the *initial capacitor voltage transformation* circuit reported in Fig. 2.1 in page 307 in [8].

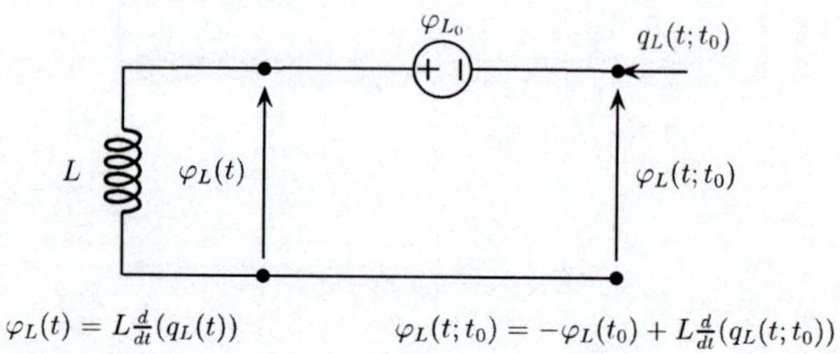

Fig. 5.8 Equivalent circuit for an ideal inductor in terms of the incremental charge $q_L(t; t_0)$ and flux $\varphi_L(t; t_0)$

$$\varphi_L(t; t_0) = -\varphi_{L_0} + L\frac{dq_L(t)}{dt}. \tag{5.25}$$

By observing that

$$L\frac{dq_L(t)}{dt} = L\frac{dq_L(t; t_0)}{dt}$$

the *CR of the ideal inductor L in terms of the incremental variables* $q_L(t; t_0)$ *and* $\varphi_L(t; t_0)$ can be expressed as

$$\varphi_L(t; t_0) = -\varphi_{L_0} + L\frac{dq_L(t; t_0)}{dt} \tag{5.26}$$

that corresponds to the equivalent circuit shown in Fig. 5.8.[5]

Again, we stress that the equivalent circuit in the (φ, q)-domain has a flux source $\varphi_{L0} = i_{L0}/L$ depending on the initial condition $i_L(t_0) = i_{L_0}$ for the state variable $i_L(t)$ in the (v, i)-domain. According to (5.26), a natural choice for the *state variable* of an inductor in the (φ, q)-domain is the incremental charge $q_L(t; t_0)$.

5.5.2.3 Ideal Resistor

Let us consider an ideal resistor R described by Ohm's law

$$v_R(t) = Ri_R(t).$$

By integrating between t_0 and $t \geq t_0$, the CR in the (φ, q)-domain results to be

$$\varphi_R(t; t_0) = Rq_R(t; t_0) \tag{5.27}$$

[5]The circuit in Fig. 5.8 represents, in the (φ, q)-domain, the dual of the *initial inductor current transformation* circuit reported in Fig. 2.1 on page 307 in [8].

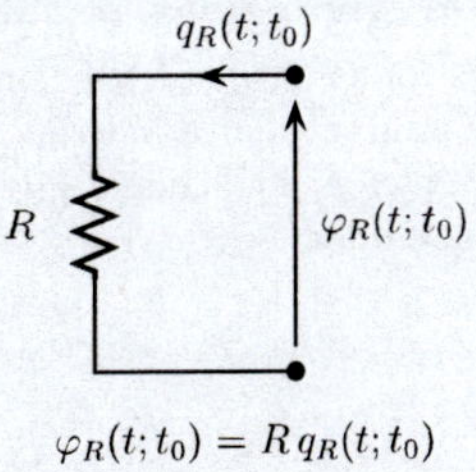

Fig. 5.9 Equivalent circuit for an ideal resistor in terms of the incremental charge $q_R(t; t_0)$ and flux $\varphi_R(t; t_0)$

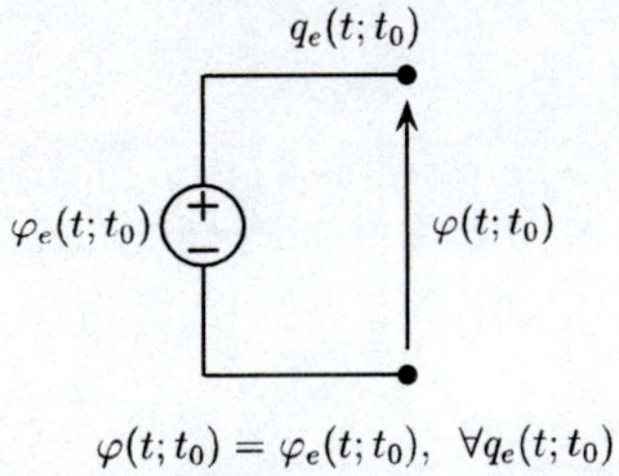

Fig. 5.10 Equivalent circuit for an ideal independent voltage source in terms of the incremental charge $q_e(t; t_0)$ and flux $\varphi_e(t; t_0)$

and the corresponding equivalent circuit is reported in Fig. 5.9. It turns out that the CR of R in the (φ, q)-domain is analogous to the usual Ohm's Law, i.e., the resistor is an adynamic element also in the (φ, q)-domain.

5.5.2.4 Ideal Independent Voltage Source

Let us consider an ideal independent voltage source

$$v(t) = e(t), \quad \forall i(t)$$

where $e(t)$ is a given function of time. By integrating between t_0 and $t \geq t_0$, the CR of the ideal independent voltage source in the (φ, q)-domain can be written as

$$\varphi(t; t_0) = \varphi_e(t; t_0) \doteq \int_{t_0}^{t} e(\tau)d\tau, \quad \forall q_e(t; t_0)$$

i.e., the incremental flux $\varphi_e(t; t_0)$ is a given function of time that is independent of the incremental charge at the source terminals. The corresponding equivalent circuit is shown in Fig. 5.10.

5.5.2.5 Ideal Independent Current Source

Let us consider an ideal independent current source

$$i(t) = a(t), \quad \forall v(t)$$

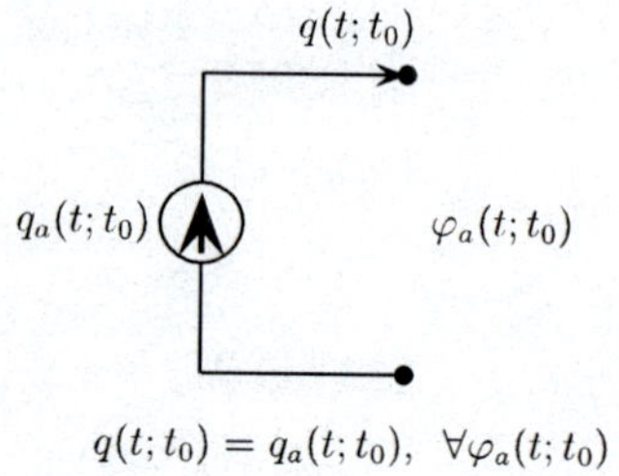

Fig. 5.11 Equivalent circuit for an ideal independent current source in terms of the incremental charge $q_a(t; t_0)$ and flux $\varphi_a(t; t_0)$

where $a(t)$ is a given function of time. By integrating between t_0 and $t \geq t_0$, the CR of the ideal independent current source in the (φ, q)-domain can be written as

$$q(t; t_0) = q_a(t; t_0) \doteq \int_{t_0}^{t} a(\tau) d\tau, \quad \forall \varphi_a(t; t_0)$$

i.e., the incremental charge $q_a(t; t_0)$ is a given function of time that is independent of the incremental voltage at the source terminals. The corresponding equivalent circuit is shown in Fig. 5.11.

5.5.2.6 Ideal Flux-Controlled Memristor

Let us consider an ideal flux-controlled memristor

$$q_M(t) = f(\varphi_M(t))$$

where $f(\cdot) : \mathbb{R} \to \mathbb{R}$ satisfies $f(0) = 0$, and let $\varphi_M(t_0) = \varphi_{M_0}$ be the initial flux of the memristor at t_0. It follows that the initial charge is $q_M(t_0) = q_{M_0} = f(\varphi_{M_0})$. By using the incremental charge and flux, we have for $t \geq t_0$

$$\begin{aligned} q_M(t; t_0) &= q_M(t) - q_M(t_0) \\ &= f(\varphi_M(t)) - f(\varphi_M(t_0)) \\ &= f(\varphi_M(t; t_0) + \varphi_M(t_0)) - f(\varphi_M(t_0)). \end{aligned}$$

As a consequence, the *CR of the ideal memristor in terms of the incremental variables* $q_M(t; t_0)$ *and* $\varphi_M(t; t_0)$ is

$$q_M(t; t_0) = f(\varphi_M(t; t_0) + \varphi_{M_0}) - f(\varphi_{M_0}). \tag{5.28}$$

Note that the memristor acts as a *nonlinear adynamic element in the* (φ, q)-*domain*. Then, we cannot associate a state variable in the (φ, q)-domain to the

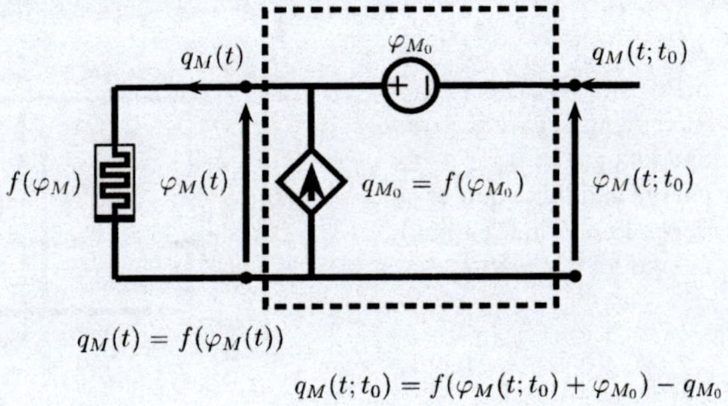

Fig. 5.12 Equivalent circuit for an ideal flux-controlled memristor in terms of the incremental charge $q_M(t; t_0)$ and flux $\varphi_M(t; t_0)$. The two charge and flux sources depend only on the initial flux $\varphi_M(t_0) = \varphi_{M_0}$

memristor. Its equivalent circuit is shown in Fig. 5.12. This includes a two-port network similar to that shown in Fig. 5.6.

It is important to stress that the equivalent circuit in the (φ, q)-domain has an independent flux source proportional to the initial condition φ_{M_0} for the memristor state variable $\varphi_M(t)$ in the (v, i)-domain. The flux-controlled charge-source $q_{M_0} = f(\varphi_{M_0})$ is also dependent on φ_{M_0}.

5.5.2.7 Ideal Charge-Controlled Memristor

Consider an ideal charge-controlled memristor

$$\varphi_M(t) = h(q_M(t))$$

where $h(\cdot) : \mathbb{R} \to \mathbb{R}$ satisfies $h(0) = 0$, and let $q_M(t_0) = q_{M_0}$ be the initial charge of the memristor at t_0. It follows that the initial flux is $\varphi_M(t_0) = \varphi_{M_0} = h(q_{M_0})$. By duality with respect to the flux-controlled memristor, the *CR of the ideal charge-controlled memristor in terms of the incremental variables $q_L(t; t_0)$ and $\varphi_L(t; t_0)$* is

$$\varphi_M(t; t_0) = h(q_M(t; t_0) + q_{M_0}) - h(q_{M_0}) \tag{5.29}$$

and the corresponding equivalent circuit is in Fig. 5.13. Again, a charge-controlled memristor is a nonlinear adynamic element in the (φ, q)-domain so that we cannot associate a state variable in the (φ, q)-domain with the memristor.

We stress once more that the equivalent circuit in the (φ, q)-domain has an independent charge source proportional to the initial condition q_{M_0} for the memristor state variable $q_M(t)$ in the (v, i)-domain. The charge-controlled flux-source $\varphi_{M_0} = h(q_{M_0})$ is also dependent on q_{M_0}.

Remark 5.6 Note that (5.29) can also be obtained from (5.28) under the assumption that function $f(\cdot)$ is globally invertible, i.e., the memristor is both flux- and charge-controlled.

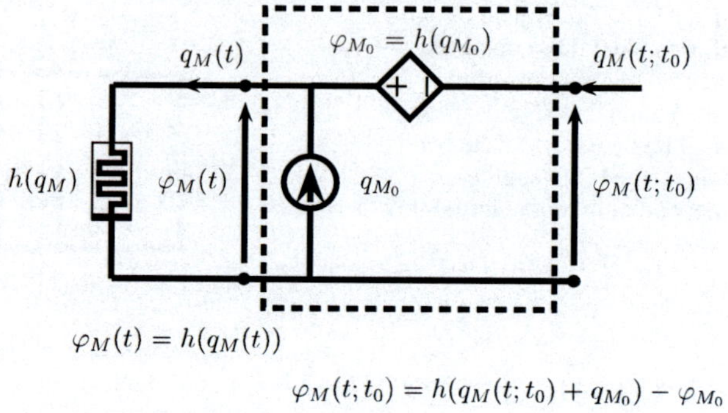

Fig. 5.13 Equivalent circuit for an ideal charge-controlled memristor in terms of the incremental charge $q_M(t; t_0)$ and flux $\varphi_M(t; t_0)$. The two charge and flux sources depend only on the initial charge $q_M(t_0) = q_{M_0}$

5.6 Extension of FCAM

FCAM as developed for the class $\mathcal{LM}$ of memristor circuits in Sect. 5.5 can be extended along several directions. In fact, FCAM can be immediately extended to include siblings of ideal flux- or charge-controlled memristors (cf. Example 5.16), since such siblings can be brought back to ideal memristors via a suitable globally 1:1 change of variables (Chap. 2). Nonlinear capacitors and inductors, as well as memcapacitors and meminductors, can also be included in the class of circuits $\mathcal{LM}$ and analyzed via FCAM. Such an extension will be treated in Chap. 11. If also nonlinear resistors are considered, then their piecewise-linear approximation allows us to exploit FCAM in each interval of linearity. Examples of this kind will be illustrated in Chap. 10.

In the next two sections we briefly address the extension of FCAM to include multiterminal and multiport elements and to deal with time varying-elements.

5.6.1 *Multiterminal and Multiport Elements*

FCAM admits an obvious extension to include linear multiport or multiterminal resistors, capacitors, and inductors, as illustrated with the next examples.

Example 5.8 A current-controlled voltage-source is a resistive two-port network satisfying in the (v, i)-domain

$$v_1(t) = 0$$
$$v_2(t) = ri_1(t)$$

where r is a control parameter with dimension of Ohm (Fig. 5.14). In the (φ, q)-domain we still have a resistive two-port networks described by

$$\varphi_1(t; t_0) = 0$$

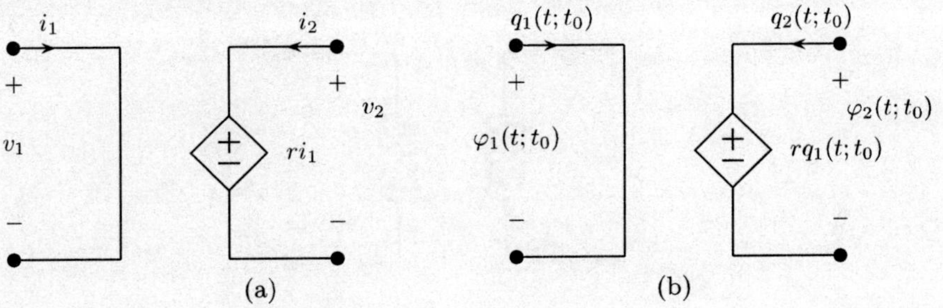

Fig. 5.14 (**a**) Current-controlled voltage-source in the (v, i)-domain and (**b**) equivalent circuit in the (φ, q)-domain

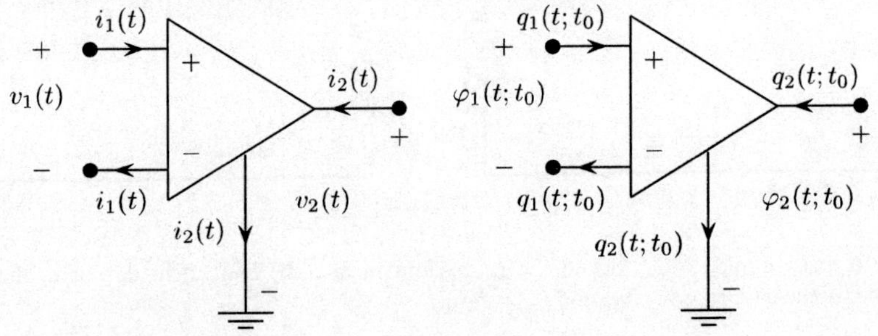

Fig. 5.15 (**a**) Ideal operational amplifier operating in the linear region in the (v, i)-domain and (**b**) equivalent circuit in the (φ, q)-domain

$$\varphi_2(t; t_0) = rq_1(t; t_0)$$

for any $t \geq t_0$.

Example 5.9 An ideal operational amplifier operating in the linear region is a resistive two-port networks satisfying in the (v, i)-domain (Fig. 5.15)

$$v_1(t) = 0$$
$$i_1(t) = 0.$$

In the (φ, q)-domain we still have a resistive two-port networks described by

$$\varphi_1(t; t_0) = 0$$
$$q_1(t; t_0) = 0$$

for any $t \geq t_0$.

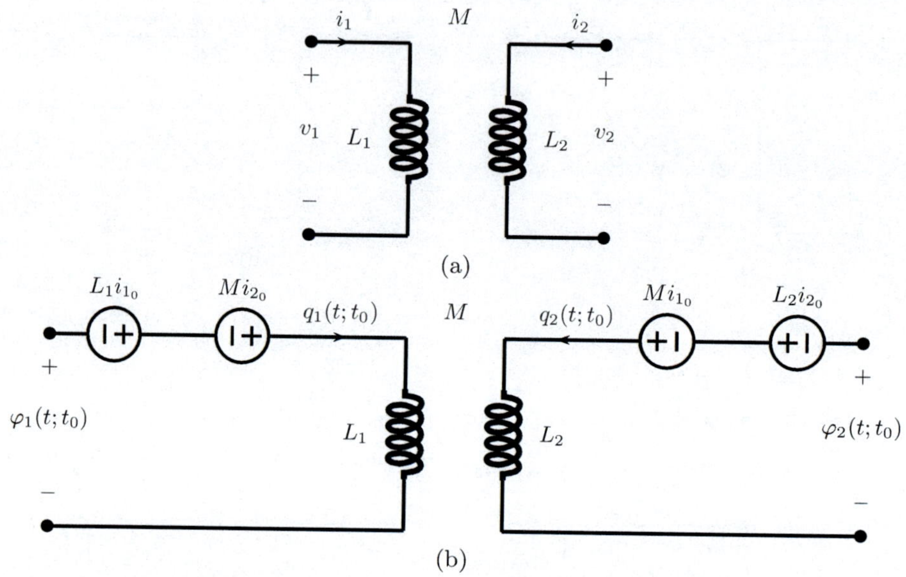

Fig. 5.16 (**a**) Coupled inductors in the (v, i)-domain and (**b**) equivalent circuit in the (φ, q)-domain. We have let $i_{1_0} = i_1(t_0)$ and $i_{2_0} = i_2(t_0)$

Example 5.10 In the (v, i)-domain, two coupled inductors satisfy (Fig. 5.16)

$$v_1 = L_1 \frac{di_1}{dt} + M \frac{di_2}{dt}$$

$$v_2 = M \frac{di_1}{dt} + L_2 \frac{di_2}{dt}$$

where M is the mutual inductance.

The CR in the (φ, q)-domain is easily obtained via integration as

$$\varphi_1(t; t_0) = L_1 \frac{dq_1(t; t_0)}{dt} + M \frac{dq_2(t; t_0)}{dt} - L_1 i_1(t_0) - M i_2(t_0)$$

$$\varphi_2(t; t_0) = M \frac{dq_1(t; t_0)}{dt} + L_2 \frac{dq_2(t; t_0)}{dt} - M i_1(t_0) - L_2 i_2(t_0)$$

for $t \geq t_0$.

The equivalent circuit in the (φ, q)-domain is depicted in Fig. 5.16.

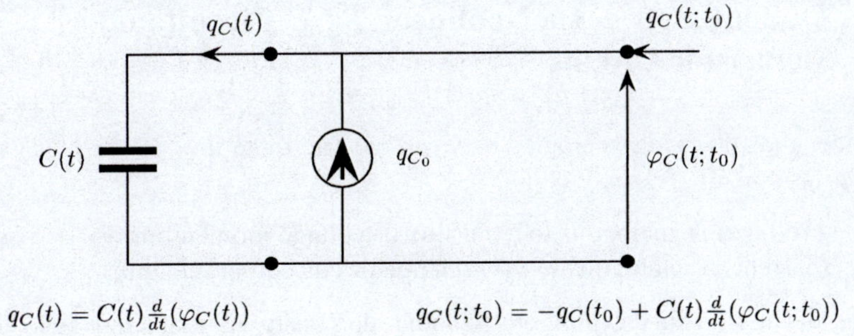

Fig. 5.17 Equivalent circuit of a time-varying capacitor in the (φ, q)-domain

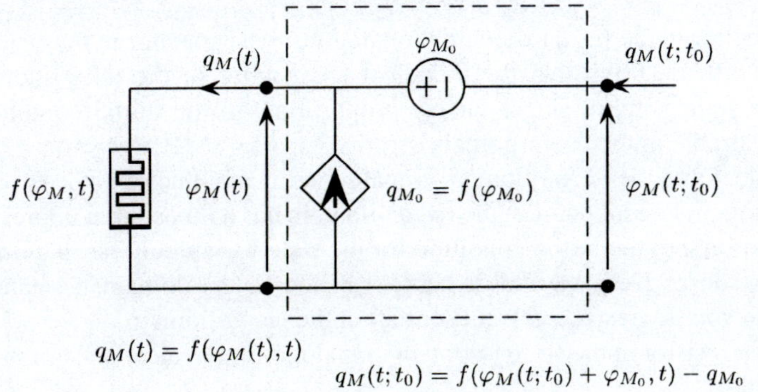

Fig. 5.18 Equivalent circuit of a time-varying flux-controlled memristor in the (φ, q)-domain

5.6.2 *Time-Varying Elements*

So far we have assumed that any element in $\mathcal{LM}$, except possibly the independent sources, is time-invariant. It is not difficult, however, to extend FCAM to the case where any element in $\mathcal{LM}$ may be time varying.

Example 5.11 Consider a time-varying capacitor defined by $q_C(t) = C(t)v_C(t)$ [8]. In the (φ, q)-domain its CR is given by

$$q_C(t;\, t_0) = C(t)\frac{d\varphi_C(t;\, t_0)}{dt} - q_{C0}$$

where $q_{C0} = C(t_0)v_C(t_0)$. The corresponding equivalent circuit is shown in Fig. 5.17.

Example 5.12 Consider a time-varying flux-controlled memristor $q_M(t) = f_M(\varphi_M(t), t)$. The CR in the (φ, q)-domain is obtained as

$$q_M(t;\, t_0) = f_M(\varphi_M(t;\, t_0) + \varphi_M(t_0), t) - f_M(\varphi_M(t_0), t_0)$$

and the equivalent circuit in the (φ, q)-domain is in Fig. 5.18.

5.7 Analogy Between a Nonlinear *RLC* Circuit and a Memristor Circuit

Consider the following analogies between electric quantities in the (v, i)-domain and (φ, q)-domain, respectively:

- $v(t)$ (voltage) is analogous to $\varphi(t)$ (flux or voltage momentum)
- $i(t)$ (current) is analogous to $q(t)$ (charge or current momentum)

The discussion in the previous sections demonstrates that there hold the following analogies between two-terminal elements in the (v, i)-domain and (φ, q)-domain.

- A capacitor in the (φ, q)-domain is analogous to a capacitor in the (v, i)-domain. It is worth recalling that the model of a capacitor in the (φ, q)-domain has a charge source impressing a charge proportional to the initial condition for the state variable in the (v, i)-domain.
- An inductor in the (φ, q)-domain is analogous to an inductor in the (v, i)-domain. The model of an inductor in the (φ, q)-domain has a flux source impressing a flux proportional to the initial condition for the state variable in the (v, i)-domain.
- A flux-source (resp., a charge source) in the (φ, q)-domain is analogous to a voltage-source (resp., a current source) in the (v, i)-domain.
- A linear resistor in the (φ, q)-domain is analogous to a linear resistor in the (v, i)-domain.

Consider a flux-controlled memristor. The CR in the (φ, q)-domain is given by an algebraic relationship between the electric quantities at its terminals $\varphi_M(t; t_0)$ and $q_M(t; t_0)$, i.e.,

$$q_M(t; t_0) = f(\varphi_M(t; t_0) + \varphi_{M_0}) - f(\varphi_{M_0}) \doteq \tilde{f}(\varphi_M(t; t_0); \varphi_{M_0}).$$

This shows that

- for any fixed φ_{M_0}, a flux-controlled memristor is analogous in the (φ, q)-domain to a voltage-controlled nonlinear resistor in the (v, i)-domain, i.e., an element defined by an algebraic relationship between voltage and current at its terminals.

It is important to stress that, different from a resistor, a flux-controlled memristor holds memory of the past history of its voltage through the source $\varphi_{M_0} = \int_{-\infty}^{t_0} v_M(\tau) d\tau$ associated with its initial condition (see Fig. 5.12). To further highlight this difference, we depict in Fig. 5.19 the nonlinear characteristic of the memristor $q_M(t; t_0) = f(\varphi_M(t; t_0)) = \tilde{f}(\varphi_M(t; t_0); 0)$ with $\varphi_{M_0} = 0$, and the characteristic $q_M(t; t_0) = \tilde{f}(\varphi_M(t; t_0); \varphi_{M_0})$ of the memristor charged at an initial flux $\varphi_{M_0} \neq 0$. Note that the latter is obtained by translating to the point $(\varphi_{M_0}, f(\varphi_{M_0}))$ the former characteristic. It is clear that the initial condition has a relevant influence on the shape of the nonlinear characteristic and behavior of the memristor within a circuit.

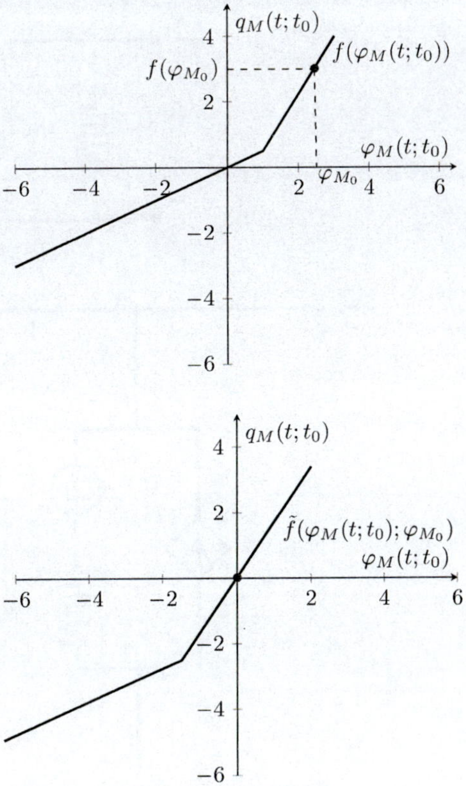

Fig. 5.19 Characteristic of a discharged memristor (upper plot) and characteristic of the same memristor charged at $\varphi_{M_0} \neq 0$ (lower plot)

Dual considerations hold for a charge-controlled memristor. Namely,

- **for any fixed q_{M_0} a flux-controlled memristor is analogous in the (φ, q)-domain** to a charge-controlled nonlinear resistor in the (v, i)-domain.

In conclusion, there is an analogy between a memristor circuit in the class $\mathcal{LM}$ in the (φ, q)-domain and a nonlinear RLC circuit in the (v, i)-domain. This simple yet fundamental observation shows that it is possible to use any method to analyze nonlinear circuits in the (v, i)-domain (Chap. 3) also to analyze memristor circuits in the class $\mathcal{LM}$, provided we take into account the analogies previously discussed. An important issue to consider is however that the memristor keeps memory of its history also in the (φ, q)-domain via the source depending on the initial condition for the memristor state variable in the (v, i)-domain.

As a first elementary application of this analogy, we study in the (φ, q)- domain some basic two-terminal elements obtained by interconnecting memristors with other elements. Then, in Sect. 5.8, we develop some general methods for writing circuit equations in the (φ, q)-domain for a circuit in $\mathcal{LM}$ starting from the corresponding methods for nonlinear RLC circuits in the (v, i)-domain (cf. Chap. 3).

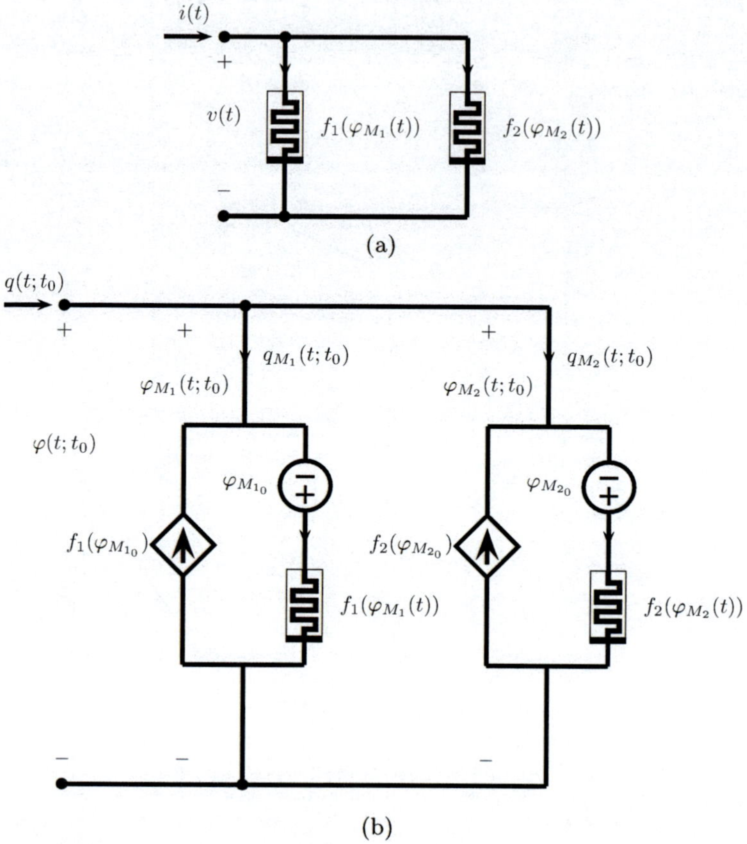

Fig. 5.20 **(a)** Parallel connection of two flux-controlled memristors and **(b)** equivalent circuit in the (φ, q)-domain

Example 5.13 (Parallel Connection of Flux-Controlled Memristors) Consider as in Fig. 5.20a the parallel connection of two flux-controlled memristors with CRs

$$q_{M_i}(t;t_0) = f_i(\varphi_{M_i}(t;t_0) + \varphi_{M_{i0}}) - f_i(\varphi_{M_{i0}}) \doteq \tilde{f}_i(\varphi_{M_i}(t;t_0);\varphi_{M_{i0}})$$

where $\varphi_{M_{i0}} = \varphi_{M_i}(t_0)$ and $i = 1, 2$.

We have from KqL

$$q(t;t_0) = q_{M_1}(t;t_0) + q_{M_2}(t;t_0)$$

while KφL yields

$$\varphi(t;t_0) = \varphi_{M_1}(t;t_0) = \varphi_{M_2}(t;t_0).$$

By substitution we obtain

$$\begin{aligned} q(t; t_0) = &-f_1(\varphi_{M_{1_0}}) - f_2(\varphi_{M_{2_0}}) \\ &+ f_1(\varphi(t; t_0) + \varphi_{M_{1_0}}) + f_2(\varphi(t; t_0) + \varphi_{M_{2_0}}). \end{aligned}$$

The parallel is thus equivalent to a flux-controlled memristor with a CR

$$q(t; t_0) = \tilde{f}(\varphi(t; t_0); \varphi_{M_{1_0}}, \varphi_{M_{2_0}})$$

where

$$\tilde{f}(\varphi(t; t_0); \varphi_{M_{1_0}}, \varphi_{M_{2_0}}) \doteq \tilde{f}_1(\varphi(t; t_0); \varphi_{M_{1_0}}) + \tilde{f}_2(\varphi(t; t_0); \varphi_{M_{2_0}}).$$

The result is analogous to the parallel connection of two nonlinear voltage-controlled resistors in the (v, i)-domain.

Example 5.14 (Parallel Connection of a Flux-Controlled Memristor and a Capacitor) Consider as in Fig. 5.21 the parallel connection of a flux-controlled memristor and a capacitor. The CRs are

$$q_M(t; t_0) = f(\varphi_M(t; t_0) + \varphi_{M_0}) - f(\varphi_{M_0})$$

where $\varphi_{M_0} = \varphi_M(t_0)$ and

$$q_C(t; t_0) = C\frac{d\varphi_C(t; t_0)}{dt} - q_{C_0}$$

where $q_{C_0} = Cv_C(t_0)$.

We have from KqL

$$q(t; t_0) = q_M(t; t_0) + q_C(t; t_0)$$

while KφL yields

$$\varphi(t; t_0) = \varphi_M(t; t_0) = \varphi_C(t; t_0).$$

By substitution we obtain the CR

$$q(t; t_0) = -f(\varphi_{M_0}) - q_{C_0} + f(\varphi(t; t_0) + \varphi_{M_0}) + C\frac{d\varphi(t; t_0)}{dt}.$$

Note that this parallel is a dynamic first-order flux-controlled two-terminal element that is analogous to a dynamic first-order voltage-controlled two-terminal element given by the parallel of a nonlinear resistor and a capacitor in the (v, i)-domain.

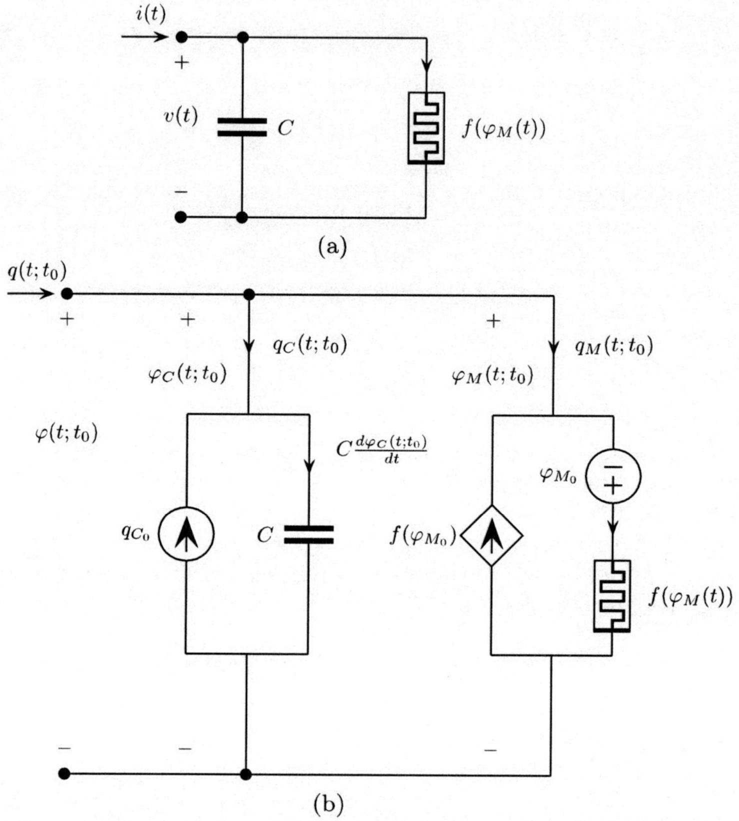

Fig. 5.21 (**a**) Parallel connection of a flux-controlled memristor and a capacitor and (**b**) equivalent circuit in the (φ, q)-domain

5.8 Formulation of Memristor Circuits Equations

In the first part of this section we derive the dynamic equations describing a circuit in $\mathcal{LM}$ in the (φ, q)-domain by exploiting: (a) the Kirchhoff laws in terms of the incremental charges and fluxes, i.e., KqL and KφL; (b) the CR in the (φ, q)-domain of each element in the class $\mathcal{LM}$. The dynamic equations are first formulated in the form of DAEs and then, under suitable additional assumptions, in the form of SEs. Memristor circuits in $\mathcal{LM}$ can be analyzed also in the traditional (v, i)-domain. In the second part of the section, we write the DAEs and SEs in the (v, i)-domain and compare the order of the SE representations in the two domains. We also discuss by means of specific examples if it is possible to pass from the SEs in a domain to the SEs in the other domain via an integration or differentiation in time. Without losing any generality, in the remaining part of this chapter, unless stated otherwise, only flux-controlled memristors are considered. All the results can be easily extended to circuits including either charge-controlled memristors or flux- and charge-controlled memristors, as illustrated via examples.

5.8.1 Differential Algebraic Equations in the Flux-Charge Domain

Following the classical tableau analysis approach (cf. Chap. 3), the set of b linear algebraic equations due to the topological constraints can be obtained by putting together the $n-1$ KqL-equations in (5.21) and the $b-n+1$ KφL-equations in (5.22), that is

$$\begin{cases} \mathbf{A}\mathbf{q}(t;t_0) = \mathbf{0} \\ \mathbf{B}\boldsymbol{\varphi}(t;t_0) = \mathbf{0}. \end{cases} \tag{5.30}$$

The collection of all CRs, one for each of the b two-terminal elements in a circuit in $\mathcal{LM}$, provides a set of algebraic and differential equations that can be written in terms of the incremental charge and flux as follows:

$$\begin{cases} \varphi_{R_s}(t;t_0) = R_s q_{R_s}(t;t_0), \ (s = 1,\ldots,n_R) \\ \varphi_w(t;t_0) = \varphi_{e_w}(t;t_0), \ \ \forall q_w(t;t_0), \ (w = 1,\ldots,n_E) \\ q_z(t;t_0) = q_{a_z}(t;t_0), \ \ \forall \varphi_z(t;t_0), \ (z = 1,\ldots,n_A) \end{cases} \tag{5.31}$$

$$\begin{cases} C_j \frac{d\varphi_{C_j}(t;t_0)}{dt} = q_{C_j}(t;t_0) + q_{C_{j0}}, \ (j = 1,\ldots,n_C) \\ L_m \frac{dq_{L_m}(t;t_0)}{dt} = \varphi_{L_m}(t;t_0) + \varphi_{L_{m0}}, \ (m = 1,\ldots,n_L) \end{cases} \tag{5.32}$$

$$q_{M_k}(t;t_0) = f_{M_k}(\varphi_{M_k}(t;t_0) + \varphi_{M_{k0}}) - f_{M_k}(\varphi_{M_{k0}}), \ (p = 1,\ldots,n_M) \tag{5.33}$$

where n_R is the number of resistors, n_E the number of independent voltage sources, n_A the number of independent current sources, n_C the number of capacitors, n_L the number of inductors, and n_M the number of flux-controlled memristors.

Equations (5.30)–(5.33) provide a system of $2b$ DAEs involving only the incremental variables $\mathbf{q}(t;t_0)$ and $\boldsymbol{\varphi}(t;t_0)$. The solution of DAEs (5.30)–(5.33), with the initial conditions

$$\mathbf{q}(t_0;t_0) = \mathbf{0}$$

and

$$\boldsymbol{\varphi}(t_0; t_0) = \mathbf{0}$$

gives the evolution of $\mathbf{q}(t; t_0)$ and $\boldsymbol{\varphi}(t; t_0)$ for $t \geq t_0$.

Remark 5.7 Although the DAEs (5.30)–(5.33) involve the incremental variables in the (φ, q)-domain, their solution depends on $\mathbf{q}_{C0}$, $\boldsymbol{\varphi}_{L0}$ and $\boldsymbol{\varphi}_{M0}$,[6] that is, the initial conditions at t_0 for the state variables in the (v, i)-domain. Indeed, in the obtained formulation, *such initial conditions appear as constant inputs in the r.h.s. of* (5.32) *and* (5.33).

5.8.2 State Equations in the Flux-Charge Domain

Circuit analysis methods based on DAEs are fundamental in numerical simulation (using, e.g., PSpice software [9]) of linear and nonlinear circuits [10]. On the other hand, *qualitative aspects of the nonlinear dynamics* (e.g., the existence of no-finite-forward-escape-time solutions, eventual boundedness of solutions, local and global asymptotic stability properties, bifurcation phenomena, etc.) are more effectively analyzed by means of the SE formulation. Moreover, as pointed out in Chap. 3, circuits that do not admit an SE description may be bad modeled from a physical viewpoint due to the presence of impasse points.

The SE description in the (φ, q)-domain of a memristor circuit in $\mathcal{LM}$, admitting it exists, can be obtained by inspection of the equivalent circuit in the (φ, q)-domain for simple low-order circuits. However, for more complex circuits, it is desirable to develop a systematic procedure for writing the SEs and to give easily checkable conditions guaranteeing that the SE representation exists.

In the following, we discuss a systematic approach for obtaining the SE description of a circuit in $\mathcal{LM}$ in the (φ, q)-domain. The formal derivation of an SE formulation of nonlinear RLC circuits in the (v, i)-domain, and the conditions for the *existence of such an SE formulation*, have been provided in Sect. 3.3.2.3 of Chap. 3. As previously observed, a memristor circuit in $\mathcal{LM}$, when described in the (φ, q)-domain, is *analogous to a nonlinear RLC circuit in the (v, i)-domain*. On the basis of this observation, we are in a position to derive the SE description in the (φ, q)-domain by exploiting the results in Chap. 3. It is worth to observe that the formulation here proposed is based on using the vector of capacitor incremental fluxes $\boldsymbol{\varphi}_C(t; t_0)$, and the vector of incremental inductor charges $\mathbf{q}_L(t; t_0)$, as state variables in the (φ, q)-domain. This is consistent with the fact that while capacitors and inductors are memory elements in the (φ, q)-domain, a memristor is instead an adynamic element in the (φ, q)-domain, hence no state variable is needed for a memristor in such domain.

[6]The DAEs (5.30)–(5.33) depend also on $\mathbf{q}_{M_{k0}}$ if charge-controlled memristors are included in the class $\mathcal{LM}$.

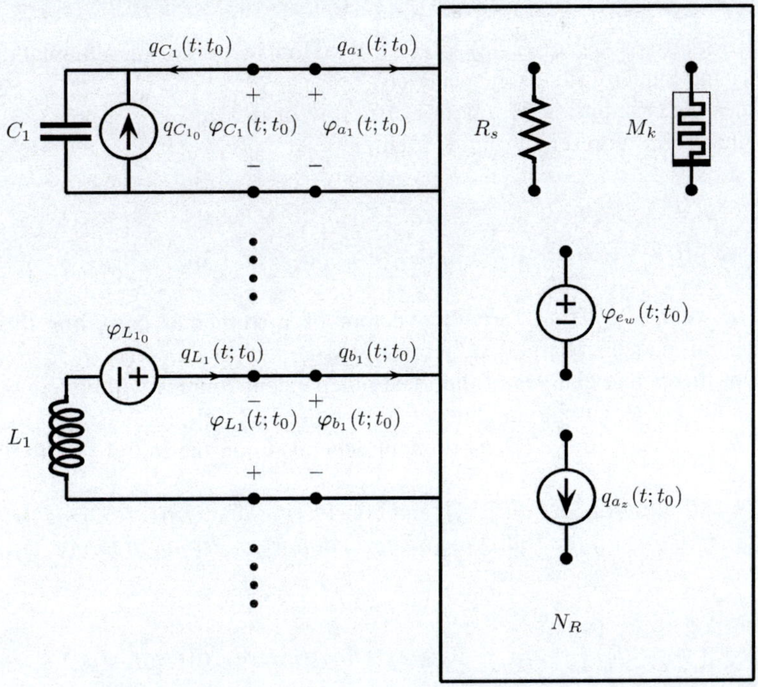

Fig. 5.22 **Circuit representation to derive the SEs in terms of the incremental fluxes $\boldsymbol{\varphi}_C(t;t_0)$ and charges $\mathbf{q}_L(t;t_0)$ in the (φ, q)-domain**

Any circuit in $\mathcal{LM}$ can be represented as shown in Fig. 5.22, where all capacitors and inductors are connected to a resistive nonlinear $(n_C + n_L)$-port N_R including ideal resistors, memristors, and ideal independent flux and charge sources.[7] Each capacitor and inductor is represented by the equivalent circuit shown in Figs. 5.7 and 5.8, respectively.

The CRs (5.32) describing the n_C capacitors and n_L inductors can be written in matrix form as follows:

$$\begin{cases} \mathbf{C}\frac{d}{dt}\Big(\boldsymbol{\varphi}_C(t;t_0)\Big) = \mathbf{q}_C(t;t_0) + \mathbf{q}_{C0} \\ \mathbf{L}\frac{d}{dt}\Big(\mathbf{q}_L(t;t_0)\Big) = \boldsymbol{\varphi}_L(t;t_0) + \boldsymbol{\varphi}_{L0} \end{cases} \tag{5.34}$$

[7] Multiport N_R is resistive because, as already pointed out, the memristor is described by an algebraic CR in the (φ, q)-domain.

where $\mathbf{C} = \text{diag}(C_1, \ldots, C_{n_C})$ and $\mathbf{L} = \text{diag}(L_1, \ldots, L_{n_L})$ are nonsingular diagonal matrices and $\mathbf{q}_C(t; t_0)$, $\boldsymbol{\varphi}_L(t; t_0)$ are the vectors of incremental capacitor charges and inductor fluxes, respectively.

Under the assumption of unique solvability of the $(n_C + n_L)$-port N_R, we can write the hybrid representation (cf. Chap. 3)

$$\begin{cases} -\mathbf{q}_C(t; t_0) = \mathbf{q}_a(t; t_0) = \mathbf{h}_a(\boldsymbol{\varphi}_C(t; t_0), \mathbf{q}_L(t; t_0), \mathbf{u}(t; t_0), \boldsymbol{\varphi}_{M_0}) \\ -\boldsymbol{\varphi}_L(t; t_0) = \boldsymbol{\varphi}_b(t; t_0) = \mathbf{h}_b(\boldsymbol{\varphi}_C(t; t_0), \mathbf{q}_L(t; t_0), \mathbf{u}(t; t_0), \boldsymbol{\varphi}_{M_0}) \end{cases} \tag{5.35}$$

where $\boldsymbol{\varphi}_C(t; t_0)$, $\mathbf{q}_L(t; t_0)$ are the vectors of incremental capacitor fluxes and incremental inductor charges, respectively. Moreover, $\mathbf{u}(t) = (\boldsymbol{\varphi}_E(t), \mathbf{q}_A(t))$ is the vector of fluxes and charges of the ideal independent sources within N_R. Note that $\mathbf{h}_a(\cdot)$ and $\mathbf{h}_b(\cdot)$ depend on the independent sources $\boldsymbol{\varphi}_E$, $\mathbf{q}_A$, and also $\boldsymbol{\varphi}_{M_0}$, since the CR of a memristor in the (φ, q)-domain depends upon the initial flux in the same memristor.[8]

By substituting (5.35) in (5.34), and considering that $\boldsymbol{\varphi}_C(t; t_0) = \boldsymbol{\varphi}_a(t; t_0)$ and $\mathbf{q}_L(t; t_0) = \mathbf{q}_b(t; t_0)$, the following *SE formulation for $t \geq t_0$ in the (φ, q)-domain for a circuit in $\mathcal{LM}$* is obtained:

$$\begin{cases} \mathbf{C}\frac{d}{dt}\Big(\boldsymbol{\varphi}_C(t; t_0)\Big) = \mathbf{q}_{C_0} - \mathbf{h}_a(\boldsymbol{\varphi}_C(t; t_0), \mathbf{q}_L(t; t_0), \mathbf{u}(t; t_0), \boldsymbol{\varphi}_{M_0}) \\ \mathbf{L}\frac{d}{dt}\Big(\mathbf{q}_L(t; t_0)\Big) = \boldsymbol{\varphi}_{L_0} - \mathbf{h}_b(\boldsymbol{\varphi}_C(t; t_0), \mathbf{q}_L(t; t_0), \mathbf{u}(t; t_0), \boldsymbol{\varphi}_{M_0}) \\ \boldsymbol{\varphi}_C(t_0; t_0) = \mathbf{0} \\ \mathbf{q}_L(t_0; t_0) = \mathbf{0}. \end{cases} \tag{5.36}$$

The structure of the SEs (5.36) suggests the following observations:

- if there are n_C capacitors and n_L inductors, the state variables of the SE representation in the (φ, q)-domain are the $n_C + n_L$ incremental variables $\boldsymbol{\varphi}_C(t; t_0)$ and $\mathbf{q}_L(t; t_0)$;
- the *order of the SE representation in the (φ, q)-domain is $n_C + n_L$*;
- the evolution of the incremental variables $\boldsymbol{\varphi}_C(t; t_0)$ and $\mathbf{q}_L(t; t_0)$ is influenced via $\mathbf{q}_{C_0}$ and $\boldsymbol{\varphi}_{L_0}$ by the initial conditions for the state variables in the (v, i)-domain and, for n_M memristors, the n_M memristors initial conditions $\boldsymbol{\varphi}_{M_0}$ as well. *The initial conditions $\mathbf{q}_{C_0}$, $\boldsymbol{\varphi}_{L_0}$ and $\boldsymbol{\varphi}_{M_0}$ indeed appear as constant inputs in the r.h.s. of (5.36);*
- the IVP for the SEs (5.36) has, by definition, zero initial conditions for the incremental variables $\boldsymbol{\varphi}_C(t; t_0)$ and $\mathbf{q}_L(t; t_0)$;

[8]Functions $\mathbf{h}_a(\cdot)$ and $\mathbf{h}_b(\cdot)$ depend also on $\mathbf{q}_{M_0}$ if charge-controlled memristors are included in $\mathcal{LM}$.

- if needed, to obtain the SEs in normal form, it is enough to multiply both sides of the first two equations (5.36) by $\mathbf{C}^{-1}$ and $\mathbf{L}^{-1}$, respectively.

Remark 5.8 Criteria ensuring the existence of the SEs (5.36) and the hybrid representation (5.35) of the adynamic (n_C+n_L)-port can be derived via the approach presented in Chap. 3.

Taking into account the analogies in Sect. 5.7, from Property 3.1 in Sect. 3.3.2.3 of Chap. 3, it is not difficult to see that a set of sufficient conditions for the existence of the SEs in the (φ, q)-domain is as follows:

1. There is no loop formed exclusively by capacitors, inductors, and/or independent flux sources. Furthermore, there is no cut-set formed exclusively by capacitors, inductors, and/or independent charge sources.
2. Each flux-controlled (but not charge-controlled) memristor is in parallel with a capacitor. Each charge-controlled (but not flux-controlled) memristor is in series with an inductor.
3. Each remaining memristor is strongly locally passive, or else it is either in parallel with a capacitor or in series with an inductor.

A verification is left to the reader. Here we have considered also the possible presence of charge-controlled memristors. A flux-controlled memristor $q_M = f(\varphi_M)$ is said to be *strongly passive* if there exist constants $0 < \gamma' < \gamma''$ such that $\gamma' \leq (f(\varphi_{M,1}) - f(\varphi_{M,2}))/(\varphi_{M,1} - \varphi_{M,2}) \leq \gamma''$ for any $\varphi_{M,1} \neq \varphi_{M,2}$. A similar definition holds for a charge-controlled memristor.

5.8.3 Differential Algebraic Equations in the Voltage-Current Domain

Given a circuit in the class $\mathcal{LM}$, the common formulation of DAEs in the (v, i)-domain can be readily derived either by differentiating the DAEs in the (φ, q)-domain, or otherwise by using KCLs, KVLs, and CRs in terms of current and voltage. The former approach is briefly discussed in this section, whereas the latter is widely reported in the literature (see for instance [11]). It turns out that both methods provide the same circuit equations.

Since

$$\frac{d\,\mathbf{q}(t;t_0)}{dt} = \frac{d\,\mathbf{q}(t)}{dt} = \mathbf{i}(t)$$

$$\frac{d\,\boldsymbol{\varphi}(t;t_0)}{dt} = \frac{d\,\boldsymbol{\varphi}(t)}{dt} = \mathbf{v}(t)$$

by differentiating (5.30)–(5.32), the "usual" KCLs and KVLs

$$\begin{cases} \mathbf{Ai}(t) = \mathbf{0} \\ \mathbf{Bv}(t) = \mathbf{0} \end{cases} \tag{5.37}$$

are obtained. Moreover, the CRs result to be

$$\begin{cases} v_{R_s}(t) = R_s\, i_{R_s}(t), \ (s = 1, \dots, n_R) \\ v_w(t) = e_w(t), \ \forall i_w(t), \ (w = 1, \dots, n_E) \\ i_z(t) = a_z(t), \ \forall v_z(t), \ (z = 1, \dots, n_A) \end{cases} \tag{5.38}$$

$$\begin{cases} C_j \dfrac{d\, v_{C_j}(t)}{dt} = i_{C_j}(t), \ (j = 1, \dots, n_C) \\ L_m \dfrac{d\, i_{L_m}(t)}{dt} = v_{L_m}(t), \ (m = 1, \dots, n_L) \\ q_{C_j}(t_0) = q_{C_{j0}} \Rightarrow v_{C_j}(t_0) = v_{C_{j0}} \\ \varphi_{L_m}(t_0) = \varphi_{L_{m0}} \Rightarrow i_{L_m}(t_0) = i_{L_{m0}}. \end{cases} \tag{5.39}$$

In addition, the following CRs of memristors in the (v, i)-domain are derived by differentiating (5.33)

$$\begin{cases} i_{M_k}(t) = G_k(\varphi_{M_k}(t)) v_{M_k}(t), \ (p = 1, \dots, n_M) \\ \dfrac{d\, \varphi_{M_k}(t)}{dt} = v_{M_k}(t) \\ \varphi_{M_k}(t_0) = \varphi_{M_{k0}}. \end{cases} \tag{5.40}$$

Once the initial conditions $\mathbf{v}_C(t_0)$, $\mathbf{i}_L(t_0)$, and $\boldsymbol{\varphi}_M(t_0)$ for the state variables in the (v, i)-domain are specified, (5.37)–(5.40) provide a system of $2b$ DAEs governing the evolution for $t \geq t_0$ of the current $\mathbf{i}(t)$ and voltage $\mathbf{v}(t)$ variables.

5.8.4 *State Equations in the Voltage-Current Domain*

To write the SEs in the (v, i)-domain of a circuit in $\mathcal{LM}$ we consider a variant of a systematic procedure proposed in [12]. The technique is sketched in the following and then further illustrated by means of some specific examples.

The state variables in the (v, i)-domain are the n_C capacitor voltages $\mathbf{v}_C$, the n_L inductor currents $\mathbf{i}_L$ and the n_M fluxes of flux-controlled memristors $\boldsymbol{\varphi}_M$.[9] First note that the CRs of circuit elements in the (v, i)-domain can be written in vector form as

$$\begin{cases} \mathbf{C}\frac{d\,\mathbf{v}_C(t)}{dt} = \mathbf{i}_C(t) \\ \mathbf{L}\frac{d\,\mathbf{i}_L(t)}{dt} = \mathbf{v}_L(t) \\ \mathbf{v}_C(t_0) = \mathbf{v}_{C0} \\ \mathbf{i}_L(t_0) = \mathbf{i}_{L0} \end{cases} \tag{5.41}$$

for capacitors and inductors and

$$\begin{cases} \mathbf{i}_M(t) = \mathbf{G}(\boldsymbol{\varphi}_M(t))\mathbf{v}_M(t) \\ \frac{d\,\boldsymbol{\varphi}_M(t)}{dt} = \mathbf{v}_M(t) \\ \boldsymbol{\varphi}_M(t_0) = \boldsymbol{\varphi}_{M0} \end{cases} \tag{5.42}$$

for flux-controlled memristors, where

$$\mathbf{G}(\boldsymbol{\varphi}_M(t)) = \mathrm{diag}(G_1(\varphi_{M_1}(t)), G_2(\varphi_{M_2}(t)), \ldots, G_{n_M}(\varphi_{M_{n_M}}(t))).$$

Any circuit in $\mathcal{LM}$ can be represented as shown in Fig. 5.23, where all capacitors and inductors are connected to a nonlinear $(n_C + n_L)$-port including ideal resistors, memristors, and ideal independent voltage and current sources. Note that, for any given value of the flux φ_{M_k}, each memristor $q_{M_k} = f_k(\varphi_{M_k})$ can be replaced by a linear conductance $G_k(\varphi_{M_k}) = f_k'(\varphi_{M_k})$. For each fixed $\boldsymbol{\varphi}_M$ the $(n_C + n_L)$-port is thus a linear adynamic network.

By following a procedure analogous to that in Chap. 3, suppose to replace each capacitor by a voltage source and each inductor by a current source. If a unique solvability assumption is satisfied *for any* $\boldsymbol{\varphi}_M$ by the linear adynamic circuit thus obtained, then we can write

$$\begin{cases} -\mathbf{i}_C(t) = \mathbf{H}_a(\mathbf{v}_C(t), \mathbf{i}_L(t), \boldsymbol{\varphi}_M(t), \mathbf{e}(t), \mathbf{a}(t)) \\ -\mathbf{v}_L(t) = \mathbf{H}_b(\mathbf{v}_C(t), \mathbf{i}_L(t), \boldsymbol{\varphi}_M(t), \mathbf{e}(t), \mathbf{a}(t)). \end{cases} \tag{5.43}$$

Solving for $\mathbf{v}_M$ the linear adynamic network we also obtain

$$\mathbf{v}_M(t) = \mathbf{H}_c(\mathbf{v}_C(t), \mathbf{i}_L(t), \boldsymbol{\varphi}_M(t), \mathbf{e}(t), \mathbf{a}(t)). \tag{5.44}$$

[9] There are also the memristor charges $\mathbf{q}_M$ for charge-controlled memristors.

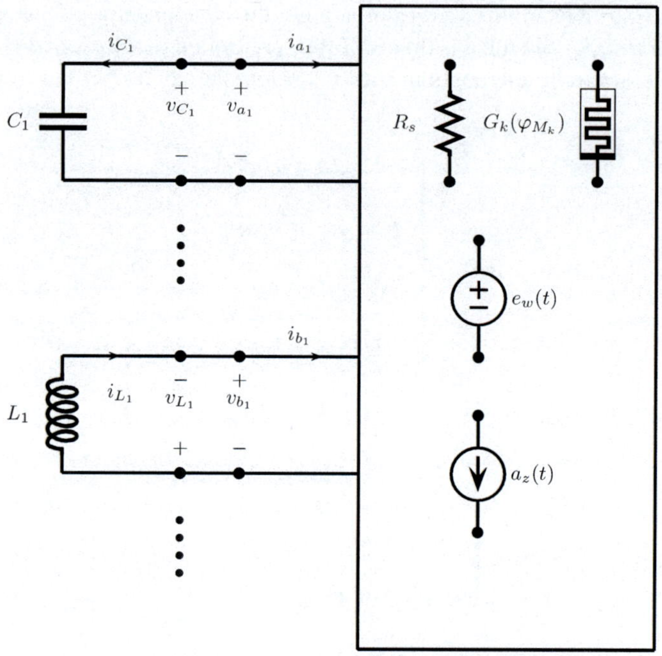

Fig. 5.23 Memristor circuit representation to derive the SEs in the (v, i)-domain

The SE formulation in the (v, i)-domain is thus derived as

$$\begin{cases} \mathbf{C}\frac{d\,\mathbf{v}_C(t)}{dt} = -\mathbf{H}_a(\mathbf{v}_C(t), \mathbf{i}_L(t), \boldsymbol{\varphi}_M(t), \mathbf{e}(t), \mathbf{a}(t)) \\ \mathbf{L}\frac{d\,\mathbf{i}_L(t)}{dt} = -\mathbf{H}_b(\mathbf{v}_C(t), \mathbf{i}_L(t), \boldsymbol{\varphi}_M(t), \mathbf{e}(t), \mathbf{a}(t)) \\ \frac{d\,\boldsymbol{\varphi}_M(t)}{dt} = \mathbf{H}_c(\mathbf{v}_C(t), \mathbf{i}_L(t), \boldsymbol{\varphi}_M(t), \mathbf{e}(t), \mathbf{a}(t)) \\ \mathbf{v}_C(t_0) = \mathbf{v}_{C0} \\ \mathbf{i}_L(t_0) = \mathbf{i}_{L0} \\ \boldsymbol{\varphi}_M(t_0) = \boldsymbol{\varphi}_{M0}. \end{cases} \tag{5.45}$$

This is a system of $n_C + n_L + n_M$ SEs in the state variables $\mathbf{v}_C$, $\mathbf{i}_L$ and $\boldsymbol{\varphi}_M$ in the (v, i)-domain.

Some important remarks on the obtained results are in order.

Remark 5.9 (Reduction of Order) The order of the SEs (5.45) in the (v, i)-domain is $n_C + n_L + n_M$ while the order of the SEs in (5.36) in the (φ, q)-domain is $n_C + n_L$. In passing from the (v, i)-domain to the (φ, q)-domain there is then an *order* *reduction equal to the number n_M of memristors*. Such a reduction is expected to yield advantages in the dynamic analysis of memristor circuits in the (φ, q)-domain.

Remark 5.10 (Smoothing Effect) The SEs (5.36) in the (φ, q)-domain are defined by a vector field containing the nonlinearities $f_{M_k}(\cdot)$ of memristors, while the SEs (5.45) in the (v, i)-domain are defined by a vector field with terms $G_k(\cdot) = f'_{M_k}(\cdot)$. This means that the SEs in the (φ, q)-domain are defined by a *smoother vector field* with respect to those in the (v, i)-domain. This represents a potential advantage in view of the numerical simulations. Another relevant consequence concerns mathematical issues related to the uniqueness of solutions for the SEs. The latter aspect will be further discussed with specific examples.

Remark 5.11 In abstract mathematical form, the correspondence between the solution of the SEs (5.36)

$$\boldsymbol{\Psi}_{\varphi,q}(t, \boldsymbol{\varphi}_C(t_0, t_0) = \mathbf{0}, \mathbf{q}_L(t_0; t_0) = \mathbf{0}) = (\boldsymbol{\varphi}_C(t; t_0), \mathbf{q}_L(t; t_0))$$

and the solution of the SEs (5.45)

$$\boldsymbol{\Psi}_{v,i}(t, \mathbf{v}_{C0}, \mathbf{i}_{L0}, \boldsymbol{\varphi}_{M0}) = (\mathbf{v}_C(t), \mathbf{i}_L(t), \boldsymbol{\varphi}_M(t))$$

is expressed by the following relations:

$$\frac{d}{dt}\Big(\boldsymbol{\varphi}_C(t; t_0)\Big) = \mathbf{v}_C(t)$$

$$\frac{d}{dt}\Big(\mathbf{q}_L(t; t_0)\Big) = \mathbf{i}_L(t)$$

$$\boldsymbol{\varphi}_M(t) = \boldsymbol{\varphi}_{M_0} + \int_{t_0}^{t} \mathbf{H}_c(\mathbf{v}_C(\tau), \mathbf{i}_L(\tau), \boldsymbol{\varphi}_M(\tau), \mathbf{e}(\tau), \mathbf{a}(\tau))d\tau.$$

Also an inverse relation can be obtained. These relationships between solutions will be further illustrated by means of specific examples.

5.8.5 Examples

We have seen in the previous section how to write the DAEs and SEs of a circuit in $\mathcal{LM}$ in the (φ, q)-domain and in the traditional (v, i)-domain. We provide here some examples to illustrate how to apply these techniques. The examples also discuss if it is possible to pass from the dynamic equations in a given domain to the corresponding dynamic equations in the other domain via differentiation or integration in time. In addition, the examples highlight some shortcomings in the techniques discussed so far for writing the SEs in the (φ, q)-domain, that will be solved later in the book in Chap. 7.

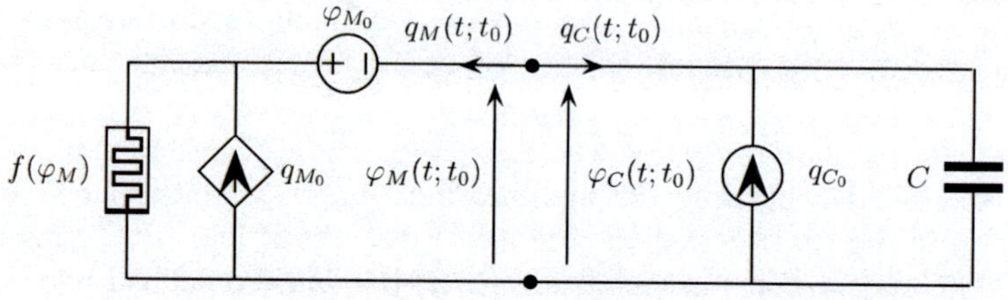

Fig. 5.24 Equivalent circuit in the (φ, q)-domain of the M–C circuit in Fig. 5.5a for $t \geq t_0$

Example 5.15 (The Memristor-Capacitor Circuit) Let us reconsider the M–C circuit in Fig. 5.5a for $t \geq t_0$. The corresponding circuit in the (φ, q)-domain (see Fig. 5.24) is obtained via FCAM by replacing the capacitor and the memristor with the equivalent circuits in Figs. 5.7 and 5.12, respectively, and connecting the two equivalent circuits via terminals with incremental variables.

Analysis by inspection permits the direct writing of the following equations, i.e., KqL, KφL, and CRs of circuit elements

$$\begin{aligned} q_C(t; t_0) + q_M(t; t_0) &= 0 \\ \varphi_C(t; t_0) - \varphi_M(t; t_0) &= 0 \\ q_M(t; t_0) &= f(\varphi_M(t; t_0) + \varphi_{M_0}) - q_{M_0} \\ C\frac{d\varphi_C(t;t_0)}{dt} &= q_C(t; t_0) + q_{C_0} \end{aligned} \tag{5.46}$$

where $q_{C_0} = q_C(t_0)$, $\varphi_{M_0} = \varphi_M(t_0)$, and $q_{M_0} = f(\varphi_{M_0})$. These correspond to the $2b$ ($b = 2$) DAEs in (5.30)–(5.33).

Since we have a flux-controlled memristor in parallel to a capacitor C, we are guaranteed that the SE description of the $M - C$ exists (cf. Sect. 5.8.2). The SE description in the (φ, q)-domain can be readily obtained from (5.46) by considering that $\varphi_C(t; t_0) = \varphi_M(t; t_0)$ and $q_C(t; t_0) = -q_M(t; t_0) = -f(\varphi_C(t; t_0) + \varphi_{M_0}) + q_{M_0}$, that is

$$\begin{cases} C\frac{d\varphi_C(t;t_0)}{dt} = -f(\varphi_C(t; t_0) + \varphi_{M_0}) + f(\varphi_{M_0}) + q_{C_0} \\ \varphi_C(t_0; t_0) = 0. \end{cases} \tag{5.47}$$

Note that the state variable is $\varphi_C(t; t_0)$ and the initial condition for the state variable is zero. Also note that the initial conditions φ_{M_0} and q_{C_0} for the state variables in the (v, i)-domain act as constant inputs. Finally, note that this is the same SE as obtained in Example 5.7 by integrating the SEs in the (v, i)-domain for the $M - C$ circuit.

The formulation of circuit equations in the (v, i)-domain can be readily derived by differentiating (5.46) with respect to time, that is

$$\begin{aligned} i_C(t) + i_M(t) &= 0 \\ v_C(t) - v_M(t) &= 0 \\ i_M(t) &= G(\varphi_M(t))v_M(t) \\ \tfrac{d\varphi_M(t)}{dt} &= v_M(t) \\ C\tfrac{dv_c(t)}{dt} &= i_C(t) \end{aligned} \tag{5.48}$$

where $G(\cdot) = f'(\cdot)$. The initial conditions are $v_C(t_0) = v_{C_0}$ and φ_{M_0}. These correspond to the DAEs in the (v, i)-domain in (5.37)–(5.40).

Finally, the SE singolare in the (v, i)-domain can be obtained by observing that $v_C(t) = v_M(t)$ implies

$$-i_C(t) = i_M(t) = G(\varphi_M(t))v_C(t) \tag{5.49}$$

and

$$\frac{d\varphi_M(t)}{dt} = v_C(t) \tag{5.50}$$

that is, functions $H_a(\cdot)$ and $H_c(\cdot)$ in (5.44) and (5.43) are

$$\begin{aligned} H_a(v_c(t), \varphi_M(t)) &= G(\varphi_M(t))v_C(t) \\ H_c(v_c(t), \varphi_M(t)) &= v_C(t). \end{aligned} \tag{5.51}$$

Note that we do not need to consider $H_b(\cdot)$ since there are no inductors and the dependency on $e(t)$ and $a(t)$ is not included because the circuit has no sources.

Hence, the SE in the (v, i)-domain for the memristor-capacitor circuit results to be

$$\begin{cases} C\frac{d\,v_C(t)}{dt} = -G(\varphi_M(t))v_C(t) \\ \frac{d\varphi_M(t)}{dt} = v_C(t) \\ v_C(t_0) = v_{C_0} \\ \varphi_M(t_0) = \varphi_{M_0}. \end{cases} \tag{5.52}$$

In the (v, i)-domain the $M - C$ circuit is described by a second-order SE in the state variables $v_C(t)$ and $\varphi_M(t)$. The initial conditions are v_{C_0} and φ_{M_0}. The IVP for the SE reported in (5.52) corresponds to (5.45) and coincides with the IVP for a second-order SE in (5.10)–(5.11).

Taking into account that $\varphi_M(t;t_0) = \varphi_C(t;t_0)$, the following relationships hold true between the solutions of (5.47) and (5.52):

- if $\varphi_C(t;t_0)$ is the solution of the IVP (5.47) for $t \geq t_0$, then the solution of the IVP (5.52) can be obtained as follows:

$$v_C(t) = \frac{d}{dt}\Big(\varphi_C(t;t_0)\Big)$$

$$\varphi_M(t) \doteq \varphi_M(t;t_0) + \varphi_{M0} = \varphi_C(t;t_0) + \varphi_{M0}$$

- if $(v_C(t), \varphi_M(t))$ is the solution of the IVP (5.52) for $t \geq t_0$, then the solution of the IVP (5.47) is given as

$$\varphi_C(t;t_0) = \varphi_M(t;t_0) = \varphi_M(t) - \varphi_{M0}$$

or else

$$\varphi_C(t;t_0) \doteq \int_{t_0}^{t} v_C(\tau)d\tau.$$

To summarize, the M–C circuit equations have been written as SEs both in the (φ, q)-domain (see (5.47)) and in the (v, i)-domain (see (5.52)). The two formulations are equivalent, but (5.47) presents a reduced number of ODEs with respect to (5.52). Moreover, (5.47) is defined by a smoother vector field (nonlinearity $f(\cdot)$), with respect to (5.52) (nonlinearity $G(\cdot) = f'(\cdot)$). We can pass from one SE formulation to the other simply by integration or differentiation in time.

Example 5.16 (Ideal Generic Memristor and Capacitor Circuit) Consider again the circuit in Fig. 5.5 but suppose the ideal memristor is replaced by an ideal generic memristor (Sect. 2.4.2 in Chap. 2)

$$i = G(x)v$$

$$\frac{dx}{dt} = w(x)v$$

where the state variables x belong to the normalized interval $[-1, 1]$ and $w(\cdot)$ is a window function such that $w(x) > 0$ for $x \in (-1, 1)$ and $w(0) = w(1) = 0$.

The circuit satisfies for $t \geq t_0$

$$C\frac{dv_C}{dt} = -G(x)v_C$$

$$\frac{dx}{dt} = w(x)v_C$$

where v_C is the capacitor voltage.

Let

$$\varphi_M \doteq h(x) = \int_0^x \frac{1}{w(\rho)} d\rho$$

for any $x \in (-1, 1)$. Since $w(x) > 0$ for $x \in (-1, 1)$, function $h(\cdot)$ is invertible in the same interval and we have $x = h^{-1}(\varphi_M)$ and

$$h'(x) = \frac{1}{w(x)}.$$

Substituting x with φ_M we obtain the SE

$$C\frac{dv_C}{dt} = -\hat{G}(\varphi_M)v_C$$
$$\frac{d\varphi_M}{dt} = v_C$$

where we have let

$$\hat{G}(\varphi_M) = G(h^{-1}(\varphi_M)).$$

Such an SE coincides with that obtained for the $M - C$ circuit studied in Example 5.15. Then, we can use the same procedure as in that example for studying the dynamics of the more general circuit with an ideal generic memristor here considered.

Example 5.17 **Consider a circuit with a flux-controlled memristor** $q_{M_1} = f(\varphi_{M_1})$ and a charge controlled memristor $\varphi_{M_2} = h(q_{M_2})$ as in Fig. 5.25a. We wish to obtain the SE representation in the (φ, q)- and (v, i)-domain using the procedure in Sect. 5.8.2. Consider the corresponding circuit in the (φ, q)-domain, extract L and C, connect them to a two-port network containing the adynamic elements in the (φ, q)-domain, and replace C and L **by a flux source and a charge source,** respectively, as shown in Fig. 5.25b. It can be checked that the hypotheses in Remark 5.8 for the existence of the hybrid representation of the two-port network **are satisfied. This representation is easily obtained as**

$$q_1(t; t_0) = f(\varphi_1(t; t_0) + \varphi_{M_{1_0}}) - f(\varphi_{M_{1_0}}) - q_a(t; t_0) - q_2(t; t_0)$$
$$\varphi_2(t; t_0) = h(q_2(t; t_0) + q_{M_{2_0}}) - h(q_{M_{2_0}}) + \varphi_1(t; t_0)$$

where the first equation derives from KqL applied to cut-set C in Fig. 5.25b and the second equation from KφL at the loop given by the source $q_2(t; t_0)$, the charge-controlled memristor and the source $\varphi_1(t; t_0)$.

The SEs in the (φ, q)-domain are given by the second-order system

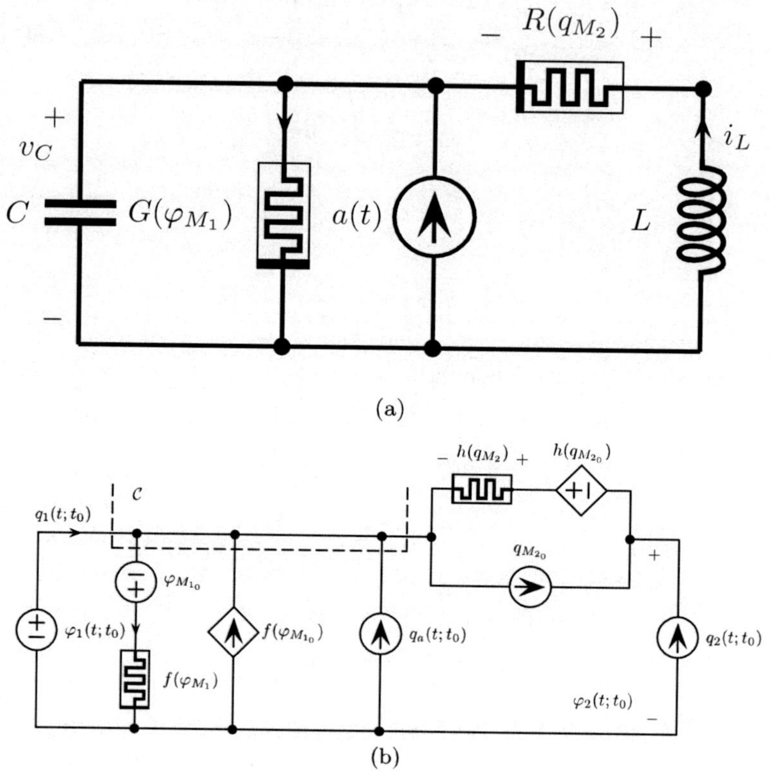

Fig. 5.25 (**a**) A circuit with a flux-controlled and a charge-controlled memristor and (**b**) equivalent circuit in the (φ, q)-domain for finding the hybrid representation. We have let $G(\cdot) = f'(\cdot)$ and $R(\cdot) = h'(\cdot)$

$$
\begin{aligned}
C\frac{d\varphi_C(t;t_0)}{dt} &= -f(\varphi_C(t;t_0) + \varphi_{M_{10}}) + f(\varphi_{M_{10}}) + q_L(t;t_0) \\
&\quad + q_a(t;t_0) + Cv_{C0} \\
L\frac{dq_L(t;t_0)}{dt} &= -h(q_L(t;t_0) + q_{M_{20}}) + h(q_{M_{20}}) - \varphi_C(t;t_0) + Li_{L0}
\end{aligned}
$$

where $v_C(t_0) = v_{C0}$, $i_L(t_0) = i_{L0}$, $\varphi_{M_1}(t_0) = \varphi_{M_{10}}$, and $q_{M_2}(t_0) = q_{M_{20}}$.

By differentiation, we obtain the SEs in the (v, i)-domain. These are given by the fourth-order system

$$
\begin{aligned}
C\frac{dv_C(t)}{dt} &= -f'(\varphi_{M_1}(t))v_C(t) + i_L(t) + a(t) \\
L\frac{di_L(t)}{dt} &= -h'(q_{M_2}(t))i_L(t) - v_C(t)
\end{aligned}
$$

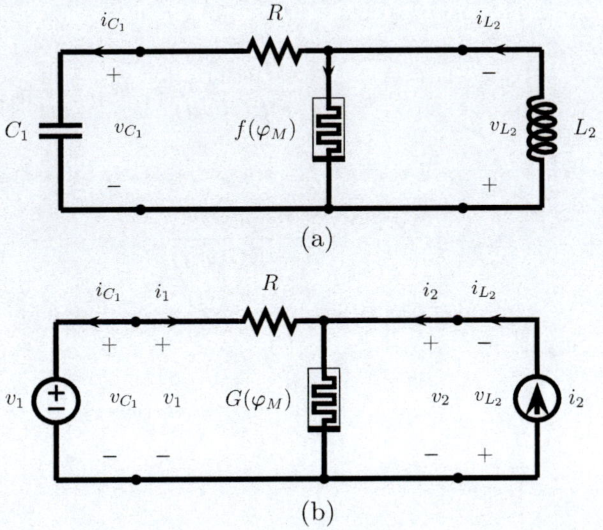

Fig. 5.26 (**a**) Circuit with a flux-controlled memristor that is not in parallel to a capacitor and (**b**) circuit for finding the hybrid representation

$$\frac{d\varphi_{M_1}(t)}{dt} = v_C(t)$$
$$\frac{dq_{M_2}(t)}{dt} = i_L(t)$$

where we have taken into account that $\varphi_C(t; t_0) + \varphi_{M_{1_0}} = \varphi_{M_1}(t; t_0) + \varphi_{M_{1_0}} = \varphi_{M_1}(t)$, $q_L(t; t_0) + q_{M_{2_0}} = q_{M_2}(t; t_0) + q_{M_{2_0}} = q_{M_2}(t)$. The initial conditions are v_{C_0}, i_{L_0}, $\varphi_{M_{1_0}}$, and $q_{M_{2_0}}$.

The same SEs in the (v, i)-domain may be also directly found via the procedure illustrated in Sect. 5.8.4. The verification is left to the reader.

As expected, the order reduction for the SEs in the (φ, q)-domain, with respect to the (v, i)-domain, is equal to the number of memristors in the circuit.

Example 5.18 Consider the circuit in Fig. 5.26a containing a flux-controlled memristor, a resistor $R > 0$, a capacitor, and an inductor. Let us first find the SEs in the (v, i)-domain by using the procedure given in Sect. 5.8.4. For any flux φ_M, let us replace the memristor with a conductance $G(\varphi_M) = f'(\varphi_M)$. Extract C and L and replace C by a voltage source v_1 and L by a current source i_2. The resulting circuit (Fig. 5.26b) is linear and the two-port network to which the sources are connected admits of the hybrid representation

$$i_1 = h_{11}(\varphi_M)v_1 + h_{12}(\varphi_M)i_2$$
$$v_2 = h_{21}(\varphi_M)v_1 + h_{22}(\varphi_M)i_2.$$

A simple analysis (details are left to the reader) shows that we have

$$h_{11} = \frac{i_1}{v_1}|_{i_2=0} = \frac{G(\varphi_M)}{1 + RG(\varphi_M)}; \quad h_{12} = \frac{i_1}{i_2}|_{v_1=0} = -\frac{1}{1 + RG(\varphi_M)}$$

and

$$h_{21} = \frac{v_2}{v_1}|_{i_2=0} = \frac{1}{1 + RG(\varphi_M)}; \quad h_{22} = \frac{v_2}{i_2}|_{v_1=0} = \frac{R}{1 + RG(\varphi_M)}.$$

This representation exists, i.e., the circuit is uniquely solvable *for any* φ_M, if and only if

$$1 + RG(\varphi_M) \neq 0$$

i.e.,

$$G(\varphi_M) > -\frac{1}{R}.$$

Note that this condition allows for the presence of an active memristor since $R > 0$.

By analyzing the same linear circuit we can express v_M as a function of v_1 and i_2 as follows:

$$v_M = \frac{1}{1 + RG(\varphi_M)} v_1 + \frac{R}{1 + RG(\varphi_M)} i_2.$$

Since $v_1 = v_C$, $i_2 = i_L$ and $i_1 = -i_C = -C dv_C/dt$, $v_2 = -v_L = -L di_L/dt$, we conclude that under the stated condition there exist the SEs in the (v, i)-domain and they are given in vector form by the third-order system

$$\begin{pmatrix} \frac{dv_C}{dt} \\ \frac{di_L}{dt} \\ \frac{d\varphi_M}{dt} \end{pmatrix} = -\frac{1}{1 + RG(\varphi_M)} \begin{pmatrix} \frac{G(\varphi_M)}{C} & -\frac{1}{C} & 0 \\ \frac{1}{L} & \frac{R}{L} & 0 \\ 1 & R & 0 \end{pmatrix} \begin{pmatrix} v_C \\ i_L \\ \varphi_M \end{pmatrix}.$$

Let us now try to find the SE representation in the (φ, q)-domain using the technique described in Sect. 5.8.2. Once more, extract L and C and connect them to an adynamic two-port network containing the memristor, that is an element with a nonlinear algebraic CR $q_M(t; t_0) = f(\varphi_M(t; t_0) + \varphi_{M_0}) - f(\varphi_{M_0}) = \tilde{f}(\varphi_M(t; t_0); \varphi_{M_0})$. If the nonlinear characteristic $\tilde{f}$ is strongly passive, then the conditions in Sect. 5.8.2 for the existence of the hybrid representation of the two-port network and hence for the existence of the SEs in the (φ, q)-domain are satisfied. One problem is that it seems not an easy task to explicitly find functions $h_u(\varphi_C(t; t_0), q_L(t; t_0), \varphi_M(t; t_0))$ and $h_i(\varphi_C(t; t_0), q_L(t; t_0), \varphi_M(t; t_0))$ as in (5.35) and hence explicitly write the SEs. We will discuss this issue again in Example 7.5 of Chap. 7.

Remark 5.12 In Example 5.18 we have been able to find the SEs in the (v, i)-domain explicitly and to show the existence in implicit form of the SEs in the (φ, q)-domain, under suitable assumptions on f. Clearly, it would be difficult to use the SEs in the (φ, q)-domain for further dynamic analysis when they are not known in explicit form. Then, we are led to look for additional conditions on a circuit in $\mathcal{LM}$ so that it admits of an SE representation and moreover such representation is explicitly known and is sufficiently simple. This problem will be overcome for a relevant subclass of circuits in $\mathcal{LM}$ in Chap. 7. For that class we will also show that we can pass from the SEs in the (φ, q)-domain to those in the (v, i)-domain, and conversely, simply by differentiation or integration in time.

We conclude this section with an example concerning a circuit not in the class $\mathcal{LM}$ containing an operational amplifier and by completing the analysis in the (φ, q)-domain of an example at the beginning of the chapter.

Example 5.19 (Memristor Circuit with Operational Amplifier) Consider a circuit with an ideal operational amplifier operating in the linear region, a flux-controlled memristor $q_M = f(\varphi_M)$, and two capacitors as shown in Fig. 5.27. The corresponding circuit in the (φ, q)-domain for finding the hybrid representation, where each capacitor has been replaced by a flux-source, is shown in Fig. 5.27b.

KqL at cut-set C and node Ⓐ, respectively, easily yield the following hybrid representation:

$$
\begin{aligned}
q_1(t;t_0) &= \frac{\varphi_1(t;t_0)-\varphi_e(t;t_0)}{R_1}+\frac{\varphi_1(t;t_0)}{R_2}\\
&\quad + f(\varphi_1(t;t_0)-\varphi_2(t;t_0)+\varphi_{M_0})-f(\varphi_{M_0})\\
q_2(t;t_0) &= \frac{\varphi_1(t;t_0)}{R_2}.
\end{aligned}
$$

Then, the SEs in the (φ, q)-domain are given by the second-order system

$$
\begin{aligned}
C_1\frac{d\varphi_{C_1}(t;t_0)}{dt} &= -\varphi_{C_1}(t;t_0)\left(\frac{1}{R_1}+\frac{1}{R_2}\right)\\
&\quad - f\left(\varphi_{C_1}(t;t_0)-\varphi_{C_2}(t;t_0)+\varphi_{M_0}\right)\\
&\quad + f(\varphi_{M_0})+\frac{\varphi_e(t;t_0)}{R_1}+C_1 v_{C_{10}}\\
C_2\frac{d\varphi_{C_2}(t;t_0)}{dt} &= -\frac{\varphi_{C_1}(t;t_0)}{R_2}+C_2 v_{C_{20}}.
\end{aligned}
$$

By differentiation we may obtain the fourth-order SEs describing the circuit in the (v, i)-domain (details are omitted).

Example 5.20 (Example 5.4 Continued) Consider again the circuit $C_1 - R - C_2$ in Example 5.4 for $t \geq 0$, which we want to analyze here in the (φ, q)-domain by

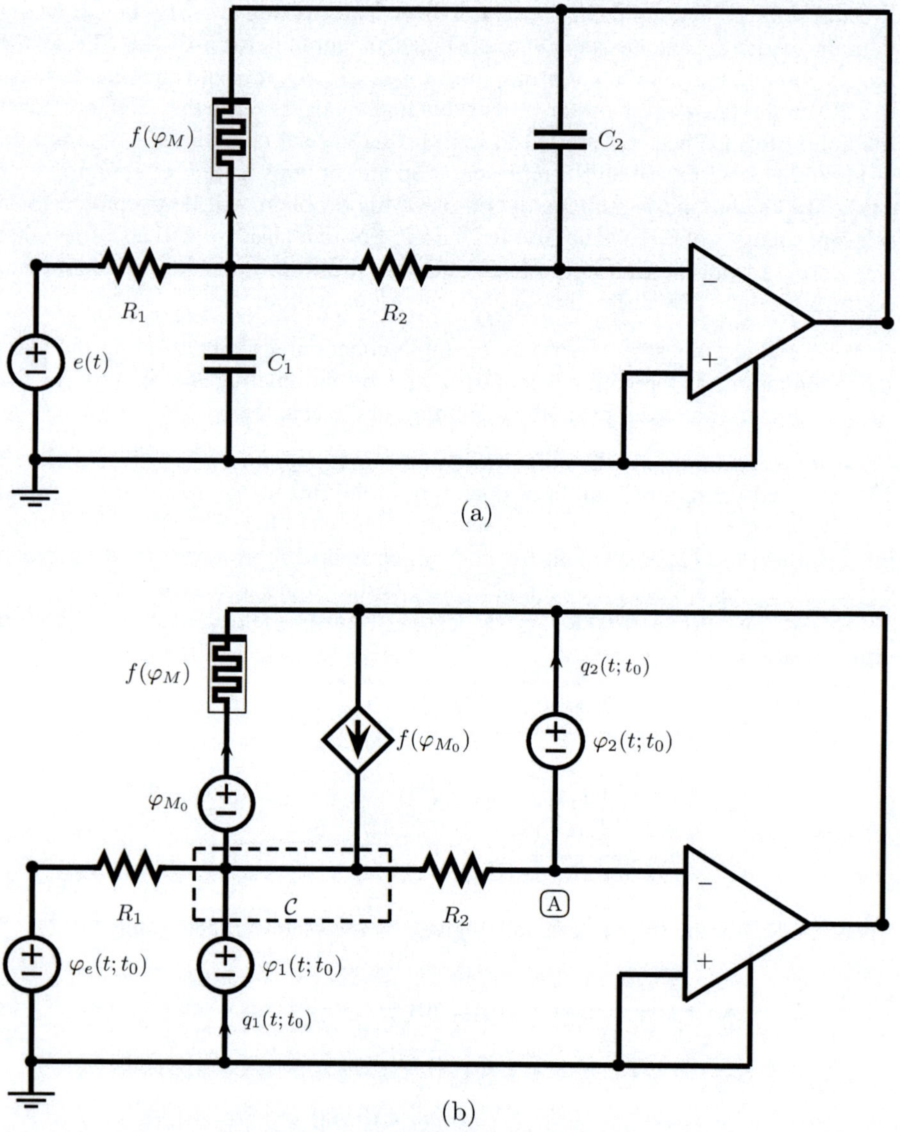

Fig. 5.27 (**a**) Memristor circuit with ideal operational amplifier and (**b**) equivalent circuit in the (φ, q)-domain for finding the hybrid representation

means of FCAM. Recall that C_1 is charged at $v_{C_1}(0) = E$ while $v_{C_2}(0) = 0$. Hence, we obtain in the (φ, q)-domain the equivalent circuit in Fig. 5.28.

Analysis by inspection yields the KqL

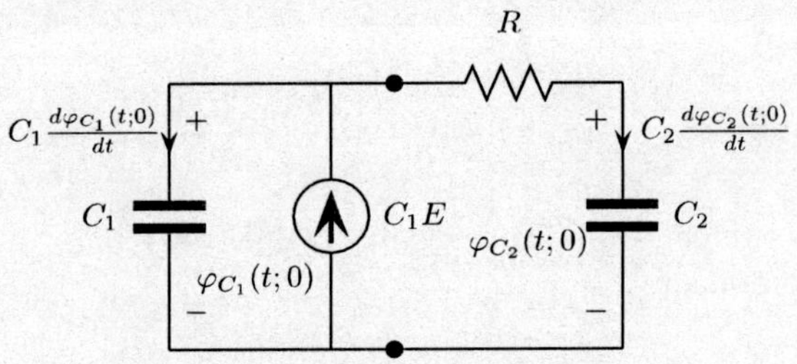

Fig. 5.28 Equivalent circuit in the (φ, q)-domain for the $C_1 - R - C_2$ circuit in Example 5.4

$$C_1\frac{d\varphi_{C_1}(t;0)}{dt} = C_1E - C_2\frac{d\varphi_{C_2}(t;0)}{dt}$$

and KφL

$$\varphi_{C_1}(t;t_0) = RC_2\frac{d\varphi_{C_2}(t;0)}{dt} + \varphi_{C_2}(t;t_0).$$

These yield the SEs

$$\frac{d\varphi_{C_1}(t;0)}{dt} = E - \frac{\varphi_{C_1}(t;0) - \varphi_{C_2}(t;0)}{RC_1}$$
$$\frac{d\varphi_{C_2}(t;0)}{dt} = \frac{\varphi_{C_1}(t;0) - \varphi_{C_2}(t;0)}{RC_2}$$

with initial conditions $\varphi_{C_1}(0;0) = 0$ and $\varphi_{C_2}(0;0) = 0$.

The solution can be easily obtained via standard techniques from linear ordinary differential equations, namely

$$\varphi_{C_1}(t;t_0) = \frac{EC_2}{C_1+C_2}t + E\tau\frac{C_1}{C_1+C_2}\left(1 - e^{-\frac{t}{\tau}}\right)$$

and

$$\varphi_{C_2}(t;0) = \frac{EC_1}{C_1+C_2}t - \frac{EC_1\tau}{C_1+C_2}\left(1 - e^{-\frac{t}{\tau}}\right)$$

where

$$\tau = R\frac{C_1C_2}{C_1+C_2}.$$

Differentiating in time we have

$$v_{C_1}(t) = \frac{d\varphi_{C_1}(t;0)}{dt} = Ee^{-\frac{t}{\tau}} + \frac{EC_2}{C_1+C_2}\left(1 - e^{-\frac{t}{\tau}}\right)$$

and

$$v_{C_2}(t) = \frac{d\varphi_{C_2}(t;0)}{dt} = \frac{EC_1}{C_1 + C_2}\left(1 - e^{-\frac{t}{\tau}}\right)$$

which is the same result obtained in Example 5.4 via an analysis in the (v, i)-domain.

5.9 Discussion

In the chapter, we started by revisiting the fundamental issue of how to write Kirchhoff laws in the (φ, q)-domain and showed that the most effective form of Kirchhoff laws is that expressed in terms of incremental fluxes and incremental charges (conservation of incremental charge in any cut-set and conservation of incremental flux around any loop). This has led to the introduction of FCAM, i.e., a method for analyzing in the (φ, q)-domain a class $\mathcal{LM}$ of nonlinear circuits with memristors, linear resistors, inductors, capacitors, and independent voltage and current sources.

One important property proved via FCAM is that a memristor circuit in $\mathcal{LM}$ is analogous in the (φ, q)-domain to a nonlinear RLC circuit in the (v, i)-domain containing nonlinear resistors and linear inductors and capacitors. Based on this analogy, general techniques for writing the DAEs and SEs describing the dynamics both in the (φ, q)-domain and in the (v, i)-domain have been obtained. We have also shown that in some relevant situations we can pass from the SEs in one domain to those in the other domain simply via a differentiation or integration in time. Despite these results, some issues concerning how to write the SEs in an effective way for subsequent dynamic analysis remain open. We will come back to these issues in Chap. 7, where under a slightly more restrictive set of assumptions we will be able to find a relevant subclass of $\mathcal{LM}$ for which the SEs can be written even in a more simple and effective form with respect to this chapter.

The main potential advantage of FCAM is that the dynamics of a memristor circuit are described in the (φ, q)-domain by a *reduced order* SE with respect to the (v, i)-domain, the reduction of order being exactly equal to the number of memristors present in the circuit. Another advantage is that the vector field describing the SEs in the (φ, q)-domain is *smoother* than that describing the SEs in the (v, i)-domain. In the next chapter, we will discuss in detail the application of FCAM to some fundamental memristor circuits in order to better highlight such advantages. In particular, we will see that the reduction of order is related to a fundamental structural property of memristor circuits, namely, the fact that the state-space in the (v, i)-domain can be foliated in a continuum of manifolds that are invariant for the dynamics.

It has been shown in the chapter that FCAM can be easily extended to include linear resistive multiport networks and also time-varying elements. Later in the book we will develop further extensions of FCAM to circuits containing nonlinear inductors and capacitors and higher-order elements as memcapacitors and meminductors (Chap. 11). We will also study the application of FCAM to circuits containing piecewise-linear approximations of nonlinear resistors (Chap. 10).

In circuit theory, conservation of flux and charge is traditionally used to analyze the possible discontinuities of state variables due to the opening or closing of ideal switches, or the application of impulsive (delta of Dirac) sources, in RLC circuits. In particular, the paper [13] has developed a systematic procedure for finding consistent initial conditions after switching, or the application of impulsive sources, in RLC circuits. This is based on finding equivalent circuits for the elements via incremental flux and charge at the terminals that hold in the interval $(0^-, 0^+)$, where $t_0 = 0$ is the critical instant. The method FCAM developed in the chapter has some relationships with the approach in [13] but also some fundamental differences. In fact, FCAM has been conceived for application to circuits containing memristors, and mem-elements in general, a situation where the fundamental advantages of the analysis in the (φ, q)-domain show themselves at their fullest extent. Moreover, FCAM is devoted to study the whole dynamics of memristor circuits starting from an initial instant t_0, not only the behavior at a certain instant.

References

1. F. Corinto, M. Forti, Memristor circuits: flux–charge analysis method. IEEE Trans. Circuits Syst. I Regul. Pap. **63**(11), 1997–2009 (2016)
2. F. Corinto, M. Forti, Memristor circuits: bifurcations without parameters. IEEE Trans. Circuits Syst. I Regul. Pap. **64**(6), 1540–1551 (2017)
3. F. Corinto, M. Forti, Memristor circuits: pulse programming via invariant manifolds. IEEE Trans. Circuits Syst. I Regul. Pap. **65**(4), 1327–1339 (2018)
4. L.O. Chua, Nonlinear circuit foundations for nanodevices. I. The four-element torus. Proc. IEEE **91**(11), 1830–1859 (2003)
5. M. Mansfield, C. O'sullivan, *Understanding Physics* (Wiley, Hoboken, 2011)
6. F. Corinto, P.P. Civalleri, L.O. Chua, A theoretical approach to memristor devices. IEEE J. Emerg. Sel. Top. Circuits Syst. **5**(2), 123–132 (2015)
7. L.O. Chua, *Introduction to Nonlinear Network Theory* (McGraw-Hill, New York, 1969)
8. L.O. Chua, C.A. Desoer, E.S. Kuh, *Linear and Nonlinear Circuits* (McGraw-Hill, New York, 1987)
9. P.W. Tuinenga, *SPICE: A Guide to Circuit Simulation and Analysis Using PSpice*, vol. 2 (Prentice Hall, Englewood Cliffs, 1995)
10. R. Riaza, DAEs in circuit modelling: a survey, in *Surveys in Differential-Algebraic Equations I* (Springer, Berlin, 2013), pp. 97–136
11. M. Itoh, L.O. Chua, Memristor oscillators. Int. J. Bifurc. Chaos **18**(11), 3183–3206 (2008)
12. M. Hasler, State equations for active circuits with memristors, in *Chaos, CNN, Memristors and Beyond: A Festschrift for Leon Chua* (World Scientific, Singapore, 2013), pp. 518–528
13. A. Opal, J. Vlach, Consistent initial conditions of nonlinear networks with switches. IEEE Trans. Circuits Syst. **38**(7), 698–710 (1991)

Chapter 6
Memristor Circuits: Invariant Manifolds, Coexisting Attractors, Extreme Multistability, and Bifurcations Without Parameters

Chapter 5 has developed the flux-charge analysis method (FCAM) for the analysis of a class $\mathcal{LM}$ of memristor circuits containing memristors, linear resistors, inductors, capacitors, and independent voltage and current sources. The formulation of circuit equations (DAEs and SEs) has been provided in the (φ, q)-domain and in the (v, i)-domain and relationships between the analysis in the two domains have been discussed. A fundamental property is that the SE description in the (φ, q)-domain of a circuit in $\mathcal{LM}$ has a lower order with respect to the corresponding description in the (v, i)-domain, so that it is expected that the dynamic analysis in the former domain is simpler. The goal of this chapter is to highlight some main advantages of FCAM by studying the dynamics of a number of low-order autonomous memristor circuits in the (φ, q)-domain, namely, circuits described by a first-order SE in the (φ, q)-domain, oscillatory circuits described by a second-order SE in the (φ, q)-domain, and chaotic circuits described by a third-order SE in the (φ, q)-domain. Such a study permits to show that memristor circuits display a number of fundamental peculiar dynamic features. First of all, the state space in the (v, i)-domain can be foliated in a continuum of *invariant manifolds*, thus implying the ***coexistence of infinitely many different reduced-order dynamics and attractors***. Moreover, such circuits display bifurcations due to changing the initial conditions for a fixed set of circuit parameters that are named *bifurcations without parameters*.

6.1 First-Order Memristor Circuits

For didactic purposes, we find it useful to first illustrate the idea of the invariant of motions, invariant manifolds, and coexisting dynamics using a simple *linear* circuit. Bifurcations without parameters are instead displayed only when the circuit is *nonlinear*, as in the case where the circuit contains also memristors.

F. Corinto et al., *Nonlinear Circuits and Systems with Memristors*,
https://doi.org/10.1007/978-3-030-55651-8_6

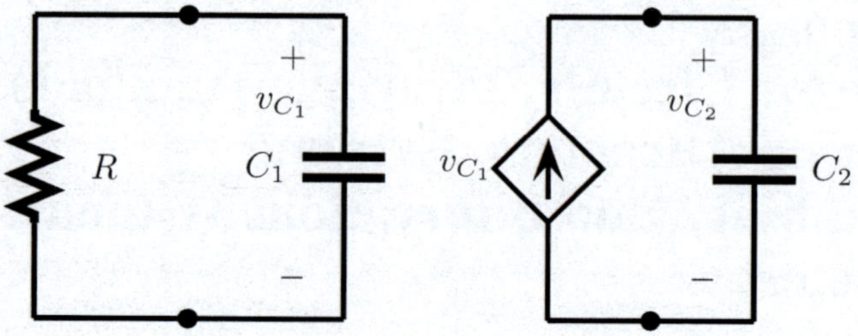

Fig. 6.1 Linear circuit with two capacitors and a voltage-controlled current-source

6.1.1 *Linear Circuits*

Example 6.1 Consider the linear circuit in Fig. 6.1 containing two capacitors, one voltage-controlled current-source and a resistor. The second-order SEs describing the circuit dynamics for $t \geq 0$ are given by the linear system

$$\begin{aligned} \dot{v}_{C_1} &= -av_{C_1} \\ \dot{v}_{C_2} &= v_{C_1} \\ v_{C_1}(0) &= v_{C_{10}} \\ v_{C_2}(0) &= v_{C_{20}} \end{aligned} \tag{6.1}$$

where we let $C_1 = C_2 = 1$ and $a = 1/RC_1 = 1/R > 0$.

By letting $\dot{v}_{C_1} = 0$ and $\dot{v}_{C_2} = 0$, we obtain that there is a continuum of non-isolated equilibrium points (EPs) coinciding with the v_{C_2}-axis, i.e.,

$$\mathcal{E} = \{(v_{C_1}, v_{C_2})^T \in \mathbb{R}^2 : v_{C_1} = 0\}.$$

It is an easy matter to verify that any EP is stable, but not asymptotically stable [1]. In fact, the definition of asymptotic stability of an EP requires that solutions starting nearby the EP converge toward the EP. Clearly, this condition cannot be met if an EP is not isolated.

The eigenvalues of the matrix defining the linear system are $\lambda_1 = -a < 0$ and $\lambda_2 = 0$ and the solution of the initial value problem (IVP) (6.1) is easily obtained as

$$(v_{C_1}(t), v_{C_2}(t)) = (v_{C_{10}} e^{-at}, v_{C_{20}} + \frac{v_{C_{10}}}{a}(1 - e^{-at})).$$

Note that any solution converges to an EP as $t \to +\infty$; moreover, the phase portrait is given by infinite parallel straight lines with slope $-a$ in the $v_{C_2} - v_{C_1}$ phase plane, as shown in Fig. 6.2.

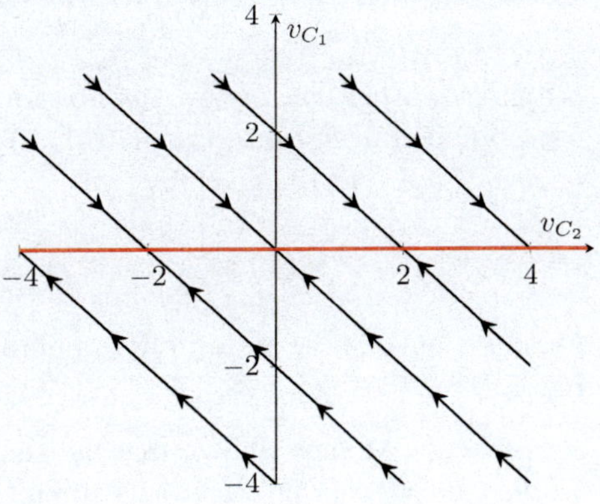

Fig. 6.2 Phase portrait of a linear circuit possessing a manifold of EPs (red line)

Integrating the previous system in $[0, t)$, where $t > 0$, yields $v_{C_1}(t) - v_{C_{1_0}} = -a(v_{C_2}(t) - v_{C_{2_0}})$, hence we can associate with (6.1) the ***reduced-order (first-order) system***

$$\begin{aligned} \dot{v}_{C_2} &= -av_{C_2} + av_{C_{2_0}} + v_{C_{1_0}} \\ v_{C_2}(0) &= v_{C_{2_0}} \end{aligned} \tag{6.2}$$

having a unique EP $v_{C_{2_0}} + v_{C_{1_0}}/a$. The solution of the IVP (6.2) is

$$v_{C_2}(t) = v_{C_{2_0}} + \frac{v_{C_{1_0}}}{a}(1 - e^{-at}). \tag{6.3}$$

Note that any solution converges to the unique EP as $t \to +\infty$.

We can verify that if $(v_{C_1}(t), v_{C_2}(t))$ is the solution of the IVP (6.1), then $v_{C_2}(t)$ is the solution of the IVP (6.2). Conversely, if $v_{C_2}(t)$ is the solution of the IVP (6.2), then $(\dot{v}_{C_2}(t), v_{C_2}(t))$ is the solution of the IVP (6.1).

Let us now consider the function of the state variables (v_{C_1}, v_{C_2}) of (6.1)

$$w(v_{C_1}, v_{C_2}) = \alpha v_{C_2} + \beta v_{C_1} - \gamma$$

where $\alpha, \beta, \gamma \in \mathbb{R}$. The time derivative of w along a solution of (6.1) is given by

$$\dot{w}(v_{C_1}, v_{C_2}) = \frac{\partial w}{\partial v_{C_1}}\dot{v}_{C_1} + \frac{\partial w}{\partial v_{C_2}}\dot{v}_{C_2} = (\alpha - a\beta)v_{C_1}.$$

If we choose $\beta = 1$, $\alpha = a$, then we have

$$\dot{w}(v_{C_1}, v_{C_2}) = 0$$

for any $t \geq 0$. This means that, for any $\gamma \in \mathbb{R}$, function

$$w(v_{C_1}, v_{C_2}) = av_{C_2} + v_{C_1} - \gamma \tag{6.4}$$

is a (non-constant) *invariant of motion* for (6.1). In the mathematical literature, (6.4) is also called a ***first integral*** of (6.1). This property implies that there exist ∞^1 *invariant sets* for (6.1) given by

$$\mathcal{M}(\gamma) = \{(v_{C_1}, v_{C_2})^T \in \mathbb{R}^2 : av_{C_2} + v_{C_1} = \gamma\}$$

for any $\gamma \in \mathbb{R}$. These are simply straight lines with slope $-a$ in the phase space (cf. Fig. 6.2).

Remark 6.1 We have shown that the state space $(v_{C_1}, v_{C_2}) \in \mathbb{R}^2$ of (6.1) can be *foliated* in ∞^1 1D invariant sets (or manifolds), where each manifold is in this case simply a straight line. In addition, the following considerations can be easily drawn:

- the SE (6.1) is second-order, while (6.2) is first-order. However, the right-hand side of the latter system *depends upon the initial conditions* $v_{C_{1_0}}$ *and* $v_{C_{2_0}}$ *for the state variables of (6.1)*
- the previous results state that the dynamics of v_{C_2} as a solution of (6.1) is ***essentially first-order***, since it satisfies the reduced-order system (6.2). However, since (6.2) depends upon the scalar term $av_{C_{2_0}} + v_{C_{1_0}}$, there are actually ∞^1 different ***coexisting first-order dynamics*** for (6.1)
- the reduced-order dynamics are all of the same type and share the same stability properties. On each manifold there is a *unique isolated EP* $v_{C_{2_0}} + v_{C_{1_0}}/a$ which is asymptotically stable for the reduced-order system (6.2). Each solution starting on a manifold converges toward the corresponding EP as $t \to \infty$.
  In the example considered, the existence of an invariant of motion is related to the existence of infinitely many (a continuum of) EPs for (6.1). The reduced-order system (6.2) has instead a unique EP depending however on the manifold.

Example 6.2 Consider the linear circuit with two capacitors and a resistor in Fig. 6.3. Note that the two capacitors form a cut-set. The SEs describing its dynamics are, assuming $C_1 = C_2 = 1$ and $R = 1$

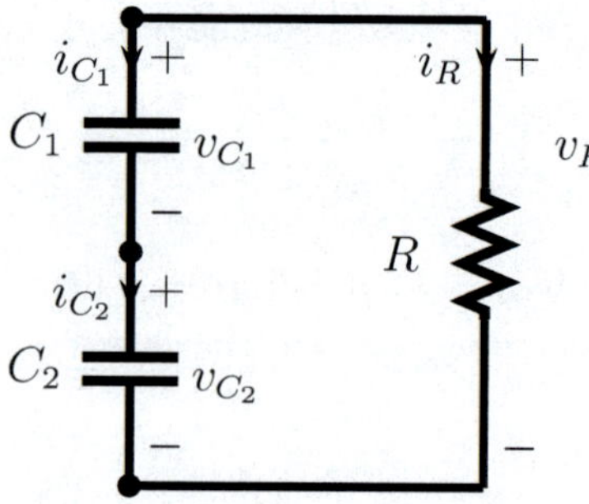

Fig. 6.3 Linear circuit with two capacitors

$$\frac{dv_{C_1}}{dt} = -v_{C_1} - v_{C_2} \tag{6.5}$$

$$\frac{dv_{C_2}}{dt} = -v_{C_1} - v_{C_2}. \tag{6.6}$$

It can easily be checked that the circuit has a continuum of non-isolated EPs given by the line

$$\mathcal{E} = \{(v_{C_1}, v_{C_2})^T \in \mathbb{R}^2 : v_{C_1} = -v_{C_2}\}.$$

Also this circuit has an invariant of motion given by

$$w(v_{C_1}, v_{C_2}) = v_{C_1} - v_{C_2} - \gamma$$

where $\gamma \in \mathbb{R}$. In fact, differentiating $w(\cdot)$ along the circuit solutions we obtain

$$\dot{w}(v_{C_1}, v_{C_2}) = \frac{\partial w}{\partial v_{C_1}}\dot{v}_{C_1} + \frac{\partial w}{\partial v_{C_2}}\dot{v}_{C_2} = -v_{C_1} - v_{C_2} + v_{C_1} + v_{C_2} = 0$$

for any t. Then, there exist ∞^1 *invariant sets* for (6.1) given by

$$\mathcal{M}(\gamma) = \{(v_{C_1}, v_{C_2})^T \in \mathbb{R}^2 : v_{C_1} - v_{C_2} = \gamma\}$$

for any $\gamma \in \mathbb{R}$. These are simply straight lines with slope 45° in the phase space (cf. Fig. 6.4). The same figure reports the global phase portrait for the circuit. Also in this case we may repeat considerations analogous to those in Remark 6.1. In particular, the phase space can be foliated in a continuum of invariant manifolds where we have a reduced-order (first-order) dynamics. On each manifold there is a unique EP which is asymptotically stable for the reduced-order dynamics.

Fig. 6.4 Phase portrait of a linear circuit with two capacitors possessing a manifold of EPs (red line)

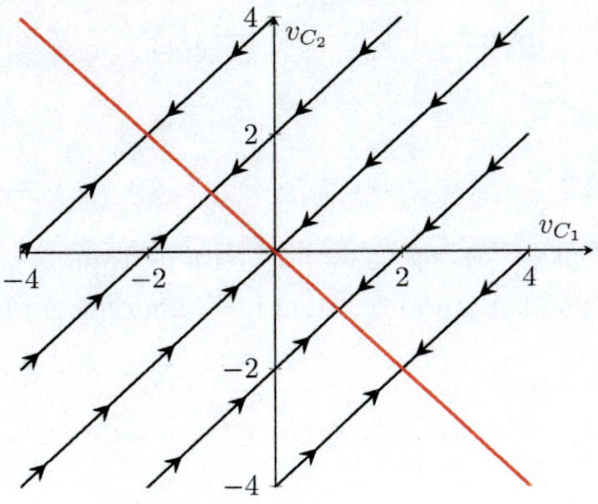

Remark 6.2 (Structural Stability of Non-isolated EPs, Invariants of Motion and Invariant Manifolds) The examples discussed so far feature the peculiar property that there exists a continuum of non-isolated EPs. Such property is *structurally stable* in the considered circuits, i.e., it persists under perturbations of the circuit parameters. Indeed, we leave to the reader the verification that any of the two circuits still displays a line of non-isolated equilibria for any value of C_1 and C_2 and the other parameters. We refer the reader to [2] and [3] for further considerations on the existence of non-isolated equilibrium points for RLC circuits and their structural stability or instability properties.

However, we stress that it is *a special property for an RLCcircuit to possess a continuum of equilibria.* For example, according to Theorem 14 in [2], it is seen that the existence of non-isolated EPs is related to the presence of particular circuit structures as cut-sets made of capacitors only (cf. Example 6.2) or loops of inductors only. Such structures also imply the existence of invariants of motion and invariant manifolds [3], that are again observable only in special classes of RLC circuits.

Different from RLC circuits, in memristor circuits, due to the presence of the new element memristor, we are *in a situation where the circuit displays a continuum of EPs in the (v, i)-domain (cf. Theorem 2.2 in Chap. 2). Moreover, this is accompanied by the presence of invariants of motion and invariant manifolds. Such properties are structurally stable, in the sense that they hold for any value of the circuit parameters.* These issues will be discussed and illustrated in detail with the memristor circuits studied in the remaining part of the book.

Remark 6.3 (Linear Systems with an Invariant of Motion) It is quite a special situation for a linear system to admit an invariant of motion. For instance, the system

$$\dot{x} = -x$$
$$\dot{y} = -y$$

does not have a continuous non-constant invariant of motion (cf. [4]). Note that such a system has a unique EP.

Instead, the linear system

$$\dot{x} = y$$
$$\dot{y} = -x$$

which can be associated with a simple conservative $L - C$ circuit, has a unique EP, and has a non-constant invariant of motion $w(x, y) = x^2 + y^2 - \gamma$, where $\gamma \in \mathbb{R}$.

6.1.2 $M - C$ Circuit

The linear circuits studied in Sect. 6.1.1 illustrate the idea of foliation of the state space, invariants of motion, invariants manifolds, and coexisting reduced-order dynamics. However, due to the linearity, each reduced-order dynamics has the same properties from a stability viewpoint and there are no bifurcations when changing the initial conditions for fixed circuit parameters. As shown next, there is a radical change from the viewpoint of stability and bifurcations when we consider a nonlinear memristor circuit.

Consider again for $t \geq t_0$, where $-\infty < t_0 < \infty$, the flux-controlled memristor and capacitor ($M - C$) circuit in the class $\mathcal{LM}$ studied in Chap. 5 (Fig. 6.5). The SEs in the (v, i)-domain are

$$\begin{cases} C\frac{d\,v_C(t)}{dt} = -G(\varphi_M(t))v_C(t) \\ \frac{d\varphi_M(t)}{dt} = v_C(t) \\ v_C(t_0) = v_{C0} \\ \varphi_M(t_0) = \varphi_{M0} \end{cases} \tag{6.7}$$

where $q_M = f(\varphi_M)$ is the memristor characteristic and $G(\varphi_M) = f'(\varphi_M)$. The $M - C$ circuit is described by a *second-order* SE in the state variables $v_C(t)$ and $\varphi_M(t)$ and the initial conditions are v_{C_0} and φ_{M_0}.

The $M - C$ has a continuum of non-isolated EPs in the (v, i)-domain coinciding with the φ_M axis, i.e.,

$$\mathcal{E} = \{(v_C, \varphi_M)^T \in \mathbb{R}^2 : v_C = 0\}.$$

Such a property is *structurally stable* and is due to the presence of the memristor that is able to store any value of the flux in steady state (cf. Theorem 2.2 in Chap. 2).

The dynamic description in the (φ, q)-domain of the $M - C$ circuit, derived via FCAM and the equivalent circuit in Fig. 6.6, is given by the ***first-order*** SE

$$\begin{cases} C\frac{d\varphi_C(t;t_0)}{dt} = -f(\varphi_C(t;t_0) + \varphi_{M0}) + f(\varphi_{M0}) + q_{C0} \\ \varphi_C(t_0;t_0) = 0. \end{cases} \tag{6.8}$$

The state variable is $\varphi_C(t; t_0)$ and the initial condition is by construction zero.

As seen in Chap. 5, if $(v_C(t), \varphi_M(t))$ is the solution of the IVP (6.7), then $\varphi_C(t; t_0) = \varphi_M(t) - \varphi_{M_0}$ is the solution of the IVP (6.8). Conversely, if $\varphi_C(t; t_0)$ is the solution of the IVP (6.8), then $(\dot{\varphi}_C(t; t_0), \varphi_C(t; t_0) + \varphi_{M_0})$ is the solution of the IVP (6.7).

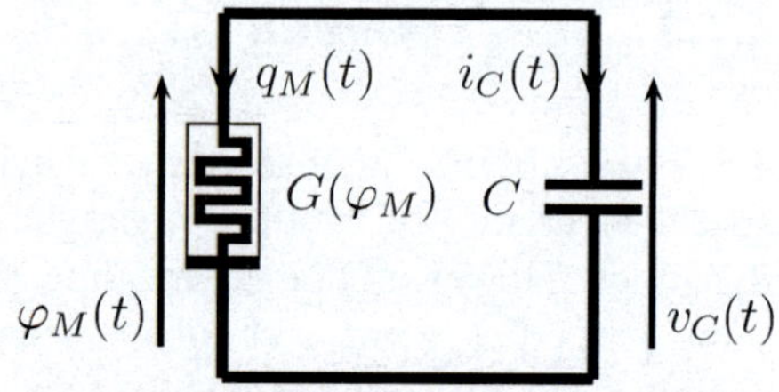

Fig. 6.5 The simplest memristor-based circuit

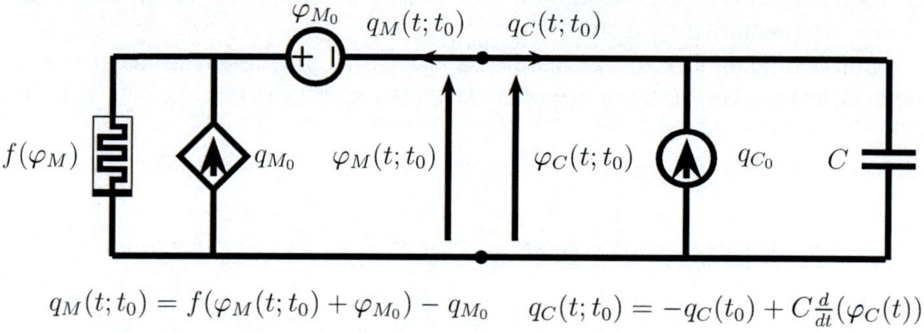

Fig. 6.6 Equivalent circuit in the (φ, q)-domain of the M–C circuit

6.1.3 Invariant Manifolds

The SE (6.8) describes the evolution in the (φ, q)-domain of the incremental capacitor flux $\varphi_C(t; t_0)$ for $t \geq t_0$ when $\varphi_C(t_0; t_0) = 0$. Such evolution depends on the initial conditions v_{C_0} and φ_{M_0} of the state variables in the (v, i)-domain that define the constant input

$$Q_0 = f(\varphi_{M_0}) + q_{C_0} = f(\varphi_{M_0}) + Cv_{C_0} \tag{6.9}$$

in the right-hand side (r.h.s.) of (6.8). The physical meaning of Q_0 is evident (by recalling $f(\varphi_{M_0}) = q_{M_0}$): it represents the *total charge* in the M–C circuit at the initial instant t_0. Let us introduce the total charge $Q(t)$ for $t \geq t_0$

$$Q(t) = f(\varphi_M(t)) + Cv_C(t). \tag{6.10}$$

Note that $Q(t)$ is given in terms of the state variables $v_C(t)$ and $\varphi_M(t)$ of the SE (6.7) in the (v, i)-domain. The following holds.

Property 6.1 The total charge $Q(t) = Q_0 =$ constant for all $t \geq t_0$.

Proof Property 6.1 is readily verified by evaluating the time-derivative of $Q(t)$ along the solutions of (6.7), i.e.,

$$\frac{d\,Q(t)}{dt} = G(\varphi_M(t))\frac{d\,\varphi_M(t)}{dt} + C\frac{d\,v_C(t)}{dt} = 0. \tag{6.11}$$

It follows that $Q(t)$ is constant in time. Hence, $Q(t)$ keeps the initial value $Q(t_0) = Q_0$ for all $t \geq t_0$. ■

Property 6.1 expresses the physical *law of conservation of charge*, i.e., the state variables in the (v, i)-domain, $v_C(t)$ and $\varphi_M(t)$, evolve according to (6.7), but the total charge $Q(t)$ remains constant to the initial value Q_0 set by v_{C_0} and φ_{M_0}.

Remark 6.4 We remark that the conservation of charge in the M–C circuit directly follows from KqL as well. Indeed, considering the equivalent circuit in the (φ, q)-domain, KqL yields

$$\begin{aligned} q_C(t; t_0) + q_M(t; t_0) &= q_C(t) - q_{C_0} + q_M(t) - q_{M_0} \\ &= (Cv_C(t) + f(\varphi_M(t))) - (Cv_{C_0} + f(\varphi_{M_0})) \\ &= Q(t) - Q(t_0) = 0 \end{aligned}$$

for any $t \geq t_0$.

Since $Q(t)$ is a function of the state variables in the (φ, q)-domain, Property 6.1 means that the total charge is an *invariant of motion* for the SEs in the (v, i)-domain. **Define, for any $Q_0 \in \mathbb{R}$, $\mathcal{M}(Q_0)$ as the subset of the phase-space in the (v, i)-domain where $Q_0 \in \mathbb{R}$ has a fixed value, that is**

$$\mathcal{M}(Q_0) = \{(v_C, \varphi_M)^T \in \mathbb{R}^2 : f(\varphi_M) + Cv_C = Q_0\}. \tag{6.12}$$

The following observations, whose verification is straightforward, make clear some salient features of $\mathcal{M}(Q_0)$.

- The set $\mathcal{M}(Q_0)$ is a one-dimensional manifold (i.e., a curve) in the phase-space $\mathbb{R}^2$ of the SEs (6.7) in the (v, i)-domain
- there are ∞^1 1D nonintersecting manifolds $\mathcal{M}(Q_0)$, spanning the whole phase-space $\mathbb{R}^2$ of the SE (6.7) in the (v, i)-domain, obtained by varying Q_0 in $\mathbb{R}$. Some of these manifolds, for different values of Q_0, are depicted in Fig. 6.7 in the case where the memristor has a cubic characteristic $q_M = f(\varphi_M) = -\varphi_M + \frac{1}{3}\varphi_M^3$ as shown in Fig. 6.9
- since $Q(t)$ is constant for any $t \geq t_0$, it follows that for any given $Q_0 \in \mathbb{R}$ the manifold $\mathcal{M}(Q_0)$ is *positively invariant* for the evolution of $(v_C(t), \varphi_M(t))$, i.e., if $(v_{C_0}, \varphi_{M_0})^T \in \mathcal{M}(Q_0)$ then the solution of (6.7) with such initial conditions belongs to $\mathcal{M}(Q_0)$ for any $t \geq t_0$
- the manifold where a solution evolves can be changed by varying Q_0, i.e., by varying the initial conditions at t_0 for the state variables in the (v, i)-domain

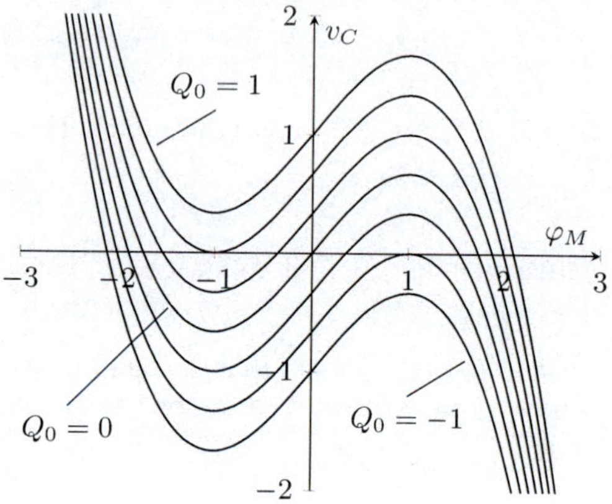

Fig. 6.7 Invariant manifolds of the $M - C$ circuit for different values of Q_0. From the lower manifold to the upper manifold we have $Q_0 = -1, -2/3, -1/3, 0, 1/3, 2/3, 1$, respectively

- the one-to-one correspondence between the solution of IVP (6.7) and IVP (6.8) implies that the dynamics on a manifold $\mathcal{M}(Q_0)$ can be also described by the first-order SE (6.8) in the (φ, q)-domain
- we have obtained for the $M - C$ circuit a *foliation of the state space in the* (v, i)*-domain* in ∞^1 manifolds $\mathcal{M}(Q_0)$ where the dynamics is first order and is described by the reduced-order system (6.7).

6.1.4 Nonlinear Dynamics and Saddle-Node Bifurcations Without Parameters

In this section, we use the concept of invariant manifolds and foliation of the state space in the (v, i)-domain to investigate nonlinear dynamics and bifurcations in the M–C circuit.

First, consider the reduced-order (first-order) SE (6.8) describing the dynamics in the (φ, q)-domain. Since $\varphi_M(t) = \varphi_C(t; t_0) + \varphi_{M_0}$, such an SE becomes

$$C\frac{d\varphi_M(t)}{dt} = -f(\varphi_M(t)) + Q_0 \tag{6.13}$$

with $\varphi_M(t_0) = \varphi_{M_0}$. Moreover, since $v_C(t) = d\varphi_M(t)/dt$, the invariant manifolds can be written as follows:

$$\mathcal{M}(Q_0) = \{(v_C, \varphi_M)^T \in \mathbb{R}^2 : v_C = \dot{\varphi}_M = \frac{1}{C}(-f(\varphi_M) + Q_0)\}. \tag{6.14}$$

Remark 6.5 Clearly, $\mathcal{M}(Q_0)$ is related to the *dynamic route* of (6.13) (cf. Chap. 4).

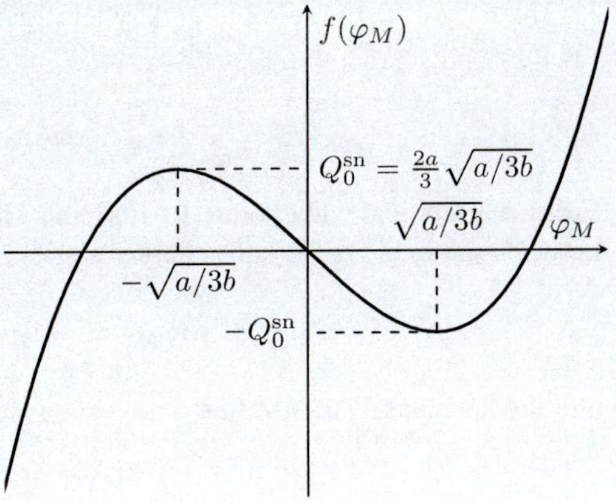

Fig. 6.8 Cubic characteristic of an active flux-controlled memristor when $a, b > 0$

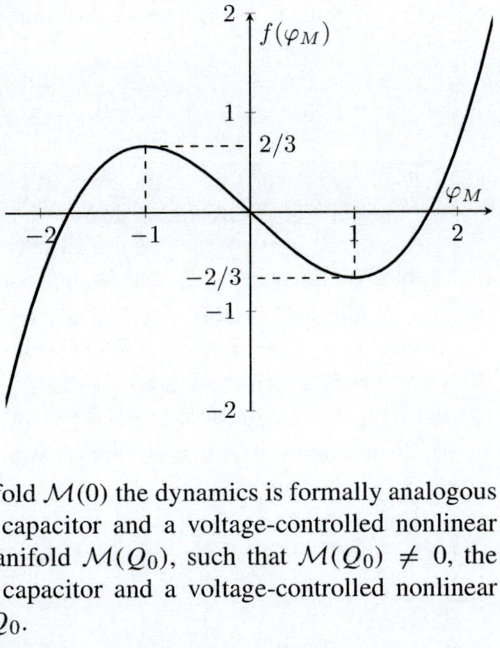

Fig. 6.9 Cubic characteristic of an active flux-controlled memristor in the case $a = 1$, $b = 1/3$

Remark 6.6 Note that on the 0-manifold $\mathcal{M}(0)$ the dynamics is formally analogous to that of a circuit composed by a capacitor and a voltage-controlled nonlinear resistor $i = f(v)$. Instead, on a manifold $\mathcal{M}(Q_0)$, such that $\mathcal{M}(Q_0) \neq 0$, the dynamics is analogous to that of a capacitor and a voltage-controlled nonlinear resistor forced with a constant term Q_0.

To further analyze the global dynamics of (6.13), assume that the CR of the memristor is defined by the cubic function

$$q_M = f(\varphi_M) = -a\varphi_M + b\varphi_M^3 \tag{6.15}$$

where $a, b > 0$. The characteristic is depicted in Fig. 6.8 for generic $a, b > 0$, while Fig. 6.9 reports the special case $a = 1$, $b = 1/3$. Note that $f(\cdot)$ is non-monotone, hence we have an active memristor (Chap. 2) which can be implemented for instance by a passive flux-controlled memristor (i.e., $b\varphi_M^3$) in parallel with an active resistor (i.e., a negative conductance $-a$).

The EPs of (6.13) are obtained by letting $d\varphi_M(t)/dt = 0$ and they correspond to the intersection of $\mathcal{M}(Q_0)$ with $v_C = 0$, i.e., they are solutions of

$$-b\varphi_M^3 + a\varphi_M + Q_0 = 0. \tag{6.16}$$

From a graphical viewpoint, to find the EPs we need to find the intersections between the cubic memristor characteristic

$$f(\varphi_M) = -a\varphi_M + b\varphi_M^3$$

and the horizontal straight line

$$\ell(\varphi_M) = Q_0.$$

From the graphic of $f(\cdot)$ (Fig. 6.8), we conclude the following. Let

$$Q_0^{\mathrm{sn}} = \frac{2a}{3}\sqrt{\frac{a}{3b}}. \tag{6.17}$$

1. If $|Q_0| < Q_0^{\mathrm{sn}}$, then there are three EPs $\bar{\varphi}_M^{\alpha} \neq \bar{\varphi}_M^{\beta} \neq \bar{\varphi}_M^{\gamma}$. The dynamic route (see Fig. 6.10 in the case $a = 1$, $b = 1/3$, hence $Q_0^{\mathrm{sn}} = 2/3$) permits to figure out the stability properties of the EPs. Note that, according to the arrowheads on the dynamic route, $\bar{\varphi}_M^{\alpha}$ and $\bar{\varphi}_M^{\gamma}$ attract solutions starting nearby, hence they are asymptotically stable. Instead, $\bar{\varphi}_M^{\beta}$ repels solutions starting nearby, i.e., it is unstable (Fig. 6.10 in the case $Q_0 = 0$ and $Q_0 = 1/3$).
2. If $Q_0 = Q_0^{\mathrm{sn}}$, then there are two EPs $\bar{\varphi}_M^{\alpha\beta}$ and $\bar{\varphi}_M^{\gamma}$. According to the dynamic route (Fig. 6.10 in the case $Q_0 = 2/3$), the EP $\bar{\varphi}_M^{\gamma}$ is asymptotically stable, while $\bar{\varphi}_M^{\alpha\beta}$ is unstable since it repels solutions starting nearby at its right.
3. If $Q_0 = -Q_0^{\mathrm{sn}}$, then there are two EPs $\bar{\varphi}_M^{\alpha}$ and $\bar{\varphi}_M^{\beta\gamma}$. The first one is asymptotically stable while the latter is unstable.
4. If $Q_0 > Q_0^{\mathrm{sn}}$, then there is a unique EP $\bar{\varphi}_M^{\gamma}$. From the dynamic route (cf. Fig. 6.10 in the case $Q_0 = 1$), it follows that $\bar{\varphi}_M^{\gamma}$ is a globally attracting asymptotically stable EP.
5. If $Q_0 < -Q_0^{\mathrm{sn}}$, then there is a unique EP $\bar{\varphi}_M^{\alpha}$ which is globally attracting.

Remark 6.7 It is also possible to analytically find the roots of a third-order algebraic equation by using some known techniques and formulas (e.g., [5]).

Let

$$\hat{\sigma} = \sqrt[3]{\frac{Q_0}{2} + \sqrt{\Delta}}$$

where

$$\Delta = \frac{Q_0^2}{4b^2} - \frac{a^3}{27b^3} = \frac{Q_0^2}{4b^2}\left[1 - \left(\frac{Q_0^{\mathrm{sn}}}{Q_0}\right)^2\right].$$

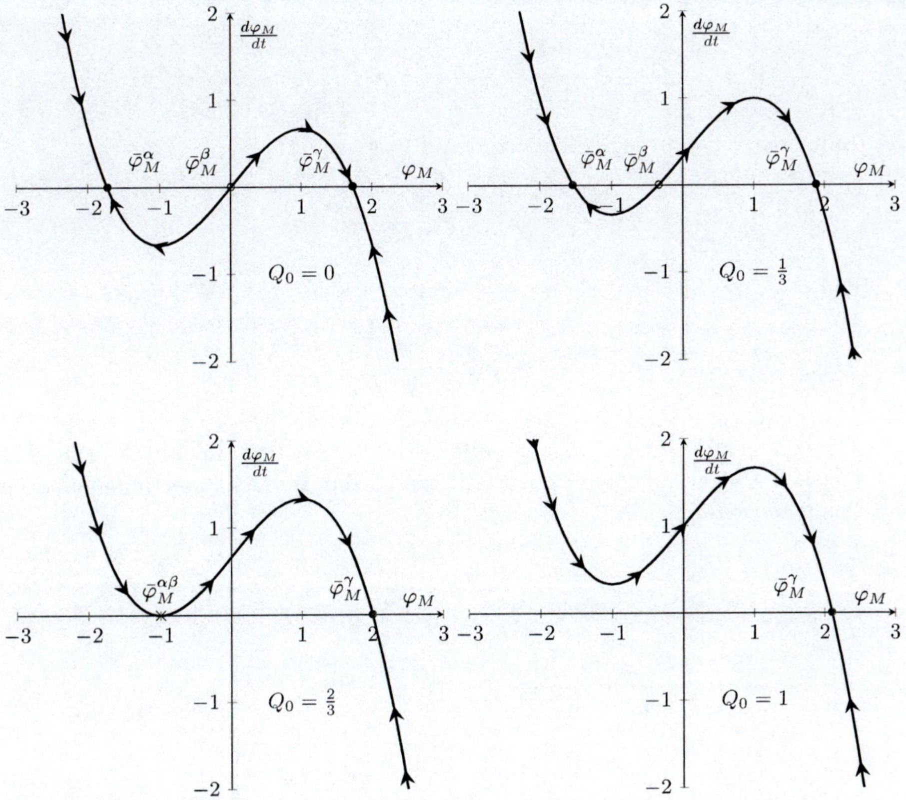

Fig. 6.10 Dynamic route of (6.13) for various values of Q_0 and the cubic nonlinearity (6.15) with $a = 1$ and $b = 1/3$. We have $Q_0^{\text{sn}} = 2/3$

1. If $Q_0 > Q_0^{\text{sn}}$, we have $\Delta > 0$ and $\hat{\sigma}$ is real. Then, there is a unique EP

$$\bar{\varphi}_M^\gamma = \frac{a}{3b\hat{\sigma}} + \hat{\sigma}.$$

2. Similarly, if $Q_0 < -Q_0^{\text{sn}}$, there is a unique EP

$$\bar{\varphi}_M^\alpha = -\frac{a}{3b\hat{\sigma}} - \hat{\sigma}.$$

3. If $Q_0 = Q_0^{\text{sn}}$, then $\Delta = 0$ and again $\hat{\sigma}$ is real. There are two EPs given by

$$\bar{\varphi}_M^{\alpha\beta} = -\sqrt{\frac{a}{3b}}$$

and

$$\bar{\varphi}_M^{\gamma} = \frac{a}{3b\hat{\sigma}} + \hat{\sigma}.$$

4. Similarly, if $Q_0 = -Q_0^{\text{sn}}$, there are two EPs given by

$$\bar{\varphi}_M^{\alpha} = -\frac{a}{3b\hat{\sigma}} - \hat{\sigma}$$

and

$$\bar{\varphi}_M^{\beta\gamma} = \sqrt{\frac{a}{3b}}.$$

5. Finally, suppose $|Q_0| < Q_0^{\text{sn}}$. In this case we have $\Delta < 0$ and so we need to consider $\sqrt{\Delta}$ in the complex sense. It can be shown via suitable manipulations that there are three EPs

$$\bar{\varphi}_M^{\gamma} = \frac{a}{3b\hat{\sigma}} + \hat{\sigma} \tag{6.18}$$

$$\bar{\varphi}_M^{\alpha} = \frac{1}{2}\left[-\bar{\varphi}_M^{\gamma} + j\sqrt{3}(-\bar{\varphi}_M^{\gamma} + 2\hat{\sigma})\right] \tag{6.19}$$

$$\bar{\varphi}_M^{\beta} = \frac{1}{2}\left[-\bar{\varphi}_M^{\gamma} - j\sqrt{3}(-\bar{\varphi}_M^{\gamma} + 2\hat{\sigma})\right] \tag{6.20}$$

and that all three EPs turn out to be real numbers.

For instance, when $Q_0 = 0$ the three EPs are simply given by

$$\hat{\varphi}_M^{\alpha} = -\sqrt{\frac{a}{b}}, \quad \hat{\varphi}_M^{\beta} = 0; \quad \hat{\varphi}_M^{\gamma} = \sqrt{\frac{a}{b}}.$$

■

Let us now study the global dynamics of (6.13). The following holds.

Property 6.2 Consider the $M - C$ circuit with the cubic nonlinearity (6.15). Any solution of the SE (6.13) describing the circuit in the (φ, q)-domain is bounded and hence defined for $t \geq t_0$; moreover, it converges to an EP as $t \to +\infty$.

Proof First, let us verify that any solution of (6.13) is defined and bounded for any $t \geq t_0$. Consider the vector field $-f(\varphi_M) + Q_0$ defining (6.13). There exists $\tilde{\varphi} > 0$ and $\epsilon > 0$ such that $-f(\varphi_M) + Q_0 < -\epsilon < 0$ for $\varphi_M \geq \tilde{\varphi}$ and $-f(\varphi_M) + Q_0 > \epsilon > 0$ for $\varphi_M \leq -\tilde{\varphi}$. Then, the set $S = [-\tilde{\varphi}, \tilde{\varphi}]$ is positively invariant for (6.13); moreover, any solution with initial condition $\varphi_M(t_0) \notin S$ reaches S in finite time and stays in S thereafter. This easily implies that any solution of (6.13) is bounded and hence defined for any $t \geq t_0$.

The state space of (6.13) is one dimensional and coincides with the φ_M-axis. Equation (6.13) is a first-order autonomous ODE with bounded solutions. Then due to known properties, any solution necessarily converges to an EP as $t \to +\infty$ (Chap. 4). ■

Remark 6.8 We have given a mathematical proof that any solution of (6.13) is bounded and convergent toward an EP. It is instructive to notice that the same properties can be immediately verified from a geometric viewpoint by using the dynamic route (cf. Fig. 6.10).

Summing up, the global dynamics in the (φ, q)-domain can be described as follows.

For any $|Q_0| < Q_0^{\text{sn}}$, the SE (6.13) has two asymptotically stable EPs $\bar{\varphi}_M^\alpha$ and $\bar{\varphi}_M^\gamma$. If the initial condition $\varphi_{M_0} \in (-\infty, \bar{\varphi}_M^\beta)$, then the corresponding solution converges to $\bar{\varphi}_M^\alpha$. If $\varphi_{M_0} \in (\bar{\varphi}_M^\beta, +\infty)$, then the solution converges to $\bar{\varphi}_M^\gamma$. The attraction basin $\mathfrak{A}(\cdot)$ of an asymptotically stable EP is defined as the set of initial conditions such that the corresponding solution converges to the considered EP. Then, we have

$$\mathfrak{A}(\bar{\varphi}_M^\alpha) = (-\infty, \bar{\varphi}_M^\beta)$$

while

$$\mathfrak{A}(\bar{\varphi}_M^\gamma) = (\bar{\varphi}_M^\beta, +\infty).$$

For these values of Q_0 the $M - C$ circuit is in *bistable* mode.

When $Q_0 < -Q_0^{\text{sn}}$ (resp., $Q_0 > Q_0^{\text{sn}}$), the SE (6.13) has only one asymptotically stable EP $\bar{\varphi}_M^\alpha$ (resp., $\bar{\varphi}_M^\gamma$) attracting all solutions. We have

$$\mathfrak{A}(\bar{\varphi}_M^\alpha) = \mathbb{R}$$

when $Q_0 < -Q_0^{\text{sn}}$, while

$$\mathfrak{A}(\bar{\varphi}_M^\gamma) = \mathbb{R}$$

when $Q_0 > Q_0^{\text{sn}}$. For all these values of Q_0 the M–C circuit is in *monostable* mode.

The special case $Q_0 = \pm Q_0^{\text{sn}}$ can be dealt with in a similar way.

Note that in any case the location of the EPs of (6.13) changes when Q_0 changes. Clearly, by varying Q_0 in $\mathbb{R}$ we obtain infinitely many different first-order dynamics for (6.13), some of which are bistable, some monostable.

The concept of foliation in invariant manifolds permits to investigate not only EPs and stability properties of (6.13), but bifurcation phenomena as well. To this end, it is convenient to identify *two logically different causes leading to bifurcations.*

On one hand, we have the variation of the circuit parameters C, a, b. On the other hand, we have the variation of Q_0, which acts as an additional parameter in the r.h.s. of the SE in the (φ, q)-domain (6.13). Note that Q_0 can be varied by varying the initial conditions v_{C_0} and φ_{M_0} for the state variables in the (v, i)-domain even when circuit parameters are held fixed.

Suppose first we fix $Q_0 = \bar{Q}_0$, i.e., the initial condition v_{C_0} and φ_{M_0} for the state variables in the (v, i)-domain are such that $v_{C_0} = \frac{1}{C}(-f(\varphi_{M_0}) + \bar{Q}_0)$. Choose for instance $\bar{Q}_0 = 0$. In this case, we may have bifurcations of EPs of (6.13) only if parameter a changes sign. These are the standard bifurcations due to changing a parameter on a *fixed invariant manifold*. However, if Q_0 changes, due to changes in the initial condition v_{C_0} and φ_{M_0}, then (6.13) can undergo bifurcations of EPs even if the circuit parameters are kept fixed. In particular, (6.13) is seen to undergo a *saddle-node bifurcation* when $\bar{\varphi}_M^{\beta}$ and $\bar{\varphi}_M^{\alpha}$ (or else $\bar{\varphi}_M^{\beta}$ and $\bar{\varphi}_M^{\gamma}$) collide and annihilate each other. The conditions $\bar{\varphi}_M^{\beta} = \bar{\varphi}_M^{\alpha}$ and $\bar{\varphi}_M^{\beta} = \bar{\varphi}_M^{\gamma}$ yield

$$Q_0 = \pm Q_0^{\mathrm{sn}} = \pm \frac{2}{3} \frac{\sqrt[3]{a^2}}{\sqrt{3b}} \tag{6.21}$$

for the values of Q_0 at which such saddle-node bifurcation occurs, which is in accordance with the results on EPs of (6.13) previously obtained.

Remark 6.9 Let us study in more detail the saddle-node bifurcations due to varying Q_0. Suppose once more $C = 1$, $a = 1$ and $b = 1/3$. Consider the critical value $Q_0 = 2/3$ for which there is a bifurcating EP at $\bar{\varphi}_M = -1$. We have

$$\frac{d\varphi_M}{dt} = \varphi_M - \frac{1}{3}\varphi_M^3 + Q_0 \doteq F(\varphi_M).$$

By expanding F in Taylor series in a neighborhood of $\bar{\varphi}_M = -1$, we obtain

$$\begin{aligned}\frac{d\varphi_M}{dt} &= F(-1) + F'(-1)(\varphi_M + 1) + \frac{1}{2}F''(-1)(\varphi_M + 1)^2 \\ &= Q_0 - \frac{2}{3} + (\varphi_M + 1)^2\end{aligned}$$

where we have neglected higher-order terms of the expansion. The situation in a small neighborhood of $\bar{\varphi}_M = -1$ is illustrated in appare prima di Fig. 6.11 for three values of Q_0, i.e., $Q_0 < 2/3$, $Q_0 = 2/3$, and $Q_0 > 2/3$. Geometrically, the situation is analogous to the saddle-node bifurcation of an EP studied in Sect. 4.4.1 of Chap. 4 (cf. Fig. 4.21 in Chap. 4 with Fig. 6.12). There is however a fundamental difference, since in that case the bifurcation was due to *varying a circuit parameter*, while in this case circuit parameters are kept fixed and the bifurcation is due to *changing the initial conditions* (Q_0).

Remark 6.10 The bifurcations due to changing the initial conditions for the state variables in the (v, i)-domain, for a fixed set of circuit parameters, are a new dynamic phenomenon peculiar to memristor circuits and are strictly related to the property of foliation of the state space in the (v, i)-domain. In the following, we will refer to them as *bifurcations without (changing circuit) parameters*, or, simply, *bifurcations without parameters*, to differentiate them from standard bifurcations due to changing the circuit parameters. By analogy, we will use the term bifurcations without parameters also when studying bifurcations due to changing initial conditions in higher-order memristor circuits in the next sections.

Remark 6.11 It is worth to remark that the mathematical concept of bifurcations without parameters has been originally introduced in [6]. We refer the reader to Appendix 1 for a discussion on the link between the bifurcations studied for the $M - C$ circuit and bifurcations without parameters introduced in [6].

Let us now study the dynamics of the $M - C$ circuit in the (v, i)-domain. We have the following.

Property 6.3 Consider the $M - C$ circuit with the cubic memristor nonlinearity (6.15). Any solution of the SEs (6.7) in the (v, i)-domain is bounded and hence defined for $t \geq t_0$, moreover it converges to an EP as $t \to +\infty$.

Proof It is quite simple to verify this result in mathematical terms. In fact, we have seen that any solution $\varphi_M(t)$ of the reduced system (6.13) is bounded (for $t \geq t_0$), hence also $v_C(t) = d\varphi_M(t)/dt = -f(\varphi_M(t)) + Q_0$ is bounded, implying that this is true also of any solution $(v_C(t), \varphi_M(t))$ of (6.7). Moreover, the reduced system (6.13) is convergent, i.e., we have $\varphi_M(t) \to \bar{\varphi}_M$ and $d\varphi_M(t)/dt \to 0$ as $t \to +\infty$, where $\bar{\varphi}_M$ is an EP of (6.13). Since $d\varphi_M(t)/dt = v_C(t)$, it follows that the corresponding solution of the SE in the (v, i)-domain tends to the EP $(0, \bar{\varphi}_M)$ as $t \to +\infty$. ■

The EPs of the SEs (6.7) in the (v, i)-domain are given by $\bar{v}_C = 0$ and $\bar{\varphi}_M \in \mathbb{R}$, i.e., there is a manifold (a continuum of EPs) given by the φ_M-axis. Note that at an EP the memristor flux is an arbitrary constant while the capacitor voltage vanishes. As remarked before, the state space (v_C, φ_M) in the (v, i)-domain can be foliated in ∞^1 1D invariant manifolds where the reduced dynamics is first order. Moreover, on any fixed manifold, the $M - C$ circuit always has isolated EPs. If $|Q_0| < Q_0^{\mathrm{sn}}$, the EPs on a given manifold $\mathcal{M}(Q_0)$ of (6.7) are $(0, \bar{\varphi}_M^{\alpha})$, $(0, \bar{\varphi}_M^{\beta})$ and $(0, \bar{\varphi}_M^{\gamma})$. Hence, due to Property 6.3, in the (v_C, φ_M) state-space any trajectory leaving from $(v_{C_0}, \varphi_{M_0}) \neq (0, \bar{\varphi}_M^{\beta})$ lies on the manifold $\mathcal{M}(Q_0)$ and converges toward one of the stable EPs, i.e., toward either $(0, \bar{\varphi}_M^{\alpha})$ or $(0, \bar{\varphi}_M^{\gamma})$. Such analytical results are confirmed by numerical simulations (see Fig. 6.11) of the SE (6.7) in the (v, i)-domain with circuit parameters $a = 1$, $b = \frac{1}{3}$ in (6.15) and $C = 1$. The dynamics for other values of Q_0 can be analyzed similarly.

Remark 6.12 We stress that although the $M - C$ circuit is described by a second-order SE in the (v, i)-domain, it never oscillates. Actually, any solution

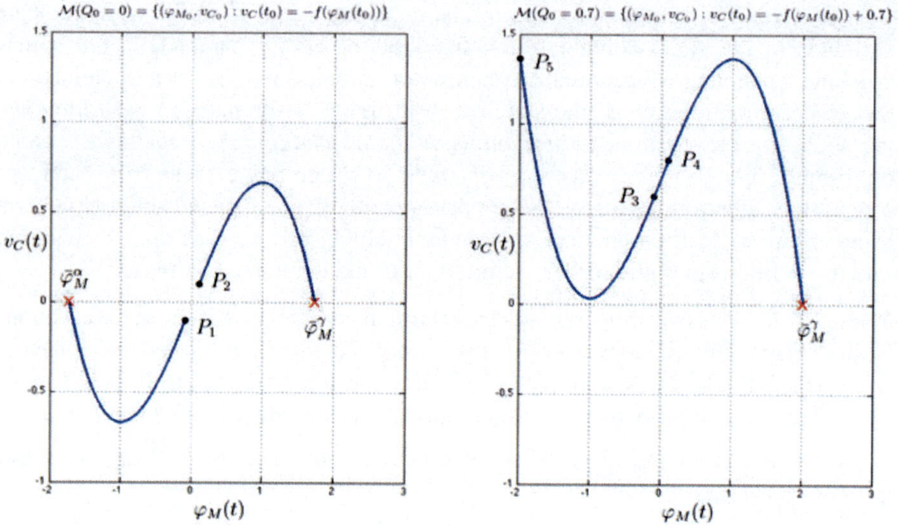

Fig. 6.11 Numerical simulations of the SE (6.7) in the (v, i)-domain with a memristor nonlinearity $-\varphi_M + (1/3)\varphi_M^3$ and $C = 1$. The left (resp., right) part shows the M–C circuit operating in bistable mode when $Q_0 = 0$ (resp., mono-stable mode when $Q_0 = 0.7 > Q_0^{sn} = \frac{2}{3}$). Initial conditions are: $P_1 = (0.1, -f(0.1))$, $P_2 = (-0.1, -f(-0.1))$ (i.e., $Q_0 = 0$ in P_1 and P_2), $P_3 = (0.1, -f(0.1) + 0.7)$, $P_4 = (-0.1, -f(-0.1) + 0.7)$ and $P_5 = (-2, -f(-2) + 0.7)$ (i.e., $Q_0 = 0.7$ in P_3, P_4 and P_5)

$(v_C(t), \varphi_M(t))$ is bounded and converges toward an EP. This result is a straightforward consequence of the principle of order reduction proved via FCAM, according to which on each manifold the dynamics is essentially first order (cf. Chap. 4). Such result would be more difficult to prove via a direct analysis of the second-order SE (6.7) in the (v, i)-domain.

Remark 6.13 (Piecewise Linear Memristors and Smoothness) We wish to briefly discuss another basic advantage of FCAM. Consider again the $M - C$ circuit but suppose the locally active flux-controlled memristor has a piecewise linear (PWL) characteristic

$$q_M = f(\varphi_M) = a\varphi_M + \frac{1}{2}(a-b)(|\varphi_M + 1| - |\varphi_M - 1|)$$

where $a < 0$ and $b > 0$ as in Fig. 6.13. This has been frequently used in the literature starting from the fundamental paper [7].

The SEs describing the dynamics in the (φ, q)-domain and (v, i)-domain are once more given by (6.8) and (6.7), respectively. Note that the r.h.s. of (6.8) is Lipschitz continuous, hence the uniqueness of the solution of the IVP problem with respect to the initial conditions is guaranteed and the dynamical analysis can be conducted via standard techniques of differential equations along the lines

Fig. 6.12 Saddle-node bifurcation of the EP $\bar{\varphi}_M = -1$ when varying Q_0 in a neighborhood of 2/3

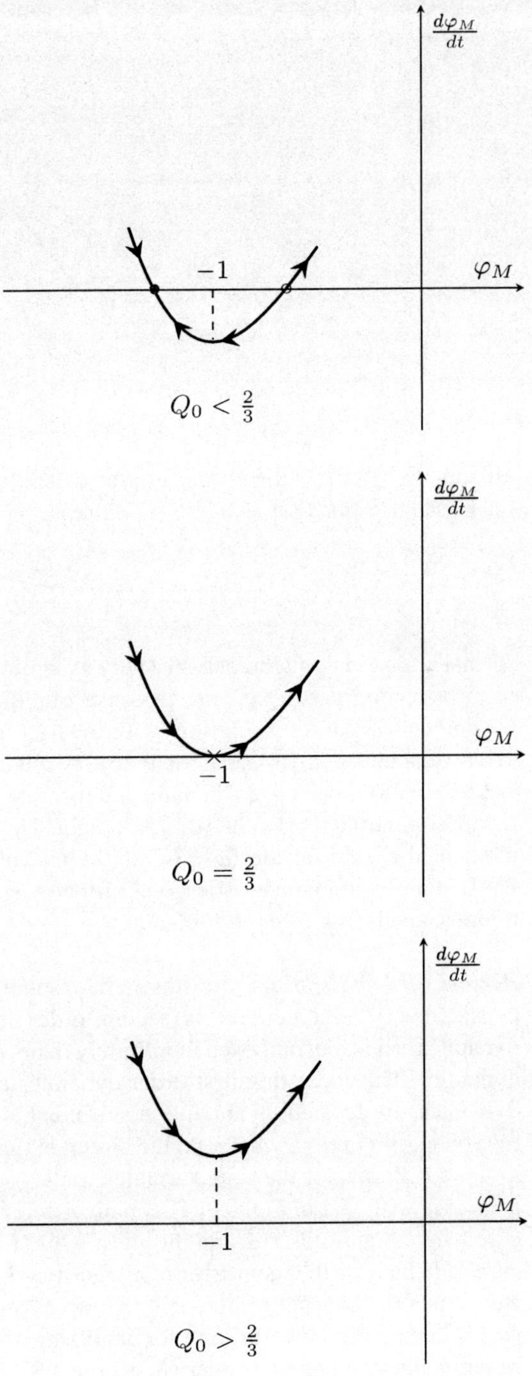

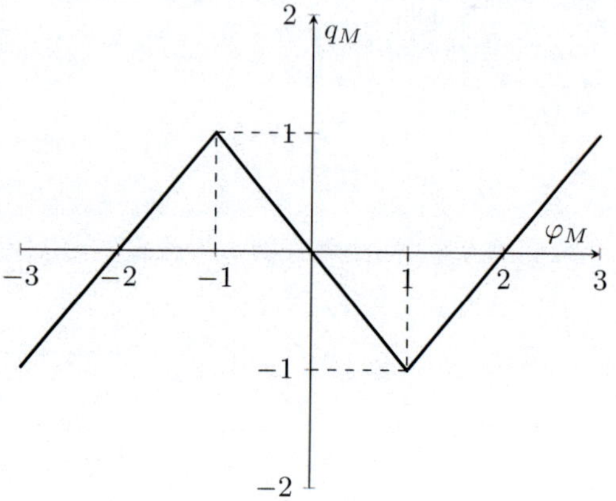

Fig. 6.13 PWL characteristic of an active memristor in the case $a = -1$ and $b = 1$

already discussed. Quite on the contrary, the r.h.s. of (6.7) contains a discontinuous memductance function

$$G(\varphi_M) = f'(\varphi_M) = \begin{cases} a, & |\varphi_M| < 1 \\ b, & |\varphi_M| > 1 \end{cases}$$

so that a rigorous mathematical analysis should rely on the concept of solutions in an extended case (e.g., in the sense of Filippov [8]). Moreover, the property of uniqueness of the solution with respect to the initial conditions is not a priori guaranteed. It is apparent that it is much easier to deal with the smoother equation (6.8) in the (φ, q)-domain and then obtain the solution in the (v, i)-domain by differentiating that in the (φ, q)-domain. The possibility to deal with a smoother dynamical system in the (φ, q)-domain, and hence easily include in the analysis PWL memristor characteristics, is a further basic advantage of FCAM with respect to the analysis in the (v, i)-domain.

Remark 6.14 It is useful to summarize some main results thus obtained. The considered $M - C$ circuit is second-order in the (v, i)-domain; however, its dynamics can be decomposed in infinitely many different reduced-order (first-order) dynamics. The coexisting first-order dynamics are smoother than the second-order dynamics, are located on suitable manifolds (curves) in $\mathbb{R}^2$, and are parameterized by a quantity Q_0 (total charge in the circuit at the initial instant t_0) depending on the initial conditions for the state variables in the (v, i)-domain. The phase portrait has analogies with that in Fig. 6.2. The main difference is that Fig. 6.2 concerns a linear case where manifolds are straight lines, while for the $M - C$ circuit manifolds are curves defined by the memristor nonlinearity. The analysis of the $M - C$ circuit has shown that the concept of invariant manifolds and foliation of the state space in the (v, i)-domain is a powerful tool for analyzing nonlinear dynamics and bifurcations. In particular, the reduced-order (first-order) SE in the (φ, q)-domain facilitates the study and permits to obtain qualitative and analytical results that cannot be easily derived from a direct analysis of the second-order SEs in the (v, i)-domain.

6.2 Second-Order Memristor Oscillators

In Sect. 6.1.2, we have considered a simple autonomous $M - C$ circuit where the memristor is flux-controlled. We have been able to write an SE describing the dynamics and, on this basis, a number of salient features have been studied. In particular, the dynamics in the (φ, q)-domain has been proved to be first-order, so that any solution is convergent toward an EP and the circuit is either mono-stable or bistable for the considered cubic nonlinearity.

In this section, we consider a variant of the $M - C$ circuit studied in Sect. 6.1.2, where we assume that the memristor is charge-controlled but not flux-controlled. This apparently simple modification leads to substantial structural changes in the dynamics. Indeed, we first observe that in the charge-controlled case we are unable to write an SE description and the circuit displays impasse points (cf. Chap. 4), i.e., points where the solution cannot be continued either forward or backward in time. We then show that we can break impasse points by considering a more realistic circuit including a parasitic element. This also shows that the circuit is intrinsically second order and is able to display relaxation oscillation. Also for this circuit we study invariant manifolds in the (v, i)-domain, coexisting dynamics and bifurcations without parameters. In particular, we show that in this case there *coexist oscillatory and convergent dynamics*; moreover, the circuit can display Hopf bifurcations without parameters.

6.2.1 *M − C Circuit with Impasse Points*

Consider for $t \geq t_0$, where $-\infty < t_0 < \infty$, a simple memristor-based circuit in the class $\mathcal{LM}$ (see the M–C circuit in Fig. 6.14), composed of one memristor M connected to a capacitor C. Suppose the memristor is charge-controlled and is defined by a smooth non-monotone function

$$\varphi_M(t) = h(q_M(t)) = -aq_M(t) + bq_M^3(t) \tag{6.22}$$

with $a, b > 0$. Note that the memristor is not flux-controlled and is (locally) active since the memristance $R(q_M(t)) = h'(q_M(t)) < 0$ for $|q_M| < \sqrt{a/3b}$.

Let $v_C(t_0) = v_{C_0}$, $q_M(t_0) = q_{M_0}$ be the initial conditions at t_0 for the state variables in the (v, i)-domain. It follows that $q_C(t_0) = q_{C_0} = Cv_{C_0}$. The corresponding circuit in the (φ, q)-domain obtained via FCAM is represented in Fig. 6.15.

Analysis by inspection permits to write the following KφL, KqL, and CRs

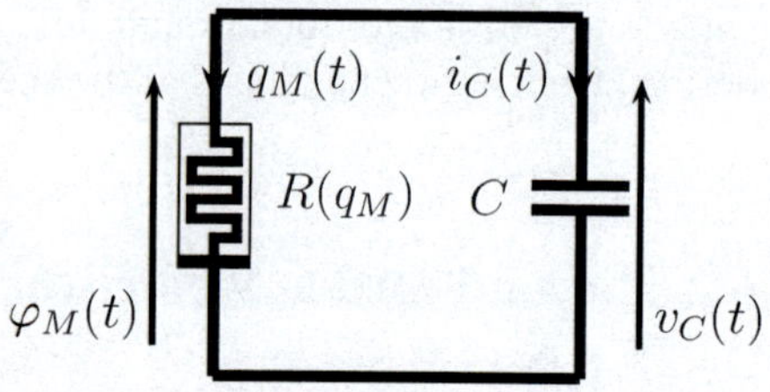

Fig. 6.14 The simplest memristor-based circuit with a charge-controlled (active) memristor

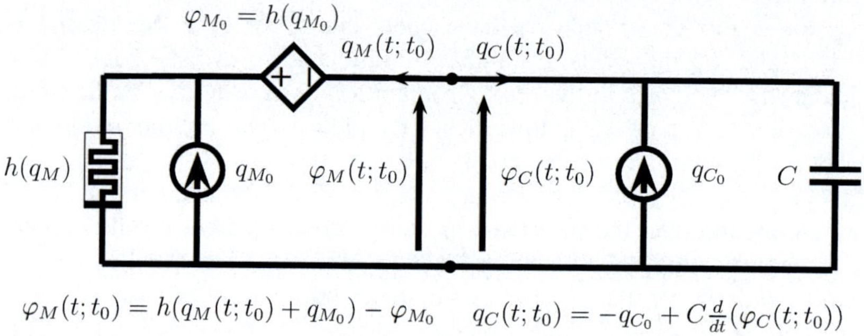

Fig. 6.15 Equivalent circuit in the (φ, q)-domain of the M–C circuit in Fig. 6.14 for $t \geq t_0$

$$\varphi_C(t;t_0) = \varphi_M(t;t_0) \tag{6.23a}$$

$$q_C(t;t_0) = -q_M(t;t_0) \tag{6.23b}$$

$$q_C(t;t_0) = -q_{C_0} + C\frac{d}{dt}(\varphi_C(t;t_0)) \tag{6.23c}$$

$$\varphi_M(t;t_0) = -h(q_{M_0}) + h(q_M(t;t_0) + q_{M_0}) \tag{6.23d}$$

for $t \geq t_0$. The state variable in the (φ, q)-domain is $\varphi_C(t;t_0)$ and the initial condition is $\varphi_C(t_0;t_0) = 0$.

It can be checked that the circuit in the (φ, q)-domain does not satisfy the conditions for the existence of the SE representation given in Property 3.1 of Chap. 3. Namely, since the nonlinear function $h(\cdot)$ is not globally invertible, it is not possible to obtain the corresponding SE in the (φ, q)-domain in terms of the state variable $\varphi_C(t;t_0)$, i.e., the SE for this circuit does not exist globally. Such a situation is analogous to that studied in Sect. 4.1.2 of Chap. 4.

On the other hand, the following DAE in the (φ, q)-domain for any $t \geq t_0$ can be readily derived from (6.23)

$$C\frac{d}{dt}\varphi_M(t;t_0) = -q_M(t;t_0) + q_{C_0} \tag{6.24a}$$

$$\varphi_M(t;t_0) = -h(q_{M_0}) + h(q_M(t;t_0) + q_{M_0}) \tag{6.24b}$$

with $\varphi_M(t_0; t_0) = 0$. This can be put in the simplified form

$$\frac{dx(t)}{dt} = -\frac{1}{C}y(t) + \frac{1}{C}Y_0 \tag{6.25a}$$

$$x(t) = h(y(t)) = -ay(t) + by^3(t) \tag{6.25b}$$

with $x(t_0) = X_0$, by letting $x(t) = \varphi_M(t; t_0) + h(q_{M0}) = \varphi_M(t)$, $y(t) = q_M(t; t_0) + q_{M0} = q_M(t)$, $X_0 = h(q_{M0})$ and

$$Y_0 = q_{M0} + q_{C0} = q_{M0} + Cv_{C0}. \tag{6.26}$$

Note that Y_0 is a term depending on the initial conditions v_{C0} and q_{M0} for the state variables in the (v, i)-domain.

It turns out that according to (6.25) the dynamics of the M–C circuit in Fig. 6.14 evolve, starting from any initial condition X_0, onto the constraint $x = h(y)$. For any Y_0 there is only one EP

$$P = (\bar{x}, \bar{y}) = (h(Y_0), Y_0). \tag{6.27}$$

The global dynamics, and the stability of the EP, can be easily assessed from the following rule (see (6.25a)):

$$\frac{dx(t)}{dt} > 0 \qquad y < Y_0$$

$$\frac{dx(t)}{dt} < 0 \qquad y > Y_0.$$

That is, for all $t > t_0$, the solution of the M–C circuit in Fig. 6.14 must follow the curve along the direction indicated by the arrowheads in the dynamic route reported in Figs. 6.16 and 6.17 for the cases $|Y_0| < \sqrt{a/3b}$ and $|Y_0| > \sqrt{a/3b}$, respectively (Fig. 6.17 reports only the case $Y_0 > \sqrt{a/3b}$, but dual considerations hold for $Y_0 < -\sqrt{a/3b}$). It is apparent in Figs. 6.16 and 6.17 that $x(t)$ is increasing in the solid part of the dynamic route (i.e., when $y < Y_0$) but decreasing in the dashed part.

The analysis of the dynamic route allows us to draw the following results:

1. if the initial conditions v_{C0}, q_{M0} for the state variables in the (v, i)-domain are such that

$$|Y_0| = |q_{M0} + q_{C0}| < \sqrt{a/3b}$$

then the unique EP P is unstable (Fig. 6.16). Moreover, there are two forward impasse points at $P_1 = \sqrt{a/3b}(-(2/3)a, 1)$ and $P_2 = -P_1$. Indeed, if we consider a solution starting at $(\hat{X}, \hat{Y})$ with $\hat{Y} \in (Y_0, \sqrt{a/3b}) \cup (\sqrt{a/3b}, \infty)$ and $\hat{X} = h(\hat{Y})$, it can be checked that the solution reaches P_1 in finite time $\hat{t}$. However, P_1 is not an EP and there is no way to prolong the solution for $t > \hat{t}$

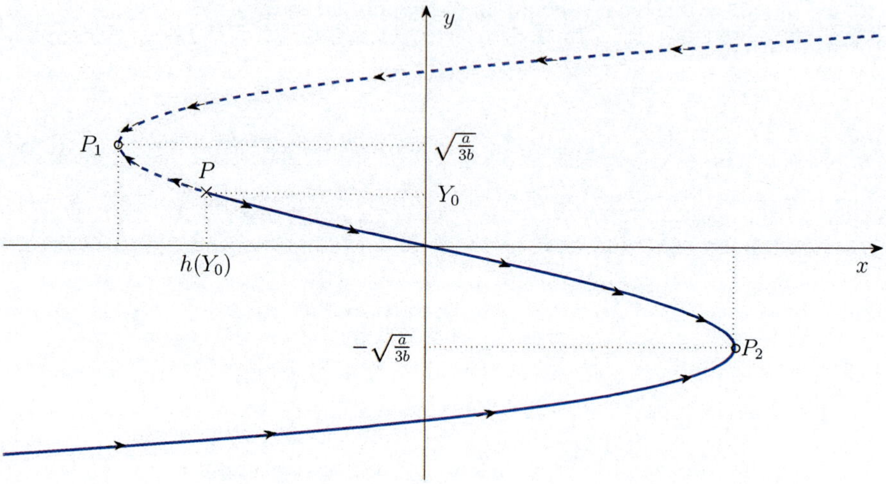

Fig. 6.16 The dynamic route of the *M–C* circuit in Fig. 6.14 when $|Y_0| < \sqrt{a/3b}$. The impasse points P_1 and P_2 are denoted by "open circle," whereas the only EP P is marked with "cross symbol"

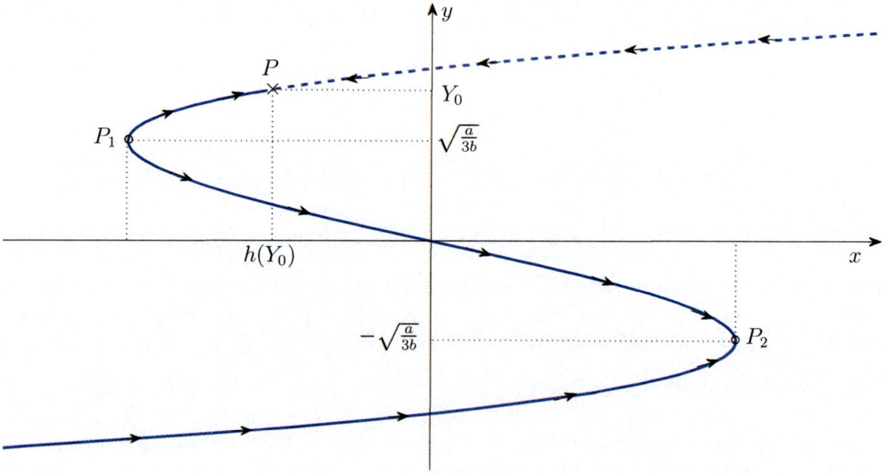

Fig. 6.17 The dynamic route of the *M–C* circuit in Fig. 6.14 when $|Y_0| > \sqrt{a/3b}$. The impasse points P_1 and P_2 are denoted by "open circle," whereas the only EP P is marked with "cross symbol"

(i.e., the solution is stuck at P_1, but the only EP is $P \neq P_1$). Similar arguments show that P_2 is a forward impasse point for a solution starting at $(\hat{X}, \hat{Y})$ with $\hat{Y} \in (Y_0, -\sqrt{a/3b}) \cup (-\sqrt{a/3b}, -\infty)$ and $\hat{X} = h(\hat{Y})$;

2. if the initial conditions v_{C_0}, q_{M_0} for the state variables in the (v, i)-domain are such that

$$Y_0 > \sqrt{a/3b}$$

then the unique EP P is asymptotically stable (Fig. 6.17). Moreover, by arguing like in the previous point, there are two impasse points at P_1 and $P_2 = -P_1$. Now, point P_1 is a backward impasse point (any backward trajectory starting at $(\hat{X}, \hat{Y})$ with $\hat{Y} \in (Y_0, \sqrt{a/3b}) \cup (\sqrt{a/3b}, -\sqrt{a/3b})$ and $\hat{X} = h(\hat{Y})$ is stuck at P_1 after a finite backward time), whereas P_2 is a forward impasse point (any forward trajectory starting at $(\hat{X}, \hat{Y})$ with $\hat{Y} \in (\sqrt{a/3b}, -\sqrt{a/3b}) \cup (-\sqrt{a/3b}, -\infty)$ and $\hat{X} = h(\hat{Y})$ is stuck at P_2 after a finite forward time);
3. the case $Y_0 < -\sqrt{a/3b}$ can be analyzed, *mutatis mutandis*, by arguing as in the previous point. Now, P_1 (resp., P_2) is a forward (resp., backward) impasse point, while P is the only asymptotically stable EP.

Such pathological situation due to the coexistence of equilibria and impasse points imply that the M–C circuit model in Fig. 6.14 is defective. The circuit needs to be remodeled by including parasitic inductances and/or capacitances at appropriate locations in order to characterize relaxation oscillations and jump phenomena that are observed in experiments with the considered circuit. This issue will be discussed in the next section.

6.2.2 *M–L–C Circuit with Relaxation Oscillations*

Any physical (memristor) circuit is of course characterized by a well-defined dynamic behavior for any $t \geq t_0$. Impasse points as those observed in the M–C circuit of the previous section represent nonphysical phenomena due to a poor circuit modeling (Chap. 4). It is known from the approach discussed in Chap. 4 that the impasse points in the M–C circuit of Fig. 6.14 can be hopefully broken by introducing a small inductance L (e.g., representing the inductance of connecting wires) connected in series with the memristor. As a result, the series M–L–C circuit in the class $\mathcal{LM}$, depicted in Fig. 6.18, is obtained. Suppose the memristor M is still defined by (6.22) and let $v_C(t_0) = v_{C_0}$, $i_L(t_0) = i_{L_0}$, $q_M(t_0) = q_{M_0}$ be the initial conditions at t_0 for the state variables in the (v, i)-domain. Then, $q_C(t_0) = q_{C_0} = Cv_{C_0}$ and $\varphi_L(t_0) = \varphi_{L_0} = Li_{L_0}$. The corresponding circuit in the (φ, q)-domain is reported in Fig. 6.19.

6.2.2.1 Formulation of the Circuit Equations

Analysis by inspection of the circuit in Fig. 6.19 permits to write the following KφL, KqLs, and CRs of circuit elements

$$\varphi_C(t; t_0) = \varphi_L(t; t_0) + \varphi_M(t; t_0) \tag{6.28a}$$

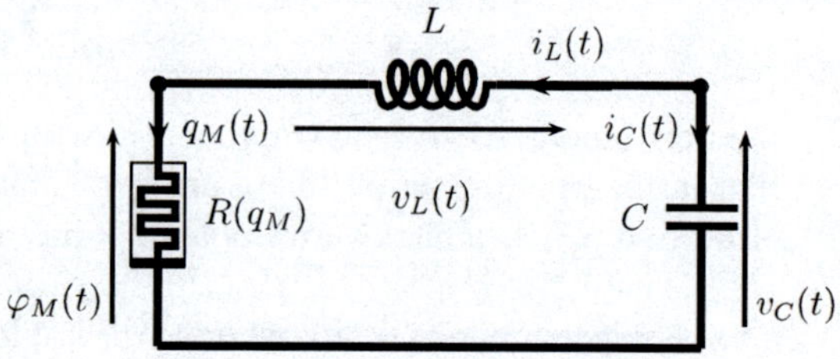

Fig. 6.18 The M–L–C obtained from the M–C circuit in Fig. 6.14 by inserting a parasitic inductance L in series with the memristor

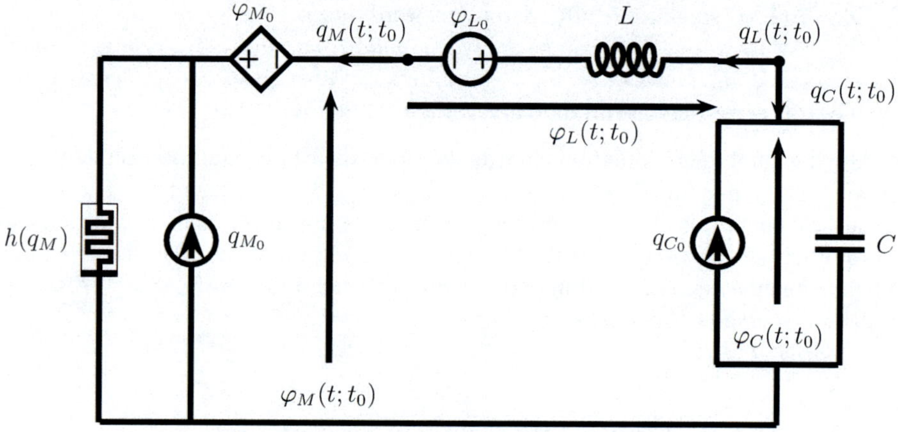

Fig. 6.19 Equivalent circuit in the (φ, q)-domain of the M–L–C circuit in Fig. 6.18 for $t \geq t_0$

$$q_L(t; t_0) = -q_C(t; t_0) \tag{6.28b}$$

$$q_L(t; t_0) = q_M(t; t_0) \tag{6.28c}$$

$$q_C(t; t_0) = -q_{C_0} + C\frac{d}{dt}\varphi_C(t; t_0) \tag{6.28d}$$

$$\varphi_L(t; t_0) = -\varphi_{L_0} + L\frac{d}{dt}q_L(t; t_0) \tag{6.28e}$$

$$\varphi_M(t; t_0) = -h(q_{M_0}) + h(q_M(t; t_0) + q_{M_0}) \tag{6.28f}$$

for $t \geq t_0$. These constitute the set of DAEs describing the circuit dynamics in the (φ, q)-domain.

It is an easy matter to verify that the circuit in Fig. 6.18 satisfies the conditions given in Property 3.1 of Chap. 5 for the existence of the SE representation. By substitution in the previous DAEs, we indeed obtain for $t \geq t_0$ the second-order SEs in the (φ, q)-domain in terms of the state variables $\varphi_C(t; t_0)$ and $q_L(t; t_0)$

$$C\frac{d}{dt}\varphi_C(t;t_0) = -q_L(t;t_0) + q_{C0} \tag{6.29a}$$

$$L\frac{d}{dt}q_L(t;t_0) = \varphi_C(t;t_0) - h(q_L(t;t_0) + q_{M0}) + h(q_{M0}) + \varphi_{L0} \tag{6.29b}$$

$$\varphi_C(t_0;t_0) = 0 \tag{6.29c}$$

$$q_L(t_0;t_0) = 0. \tag{6.29d}$$

The change of variables $y(t) = q_L(t;t_0) + q_{M_0} = q_M(t)$, $x(t) = \varphi_C(t;t_0) + h(q_{M_0}) + \varphi_{L0} = \varphi_L(t) + h(q_M(t))$ permits to rewrite (6.29) in the following simplified form:

$$\frac{dx(t)}{dt} = \frac{-y(t) + Q_0}{C} \tag{6.30a}$$

$$\frac{dy(t)}{dt} = \frac{x(t) - h(y(t))}{L} \tag{6.30b}$$

for $t \geq t_0$, where $x(t_0) = h(q_{M_0}) + \varphi_{L0}$, $y(t_0) = q_{M_0}$ and quantity

$$Q_0 = q_{C_0} + q_{M_0} = Cv_{C_0} + q_{M_0} \tag{6.31}$$

depends on the initial conditions for the state variables v_{C_0} and q_{M_0} in the (v, i)-domain.

Remark 6.15 It is important to remark that when $Q_0 = 0$, the second-order system (6.30) is *analogous to that describing a Van der Pol oscillator* (see Sect. 4.2 in Chap. 4). This enables to use the bulk of results for Van der Pol oscillators for studying the dynamics of the $M-L-C$ circuit in the (φ, q)-domain. When $Q_0 \neq 0$, the system describes a Van der Pol oscillator with a *constant forcing term* Q_0, which will play a crucial role in the bifurcation phenomena of the $M - L - C$ circuit.

The circuit in Fig. 6.18 admits of the common formulation of circuit equations in the (v, i)-domain in terms of the state variables $(v_C(t), i_L(t), q_M(t))$. The SEs in the (v, i)-domain can be readily derived by differentiating (6.29) with respect to time (Chap. 5)

$$C\frac{d}{dt}v_C(t) = -i_L(t) \tag{6.32a}$$

$$L\frac{d}{dt}i_L(t) = v_C(t) - h'(q_M(t))i_M(t) \tag{6.32b}$$

$$\frac{d}{dt}q_M(t) = i_M(t) \tag{6.32c}$$

$$v_C(t_0) = v_{C0} \tag{6.32d}$$

$$i_L(t_0) = i_{L0} \tag{6.32e}$$

$$q_M(t_0) = q_{M_0} \tag{6.32f}$$

where $h'(q_M(t)) = -a + 3bq_M^2(t)$. In particular, it can be easily checked that if $\varphi_C(t;t_0), q_L(t;t_0)$ is the unique solution of the IVP given in (6.29), then $v_C(t) = d\varphi_C(t;t_0)/dt$, $i_L(t) = dq_L(t;t_0)/dt$, $q_M(t) = q_L(t;t_0) + q_{M_0}$, is the unique solution of the IVP (6.32) for $t \geq t_0$. Conversely, if $v_C(t), q_L(t), \varphi_M(t)$ is the solution of the IVP (6.32) for $t \geq t_0$, then $\varphi_C(t;t_0) = \int_{t_0}^{t} v_C(\tau)d\tau$, $q_L(t;t_0) = \int_{t_0}^{t} i_L(\tau)d\tau$ is the solution of the IVP (6.29) for $t \geq t_0$.

Note that (6.32) is an IVP for a *third-order* system in the (v,i)-domain whereas (6.29) is an IVP for a *second-order* system in the (φ,q)-domain, i.e., FCAM leads to formulate the circuit equations by means of a reduced-order SE in the (φ,q)-domain. It is also worth noting that the third-order oscillator (6.32) would be difficult to analyze directly in the (v,i)-domain since its equations do not resemble those of any typical oscillator.

6.2.3 Invariant Manifolds

The KqLs (6.28b) and (6.28c) imply the law of conservation of incremental charge $q_C(t;t_0) + q_M(t;t_0) = 0$, i.e., we have

$$Q(t) \doteq q_C(t) + q_M(t) = q_{C_0} + q_{M_0} = Q_0$$

for any $t \geq t_0$, where Q_0 is given in (6.31). This is equivalent to saying that $Q(t)$ is an invariant of motion for the SEs (6.32) and so we can define the positively invariant manifolds

$$\mathcal{M}(Q_0) = \{(v_C(t), i_L(t), q_M(t))^T \in \mathbb{R}^3 : Cv_C(t) + q_M(t) = Q_0\} \tag{6.33}$$

for the circuit dynamics in the (v,i)-domain. Note that each manifold is simply a plane in the state-space $(v_C(t), i_L(t), q_M(t))^T \in \mathbb{R}^3$ in the (v,i)-domain. Since each manifold is identified by $Q_0 \in \mathbb{R}$, there are ∞^1 of such manifolds. Moreover, it can be easily checked that they are non-intersecting and that the whole state-space in the (v,i)-domain is covered by such manifolds by varying Q_0 in $\mathbb{R}$. In this way we obtained a foliation of the phase-space $(v_C(t), i_L(t), q_M(t))$ in ∞^1 invariant manifolds $\mathcal{M}(Q_0)$ and on each manifold the dynamics is described in the (φ,q)-domain by the second-order system (6.30).

Note that the vector field defining the SEs (6.30) depends on the circuit parameters (L, C, a, b) and on Q_0 as well, where Q_0 in turn depends on the initial conditions v_{C_0}, i_{L_0} and q_{M_0} for the state variables in the (v,i)-domain.

6.2.4 Nonlinear Dynamics and Hopf Bifurcations Without Parameters

Let us study the dynamics of the $M-C-L$ circuit in the (φ, q)-domain. The unique EP of (6.30) is given by $\bar{x} = h(Q_0)$ and $\bar{y} = Q_0$. The Jacobian of the vector field defining (6.30) at the EP is

$$\mathbf{J}_{(\bar{x},\bar{y})} = \begin{pmatrix} 0 & -\frac{1}{C} \\ \frac{1}{L} & -\frac{h'(Q_0)}{L} \end{pmatrix} = \begin{pmatrix} 0 & -\frac{1}{C} \\ \frac{1}{L} & \frac{a-3bQ_0^2}{L} \end{pmatrix}. \tag{6.34}$$

By denoting with

$$\alpha = \text{tr}\mathbf{J}_{(\bar{x},\bar{y})} = \frac{a - 3bQ_0^2}{L}$$

$$\Delta = \text{det}\mathbf{J}_{(\bar{x},\bar{y})} = \frac{1}{LC} > 0$$

the eigenvalues of $\mathbf{J}_{(\bar{x},\bar{y})}$ are

$$\lambda_{1,2} = \frac{-\alpha \pm \sqrt{\alpha^2 - 4\Delta}}{2}. \tag{6.35}$$

For $\alpha = 0$, i.e., $Q_0 = \pm\sqrt{a/3b}$, the vector field defining (6.30) is in Normal Form and the linearization at the EP exhibits a center (i.e., a pair of purely imaginary eigenvalues $\lambda_{1,2} = \pm j\sqrt{\Delta}$). By evaluating the real and imaginary part of the eigenvalues $\lambda_{1,2}$, we can easily conclude the following.

1. If $\alpha < 0$, i.e.,

$$|Q_0| < \sqrt{a/3b}$$

then the unique EP is unstable ($\mathbf{J}_{(\bar{x},\bar{y})}$ has two eigenvalues with positive real part) and it is seen that (6.30) presents persistent oscillations. For verification, Fig. 6.20 reports the real and imaginary parts of the eigenvalues $\lambda_{1,2}$ in (6.35) as a function of Q_0 when the circuit parameters are $C = 9/2$, $L = 1$, $a = 1$, and $b = 1/3$.

2. If $\alpha > 0$, i.e.,

$$|Q_0| > \sqrt{a/3b}$$

then the unique EP is asymptotically stable ($\mathbf{J}_{(\bar{x},\bar{y})}$ has two eigenvalues with negative real part—see Fig. 6.20). It is seen that (6.30) exhibits no oscillatory behavior and the only EP is also globally attracting for the trajectories.

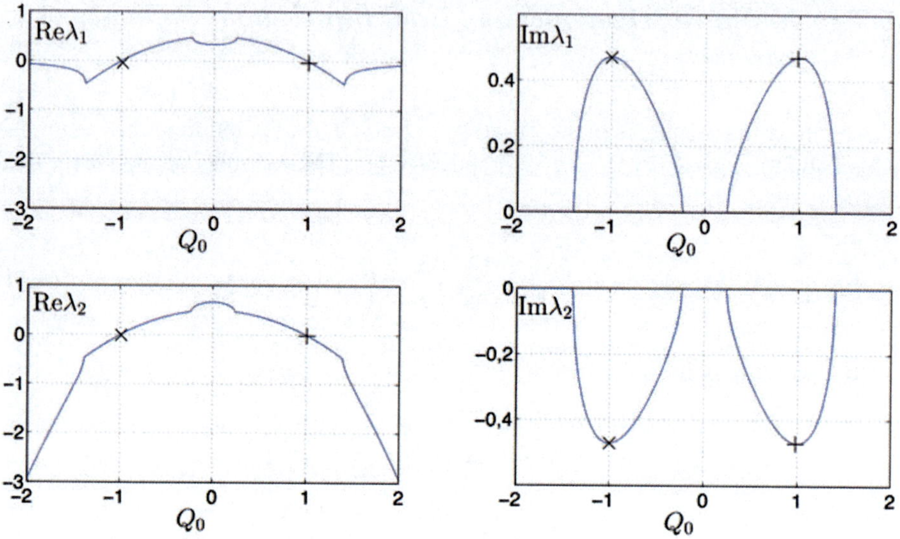

Fig. 6.20 The real and imaginary parts of the eigenvalues $\lambda_{1,2}$ in (6.35) as a function of Q_0. The circuit parameters are: $C = 9/2$, $L = 1$, $a = 1$, and $b = 1/3$. The values of $\lambda_{1,2}$ for $Q_0 = -1$ and $Q_0 = +1$ are marked with "×" and "+," respectively

3. If $\alpha = 0$, i.e., the initial conditions v_{C0} and q_{M0} are such that $Q_0 = \pm\sqrt{a/3b}$, then $\mathbf{J}_{(\bar{x},\bar{y})}$, as already noted, has two purely imaginary eigenvalues (see Fig. 6.20).

Concerning bifurcations, it is possible to distinguish the following two main case studies:

- if Q_0 is fixed (this case is referred to as *fixed-invariant manifold*), then qualitative changes in the phase-portrait and bifurcations of (6.30) might occur only if the circuit parameters (L, C, a, b) are varied. For example, if $Q_0 = 0$, the bifurcation scenario is similar to that of the well-known standard Van der Pol oscillator;
- if Q_0 is varied, then qualitative changes and bifurcations in the phase-portrait of (6.30) might occur even if the circuit parameters (L, C, a, b) are fixed, a case referred to as *bifurcations without parameters*. In particular, a *supercritical Hopf bifurcation without parameters* occurs at

$$Q_0 = -\sqrt{a/3b}$$

while a reverse supercritical Hopf bifurcation without parameters occurs at[1]

[1] The presence of such Hopf bifurcations can be justified rigorously from a mathematical viewpoint using the techniques in [9]. We omit the details to avoid excessive technicalities.

$$Q_0 = \sqrt{a/3b}$$

- since we have an explicit expression of Q_0 as a function of initial conditions, i.e., $Q_0 = Cv_{C_0} + q_{M_0}$, we can easily develop a simulation scheme for obtaining Hopf bifurcations due to varying the initial conditions. Clearly, in order to vary Q_0, we can vary either q_{M_0} for fixed v_{C_0}, or conversely. Note that varying i_{L_0} has instead no influence on the bifurcations. As an example, Figs. 6.21 and 6.22 show how variations of q_{M_0} brings about the Hopf bifurcations when $i_{L_0} = 5$, $v_{C_0} = 0.5$ and the following circuit parameters are fixed at $a = 1$, $b = 1/3$, $L = 1$, $C = 9/2$. The Hopf bifurcations without parameters take place onto the grey planes in Figs. 6.21 and 6.22 defined by $Q_0 = \pm 1$. The trajectories of (6.30) for $Q_0 \neq \pm 1$ confirm the nonlinear analysis and the stability properties derived from the study of (6.35).

Remark 6.16 The scenario illustrated in Figs. 6.21 and 6.22 is analogous to that of the Hopf bifurcation of an EP studied in Sect. 4.4.2 of Chap. 4. However, one fundamental difference is that in that case the Hopf bifurcation is due to changing a circuit parameter, while for the $M - L - C$ memristor circuit the bifurcation is due to changing the initial conditions for the state variables in the (v, i)-domain for a fixed set of circuit parameters.

Remark 6.17 From the previous analysis, and the relationship between the solutions in the (v, i)-domain and (φ, q)-domain, it can be concluded that there coexist infinitely many different second-order dynamics and attractors, one for each manifold, for the third-order system (6.32). In particular, we have coexistence of a continuum of stable EPs and a continuum of different periodic orbits for the dynamics in the (v, i)-domain (cf. Figs. 6.21 and 6.22). The coexistence of infinitely many different attractors is a property which is sometimes referred to in the mathematical and physical literature as *extreme multistability* [10].

Remark 6.18 A relevant consequence of the principle of reduction of order of FCAM and the foliation of the state space is that the $M - L - C$ circuit cannot display complex dynamics. This is again a consequence of the fact that the dynamics is essentially second order and autonomous (cf. Chap. 4). It would have been a more complex task to rule out complex dynamics by a direct analysis in the (v, i)-domain of the third-order oscillator (6.32).

Remark 6.19 The advantages of FCAM for the $M - L - C$ circuit are even more evident than for the $M - C$ circuit. In fact, via FCAM and the principle of reduction of order we have been able to study the third-order circuit $M - L - C$ by bringing back the analysis to that of a second-order (planar) Van der Pol oscillator. As already noticed, it would have been a much more difficult task to directly analyze the third-order oscillator in the (v, i)-domain.

Remark 6.20 We refer the reader to the articles [11–13], and references therein, for a study on the existence of invariant manifolds and coexisting oscillatory dynamics

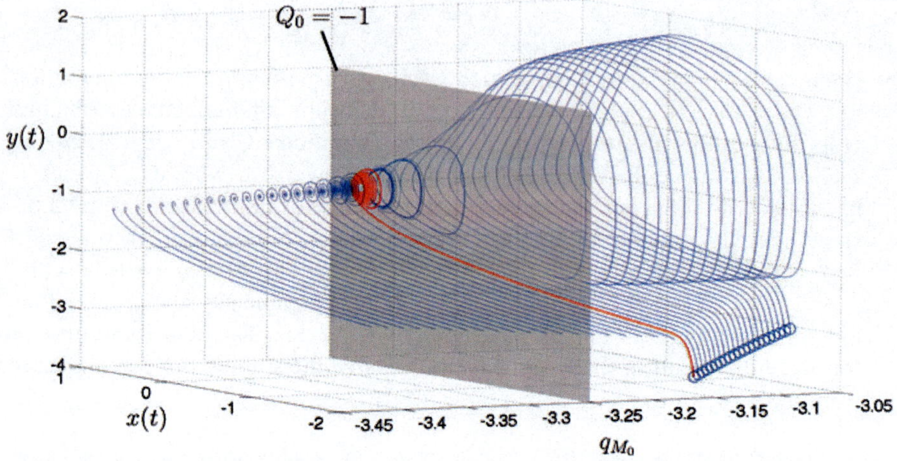

Fig. 6.21 Supercritical Hopf bifurcation ($Q_0 = -1$, $a = 1$, and $b = 1/3$) in the M–L–C of Fig. 6.18. The initial condition q_{M_0} is such that, according to (6.31), $Q_0 = Cv_{C_0} + q_{M_0} = -1$ for $q_{M_0} = -3.25$ (being $C = 9/2$ and $v_{C_0} = 0.5$)

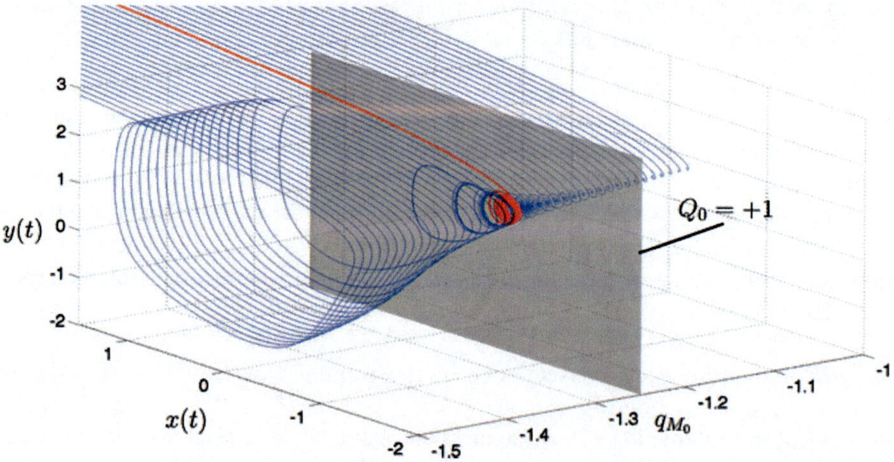

Fig. 6.22 Inverse supercritical Hopf bifurcation ($Q_0 = +1$, $a = 1$, and $b = 1/3$) in the M–L–C of Fig. 6.18. The initial condition q_{M_0} is such that, according to (6.31), $Q_0 = Cv_{C_0} + q_{M_0} = +1$ for $q_{M_0} = -1.25$ (being $C = 9/2$ and $v_{C_0} = 0.5$)

in other classes of circuits that are described by third-order systems in the (v, i)-domain and contain a capacitor, an inductor, and a memristor.

Example 6.3 Consider a variant of the $M - L - C$ circuit containing a flux-controlled memristor $q_M = f(\varphi_M)$ and a linear resistor in addition to a capacitor and an inductor (Fig. 6.23). Such a circuit has been frequently considered in the

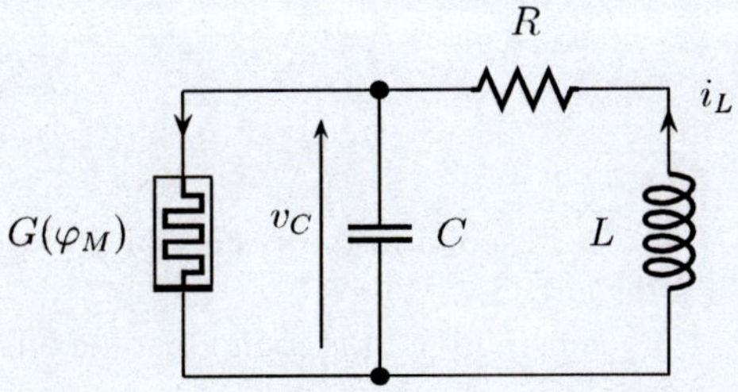

Fig. 6.23 Circuit with **flux-controlled memristor,** resistor, inductor, and capacitor

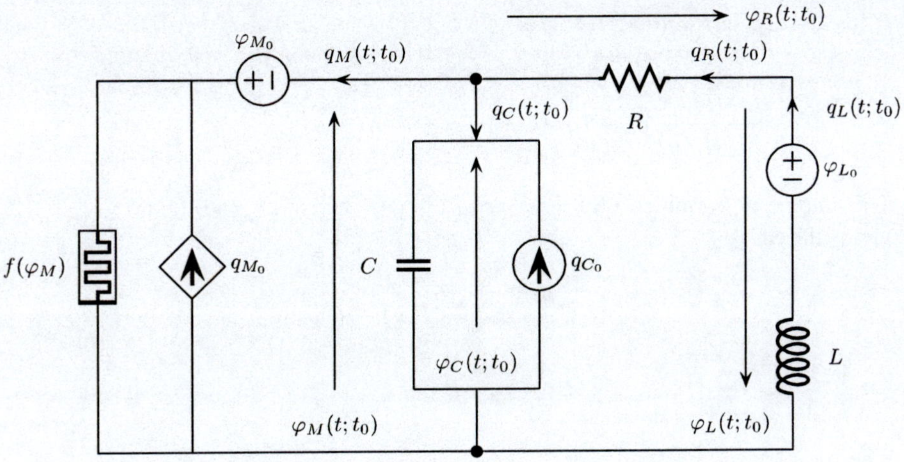

Fig. 6.24 Equivalent circuit in the **(φ, q)**-domain

literature as a prototypical circuit for studying dynamic phenomena in memristor circuits [7, 12, 14].

The equivalent circuit in the (φ, q)-domain is shown in Fig. 6.24. Applying KqL and KφL, and then listing the CRs, we obtain

$$
\begin{aligned}
q_L(t; t_0) &= q_C(t; t_0) + q_M(t; t_0) \\
q_L(t; t_0) &= q_R(t; t_0) \\
\varphi_R(t; t_0) + \varphi_L(t; t_0) + \varphi_M(t; t_0) &= 0 \\
\varphi_C(t; t_0) &= \varphi_M(t; t_0)
\end{aligned}
$$

and

$$
\begin{aligned}
\varphi_R(t; t_0) &= R q_R(t; t_0) \\
q_C(t; t_0) &= C\frac{d\varphi_C(t; t_0)}{dt} - q_{C_0}
\end{aligned}
$$

$$\varphi_L(t;t_0) = L\frac{dq_L(t;t_0)}{dt} - \varphi_{L_0}$$

$$q_M(t;t_0) = f(\varphi_M(t;t_0) + \varphi_{M_0}) - f(\varphi_{M_0}).$$

These correspond to the system of $2b = 8$ DAEs describing the tableau equations for the circuit.

By substitution, it is found that the circuit dynamics is described by the second-order SE

$$C\frac{d\varphi_C(t;t_0)}{dt} = q_L(t;t_0) - f(\varphi_C(t;t_0) + \varphi_{M_0}) + f(\varphi_{M_0}) + q_{C_0}$$

$$L\frac{dq_L(t;t_0)}{dt} = -Rq_L(t;t_0) - \varphi_C(t;t_0) + \varphi_{L_0}.$$

The change of variables $x(t) = \varphi_C(t;t_0) + \varphi_{M_0}$ and $y(t) = q_L(t;t_0) - \frac{\varphi_{M_0}+\varphi_{L_0}}{R}$ yields the SE

$$C\frac{dx}{dt} = y - f(x) + Q_0$$

$$L\frac{dy}{dt} = -Ry - x$$

where we let

$$Q_0 = q_{C_0} + \frac{\varphi_{M_0} + \varphi_{L_0}}{R} + f(\varphi_{M_0}).$$

This is in a form analogous to a Van der Pol oscillator with a constant forcing term Q_0.

By differentiation in time, the SEs in the (v, i)-domain are given by the third-order system

$$C\frac{dv_C}{dt} = i_L - G(\varphi_M)v_C$$

$$L\frac{di_L}{dt} = -Ri_L - v_C$$

$$\frac{d\varphi_M}{dt} = v_C$$

where $G(\varphi_M) = f'(\varphi_M)$.

We leave to the reader the verification that function

$$Q(t) = Cv_C(t) + \frac{\varphi_M(t) + Li_L(t)}{R} + f(\varphi_M(t))$$

is an invariant of motion for the SEs in the (v, i)-domain, i.e., the state space $\mathbb{R}^3$ can be foliated in a continuum of 2D positively invariant manifolds

$$\mathcal{M}(Q_0) = \{(v_C(t), i_L(t), \varphi_M(t))^T \in \mathbb{R}^3 : Q(t) = Q_0\}$$

where $Q_0 \in \mathbb{R}$.

On the basis of these formulas, we can study coexisting dynamics and bifurcations without parameters in the circuit along lines similar to the $M - L - C$ circuit.

6.3 Third-Order Memristor Chaotic Circuits

This section focuses on memristor-based oscillators in the class $\mathcal{LM}$ that exhibit a wide range of complex nonlinear dynamical behaviors (e.g., coexistence of stable EPs, periodic oscillations, and chaotic attractors). In particular, the Memristor-based Chaotic Circuit (MCC) in Fig. 6.25 is considered. The circuit is made of two passive resistors r and R, two passive capacitors C_1 and C_2, one passive inductor L, and an **active flux-controlled memristor** M. It is noticed that MCC is simply obtained by *replacing the nonlinear resistor of a Chua's oscillator (Sect. 4.3.1 in Chap. 4) with a flux-controlled memristor.*

Let us assume once more that the active flux-controlled memristor has a CR like (6.22), that is

$$q_M(t) = f(\varphi_M(t)) = -a\varphi_M(t) + b\varphi_M^3(t) \tag{6.36}$$

with $a, b > 0$. The memristor is (locally) active since the memductance $G(\varphi_M(t)) = f'(\varphi_M(t)) < 0$ for $|\varphi_M| < \sqrt{a/3b}$.

Let $v_{C_1}(t_0) = v_{C_{10}}$, $v_{C_2}(t_0) = v_{C_{20}}$, $i_L(t_0) = i_{L0}$, $\varphi_M(t_0) = \varphi_{M0}$ be the initial conditions at t_0 for the state variables in the (v, i)-domain. It follows that

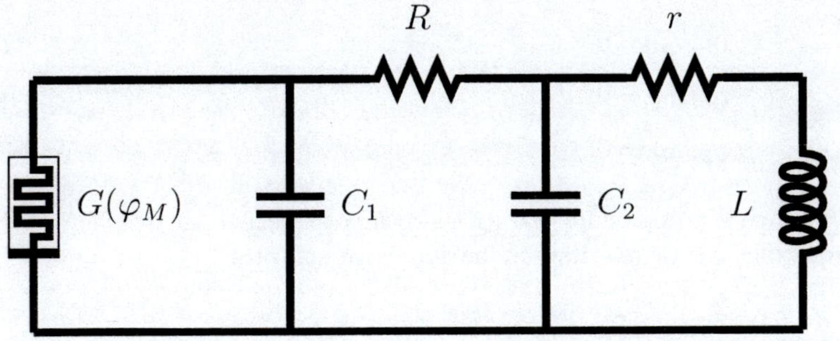

Fig. 6.25 Memristor-based chaotic circuit

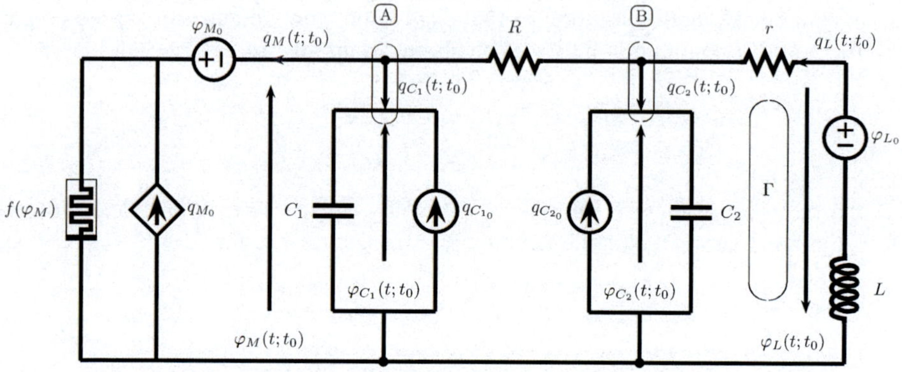

Fig. 6.26 Equivalent circuit of the MCC in the (φ, q)-domain

$q_{C_1}(t_0) = q_{C_{1_0}} = C_1 v_{C_{1_0}}$, $q_{C_2}(t_0) = q_{C_{2_0}} = C_2 v_{C_{2_0}}$, and $\varphi_L(t_0) = \varphi_{L_0} = L i_{L_0}$. The corresponding circuit in the (φ, q)-domain is reported in Fig. 6.26.

It can be easily checked that MCC satisfies the conditions in Chap. 5 for the existence of the SE representation. Analysis of the MCC in Fig. 6.26 permits to write, by means of the KφL at the loop Γ and the KqLs at the nodes (A) and (B), the following SEs in the (φ, q)-domain in terms of the state variables $\varphi_{C_1}(t; t_0)$, $\varphi_{C_2}(t; t_0)$ and $q_L(t; t_0)$. We have for any $t \geq t_0$

$$C_1 \frac{d\varphi_{C_1}(t; t_0)}{dt} = \frac{1}{R}(\varphi_{C_2}(t; t_0) - \varphi_{C_1}(t; t_0)) - f(\varphi_{C_1}(t; t_0) + \varphi_{M_0}) + f(\varphi_{M_0}) + q_{C_{1_0}} \tag{6.37a}$$

$$C_2 \frac{d\varphi_{C_2}(t; t_0)}{dt} = -\frac{1}{R}(\varphi_{C_2}(t; t_0) - \varphi_{C_1}(t; t_0)) + q_L(t; t_0) + q_{C_{2_0}} \tag{6.37b}$$

$$L \frac{dq_L(t; t_0)}{dt} = -r q_L(t; t_0) - \varphi_{C_2}(t; t_0) + \varphi_{L_0} \tag{6.37c}$$

$$\varphi_{C_1}(t_0; t_0) = 0 \tag{6.37d}$$

$$\varphi_{C_2}(t_0; t_0) = 0 \tag{6.37e}$$

$$q_L(t_0; t_0) = 0. \tag{6.37f}$$

By following a procedure analogous to that in Sect. 4.3.1 of Chap. 4, the first three equations can be rewritten in the adimensional form

$$\frac{d\varphi_{C_1}(\tau; t_0)}{d\tau} = \alpha \Big[(\varphi_{C_2}(\tau; t_0) - \varphi_{C_1}(\tau; t_0))$$

Table 6.1 The normalization values for circuit elements and electrical variables of the MCC in Fig. 6.26 are: R_0 for resistances, C_0 for capacitances, L_0 for inductances, V_0 for voltages, I_0 for currents, Q_0 for charges, Φ_0 **for fluxes, and** T_0 for time

R_0	1 kΩ	V_0	1 V
C_0	1 nF	I_0	1 mA
L_0	1 mH	Q_0	1 nAs
T_0	1 µs	Φ_0	1 µVs

$$-Rf(\varphi_{C_1}(\tau;t_0)+\varphi_{M_0})+Rf(\varphi_{M_0})+Rq_{C_{1_0}}\Big] \tag{6.38a}$$

$$\frac{d\varphi_{C_2}(\tau;t_0)}{d\tau}=-(\varphi_{C_2}(\tau;t_0)-\varphi_{C_1}(\tau;t_0))$$
$$+(Rq_L(\tau;t_0))+Rq_{C_{2_0}} \tag{6.38b}$$

$$\frac{d(Rq_L(\tau;t_0))}{d\tau}=-\gamma(Rq_L(\tau;t_0))-\beta\varphi_{C_2}(\tau;t_0)+\beta\varphi_{L_0} \tag{6.38c}$$

where we used the normalization values in Table 6.1 and introduced the parameters

$$\tau=\frac{t}{RC_2},\quad \alpha=\frac{C_2}{C_1},\quad \beta=\frac{R^2C_2}{L},\quad \gamma=\frac{RrC_2}{L}.$$

For the sake of simplicity let us denote τ with t in (6.38). The change of variables (details are reported in Appendix 2)

$$x(t)=\varphi_{C_1}(t;t_0)+\varphi_{M_0} \tag{6.39a}$$

$$y(t)=\varphi_{C_2}(t;t_0)+\frac{\gamma}{\beta+\gamma}\varphi_{M_0}-\frac{\beta}{\beta+\gamma}\varphi_{L_0}-\frac{\gamma}{\beta+\gamma}Rq_{C_{2_0}} \tag{6.39b}$$

$$z(t)=Rq_L(t;t_0)+\frac{\beta}{\beta+\gamma}\left(-\varphi_{M_0}-\varphi_{L_0}+Rq_{C_{2_0}}\right) \tag{6.39c}$$

allows us to cast (6.37) into the third-order system

$$\frac{d\,x(t)}{dt}=\alpha\left[-x(t)+y(t)-n(x(t))+X_0\right] \tag{6.40a}$$

$$\frac{d\,y(t)}{dt}=x(t)-y(t)+z(t) \tag{6.40b}$$

$$\frac{d\,z(t)}{dt}=-\beta y(t)-\gamma z(t) \tag{6.40c}$$

for $t\geq t_0$, where $x(t_0)=\varphi_{M_0}$, $y(t_0)=\frac{\gamma}{\beta+\gamma}\varphi_{M_0}-\frac{\beta}{\beta+\gamma}\varphi_{L_0}-\frac{\gamma}{\beta+\gamma}Rq_{C_{2_0}}$, $z(t_0)=\frac{\beta}{\beta+\gamma}\left(-\varphi_{M_0}-\varphi_{L_0}+Rq_{C_{2_0}}\right)$ and the nonlinearity

$$n(x(t)) = Rf(x(t)) = -Rax + Rbx^3 = -m_0x(t) + m_1x(t)^3. \tag{6.41}$$

Moreover,

$$X_0 = \frac{\beta}{\beta+\gamma}\varphi_{M_0} + \frac{\beta}{\beta+\gamma}Li_{L_0} + \frac{\gamma}{\beta+\gamma}RC_2v_{C_{2_0}} + n(\varphi_{M_0}) + RC_1v_{C_{1_0}} \tag{6.42}$$

which is a term depending on the initial conditions for the state variables $v_{C_{1_0}}$, $v_{C_{2_0}}$, i_{L_0}, and φ_{M_0} of the MCC in the (v, i)-domain.

Remark 6.21 It is stressed that (6.40) with $X_0 = 0$ describes the dynamics of the well-known *canonical Chua's oscillator* (Chap. 4). Instead, when $X_0 \neq 0$, (6.40) describes the dynamics of a Chua's oscillator with a constant forcing term. Again, this enables to use the bulk of results for Chua's oscillator in the analysis of MCC in the (φ, q)-domain.

Invariant Manifolds
The SE describing the MCC in the (v, i)-domain can be obtained by differentiating the SE (6.37). This yields the fourth-order SE

$$C_1\frac{dv_{C_1}(t)}{dt} = \frac{1}{R}(v_{C_2}(t) - v_{C_1}(t)) - G(\varphi_M(t))v_{C_1}(t) \tag{6.43a}$$

$$C_2\frac{dv_{C_2}(t)}{dt} = -\frac{1}{R}(v_{C_2}(t) - v_{C_1}(t)) + i_L(t) \tag{6.43b}$$

$$L\frac{di_L(t)}{dt} = -ri_L(t) - v_{C_2}(t) \tag{6.43c}$$

$$\frac{d\varphi_M(t)}{dt} = v_{C_1}(t) \tag{6.43d}$$

where we have taken into account that $f'(\cdot) = G(\cdot)$ and $\varphi_C(t; t_0) + \varphi_{M_0} = \varphi_M(t; t_0) + \varphi_{M_0} = \varphi_M(t)$.

Consider the function of the state variables $v_{C_1}(t)$, $v_{C_2}(t)$, $i_L(t)$, and $\varphi_M(t)$ in the (v, i)-domain

$$X(t) = \frac{\beta}{\beta+\gamma}\varphi_M(t) + \frac{\beta}{\beta+\gamma}Li_L(t) + \frac{\gamma}{\beta+\gamma}RC_2v_{C_2}(t) + n(\varphi_M(t)) + RC_1v_{C_1}(t).$$

It can be checked that the time derivative of $X(\cdot)$ along the solutions of the SE (6.43) is equal to 0. To see this, note that we have (omitting dependence on t)

$$\frac{X}{R} = \frac{1}{R+r}(\varphi_M + Li_L + rC_2v_{C_2}) + f(\varphi_M) + C_1v_{C_1}.$$

Taking its derivative along the solutions of (6.43) we obtain

$$\frac{1}{R}\dot{X} = \frac{1}{R+r}(\dot{\varphi}_M + L\frac{di_L}{dt} + rC_2\dot{v}_{C_2}) + \frac{df(\varphi_M)}{dt} + C_1\dot{v}_{C_1} \tag{6.44}$$

$$= \frac{1}{R+r}(v_{C_1} - ri_L - v_{C_2} - \frac{r}{R}v_{C_2} + \frac{r}{R}v_{C_1} - \frac{r}{R}i_L) \tag{6.45}$$

$$+\frac{1}{R}(v_{C_2} - v_{C_1}) - G(\varphi_M)v_{C_1} + G(\varphi_M)v_{C_1} = 0 \tag{6.46}$$

for any $t \geq t_0$.

This implies that $X(t)$ is an invariant of motion for the SEs (6.43) in the (v, i)-domain. As a consequence, we can define ∞^1 three-dimensional manifolds

$$\mathcal{M}(X_0) = \{(v_{C_1}(t), v_{C_2}(t), i_L(t), q_M(t))^T \in \mathbb{R}^4 : \frac{\beta}{\beta+\gamma}\varphi_M(t) + \frac{\beta}{\beta+\gamma}Li_L(t) + \frac{\gamma}{\beta+\gamma}RC_2v_{C_2}(t) + n(\varphi_M(t)) + RC_1v_{C_1}(t) = X_0\} \tag{6.47}$$

where each manifold is identified by $X_0 \in \mathbb{R}$, which are positively invariant for the dynamics of (6.43).

6.3.1 *Nonlinear Dynamics and Period-Doubling Bifurcations Without Parameters*

Assume that $\gamma = 0$ (i.e., $r = 0$) and $R = 1$ for the sake of simplicity. The results reported in this section are similar, *mutatis mutandis*, to those derived from (6.40) with $\gamma \neq 0$ and $R \neq 1$. It follows that (6.40), (6.42), and (6.47) have the following simplified expressions:

$$\frac{dx(t)}{dt} = \alpha[-x(t) + y(t) - n(x(t)) + X_0] \tag{6.48a}$$

$$\frac{dy(t)}{dt} = x(t) - y(t) + z(t) \tag{6.48b}$$

$$\frac{dz(t)}{dt} = -\beta y(t) \tag{6.48c}$$

with $x(t_0) = \varphi_{M_0}$, $y(t_0) = \varphi_{L_0}$, $z(t_0) = -\varphi_{M_0} + \varphi_{L_0} + q_{C_{2_0}}$ and where we let

$$X_0 = \varphi_{M_0} + Li_{L_0} + n(\varphi_{M_0}) + C_1v_{C_{1_0}} \tag{6.49a}$$

$$\mathcal{M}(X_0) = \{(v_{C_1}(t), v_{C_2}(t), i_L(t), \varphi_M(t)) \in \mathbb{R}^4 : \varphi_M(t) + Li_L(t) + n(\varphi_M(t)) + C_1v_{C_1}(t) = X_0\}. \tag{6.49b}$$

The EPs of (6.48) are in the form $P = (\bar{x}, 0, -\bar{x})$ where $\bar{x}$ is the solution of the algebraic equation

$$\bar{x} + n(\bar{x}) = m_1\bar{x}^3 - (m_0 - 1)\bar{x} = X_0. \tag{6.50}$$

The graphical intersection between constant X_0 and curve $m_1\bar{x}^3 - (m_0 - 1)\bar{x}$ permits to derive that:

- there exist three EPs iff

$$|X_0| < \frac{2}{3}(m_0 - 1)\sqrt{\frac{m_0 - 1}{3m_1}} \tag{6.51}$$

- there exist two EPs iff

$$|X_0| = \frac{2}{3}(m_0 - 1)\sqrt{\frac{m_0 - 1}{3m_1}} \tag{6.52}$$

- there exists only one EP iff

$$|X_0| > \frac{2}{3}(m_0 - 1)\sqrt{\frac{m_0 - 1}{3m_1}}. \tag{6.53}$$

If $X_0 = 0$, the dynamics evolve on manifold $\mathcal{M}(0)$ and is the same as that of Chua's oscillator. In particular, if $\beta = 15$, $m_0 = 8/7$, and $m_1 = 4/63$, then the EPs $P_+ = (3/2, 0, -3/2)$, $P_- = P_+$ are asymptotically stable if $\alpha < 7$ (a Hopf bifurcation occurs at $\alpha = 7$), whereas $P_0 = (0, 0, 0)$ is always an unstable saddle point (see for instance the analysis in [15]). On the other hand, changing X_0 implies that the EPs change their location in the phase-space, as well as their stability properties, with respect to Chua's oscillator.

Numerical simulations confirm the analysis described above and provide insights into the complex behavior in the MCC of Fig. 6.26. Hereinafter the following values are assumed to carry out the numerical study: $\beta = 15$, $m_0 = 8/7$, and $m_1 = 4/63$. The circuit parameter α and the constant X_0 depending on the initials conditions are varied to make clear bifurcation phenomena. Two main cases are identified:

- *bifurcations on a fixed manifold*: the initial conditions $(v_{C_{1_0}}, v_{C_{2_0}}, i_{L_0}, \varphi_{M_0})$ are such that X_0 in (6.49) is *fixed*; qualitative changes in the global phase portrait and bifurcations of the SEs (6.48) occur due to changes of the circuit parameter α;
- *bifurcations without parameters*: the circuit parameter α is kept constant, whereas the initial conditions $(v_{C_{1_0}}, v_{C_{2_0}}, i_{L_0}, \varphi_{M_0})$ are changed in such a way that X_0 varies and the variation of X_0 in turn gives rise to qualitative changes in the global phase portrait and bifurcations of the SEs (6.48).

In the next sections, bifurcations on a fixed manifold and bifurcations without parameters for the SEs (6.48) are presented. The quantity X_0 is varied by changing the initial flux in the memristor φ_{M_0}, whereas the remaining initial conditions are set to $q_{C_{1_0}} = C_1 v_{C_{1_0}} = 0$, $q_{C_{2_0}} = C_2 v_{C_{2_0}} = 1$, and $\varphi_{L_0} = L i_{L_0} = 0$ in all numerical simulations.

6.3.1.1 Bifurcations on a Fixed Manifold

It is apparent that the nonlinear dynamics of the SEs (6.48) on the fixed manifold $\mathcal{M}(0)$ present the same bifurcations (exactly for the same values of α) of the canonical Chua's oscillator. Let us then consider the fixed manifold $\mathcal{M}(X_0)$ in (6.49b) with $\varphi_{M_0} = 0.02$, hence (6.49a) gives $X_0 = -0.0029$. The aim is to show bifurcation phenomena on $\mathcal{M}(-0.0029)$ due to changes of the circuit parameter α. Figure 6.27a, b, and c shows that a period-doubling cascade takes place by increasing α from the value 8.8 to 8.9 and then to 9.0. A further increase of α leads to chaotic behavior. The same period-doubling cascade occurs in the canonical Chua's oscillator (i.e., for $X_0 = 0$), but for slightly different values of α. Numerical simulations show that the canonical Chua's oscillator exhibits a similar form for the projection of limit cycles as in Fig. 6.27a, b, and c approximately for $\alpha = 8.79$, $\alpha = 9.0$, and $\alpha = 9.01$, respectively. The only difference between the projection of the limit cycles of the canonical Chua's oscillator and those reported in Fig. 6.27a, b, and c is the position in the phase-space because the limit cycles are embedded into different invariant manifolds ($\mathcal{M}(0)$ for the canonical Chua's oscillator and $\mathcal{M}(-0.0029)$ for the memristor-based chaotic circuit in Fig. 6.25 described by (6.48)). We can conclude that: *on the fixed invariant manifold* $\mathcal{M}(0)$, *and on manifolds* $\mathcal{M}(X_0)$ *with* X_0 *close to 0, the nonlinear dynamics in the memristor-based chaotic circuit of Fig. 6.25 are analogous to those of the canonical Chua's oscillator*, i.e., nonlinear attractors and bifurcations are of a very similar type, but they are displayed for slightly different circuit parameters.

6.3.1.2 Period-Doubling Bifurcations Without Parameters

Let us pick $\alpha = 8.7$ in the SEs (6.48), in which case the MCC in Fig. 6.25 displays periodic oscillations. The aim is to show that by changing X_0, for example by means of φ_{M_0}, gives rise to bifurcation phenomena. Figure 6.28 presents the projection on the (x, y) plane of one of the limit cycles—in the circuit described by (6.48)—when $X_0 = 0$, $X_0 = 0.0103$, and $X_0 = 0.0281$, i.e., according to (6.49b), $\varphi_{M_0} = 0$, $\varphi_{M_0} = -0.0725$, and $\varphi_{M_0} = -0.2$, respectively. It turns out that the period-doubling bifurcations of the limit cycle take place without changing the circuit parameters, but only the initial condition φ_{M_0}.

Similar results can be obtained for $\alpha = 9.5$ in (6.48), in which case the MCC in Fig. 6.25 displays chaotic attractors. Figure 6.29a, b, and c show how the double-

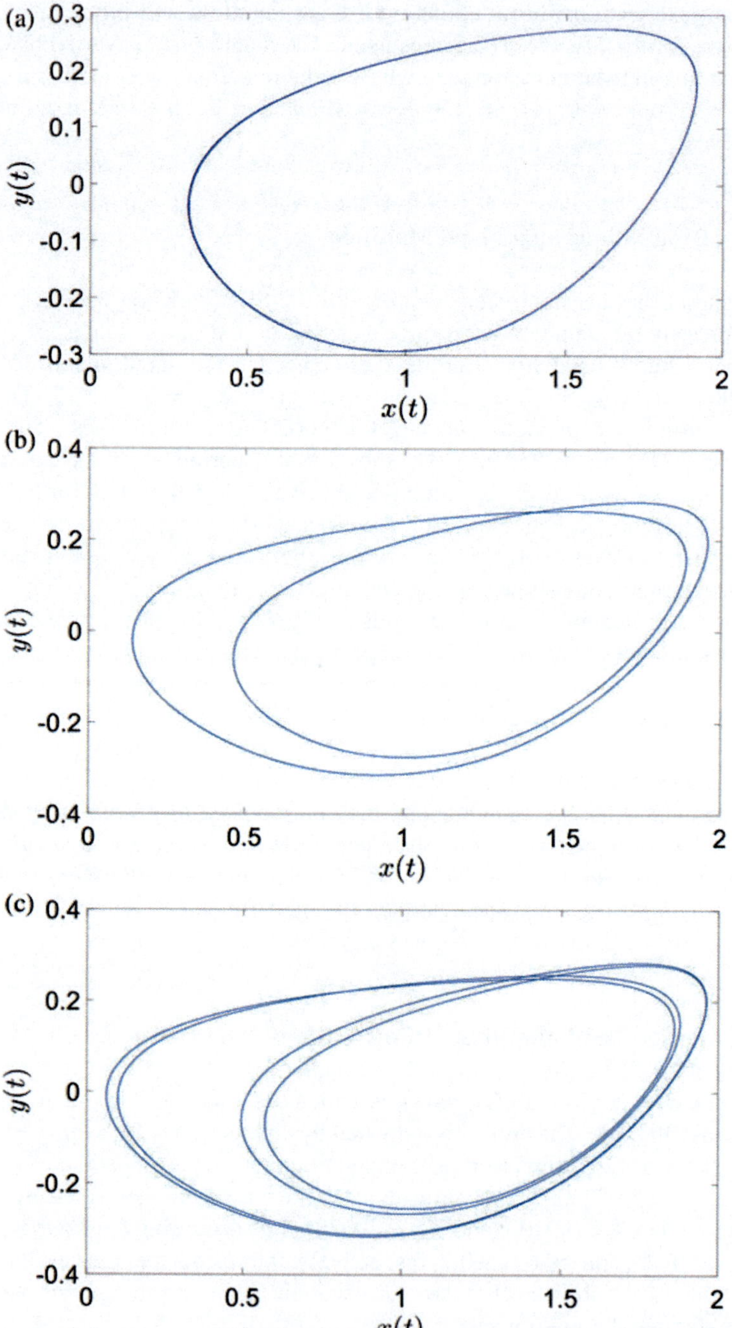

Fig. 6.27 Bifurcations on the fixed manifold $\mathcal{M}(X_0)$ specified by $X_0 = -0.0029$. The period-doubling cascade is induced by varying the circuit parameter α. (**a**) $\alpha = 8.8$. (**b**) $\alpha = 8.9$. (**c**) $\alpha = 9.0$

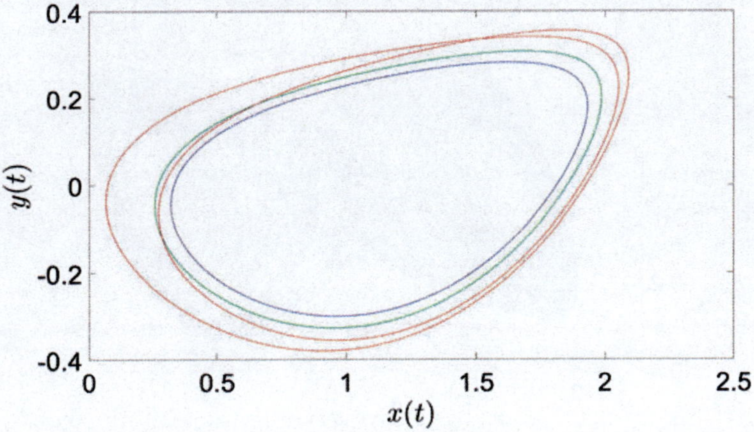

Fig. 6.28 Bifurcations without parameters of a limit cycle in the MCC of Fig. 6.25 with the circuit parameter fixed at $\alpha = 8.7$. The projection of the limit cycle for $X_0 = 0$ (blue curve) turns into a different limit cycle (green curve) for $X_0 = 0.0103$ and then into a period-two limit cycle for $X_0 = 0.0281$. The period-doubling bifurcation of the limit cycle takes place without changing the circuit parameters, but only the initial condition φ_{M_0}

scroll chaotic attractor[2] for $X_0 = 0$ (blue curve) turns into a different double-scroll chaotic attractor (green curve) for $X_0 = -0.0103$ and then into a spiral chaotic attractor for $X_0 = -0.0281$.

This shows that the MCC in Fig. 6.25 exhibits, for fixed circuit parameters, complex bifurcation phenomena due to varying initial conditions, in particular it displays *period-doubling bifurcations without parameters* leading to the birth or disappearance of chaotic attractors.

Remark 6.22 This discussion shows that for the MCC there is coexistence of infinitely many third-order dynamics, one for each invariant manifold. In particular, there is coexistence of a continuum of EPs, of periodic attractors (cycles of period one or of multiple period), and also of a continuum of different complex chaotic attractors, for a fixed set of circuit parameters. Such an extremely rich and complex dynamic scenario corresponds to the so-called property of extreme multistability.

Remark 6.23 We refer the reader to [16] where use is made of the harmonic balance method (HBM) in combination with FCAM to analytically predict period-doubling bifurcations without parameters in MCC. The article [17] instead uses HBM to study analogous bifurcations in a class of harmonically forced second-order memristor oscillators in the (φ, q)-domain.

[2]The double-scroll attractor for $X_0 = 0$ coincides with that of the canonical Chua's oscillator.

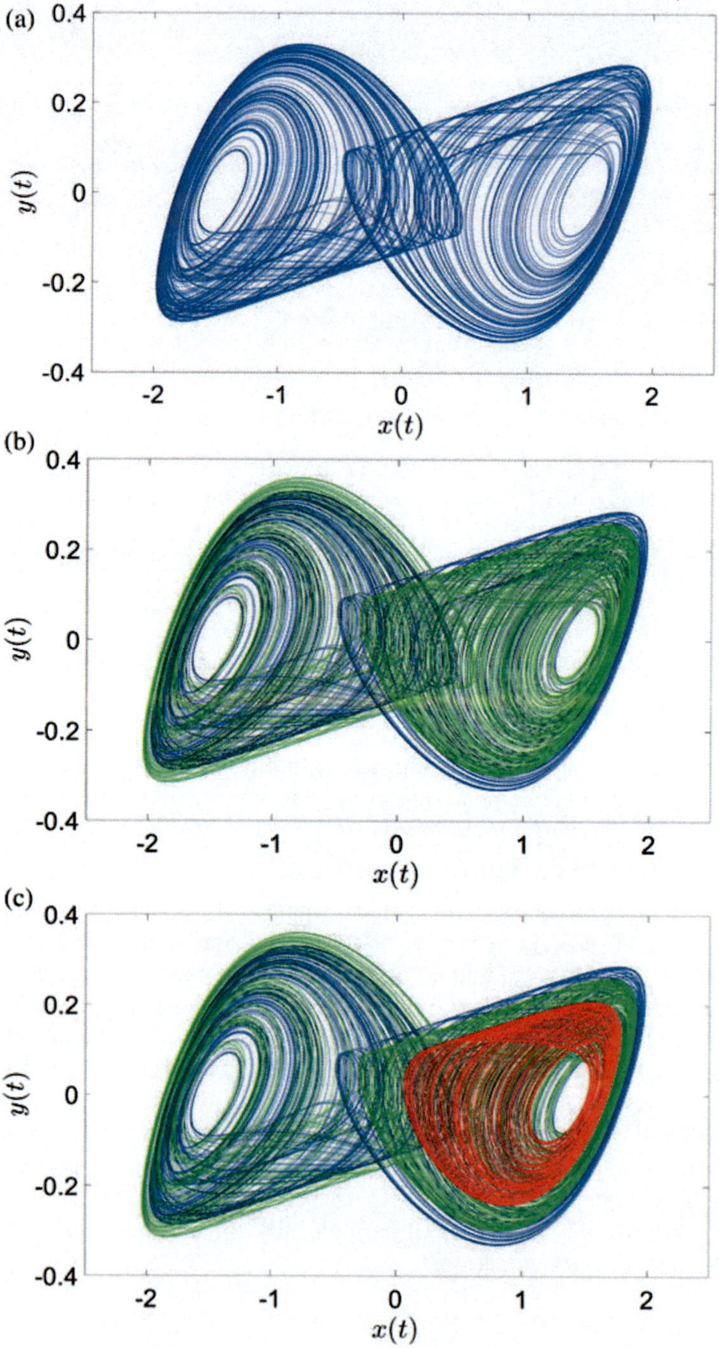

Fig. 6.29 Bifurcations without parameters of the chaotic attractor in the MCC of Fig. 6.25 with the circuit parameter α fixed at $\alpha = 9.5$. The double-scroll chaotic attractor for (**a**) $X_0 = 0$ (blue curve) turns into a different double-scroll chaotic attractor (green curve) for (**b**) $X_0 = -0.0103$ and then into a spiral chaotic attractor (red curve) for (**c**) $X_0 = -0.0281$

6.4 Discussion

Here, we collect some concluding remarks concerning the results obtained in the chapter.

1. Several papers in the literature have studied the dynamics and bifurcations of specific memristor circuits in the class $\mathcal{LM}$. A summary of some main contributions is reported next. The papers [11, 12] studied via simulations the influence of initial conditions on the oscillatory behavior of a circuit in the class $\mathcal{LM}$ pointing out the coexistence of several different attractors for the same set of circuit parameters. A class of Chua's oscillators containing a memristor is studied in [18] and coexistence of different attractors, initial-condition dependent bifurcations, and complex dynamics are highlighted by means of simulations. In the article [19], a class of memristor cellular neural networks is considered which is able to generate Turing patterns. The paper highlights by numerical means how the initial conditions in the memristors influence the onset of Turing patterns and also the types of Turing patterns generated by the network. The paper [20] has experimentally studied the role of initial conditions in relation to a type of bifurcations, named grazing bifurcations, for a fixed set of parameters, in a memristor circuit with a nonsmooth memristor characteristic. Numerical studies in the (v, i)-domain on the influence of initial conditions on dynamics and bifurcations have been carried out also in [21–23]. The articles [24, 25] developed a method in the (v, i)-domain to study bifurcations, and in particular bifurcations without parameters, in cases where the DAE description of memristor circuits is available but the SE description is not. However, it is not analytically investigated how initial conditions are related to such bifurcations. Stability properties of attractors, local and global bifurcations, and the role of the initial conditions have been extensively investigated in [26] as well. We refer the reader to [27, 28], and references therein, for other contributions along this line.

   Overall, the quoted papers aim at highlighting peculiar phenomena observable in memristor circuits as the coexistence of a huge number of different attractors for a fixed set of circuit parameters [10]. Moreover, bifurcations due to changing initial conditions, for a fixed set of circuit parameters, are highlighted. One fundamental shortcoming of these contributions is that such phenomena are investigated basically by numerical or experimental means, while an analytic explanation is lacking.

   In this chapter we have shown, by discussing some significant applications, that FCAM is effective to analytically explain these initial-condition related phenomena and is especially well suited to rigorously prove the coexistence of infinitely many different attractors and the presence of bifurcations without parameters in memristor circuits. Such an extremely rich and complex dynamic scenario corresponds to a case referred to in the literature as extreme multistability. The reader is referred to [10] for other classes of nonlinear dynamical systems in physics and engineering featuring extreme multistability.

2. *Foliation of state space, invariant manifolds, reduction of order, and smoother dynamics:* in the chapter we used FCAM to analyze the dynamics of a M–C circuit, a M–L–C circuit, and a Memristor-based Chaotic Circuit (MCC) in the class $\mathcal{LM}$. In all cases FCAM has enabled to obtain a foliation of the phase-space in the (v, i)-domain in invariant manifolds where the circuit dynamics is described by a lower-order system of SEs in the (φ, q)-domain. Dealing with a lower-order system is of course advantageous and also reveals dynamical aspects that are difficult to grasp when dealing with a higher-order system. For instance, the $M - L - C$ circuit is described in the (v, i)-domain by a third-order system whose dynamics is not easy to analyze directly. Instead, in the flux-charge domain it is described on each manifold by a second-order system that can be brought back to a forced Van Der Pol oscillator and analyzed via quite standard techniques. Analogous considerations apply to the MCC, which is equivalent in the (φ, q)-domain to a forced Chua's oscillator. It is worth to remark that a further advantage of FCAM is that the vector fields defining the SEs in the (φ, q)-domain contain the memristor nonlinearity $f(\cdot)$ (or $h(\cdot)$), which is smoother than the nonlinearity $f'(\cdot)$ (or $h'(\cdot)$) in the vector fields defining the SEs in the (v, i)-domain. As a consequence, numerical problems in the simulations (cf. for instance [29]) of memristor circuits are expected to be less relevant in the (φ, q)-domain than in the (v, i)-domain.
3. *Coexisting dynamics:* the principle of foliation of the state space naturally explains the coexistence of different dynamics and attractors for a memristor circuit. We have shown in particular that for the M–C circuit we may have coexistence of monostable and bistable dynamics, while for the M–L–C circuit there coexist both convergent and oscillatory dynamics. The MCC circuit displays a much reacher scenario with the coexistence of convergent dynamics, periodic dynamics, and complex (chaotic) dynamics.
4. *Bifurcations without parameters:* Given the initial conditions for the state variables in the (v, i)-domain, we can explicitly find via FCAM the corresponding invariant manifold where the memristor circuit dynamics evolve. As a consequence, bifurcations induced by varying these initial conditions for fixed circuit parameters (i.e., bifurcations without parameters) can be analytically investigated. For the M–L–C circuit we were able to thoroughly analyze Hopf bifurcations without parameters originating nonlinear oscillations whereas for the MCC we investigated period-doubling cascades induced by varying the initial conditions that lead to the birth or disappearance of a chaotic attractor. To the best of authors' knowledge, the M–L–C oscillatory circuit in Fig. 6.18 and the MCC in Fig. 6.25 represent the first examples in the literature where a wide gamut of bifurcations without parameters have been rigorously shown.
5. Overall, via FCAM we have shown that *the presence of a continuum of EPs, the existence of invariants of motion, and invariant manifolds are structurally stable properties for a memristor circuit. Namely, they hold for any value of the circuit parameters and are due only to the presence of the special element memristor. The same holds for the feature of coexisting attractors and extreme multistability.*

Appendix 1: Bifurcations Without Parameters of Equilibrium Points

We want to verify that the bifurcation of EPs observed in the $M - C$ studied in Sect. 6.1.2 exactly corresponds to one type of bifurcations without parameters introduced theoretically in [6].

Suppose the memristor has a CR $q_M = -\varphi_M + \frac{1}{3}\varphi_M^3$. In the (v, i)-domain the $M - C$ circuit satisfies

$$\dot{\varphi}_M(t) = v_C(t)$$
$$C\dot{v}_C(t) = (1 - \varphi_M^2(t))v_C(t).$$

There exists a manifold (a line) of EPs coinciding with the φ_M axis. By letting $Y = \varphi_M$ and $Z = v_C$, assuming $C = 1$ F, and omitting dependence on t we can rewrite the system as

$$\dot{Y} = F^Y(Y, Z) = Z$$
$$\dot{Z} = F^Z(Y, Z) = (1 - Y^2)Z$$

which is in the form of system (10) in [6]. Let $\mathbf{F} = (F^Y, F^Z)$. We have $\mathbf{F}(Y, 0) = 0$ for any $Y \in \mathbb{R}$, i.e., the Y-axis is a line of EPs.

To find the EPs that are candidates for a bifurcation without parameter we impose condition (cf. (11) in [6])

$$\partial_Z F^Z(Y, 0) = 0.$$

This yields

$$\partial_Z F^Z(Y, 0) = 1 - Y^2 = 0$$

hence the EPs candidate for a bifurcation without parameter are $Y_e = \pm 1$.

Let us consider $Y_e = -1$ (a similar analysis holds for $Y_e = 1$). The change of variables $y = Y - Y_e = Y + 1$ and $z = Z$ yields

$$\dot{y} = f^y(y, z) = z$$
$$\dot{z} = f^z(y, z) = (2y - y^2)z.$$

Let $\mathbf{f} = (f^y, f^z)$. We have $\mathbf{f}(y, 0) = 0$ for $y \in \mathbb{R}$, hence the y-axis is a line of EPs. Moreover, we have

$$\partial_z f^z(0, 0) = 0$$
$$\partial_{yz} f^z(0, 0) = 2 \neq 0$$

$$\partial_z f^y(0,0) = 1 \neq 0$$

hence conditions (11) and (12) in [6] guaranteeing the existence of a bifurcation without parameter at the EP $y_e = 0$ are satisfied.

It is worth noting that for small $|y|$ the previous system simplifies as

$$\dot{y} = f^y(y,z) = z$$
$$\dot{z} = f^z(y,z) = 2y(1 - \frac{y}{2})z \simeq 2yz$$

which is the normal form of a transcritical bifurcation without parameters according to Theorem 1.1 in [6].

We conclude that the capacitor and flux-controlled memristor originate a bifurcation without parameters as in the theory developed in [6].

Finally, it is noted that the condition $\partial_Z F^Z(0, Z) = 0$ that is used for finding EPs that are candidates for bifurcation without parameters corresponds to the loss of normal hyperbolicity of the manifold of EPs, as briefly discussed next. In the considered $M - C$ circuit, by evaluating the Jacobian of F at the generic EP $(Y, 0)$ we obtain

$$J_F(Y,0) = \begin{pmatrix} 0 & 1 \\ 0 & 1 - Y^2 \end{pmatrix}.$$

The Jacobian has an eigenvalue 0 due to the fact that there is a line of EPs (the Y-axis). An EP $(Y, 0)$ is said to be normally hyperbolic if the remaining eigenvalue $1 - Y^2$ of the Jacobian does not vanish. Now, it can be checked that the condition $\partial_Z F^Z(Y, 0) = 1 - Y^2 = 0$ is equivalent to requiring that the remaining eigenvalue of the Jacobian vanishes, i.e., the corresponding EP loses normal hyperbolicity.

Appendix 2: Change of Variable for Invariant Manifolds

This appendix summarizes the method to derive (6.40) from (6.38). Let us write the change of variables (6.39) in the general form

$$x(t) = \varphi_{C_1}(t;t_0) + k_x \tag{6.54a}$$

$$y(t) = \varphi_{C_2}(t;t_0) + k_y \tag{6.54b}$$

$$z(t) = Rq_L(t;t_0) + k_z \tag{6.54c}$$

where constants k_x, k_y, and k_z have to be determined in order to rewrite (6.38) as (6.40). By taking the time derivative of (6.54), it is apparent that the l.h.s. of (6.38) and (6.40) are identical. The substitution of (6.54) into (6.38) yields

$$\frac{dx(t)}{d\tau} = \alpha\Big[y(t) - k_y - x(t) + k_x$$
$$-Rf(x(t) - k_x + \varphi_{M_0}) + \boldsymbol{Rf}(\varphi_{M_0}) + Rq_{C_{1_0}}\Big]$$
$$\frac{dy(t)}{d\tau} = -(y(t) - k_y - x(t) + k_x) + (z(t) - k_z) + Rq_{C_{2_0}}$$
$$\frac{dz(t)}{d\tau} = -\gamma(z(t) - k_z) - \beta(y(t) - k_y) + \beta\varphi_{L_0}$$

that is

$$\frac{dx(t)}{d\tau} = \alpha\Big[y(t) - x(t) - Rf(x(t) - k_x + \varphi_{M_0})$$
$$+(k_x - k_y + \boldsymbol{Rf}(\varphi_{M_0}) + Rq_{C_{1_0}})\Big] \tag{6.55a}$$
$$\frac{dy(t)}{d\tau} = x(t) - y(t) + z(t) + (-k_x + k_y - k_z + Rq_{C_{2_0}}) \tag{6.55b}$$
$$\frac{dz(t)}{d\tau} = -\gamma z(t) - \beta y(t) + (\gamma k_y + \beta k_z + \beta\varphi_{L_0}). \tag{6.55c}$$

It turns out that the Eqs. (6.55) assume the form of the SEs (6.40) if and only if k_x, k_y, and k_z are such that

$$-k_x + \varphi_{M_0} = 0 \tag{6.56a}$$
$$-k_x + k_y - k_z + Rq_{C_{2_0}} = 0 \tag{6.56b}$$
$$\beta k_y + \gamma k_z + \beta\varphi_{L_0} = 0 \tag{6.56c}$$

and

$$X_0 = k_x - k_y + \boldsymbol{Rf}(\varphi_{M_0}) + Rq_{C_{1_0}}. \tag{6.57}$$

The solution of (6.56) permits to obtain

$$k_x = \varphi_{M_0} \tag{6.58a}$$
$$k_y = \frac{\gamma}{\beta+\gamma}\varphi_{M_0} - \frac{\beta}{\beta+\gamma}\varphi_{L_0} - \frac{\gamma}{\beta+\gamma}Rq_{C_{2_0}} \tag{6.58b}$$
$$k_z = \frac{\beta}{\beta+\gamma}\left(-\varphi_{M_0} + \varphi_{L_0} + Rq_{C_{2_0}}\right). \tag{6.58c}$$

The expressions (6.39) and (6.42) are readily obtained by inserting (6.58) in (6.54) and (6.57), respectively.

References

1. H.K. Khalil, *Nonlinear Systems* (Prentice Hall, Englewood Cliffs, 2002)
2. L.O. Chua, Dynamic nonlinear networks: state-of-the-art. IEEE Trans. Circuits Syst. **27**(11), 1059–1087 (1980)
3. T. Matsumoto, L. Chua, A. Makino, On the implications of capacitor-only cutsets and inductor-only loops in nonlinear networks. IEEE Trans. Circuits Syst. **26**(10), 828–845 (1979)
4. V.I. Arnold, *Geometrical Methods in the Theory of Ordinary Differential Equations* (Springer, New York, 1988)
5. O. Merino, A short history of complex numbers. University of Rhode Island, Kingston (2006)
6. B. Fiedler, S. Liebscher, J.C. Alexander, Generic Hopf bifurcation from lines of equilibria without parameters: I. Theory. J. Differ. Equ. **167**(1), 16–35 (2000)
7. M. Itoh, L.O. Chua, Memristor oscillators. Int. J. Bifurc. Chaos **18**(11), 3183–3206 (2008)
8. J.P. Aubin, A. Cellina, *Differential Inclusions. Set-Valued Maps and Viability Theory* (Springer, Berlin, 1984)
9. A. Mees, L. Chua, The Hopf bifurcation theorem and its applications to nonlinear oscillations in circuits and systems. IEEE Trans. Circuits Syst. **26**(4), 235–254 (1979)
10. C.R. Hens, R. Banerjee, U. Feudel, S.K. Dana, How to obtain extreme multistability in coupled dynamical systems. Phys. Rev. E **85**(3), 035202 (2012)
11. M. da Cruz Scarabello, M. Messias, Bifurcations leading to nonlinear oscillations in a 3D piecewise linear memristor oscillator. Int. J. Bifurc. Chaos **24**(1), 1430001 (2014)
12. M. Messias, C. Nespoli, V.A. Botta, Hopf bifurcation from lines of equilibria without parameters in memristor oscillators. Int. J. Bifurcation Chaos **20**(2), 437–450 (2010)
13. A. Amador, E. Freire, E. Ponce, J. Ros, On discontinuous piecewise linear models for memristor oscillators. Int. J. Bifurcation Chaos **27**(6), 1730022 (2017)
14. F. Corinto, A. Ascoli, M. Gilli, Nonlinear dynamics of memristor oscillators. IEEE Trans. Circuits Syst. I: Regul. Pap. **58**(6), 1323–1336 (2011)
15. M. Gilli, F. Corinto, P. Checco, Periodic oscillations and bifurcations in cellular nonlinear networks. IEEE Trans. Circ. Syst. I: Regul. Pap. **51**(5), 948–962 (2004)
16. M. Di Marco, M. Forti, G. Innocenti, A. Tesi, Harmonic balance method to analyze bifurcations in memristor oscillatory circuits. Int. J. Circuit Theory Appl. **46**(1), 66–83 (2018)
17. G. Innocenti, M. Di Marco, M. Forti, A. Tesi, Prediction of period doubling bifurcations in harmonically forced memristor circuits. Nonlinear Dyn. **96**(2), 1169–1190 (2019)
18. B.-C. Bao, Q. Xu, H. Bao, M. Chen, Extreme multistability in a memristive circuit. Electron. Lett. **52**(12), 1008–1010 (2016)
19. A. Buscarino, C. Corradino, L. Fortuna, M. Frasca, L.O. Chua, Turing patterns in memristive cellular nonlinear networks. IEEE Trans. Circuits Syst. I: Regul. Pap. **63**(8), 1222–1230 (2016)
20. A.I. Ahamed, M. Lakshmanan, Nonsmooth bifurcations, transient hyperchaos and hyperchaotic beats in a memristive Murali–Lakshmanan–Chua circuit. Int. J. Bifurcation Chaos **23**(6), 1350098 (2013)
21. V.-T. Pham, S. Vaidyanathan, C.K. Volos, S. Jafari, N.V. Kuznetsov, T.M. Hoang, A novel memristive time-delay chaotic system without equilibrium points. Eur. Phys. J. Spec. Top. **225**(1), 127–136 (2016)
22. B. Bao, Z. Ma, J. Xu, Z. Liu, Q. Xu, A simple memristor chaotic circuit with complex dynamics. Int. J. Bifurc. Chaos **21**(9), 2629–2645 (2011)
23. Q. Li, S. Hu, S. Tang, G. Zeng, Hyperchaos and horseshoe in a 4D memristive system with a line of equilibria and its implementation. Int. J. Circuit Theory Appl. **42**(11), 1172–1188 (2014)
24. R. Riaza, C. Tischendorf, Semistate models of electrical circuits including memristors. Int. J. Circuit Theory Appl. **39**(6), 607–627 (2011)
25. R. Riaza, Manifolds of equilibria and bifurcations without parameters in memristive circuits. SIAM J. Appl. Math. **72**(3), 877–896 (2012)

26. F. Corinto, A. Ascoli, M. Gilli, Analysis of current–voltage characteristics for memristive elements in pattern recognition systems. Int. J. Circuit Theory Appl. **40**(12), 1277–1320 (2012)
27. B.C. Bao, H. Bao, N. Wang, M. Chen, Q. Xu, Hidden extreme multistability in memristive hyperchaotic system. Chaos Solitons Fractals **94**, 102–111 (2017)
28. H. Bao, N. Wang, B. Bao, M. Chen, P. Jin, G. Wang, Initial condition-dependent dynamics and transient period in memristor-based hypogenetic jerk system with four line equilibria. Commun. Nonlinear Sci. Numer. Simul. **57**, 264–275 (2018)
29. A. Ascoli, R. Tetzlaff, Z. Biolek, Z. Kolka, The art of finding accurate memristor model solutions. IEEE J. Emerg. Sel. Top. Circuits Syst. **5**(2), 133–142 (2015)

Chapter 7
Pulse Programming of Memristor Circuits

7.1 Introduction

Chapter 5 introduced the "Flux-Charge Analysis Method" (FCAM) for the dynamic analysis of memristor circuits in the (φ, q)-domain. FCAM is based on using Kirchhoff flux and charge laws, and CRs of circuit elements, directly expressed in the (φ, q)-domain and, as such, it basically differs from other traditional approaches for analyzing memristor circuits described in the (v, i)-domain [1–3].

Then, in Chap. 6, the application of FCAM to specific low-order autonomous circuits with one memristor has shown that the state space in the (v, i)-domain can be foliated in a continuum of positively invariant manifolds and that on each manifold a memristor circuit is characterized by a different reduced-order dynamics and attractors. A relevant consequence is that *there coexist different nonlinear dynamics and regimes for the same set of circuit parameters*. Moreover, bifurcations may be induced in two different ways. In the classical way, where the circuit parameters are changed for initial conditions (ICs) belonging to a fixed manifold (standard bifurcations). Otherwise, by changing invariant manifold via the variation of ICs for fixed circuit parameters (also named bifurcations without parameters). In particular, in Chap. 5, FCAM is used to analyze saddle-node bifurcations without parameters for the simplest memristor circuit composed by a capacitor and a flux-controlled memristor, and the Hopf and period-doubling bifurcations without parameters for some memristor oscillatory circuits. Overall, these results have provided an analytic explanation of novel and intriguing dynamic phenomena displayed by memristor circuits, as the coexistence of several attractors for the same set of parameters and the strong dependence upon ICs of the dynamic behavior of solutions. We stress that such phenomena have been investigated elsewhere in the literature mainly via numerical, or experimental means, see, e.g., [4–11], and references therein.

In Chap. 6, only specific low-order autonomous circuits with one memristor have been considered for which the state equations (SEs) are written via FCAM

F. Corinto et al., *Nonlinear Circuits and Systems with Memristors*,
https://doi.org/10.1007/978-3-030-55651-8_7

by inspection. Furthermore, invariant manifolds and the dynamics on manifolds are obtained by relying on ad hoc mathematical manipulations of these SEs. Due to the huge variety and complexity of memristor circuits encountered in different applications (e.g., memristor synapses for neuromorphic systems, memristor-based chaotic circuits for cryptography, memristor-based biosensors, etc.), it is desirable to extend the results in Chap. 6 in order that they are applicable to a large class of memristor circuits used in the technical applications.

The main contributions in this chapter are as follows:

(a) we identify a wide class of memristor circuits, of any order and with any number of flux- or charge-controlled memristors, and introduce a systematic method for writing in an explicit way the SEs, both in the (φ, q) and in the (v, i)-domain. The conditions for the existence of the SEs for such class are easily checkable (usually by inspection), since they are couched in topological terms. The techniques overcome drawbacks of the analogous technique described in Chap. 5, that was able in general only to yield the SE formulation in implicit form.

(b) The obtained SEs have a relatively simple mathematical structure and, in the *autonomous* case, under certain assumptions, a systematic method is introduced to identify and write analytically the invariant manifolds, to show the coexistence of different dynamics and to find the reduced-order dynamics on each invariant manifold.

The results in (a) and (b) are then used and extended for addressing the *nonautonomous* case, i.e., the case where time-varying independent sources are present in the memristor circuit. In this regard, the chief contribution in the chapter is as follows:

(c) we develop analytic results showing how invariant manifolds, different regimes and reduced-order dynamics, and attractors of nonautonomous memristor circuits, can be easily and effectively *programmed* by applying suitable charge and/or flux *pulses* via time-varying current and/or voltage sources with finite time duration.

For example, we have seen in Chap. 6 that in a given Memristor Chaotic Circuit (MCC) built upon Chua's oscillator there coexist stationary, periodic, and complex attractors for the same set of parameters. We show in the chapter that, by means of a *single pulse source*, it is possible to set in an effective way any desired stationary or oscillatory regime in the MCC. This represents a significant practical improvement with respect to results in Chap. 6, since in that case manifolds and reduced-order dynamics were selected by imposing the whole set of four initial conditions for the state variables of MCC in the (v, i)-domain.

Although we are dealing in this chapter with general circuits with an arbitrary number of memristors, we find it useful for pedagogical reasons to first illustrate the basic idea of pulse programming via a simple memristor circuit.

7.2 Motivating Example

We have seen in Example 2.9 of Chap. 2 how it is possible to set a desired value of flux or memductance in a flux-controlled memristor by applying a suitable voltage pulse by means of a battery. This is basically a static example since the memristor flux remains unchanged once the pulse is over. In the next example, we consider a simple dynamic memristor circuit and study how it is possible to effectively programme its dynamics via a pulse generator.

Example 7.1 Consider the simple $M - C$ circuit with a capacitor and a flux-controlled memristor $q_M = f(\varphi_M)$ studied in Sect. 6.1.2 of Chap. 6. The $M - C$ circuit has an invariant of motion

$$Q(t) = Cv_C(t) + f(\varphi_M(t))$$

given by the total charge in the circuit. This means that the dynamics of the $M - C$ circuit evolves on one of the invariant manifolds

$$\mathcal{M}(Q) = \{(v_C, \varphi_M)^T \in \mathbb{R}^2 : Cv_C + f(\varphi_M) = Q\}$$

where $Q \in \mathbb{R}$.

Suppose we have $v_C(t_0) = v_{C_0}$ and $\varphi_M(t_0) = \varphi_{M_0}$, hence $(v_{C_0}, \varphi_{M_0})^T \in \mathcal{M}(Q_0)$, where $Q_0 = Cv_{C_0} + f(\varphi_{M_0})$. Consider two instants $t_2 > t_1 > t_0$ and suppose we wish that the solution of the $M - C$ circuit evolves on the manifold $\mathcal{M}(Q_0)$ in $[t_0, t_1]$, while it evolves on manifold $\mathcal{M}(\bar{Q})$ for $t \geq t_2$, where $\bar{Q} \neq Q_0$. This means that we want the *solution to switch between the manifolds* $\mathcal{M}(Q_0)$ *and* $\mathcal{M}(\bar{Q})$ *in* $[t_1, t_2]$. This can be achieved in two different ways.

The first possibility is to use additional circuitry and switches in the interval $[t_1, t_2]$ to set the state $(v_C(t_2), \varphi_M(t_2))$ at t_2 in a way that $Q(t_2) = Cv_C(t_2) + f(\varphi_M(t_2)) = \bar{Q}$. This would require two additional auxiliary networks, one to set the state $v_C(t_2)$ of the capacitor and the other to set the state $\varphi_M(t_2)$ of the memristor, which is not easy to implement.

There is also a second simpler and more effective way to solve the same problem. Suppose to add a current source as in Fig. 7.1 applying to the circuit a rectangular current pulse in $[t_1, t_2]$, i.e.,

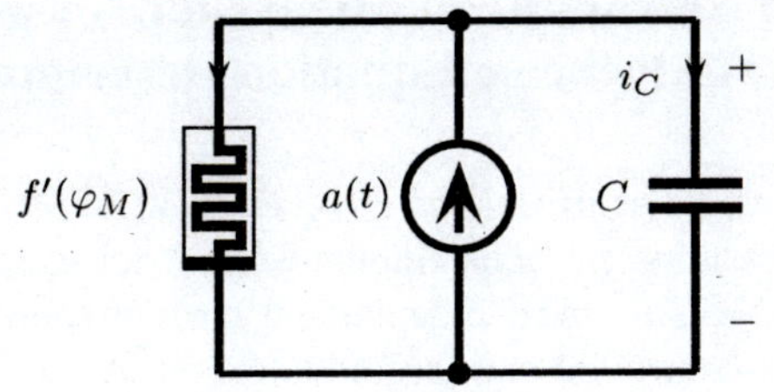

Fig. 7.1 A simple memristor circuit with a current-source introduced for programming purposes

$$a(t) = \begin{cases} 0, & t \in [t_0, t_1) \\ I_0, & t \in [t_1, t_2] \\ 0, & t > t_2. \end{cases}$$

To see the effect of the current source note that the KCL yields, for $t \geq t_0$, $i_M(t) + i_C(t) = a(t)$. Since $i_M(t) = dq_M(t)/dt$, we obtain

$$\frac{d(f(\varphi_M(t)))}{dt} + \frac{d(Cv_C(t))}{dt} = \frac{dQ(t)}{dt} = a(t).$$

By integrating between t_2 and t_1 we obtain

$$Q(t_2) - Q(t_0) = \int_{t_1}^{t_2} a(\tau)d\tau = I_0(t_2 - t_1).$$

Hence, if we choose $I_0 = (\bar{Q} - Q_0)/(t_2 - t_1)$, we obtain $Q(t_2) = \bar{Q}$. Since $a(t) = 0$ for $t \geq t_2$, it follows that the dynamics of the $M - C$ circuit evolves on manifold $\mathcal{M}(\bar{Q})$ for any $t \geq t_2$. Finally note that, as required, the dynamics evolves on $\mathcal{M}(Q_0)$ in $[t_0, t_1]$ since $a(t) = 0, t \in [t_0, t_1)$, implies $Q(t) = Q_0, t \in [t_0, t_1]$.

It is worth noting that a *single current source and a single pulse* is able to switch the dynamics of the $M - C$ circuit between manifolds $\mathcal{M}(Q_0)$ and $\mathcal{M}(\bar{Q})$. It is also clear that what counts is the area of the pulse impressed by the current source, not the shape of $a(t)$. In other words, any current pulse with area $\int_{t_1}^{t_2} a(\tau)d\tau = \bar{Q} - Q_0$ would result in the $M - C$ circuit evolving on $\mathcal{M}(\bar{Q})$ for $t \geq t_2$.

Switching the dynamics of the $M - C$ circuit between different manifolds is of potential practical interest since, as we have seen in Chap. 6, the dynamics on different manifolds may be qualitatively different. In fact, in the $M - C$ circuit in Example 7.1 there coexist both mono-stable dynamics and bistable dynamics on different manifolds for a cubic memristor nonlinearity $q = -\varphi + \frac{1}{3}\varphi^3$ and so via a pulse we can easily switch between these two different types of dynamics.

The idea in this simple example will be greatly generalized in this chapter by showing how we can programme the dynamics of a large class of memristor circuits via the application of suitable finite-duration impulsive current or voltage sources.

7.3 Structure of Memristor Network and Differential Algebraic Equation Description

Consider as in Chap. 5 the class $\mathcal{LM}$ of *memristor circuits* constituted by n_C ideal capacitors, n_L ideal inductors, n_R ideal resistors, n_E ideal independent voltage sources, n_A ideal independent current sources, and n_M memristors that are either flux-controlled or charge-controlled. Assume that any flux-controlled memristor is described by a CR of the type $q(t) = f(\varphi(t))$, where $\varphi(t)$ and $q(t)$ are the flux and charge in the memristor (i.e., the voltage and current momenta), respectively. Moreover, each charge-controlled memristor has a CR of the type $\varphi(t) = h(q(t))$.

We wish to analyze the nonlinear dynamics and bifurcations in a memristor network $N \in \mathcal{LM}$ for $t \geq t_0$, where t_0 is a given finite time instant. For any two-terminal circuit element in N, in addition to the voltage $v(t)$, current $i(t)$, flux $\varphi(t)$, and charge $q(t)$, we consider also the *incremental* flux and charge which we have defined as

$$\varphi(t; t_0) = \varphi(t) - \varphi(t_0) = \int_{t_0}^{t} v(\tau) d\tau \tag{7.1a}$$

$$q(t; t_0) = q(t) - q(t_0) = \int_{t_0}^{t} i(\tau) d\tau. \tag{7.1b}$$

Note that the following properties hold: $\varphi(t) = \varphi(t; t_0) + \varphi(t_0)$; $q(t) = q(t; t_0) + q(t_0)$; $\varphi(t_0; t_0) = 0$; $q(t_0; t_0) = 0$; $\dot{\varphi}(t; t_0) = \dot{\varphi}(t) = v(t)$; $\dot{q}(t; t_0) = \dot{q}(t) = i(t)$; $\dot{\varphi}(t_0; t_0) = \dot{\varphi}(t_0) = v(t_0)$; $\dot{q}(t_0; t_0) = \dot{q}(t_0) = i(t_0)$.

Hereinafter, N is described by means of FCAM in the (φ, q)-domain (Chap. 5). The goal is first to obtain via FCAM a DAE description of N and, on this basis, to obtain the SE description for a large subclass of memristor networks $N \in \mathcal{LM}$. We stress that the knowledge of the SEs is a fundamental prerequisite for studying basic qualitative aspects such as identifying the invariant manifolds, and the dynamics on manifolds, and study how dynamics through different manifolds can be controlled by external inputs. The importance to derive the explicit SEs for any dynamic network has been discussed also in Chap. 6, where it has been noticed that network models that do not admit an SE representation may be ill defined due to the presence of singular points (a.k.a. impasse points) where solutions cannot be continued forward or backward in time.

To obtain a DAE description we note that, without losing generality, any network $N \in \mathcal{LM}$ can be decomposed into a dynamic part N_D and an adynamic part A_D.

Subnetwork N_D contains: n_C (resp., n_L) nonlinear dynamic flux-controlled (resp., charge-controlled) two-terminal elements $\mathcal{D}^{\varphi}$ (resp., $\mathcal{D}^{q}$) connected to the ports of N_A, as described next:[1]

- γ_M two-terminal elements $\mathcal{D}_M^{\varphi} \in N_D$ made of a capacitor in parallel with a flux-controlled memristor
- γ_G two-terminal elements $\mathcal{D}_G^{\varphi} \in N_D$ made of a capacitor in parallel with a negative resistor
- γ_C two-terminal elements $\mathcal{D}_C^{\varphi} \in N_D$ made of just a capacitor

[1]Mutual coupling between the elements $\mathcal{D}^{\varphi}$ (and elements $\mathcal{D}^{q}$) is excluded for simplicity in the treatment since by assumption $\mathcal{LM}$ is constituted of two-terminal elements, only. However, it is worth to remark that it would be possible to extend the treatment also to coupled capacitors, inductors, and resistors starting from the theoretic results in [12, Sect. III].

- λ_M two-terminal elements $\mathcal{D}_M^q \in N_D$ made of an inductor in series with a charge-controlled memristor
- λ_R two-terminal elements $\mathcal{D}_R^q \in N_D$ made of an inductor in series with a negative resistor
- $\lambda_L = n_L - \lambda_M - \lambda_R$ two-terminal elements $\mathcal{D}_L^q \in N_D$ made of just an inductor

Note that:

- $\mathcal{D}^q = \mathcal{D}_M^q \bigcup \mathcal{D}_R^q \bigcup \mathcal{D}_L^q$ and $n_L = \lambda_M + \lambda_R + \lambda_L$
- $\mathcal{D}^\varphi = \mathcal{D}_M^\varphi \bigcup \mathcal{D}_G^\varphi \bigcup \mathcal{D}_C^\varphi$ and $n_C = \gamma_M + \gamma_G + \gamma_C$.

Subnetwork N_A denotes the nonlinear adynamic $(n_C + n_L)$-port network with no capacitors and inductors. The conceptual decomposition of N into N_A and N_D is shown in Fig. 7.2. Note that $N_A \bigcup N_D$ results to be N.

Concerning the structure of N_A, let us assume that it contains:

- μ_F flux-controlled memristors $\mathcal{A}_M^\varphi \in N_A$
- μ_Q charge-controlled memristors $\mathcal{A}_M^q \in N_A$
- ρ_G negative conductances $\mathcal{A}_G^\varphi \in N_A$
- ρ_R negative resistors $\mathcal{A}_R^q \in N_A$.

The structure of N_A is essential to develop a general methodology to study DAEs and SEs. In this regard, it is convenient to further decompose N_A, i.e., to extract from N_A the linear negative resistors and the memristors. Such negative resistors and memristors are not in parallel to a capacitor or in series with an inductor.

Note that

$$n_M = \mu_F + \mu_Q + \gamma_M + \lambda_M$$

is the total number of memristors in N. The extraction of $\mathcal{A}_M^\varphi$, $\mathcal{A}_G^\varphi$, $\mathcal{A}_M^q$, and $\mathcal{A}_R^q$ from N_A yields a linear network N_R with $(n_C + n_L + \mu_F + \mu_Q + \rho_G + \rho_R)$-ports having *only positive (linear) resistors and independent sources.* We suppose that there are:

- n_E ideal independent voltage sources $\mathbf{e}(t)$ such that $\boldsymbol{\varphi}_e(t; t_0) = \int_{t_0}^t \mathbf{e}(\tau)d\tau$
- n_A ideal independent current sources $\mathbf{a}(t)$ such that $\mathbf{q}_a(t; t_0) = \int_{t_0}^t \mathbf{a}(\tau)d\tau$.

7.3.1 Constitutive Relations of Two-Terminal Elements

Here, we provide a systematic description of the different types of two-terminal elements in N_D, N_A, and N_R. For each two-terminal element connected to N_R we use coordinated reference directions for incremental flux and charge as shown in Fig. 7.2.

We start with simple examples illustrating how we can find the CR of an element $\mathcal{D}_M^\varphi \in N_D$ and $\mathcal{D}_M^q \in N_D$.

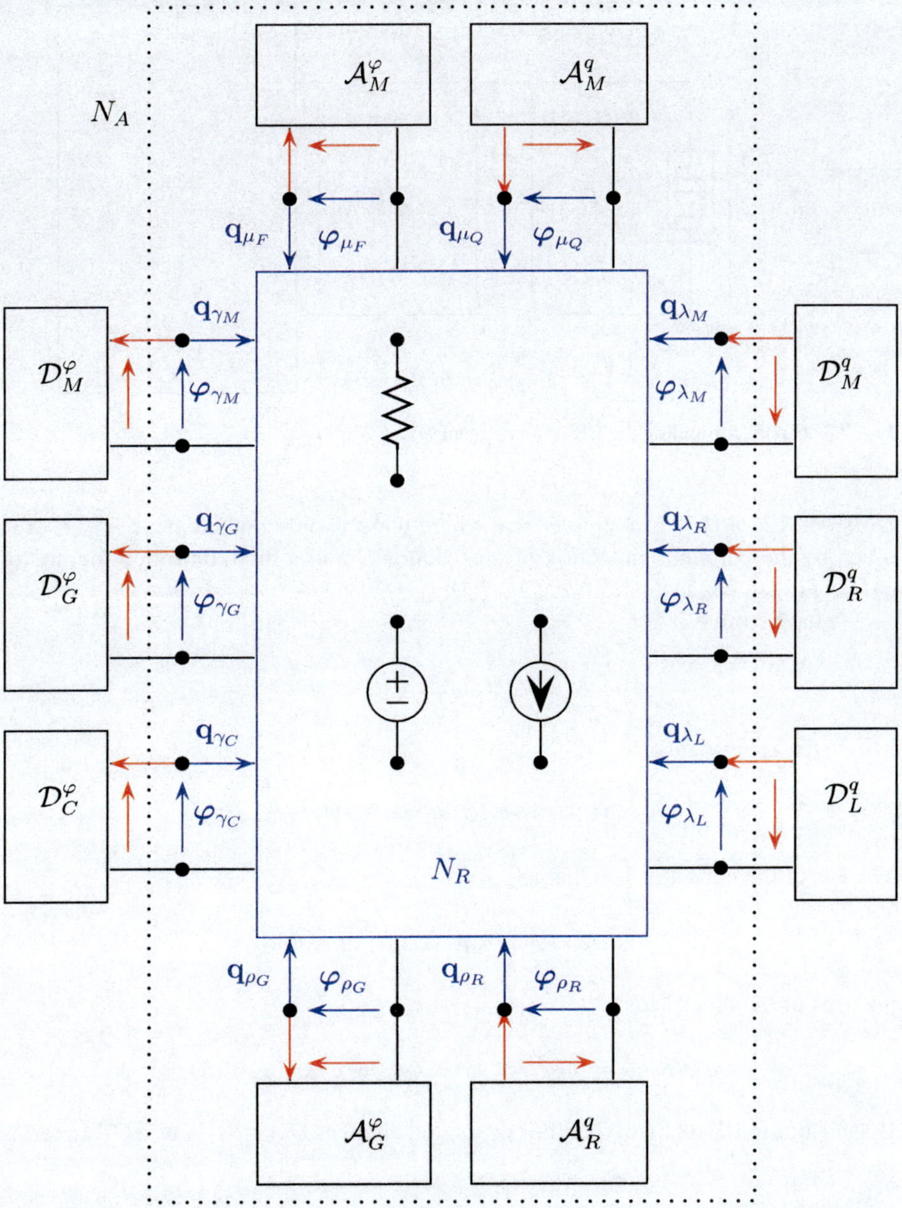

Fig. 7.2 Decomposition of N into N_A and N_D. The network $N_D = \mathcal{D}^{\varphi} \bigcup \mathcal{D}^{q}$ contains all two-terminal elements external to N_A. The n_C elements $\mathcal{D}^{\varphi}$, n_L elements $\mathcal{D}^{q}$, and the linear network N_R are highlighted. Reference directions for circuit elements external to N_R are in red

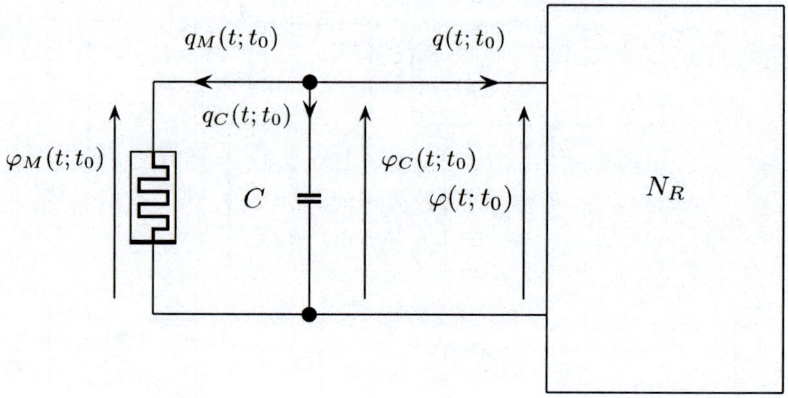

Fig. 7.3 Circuit for finding the CR of an element $\mathcal{D}_M^{\varphi} \in N_D$

Example 7.2 Consider a simple case where there is only one element $\mathcal{D}_M^{\varphi} \in N_D$, given by the parallel connection of a capacitor C and a flux-controlled memristor $q_M = f(\varphi_M)$, connected to N_R (Fig. 7.3).

We have from Kφ L

$$q(t;t_0) + q_C(t;t_0) + q_M(t;t_0) = 0.$$

Moreover, KφL yields

$$\varphi(t;t_0) = \varphi_C(t;t_0) = \varphi_M(t;t_0).$$

The CR of the capacitor is (Chap. 5)

$$q_C(t;t_0) = C\dot{\varphi}_C(t;t_0) - q_C(t_0)$$

and that of the memristor is

$$q_M(t;t_0) = f(\varphi_M(t;t_0) + \varphi_M(t_0)) - f(\varphi_M(t_0)).$$

By substitution, we obtain that the considered element $\mathcal{D}_M^{\varphi} \in N_D$ has a CR in terms of $q(t;t_0)$ and $\varphi(t;t_0)$ given by

$$-q(t;t_0) = C\dot{\varphi}(t;t_0) - q_C(t_0) + f(\varphi(t;t_0) + \varphi_M(t_0)) - f(\varphi_M(t_0)).$$

Example 7.3 Consider now the case where there is only one element $\mathcal{D}_M^{q} \in N_D$, given by the series connection of an inductor L and a charge-controlled memristor $\varphi_M = h(q_M)$, connected to N_R (Fig. 7.4).

We have from KφL

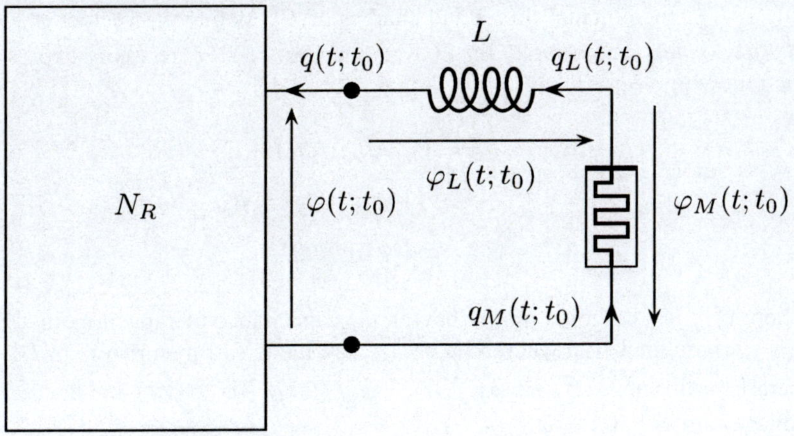

Fig. 7.4 Circuit for finding the CR of an element $\mathcal{D}_M^q \in N_D$

$$\varphi(t;t_0) + \varphi_L(t;t_0) + \varphi_M(t;t_0) = 0.$$

Moreover, KqL yields

$$q(t;t_0) = q_L(t;t_0) = q_M(t;t_0).$$

The CR of the inductor is (Chap. 5)

$$\varphi_L(t;t_0) = L\dot{q}_L(t;t_0) - \varphi_L(t_0)$$

and that of the memristor is

$$\varphi_M(t;t_0) = h(q_M(t;t_0) + q_M(t_0)) - h(q_M(t_0)).$$

By substitution, we obtain that the considered element $\mathcal{D}_M^q \in N_D$ has a CR in terms of $q(t;t_0)$ and $\varphi(t;t_0)$ given by

$$-\varphi(t;t_0) = L\dot{q}(t;t_0) - \varphi_L(t_0) + h(q(t;t_0) + q_M(t_0)) - h(q_M(t_0)).$$

The CRs in the (φ, q)-domain of elements $\mathcal{D}^\varphi$ and $\mathcal{D}^q$ in N_D can be expressed in vector form as follows.[2] We leave to the reader the rather straightforward verification of these formula using KqL, KφL and the CRs of basic circuit elements obtained in Chap. 5.

[2]A negative resistor in parallel with a flux-controlled memristor or in series with a charge-controlled memristor can be embedded in the memristor without modifying the results here provided.

- $\mathcal{D}_M^{\varphi} \in N_D$. Incremental fluxes and charges of $\mathcal{D}_M^{\varphi}$ define the vectors $\boldsymbol{\varphi}_{\gamma_M}(t; t_0)$ and $\mathbf{q}_{\gamma_M}(t; t_0)$, respectively. By FCAM, the CRs of $\mathcal{D}_M^{\varphi}$ are easily expressed in the following vector form:[3]

$$
\begin{aligned}
-\mathbf{q}_{\gamma_M}(t; t_0) = \mathbf{C}_{\gamma_M}\dot{\boldsymbol{\varphi}}_{\gamma_M}(t; t_0) - \mathbf{q}_{C_{\gamma_M}}(t_0) \\
+\mathbf{f}(\boldsymbol{\varphi}_{\gamma_M}(t; t_0) + \boldsymbol{\varphi}_{M_{\gamma_M}}(t_0)) \\
-\mathbf{f}(\boldsymbol{\varphi}_{M_{\gamma_M}}(t_0))
\end{aligned}
\tag{7.2}
$$

where $\mathbf{C}_{\gamma_M}$ is a diagonal matrix having the capacitance of capacitors in $\mathcal{D}_M^{\varphi}$ and $\mathbf{q}_{C_{\gamma_M}}(t_0)$ are their ICs, whereas $\boldsymbol{\varphi}_{M_{\gamma_M}}(t_0)$ are the ICs of memristors in $\mathcal{D}_M^{\varphi}$. The memristor fluxes in $\mathcal{D}_M^{\varphi}$ are $\boldsymbol{\varphi}_{M_{\gamma_M}}(t) = \boldsymbol{\varphi}_{\gamma_M}(t; t_0) + \boldsymbol{\varphi}_{M_{\gamma_M}}(t_0)$ and the capacitor voltages are $\mathbf{v}_{C_{\gamma_M}}(t) = \dot{\boldsymbol{\varphi}}_{\gamma_M}(t; t_0)$.
- $\mathcal{D}_G^{\varphi} \in \gamma_G$. Fluxes and charges in such $\mathcal{D}_G^{\varphi}$ define the vectors $\boldsymbol{\varphi}_{\gamma_G}(t; t_0)$ and $\mathbf{q}_{\gamma_G}(t; t_0)$, respectively. The CRs of the elements $\mathcal{D}_G^{\varphi}$ are

$$
-\mathbf{q}_{\gamma_G}(t; t_0) = \mathbf{C}_{\gamma_G}\dot{\boldsymbol{\varphi}}_{\gamma_G}(t; t_0) - \mathbf{q}_{C_{\gamma_G}}(t_0) + \mathbf{G}_{\gamma_G}\boldsymbol{\varphi}_{\gamma_G}(t; t_0) \tag{7.3}
$$

where $\mathbf{C}_{\gamma_G}$ is a diagonal matrix having the capacitance of capacitors in $\mathcal{D}_G^{\varphi}$ and $\mathbf{q}_{C_{\gamma_M}}(t_0)$ are their ICs.
- $\mathcal{D}_C^{\varphi} \in N_D$. Fluxes and charges in such capacitors define the vectors $\boldsymbol{\varphi}_{\gamma_C}(t; t_0)$ and $\mathbf{q}_{\gamma_C}(t; t_0)$, respectively. The CRs of the elements $\mathcal{D}_C^{\varphi}$ are

$$
-\mathbf{q}_{\gamma_C}(t; t_0) = \mathbf{C}_{\gamma_C}\dot{\boldsymbol{\varphi}}_{\gamma_C}(t; t_0) - \mathbf{q}_{C_{\gamma_C}}(t_0) \tag{7.4}
$$

where $\mathbf{C}_{\gamma_C}$ is a diagonal matrix having the capacitance of capacitors in $\mathcal{D}_C^{\varphi}$ and $\mathbf{q}_{C_{\gamma_C}}(t_0)$ are their ICs.

We can use dual notations for $\mathcal{D}_M^{q} \in N_D$ as follows.

- $\mathcal{D}_M^{q} \in N_D$. The CRs are

$$
\begin{aligned}
-\boldsymbol{\varphi}_{\lambda_M}(t; t_0) = \mathbf{L}_{\lambda_M}\dot{\mathbf{q}}_{\lambda_M}(t; t_0) - \boldsymbol{\varphi}_{L_{\lambda_M}}(t_0) \\
+\mathbf{h}(\mathbf{q}_{\lambda_M}(t; t_0) + \mathbf{q}_{M_{\lambda_M}}(t_0)) \\
-\mathbf{h}(\mathbf{q}_{M_{\lambda_M}}(t_0))
\end{aligned}
\tag{7.5}
$$

where $\boldsymbol{\varphi}_{L_{\lambda_M}}(t_0)$ (resp., $\mathbf{q}_{M_{\lambda_M}}(t_0)$) are the ICs for inductors (resp., memristors) in $\mathcal{D}_M^{q}$. The memristor charges in $\mathcal{D}_M^{q}$ are $\mathbf{q}_{M_{\lambda_M}}(t) = \mathbf{q}_{\lambda_M}(t; t_0) + \mathbf{q}_{M_{\lambda_M}}(t_0)$ and the inductor currents are $\mathbf{i}_{L_{\lambda_M}}(t) = \dot{\mathbf{q}}_{\lambda_M}(t; t_0)$.

[3] The active sign convention is used for any element $\mathcal{D}^{\varphi}$ and $\mathcal{D}^{q}$ in N_D.

- $\mathcal{D}_R^q \in N_D$. The CRs are

$$-\boldsymbol{\varphi}_{\lambda_R}(t;t_0) = \mathbf{L}_{\lambda_R}\dot{\mathbf{q}}_{\lambda_R}(t;t_0) - \boldsymbol{\varphi}_{L_{\lambda_R}}(t_0) + \mathbf{R}_{\lambda_R}\mathbf{q}_{\lambda_R}(t;t_0) \tag{7.6}$$

  where $\boldsymbol{\varphi}_{L_{\lambda_R}}(t_0)$ are the ICs of inductors in $\mathcal{D}_R^q$.
- $\mathcal{D}_L^q \in N_D$. The CRs are

$$-\boldsymbol{\varphi}_{\lambda_L}(t;t_0) = \mathbf{L}_{\lambda_L}\dot{\mathbf{q}}_{\lambda_L}(t;t_0) - \boldsymbol{\varphi}_{L_{\lambda_L}}(t_0) \tag{7.7}$$

  where $\boldsymbol{\varphi}_{L_{\lambda_L}}(t_0)$ are the ICs of inductors in $\mathcal{D}_L^q$.

  Concerning the CRs of elements in N_A we have the following.

- $\mathcal{A}_M^{\varphi} \in N_A$ have CRs

$$-\mathbf{q}_{\mu_F}(t;t_0) = \mathbf{f}(\boldsymbol{\varphi}_{\mu_F}(t;t_0) + \boldsymbol{\varphi}_{M_{\mu_F}}(t_0)) - \mathbf{f}(\boldsymbol{\varphi}_{M_{\mu_F}}(t_0)) \tag{7.8}$$

  where $\boldsymbol{\varphi}_{M_{\mu_F}}(t_0)$ are the ICs of memristors in $\mathcal{A}_M^{\varphi}$.
- $\mathcal{A}_M^q \in N_A$ have CRs

$$-\boldsymbol{\varphi}_{\mu_Q}(t;t_0) = \mathbf{h}(\mathbf{q}_{\mu_Q}(t;t_0) + \mathbf{q}_{M_{\mu_Q}}(t_0)) - \mathbf{h}(\mathbf{q}_{M_{\mu_Q}}(t_0)) \tag{7.9}$$

  where $\mathbf{q}_{M_{\mu_Q}}(t_0)$ are the ICs of memristors in $\mathcal{A}_M^q$.
- $\mathcal{A}_G^{\varphi} \in N_A$ have CRs

$$-\mathbf{q}_{\rho_G}(t;t_0) = \mathbf{G}_{\rho_G}\boldsymbol{\varphi}_{\rho_G}(t;t_0) \tag{7.10}$$

- $\mathcal{A}_R^q \in N_A$ have CRs

$$-\boldsymbol{\varphi}_{\rho_R}(t;t_0) = \mathbf{R}_{\rho_R}\mathbf{q}_{\rho_R}(t;t_0). \tag{7.11}$$

7.3.2 Hybrid Representation of N_R

Among the different representations of the linear (n_C+n_L)-port N_R, it is convenient to exploit its hybrid description (cf. Remark 3.8 in Chap. 4)

$$\begin{pmatrix} \mathbf{q}_{\gamma_M}(t; t_0) \\ \boldsymbol{\varphi}_{\lambda_M}(t; t_0) \\ \mathbf{q}_{\gamma_G}(t; t_0) \\ \boldsymbol{\varphi}_{\lambda_R}(t; t_0) \\ \mathbf{q}_{\gamma_C}(t; t_0) \\ \boldsymbol{\varphi}_{\lambda_L}(t; t_0) \\ \boldsymbol{\varphi}_{\mu_Q}(t; t_0) \\ \mathbf{q}_{\mu_F}(t; t_0) \\ \boldsymbol{\varphi}_{\rho_R}(t; t_0) \\ \mathbf{q}_{\rho_G}(t; t_0) \end{pmatrix} = \mathbf{H}_R \begin{pmatrix} \boldsymbol{\varphi}_{\gamma_M}(t; t_0) \\ \mathbf{q}_{\lambda_M}(t; t_0) \\ \boldsymbol{\varphi}_{\gamma_G}(t; t_0) \\ \mathbf{q}_{\lambda_R}(t; t_0) \\ \boldsymbol{\varphi}_{\gamma_C}(t; t_0) \\ \mathbf{q}_{\lambda_L}(t; t_0) \\ \mathbf{q}_{\mu_Q}(t; t_0) \\ \boldsymbol{\varphi}_{\mu_F}(t; t_0) \\ \mathbf{q}_{\rho_R}(t; t_0) \\ \boldsymbol{\varphi}_{\rho_G}(t; t_0) \end{pmatrix} + \mathbf{u}(t; t_0) \tag{7.12}$$

where

- $\mathbf{u}(t; t_0)$ takes into account the effects due to the sources $\boldsymbol{\varphi}_e(t; t_0)$ and $\mathbf{q}_a(t; t_0)$ within N_R
- the independent port variables in N_R are the fluxes in the two-terminal flux-controlled elements $\mathcal{D}^{\varphi}$, $\mathcal{A}_M^{\varphi}$, and $\mathcal{A}_G^{\varphi}$ and the charges in the two-terminal charge-controlled elements $\mathcal{D}^q$, $\mathcal{A}_M^q$, and $\mathcal{A}_R^q$
- the dependent port variables in N_R are the charges in the two-terminal flux-controlled elements and fluxes in the two-terminal charge-controlled elements.

Since N_R contains only positive resistors and independent sources, according to Remark 3.8 in Chap. 4, the hybrid description (7.12) of N_R exists if and only if the following fundamental topological assumption (hereinafter named *Assumption 1*, or (A1) for short) is satisfied

(A1) *The memristor circuit N is such that*

- *there exist no loops made of flux-controlled two-terminal elements* $\mathcal{D}^{\varphi}$, $\mathcal{A}_M^{\varphi}$, *and* $\mathcal{A}_G^{\varphi}$ *and/or flux sources* $\boldsymbol{\varphi}_e(t; t_0)$
- *there exist no cut-sets made of charge-controlled elements* $\mathcal{D}^q$, $\mathcal{A}_M^q$, *and* $\mathcal{A}_R^q$ *and/or charge sources* $\mathbf{q}_a(t; t_0)$.

Remark 7.1 As discussed in [13], the hybrid description of N_R is as general as that obtained via the approach based on the incidence matrix or the tableau method (Chap. 3).

7.3.3 Differential Algebraic Equations in the Flux-Charge Domain

The set of equations from (7.2) to (7.12) describe any memristor circuit $N \in \mathcal{LM}$ that satisfies (A1) in terms of a system of DAEs in the (φ, q)-domain.

The next example shows that the derivation of the SEs from the DAEs requires further assumptions on N. In particular, it shows that two-terminal elements in

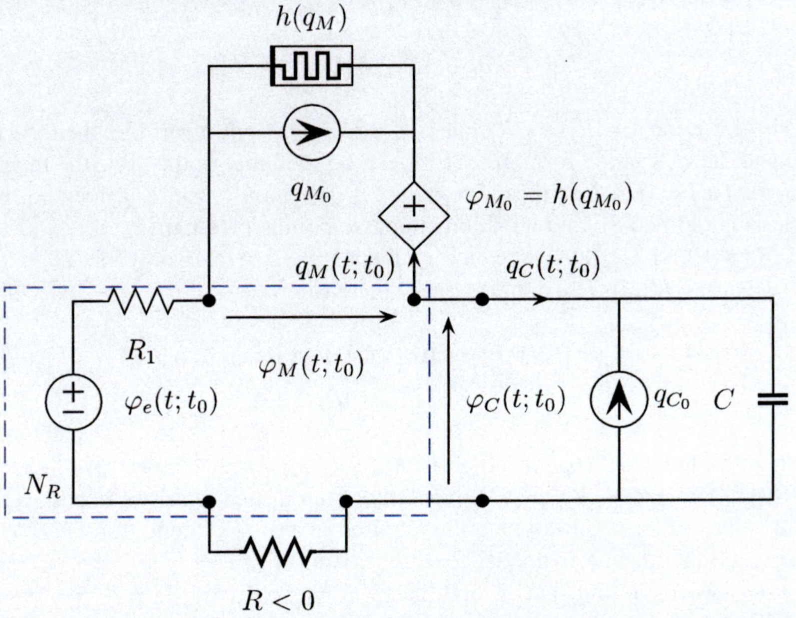

Fig. 7.5 Example of decomposition of a network $N \in \mathcal{LM}$ into N_A and N_D. The network N_D contains only a capacitor. The network N_A has a linear network N_R, made up of a resistor $R_1 \geq 0$ and a flux source $\varphi_e(t; t_0)$, connected to a charge-controlled memristor $\varphi_M = h(q_M)$ and a negative resistor $R < 0$

$N_A \backslash N_R$, such as negative resistors and/or charge-controlled memristors not in series with an inductor and, by duality, negative conductances and/or flux-controlled memristors not in parallel with a capacitor, might cause problems for the existence of the SE description.

Example 7.4 Consider the circuit with a charge-controlled memristor and a negative resistor $R < 0$ in Fig. 7.5. The DAE description in the (φ, q)-domain is obtained as

$$C\dot{\varphi}_C(t; t_0) = -q_M(t; t_0) + q_C(t_0) \tag{7.13a}$$

$$h(q_M(t)) + (R_1 + R)q_M(t) = \varphi_C(t; t_0) - \varphi_e(t; t_0) + h(q_M(t_0)) + (R_1 + R)q_M(t_0) \tag{7.13b}$$

for $t \geq t_0$, where $q_C(t_0) = Cv_C(t_0)$ and $q_M(t_0)$ are the ICs for the state variables in the (v, i)-domain. The SE for the memristor circuit in Fig. 7.5 can be derived from the DAEs (7.13) only under additional assumptions on the negative resistor R and the nonlinearity of the charge-controlled memristor.

In particular, the following cases can take place:

1. if $R = -R_1$ and $h(\cdot)$ is strictly increasing, hence we can write $q_M(t) = h^{-1}(\varphi_C(t; t_0) - \varphi_e(t; t_0) + h(q_M(t_0)))$, then the SE is

$$C\dot{\varphi}_C(t;t_0) = -h^{-1}(\varphi_C(t;t_0) - \varphi_e(t;t_0) + h(q_M(t_0))) \\ + q_C(t_0) + q_M(t_0). \tag{7.14}$$

On the contrary, if $R = -R_1$ but $\varphi_M = h(q_M)$ is not invertible, then the DAEs cannot be cast in the form of a SE. It can be seen that in this case the memristor circuit in Fig. 7.5 describes a *nonphysical situation* because it does not have a globally defined SE and it exhibits impasse points (cf. Chap. 4).

2. If $R \neq -R_1$ the DAEs reduce to the next SE if and only if $H(q_M(t)) = h(q_M(t)) + (R_1 + R)q_M(t)$ is strictly increasing

$$C\dot{\varphi}_C(t;t_0) = -\mathcal{H}(\varphi_C(t;t_0), \varphi_e(t;t_0), q_M(t_0)) \\ + q_C(t_0) + q_M(t_0) \tag{7.15}$$

where we have let $\mathcal{H}(\varphi_C(t;t_0), \varphi_e(t;t_0), q_M(t_0)) = H^{-1}(\varphi_C(t;t_0) - \varphi_e(t;t_0) + h(q_M(t_0)) + (R_1 + R)q_M(t_0))$. Note that, even if the memristor is passive, there are values of $R < 0$ for which the required invertibility condition on $H(q_M(t))$ fails, so that once more the SE does not exist.

It is worth noting that in both cases $R = -R_1$, or $R \neq -R_1$, as shown in Chap. 5, a technique that guarantees the explicit derivation of the SE for the memristor circuit in Fig. 7.5 is the insertion of an inductor in series with the charge-controlled memristor and the negative resistor.

7.4 State Equations in the Flux-Charge Domain

We consider henceforth memristor circuits $N \in \mathcal{LM}$ satisfying (A1). Consider also the following assumptions—denoted as *Assumption 2*, i.e., (A2), and *Assumption 3*, i.e., (A3)—enabling to specify all possible configurations of $N = N_D \bigcup N_A$ according to the structure of N_A:

(A2) *The subnetwork N_A of N has no memristors.*
(A3) *The subnetwork N_A of N has no negative resistors.*

Clearly, N admits of only these four configurations:

(i) if both (A2) and (A3) are fulfilled, then $\mu_F = \mu_Q = 0$ and $\rho_G = \rho_R = 0$, that is $N_A = N_R$ and $N = N_D \bigcup N_R$
(ii) if only (A2) is fulfilled, then $\mu_F = \mu_Q = 0$, $\rho_G \neq 0$ and $\rho_R \neq 0$, that is $N = N_D \bigcup N_R \bigcup \mathcal{A}_G^{\varphi} \bigcup \mathcal{A}_R^{q}$
(iii) if only (A3) is fulfilled, then $\mu_F \neq 0$, $\mu_Q \neq 0$ and $\rho_G = \rho_R = 0$, that is $N = N_D \bigcup N_R \bigcup \mathcal{A}_M^{\varphi} \bigcup \mathcal{A}_M^{q}$

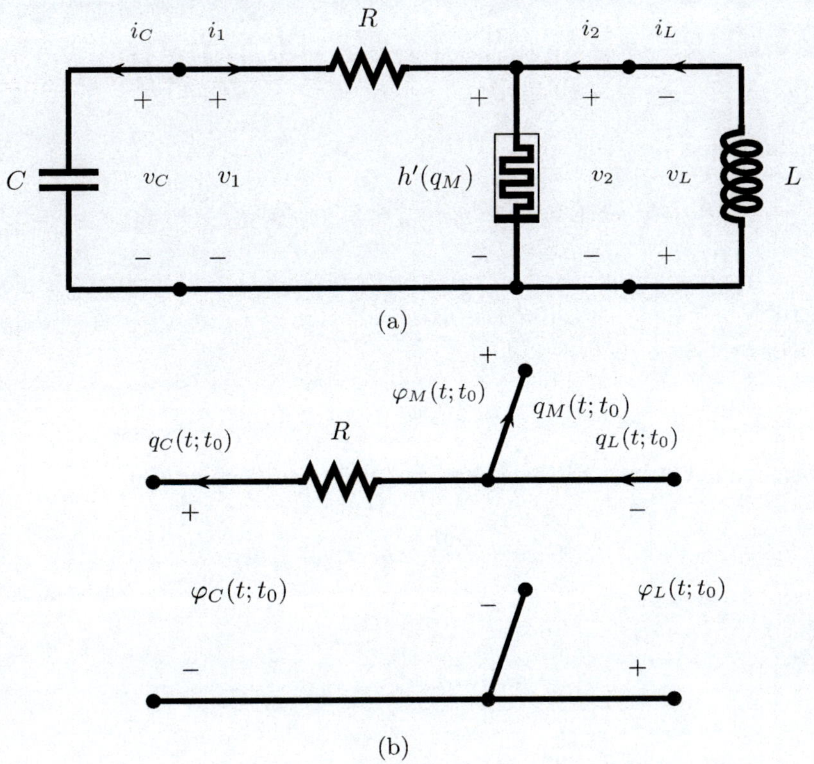

Fig. 7.6 (**a**) Circuit with a memristor which is both flux- and charge-controlled and (**b**) corresponding three-port network for finding the SEs

(iv) if both (A2) and (A3) are not satisfied, then $\mu_F \neq 0$, $\mu_Q \neq 0$, $\rho_G \neq 0$ and/or $\rho_R \neq 0$, that is $N = N_D \bigcup N_R \bigcup \mathcal{A}_M^{\varphi} \bigcup \mathcal{A}_M^{q} \bigcup \mathcal{A}_G^{\varphi} \bigcup \mathcal{A}_R^{q}$.

The first case (i) is thoroughly investigated in this chapter. The other cases can be discussed, *mutatis mutandis*, in a similar way and their treatment is left to the reader.

Example 7.5 Consider again the circuit studied in Example 5.18 of Chap. 5, where the memristor is both flux- and charge-controlled. Let us use the charge-controlled representation $\varphi_M(t; t_0) = h(q_M(t; t_0) + q_{M_0}) - h(q_{M_0}) = \tilde{h}(q_M(t; t_0); q_{M_0})$.

We have seen that it is not an easy matter to write the SEs in the (φ, q)-domain using the procedure in Sect. 5.8.2 of Chap. 5. We wish now to address again this problem using the hybrid representation and technique described in this chapter.

By extracting L, C and the memristor, we obtain a three-port network as in Fig. 7.6 for which we can write

$$-\begin{pmatrix} q_C(t;t_0) \\ \varphi_L(t;t_0) \\ \varphi_M(t;t_0) \end{pmatrix} = \begin{pmatrix} 0 & -1 & -1 \\ 1 & R & R \\ 1 & R & R \end{pmatrix} \begin{pmatrix} \varphi_C(t;t_0) \\ q_L(t;t_0) \\ q_M(t;t_0) \end{pmatrix}.$$

The third equation yields

$$Rq_M(t;t_0) + \tilde{h}(q_M(t;t_0); q_{M_0}) = -\varphi_C(t;t_0) - Rq_L(t;t_0).$$

Define function $s(q_M(t;t_0)) = Rq_M(t;t_0) + \tilde{h}(q_M(t;t_0); q_{M_0})$. If $s(\cdot)$ is globally invertible in $\mathbb{R}$ (this is true if $R+\tilde{h}'(q_M(t;t_0); q_{M_0}) = R+h'(q_M(t;t_0)+q_{M_0}) > 0$ for any $q_M(t;t_0)$), we have

$$q_M(t;t_0) = s^{-1}(-\varphi_C(t;t_0) - Rq_L(t;t_0)).$$

Substituting in the first two equations, and taking into account that

$$C\frac{d\varphi_C(t;t_0)}{dt} = q_C(t;t_0) + q_{C_0}$$

and

$$L\frac{dq_L(t;t_0)}{dt} = q_L(t;t_0) + \varphi_{L_0}$$

we are able to derive the SE representation

$$\begin{aligned} C\frac{d\varphi_C(t;t_0)}{dt} &= q_L(t;t_0) + s^{-1}(-\varphi_C(t;t_0) - Rq_L(t;t_0)) + q_{C_0} \\ L\frac{dq_L(t;t_0)}{dt} &= -\varphi_C(t;t_0) - Rq_L(t;t_0) - Rs^{-1}(-\varphi_C(t;t_0) \\ &\quad -Rq_L(t;t_0)) + \varphi_{L_0}. \end{aligned}$$

Note that in this representation $s(\cdot)$ depends also on q_{M_0}. It is also noted that although the SE representation exists, it seems difficult to find the invariant manifolds through the SE. Anyway, by inspection it can be easily checked that the considered circuit has an invariant of motion $\varphi_M(t) - Li_L(t) = h(q_M(t)) - Li_L(t)$, from which ∞^1 invariant manifolds for the dynamics in the (v,i)-domain can be found.

7.4.1 Class $N_D \bigcup N_R$ of Memristor Circuits

Suppose that (A1)–(A3) are satisfied, hence the memristor circuit N is made of the $(n_C + n_L)$-port N_R connected to the n_C elements $\mathcal{D}^{\varphi}$ and to the n_L elements $\mathcal{D}^{q}$. In this case, (A1) is equivalent to the following topological assumption that can be easily checked by inspection: *the memristor network N has no loops made by capacitors and/or flux sources and no cut-sets made by inductors and/or charge sources.*

Summing up, the case we are going to analyze in detail is that where any flux-controlled (resp., charge-controlled) memristor of N is in parallel to a capacitor (resp., in series with an inductor); moreover, any negative resistor is either in parallel to a capacitor or in series with an inductor. Recall that, by construction, the adynamic network N_R contains only positive resistors and/or independent flux or charge sources.

The use of (7.2)–(7.12) directly yields the SE representation[4] of N in the (φ, q)-domain

$$\mathbf{M}\begin{pmatrix}\dot{\boldsymbol{\varphi}}_{\gamma_M}(t;t_0)\\ \dot{\mathbf{q}}_{\lambda_M}(t;t_0)\\ \dot{\boldsymbol{\varphi}}_{\gamma_G}(t;t_0)\\ \dot{\mathbf{q}}_{\lambda_R}(t;t_0)\\ \dot{\boldsymbol{\varphi}}_{\gamma_C}(t;t_0)\\ \dot{\mathbf{q}}_{\lambda_L}(t;t_0)\end{pmatrix} = -(\mathbf{H}_R+\mathbf{G})\begin{pmatrix}\boldsymbol{\varphi}_{\gamma_M}(t;t_0)\\ \mathbf{q}_{\lambda_M}(t;t_0)\\ \boldsymbol{\varphi}_{\gamma_G}(t;t_0)\\ \mathbf{q}_{\lambda_R}(t;t_0)\\ \boldsymbol{\varphi}_{\gamma_C}(t;t_0)\\ \mathbf{q}_{\lambda_L}(t;t_0)\end{pmatrix}$$
$$-\begin{pmatrix}\mathbf{f}(\boldsymbol{\varphi}_{\gamma_M}(t;t_0)+\boldsymbol{\varphi}_{M_{\gamma_M}}(t_0))\\ \mathbf{h}(\mathbf{q}_{\lambda_M}(t;t_0)+\mathbf{q}_{M_{\lambda_M}}(t_0))\\ \mathbf{0}\\ \mathbf{0}\\ \mathbf{0}\\ \mathbf{0}\end{pmatrix}$$
$$+\begin{pmatrix}\mathbf{f}(\boldsymbol{\varphi}_{M_{\gamma_M}}(t_0))\\ \mathbf{h}(\mathbf{q}_{M_{\gamma_M}}(t_0))\\ \mathbf{0}\\ \mathbf{0}\\ \mathbf{0}\\ \mathbf{0}\end{pmatrix}+\begin{pmatrix}\mathbf{q}_{C_{\gamma_M}}(t_0)\\ \boldsymbol{\varphi}_{L_{\lambda_M}}(t_0)\\ \mathbf{q}_{C_{\gamma_G}}(t_0)\\ \boldsymbol{\varphi}_{L_{\lambda_R}}(t_0)\\ \mathbf{q}_{C_{\gamma_C}}(t_0)\\ \boldsymbol{\varphi}_{L_{\lambda_L}}(t_0)\end{pmatrix}+\mathbf{u}(t;t_0) \quad (7.16)$$

for $t \geq t_0$ where

[4] Since $\mathbf{M}$ is nonsingular, the SEs in normal form may be obtained by multiplying both sides of (7.16) by $\mathbf{M}^{-1}$.

- $\mathbf{M}$ and $\mathbf{G}$ are diagonal matrices defined as

$$\mathbf{M} = \text{diag}(\mathbf{C}_{\gamma_M}, \mathbf{L}_{\lambda_M}, \mathbf{C}_{\gamma_G}, \mathbf{L}_{\lambda_R}, \mathbf{C}_{\gamma_C}, \mathbf{L}_{\lambda_L})$$

and

$$\mathbf{G} = \text{diag}(\mathbf{0}, \mathbf{0}, \mathbf{G}_{\gamma_G}, \mathbf{R}_{\lambda_R}, \mathbf{0}, \mathbf{0}).$$

In particular, we have the following:

- $\mathbf{M}$ is nonsingular;
- the state variables in the (φ, q)-domain are the n_C fluxes across the capacitors (i.e., $\boldsymbol{\varphi}_{\gamma_M}(t; t_0)$, $\boldsymbol{\varphi}_{\gamma_G}(t; t_0)$ and $\boldsymbol{\varphi}_{\gamma_C}(t; t_0)$) and n_L charges through the inductors (i.e., $\mathbf{q}_{\lambda_M}(t; t_0)$, $\mathbf{q}_{\lambda_R}(t; t_0)$ and $\mathbf{q}_{\lambda_L}(t; t_0)$); the ICs at t_0 are by construction null;
- (7.16) is a system of $n = n_C + n_L$ SEs where the first $n_M = \gamma_M + \lambda_M$ equations are nonlinear, whereas the other $(n_C + n_L - n_M)$ equations are linear;
- since N_R is linear, vector $\mathbf{u}(t, t_0)$ in (7.12) can be explicitly written in terms of the internal sources as

$$\mathbf{u}(t; t_0) = \mathbf{B} \begin{pmatrix} \boldsymbol{\varphi}_e(t; t_0) \\ \mathbf{q}_a(t; t_0) \end{pmatrix}$$

with $\mathbf{B} \in \mathbb{R}^{(n_C+n_L)\times(n_E+n_A)}$.[5]

It is useful to write the SEs in a more compact form by introducing the following notations, vectors, and matrices:

$$n_x = \gamma_M + \lambda_M = n_M \tag{7.17a}$$

$$n_y = (n_C + n_L) - n_x = n - n_x \tag{7.17b}$$

$$\mathbf{x}(t) = (\boldsymbol{\varphi}_{\gamma_M}(t; t_0), \mathbf{q}_{\lambda_M}(t; t_0)) \in \mathbb{R}^{n_x} \tag{7.17c}$$

$$\mathbf{y}(t) = (\boldsymbol{\varphi}_{\gamma_G}(t; t_0), \mathbf{q}_{\lambda_R}(t; t_0), \boldsymbol{\varphi}_{\gamma_C}(t; t_0), \mathbf{q}_{\lambda_L}(t; t_0)) \in \mathbb{R}^{n_y} \tag{7.17d}$$

$$\mathbf{x}_M(t) = (\boldsymbol{\varphi}_{M_{\gamma_M}}(t), \mathbf{q}_{M_{\lambda_M}}(t)) \in \mathbb{R}^{n_x} \tag{7.17e}$$

$$\mathbf{M}_x = \text{diag}(\mathbf{C}_{\gamma_M}, \mathbf{L}_{\lambda_M}) \in \mathbb{R}^{n_x \times n_x} \tag{7.17f}$$

$$\mathbf{M}_y = \text{diag}(\mathbf{C}_{\gamma_G}, \mathbf{L}_{\lambda_R}, \mathbf{C}_{\gamma_C}, \mathbf{L}_{\lambda_L}) \in \mathbb{R}^{n_y \times n_y} \tag{7.17g}$$

$$\mathbf{M} = \text{diag}(\mathbf{M}_x, \mathbf{M}_y) \in \mathbb{R}^{n \times n} \tag{7.17h}$$

$$\mathbf{F}(\cdot) = (\mathbf{f}(\cdot), \mathbf{h}(\cdot)) : \mathbb{R}^{n_x} \to \mathbb{R}^{n_x} \tag{7.17i}$$

$$\mathbf{u}(t; t_0) = \begin{pmatrix} \mathbf{u}_x(t) \\ \mathbf{u}_y(t) \end{pmatrix} = \begin{pmatrix} \mathbf{B}_{11} & \mathbf{B}_{12} \\ \mathbf{B}_{21} & \mathbf{B}_{22} \end{pmatrix} \begin{pmatrix} \boldsymbol{\varphi}_e(t; t_0) \\ \mathbf{q}_a(t; t_0) \end{pmatrix} \tag{7.17j}$$

[5] Matrix $\mathbf{B}$ should not be confused with the fundamental loop matrix introduced in Chap. 3.

$$\mathbf{H} = \mathbf{H}_R + \mathbf{G} = \begin{pmatrix} \mathbf{H}_{11} & \mathbf{H}_{12} \\ \mathbf{H}_{21} & \mathbf{H}_{22} \end{pmatrix} \tag{7.17k}$$

where $\mathbf{H}_{11} \in \mathbb{R}^{n_x \times n_x}$, $\mathbf{H}_{12} \in \mathbb{R}^{n_x \times n_y}$, $\mathbf{H}_{21} \in \mathbb{R}^{n_y \times n_x}$, and $\mathbf{H}_{22} \in \mathbb{R}^{n_y \times n_y}$, whereas $\mathbf{B}_{11} \in \mathbb{R}^{n_x \times n_E}$, $\mathbf{B}_{12} \in \mathbb{R}^{n_x \times n_A}$, $\mathbf{B}_{21} \in \mathbb{R}^{n_y \times n_E}$, and $\mathbf{B}_{22} \in \mathbb{R}^{n_y \times n_A}$.

Note that $\mathbf{x}_M(t)$ is the vector of fluxes and charges in the memristors and we have

$$\mathbf{x}_M(t) = \mathbf{x}(t) + \mathbf{x}_M(t_0) \Rightarrow \dot{\mathbf{x}}_M(t) = \dot{\mathbf{x}}(t), \quad \forall t \geq t_0. \tag{7.18}$$

In addition, the following relationships hold for any $t \geq t_0$:

$$\mathbf{M}_x \dot{\mathbf{x}}(t) = \mathbf{M}_x \begin{pmatrix} \mathbf{v}_{C_{\gamma M}}(t) \\ \mathbf{i}_{L_{\lambda M}}(t) \end{pmatrix} = \begin{pmatrix} \mathbf{q}_{C_{\gamma M}}(t) \\ \boldsymbol{\varphi}_{L_{\lambda M}}(t) \end{pmatrix} \tag{7.19a}$$

$$\mathbf{M}_y \dot{\mathbf{y}}(t) = \mathbf{M}_y \begin{pmatrix} \mathbf{v}_{C_{\gamma G}}(t) \\ \mathbf{i}_{L_{\lambda R}}(t) \\ \mathbf{v}_{C_{\gamma C}}(t) \\ \mathbf{i}_{L_{\lambda L}}(t) \end{pmatrix} = \begin{pmatrix} \mathbf{q}_{C_{\gamma G}}(t) \\ \boldsymbol{\varphi}_{L_{\lambda R}}(t) \\ \mathbf{q}_{C_{\gamma C}}(t) \\ \boldsymbol{\varphi}_{L_{\lambda L}}(t) \end{pmatrix} \tag{7.19b}$$

$$\mathbf{M}_x \ddot{\mathbf{x}}(t) = \mathbf{M}_x \begin{pmatrix} \dot{\mathbf{v}}_{C_{\gamma M}}(t) \\ \frac{d}{dt}\mathbf{i}_{L_{\lambda M}}(t) \end{pmatrix} = \begin{pmatrix} \mathbf{i}_{C_{\gamma M}}(t) \\ \mathbf{v}_{L_{\lambda M}}(t) \end{pmatrix} \tag{7.19c}$$

$$\mathbf{M}_y \ddot{\mathbf{y}}(t) = \mathbf{M}_y \begin{pmatrix} \dot{\mathbf{v}}_{C_{\gamma G}}(t) \\ \frac{d}{dt}\mathbf{i}_{L_{\lambda R}}(t) \\ \dot{\mathbf{v}}_{C_{\gamma C}}(t) \\ \frac{d}{dt}\mathbf{i}_{L_{\lambda L}}(t) \end{pmatrix} = \begin{pmatrix} \mathbf{i}_{C_{\gamma G}}(t) \\ \mathbf{v}_{L_{\lambda R}}(t) \\ \mathbf{i}_{C_{\gamma C}}(t) \\ \mathbf{v}_{L_{\lambda L}}(t) \end{pmatrix}. \tag{7.19d}$$

We can summarize the results obtained so far as follows.

Theorem 7.1 *If the memristor circuit N satisfies (A1)–(A3), then it has the SE representation in the (φ, q)-domain for $t \geq t_0$*

$$\begin{aligned} \begin{pmatrix} \mathbf{M}_x \dot{\mathbf{x}}(t) \\ \mathbf{M}_y \dot{\mathbf{y}}(t) \end{pmatrix} &= -\begin{pmatrix} \mathbf{H}_{11} & \mathbf{H}_{12} \\ \mathbf{H}_{21} & \mathbf{H}_{22} \end{pmatrix} \begin{pmatrix} \mathbf{x}(t) \\ \mathbf{y}(t) \end{pmatrix} \\ &\quad - \begin{pmatrix} \mathbf{F}(\mathbf{x}(t) + \mathbf{x}_M(t_0)) \\ \mathbf{0} \end{pmatrix} - \begin{pmatrix} \mathbf{u}_x(t) \\ \mathbf{u}_y(t) \end{pmatrix} \\ &\quad + \begin{pmatrix} \mathbf{F}(\mathbf{x}_M(t_0)) \\ \mathbf{0} \end{pmatrix} + \begin{pmatrix} \mathbf{M}_x \dot{\mathbf{x}}(t_0) \\ \mathbf{M}_y \dot{\mathbf{y}}(t_0) \end{pmatrix}. \end{aligned} \tag{7.20}$$

This is a system of $n_C + n_L$ SEs in the state variables $(\mathbf{x}(t), \mathbf{y}(t))$ in the (φ, q)-domain. The initial conditions at t_0 are

$$(\mathbf{x}(t_0), \mathbf{y}(t_0)) = (\mathbf{0}, \mathbf{0}). \tag{7.21}$$

Proof It suffices to rewrite the SEs (7.16) with the introduced notations. ■

7.4.2 State Equations in the Voltage-Current Domain

By time-differentiation of (7.20), the following system of second-order ODEs is obtained (being $\dot{\mathbf{x}}(t) = \dot{\mathbf{x}}_M(t)$):

$$\begin{aligned}\begin{pmatrix}\mathbf{M}_x\ddot{\mathbf{x}}(t)\\ \mathbf{M}_y\ddot{\mathbf{y}}(t)\end{pmatrix} &= -\begin{pmatrix}\mathbf{H}_{11} & \mathbf{H}_{12}\\ \mathbf{H}_{21} & \mathbf{H}_{22}\end{pmatrix}\begin{pmatrix}\dot{\mathbf{x}}(t)\\ \dot{\mathbf{y}}(t)\end{pmatrix}\\ &\quad -\begin{pmatrix}\mathbf{J}_\mathbf{F}(\mathbf{x}_M(t))\dot{\mathbf{x}}(t)\\ \mathbf{0}\end{pmatrix} - \begin{pmatrix}\dot{\mathbf{u}}_x(t)\\ \dot{\mathbf{u}}_y(t)\end{pmatrix}\end{aligned} \tag{7.22}$$

where $\mathbf{J}_\mathbf{F}(\cdot) \in \mathbb{R}^{n_x \times n_x}$ is the Jacobian of $\mathbf{F}(\cdot)$ (i.e., $\mathbf{J}_\mathbf{F}(\cdot)$ is defined by memristances and memconductances of the $n_x = n_M$ memristors).

Theorem 7.2 *If the memristor circuit N satisfies (A1)–(A3), then it has the SE representation in the* (v, i)*-domain for* $t \geq t_0$

$$\dot{\mathbf{x}}_M(t) = \begin{pmatrix}\dot{\boldsymbol{\varphi}}_{M_{\gamma M}}(t)\\ \dot{\mathbf{q}}_{M_{\lambda M}}(t)\end{pmatrix} = \begin{pmatrix}\mathbf{v}_{C_{\gamma M}}(t)\\ \mathbf{i}_{L_{\lambda M}}(t)\end{pmatrix} \tag{7.23a}$$

$$\begin{aligned}\mathbf{M}_x\begin{pmatrix}\dot{\mathbf{v}}_{C_{\gamma M}}(t)\\ \frac{d}{dt}\mathbf{i}_{L_{\lambda M}}(t)\end{pmatrix} &= -(\mathbf{H}_{11} + \mathbf{J}_\mathbf{F}(\mathbf{x}_M(t)))\begin{pmatrix}\mathbf{v}_{C_{\gamma M}}(t)\\ \mathbf{i}_{L_{\lambda M}}(t)\end{pmatrix}\\ &\quad - \mathbf{H}_{12}\begin{pmatrix}\mathbf{v}_{C_{\gamma G}}(t)\\ \mathbf{i}_{L_{\lambda R}}(t)\\ \mathbf{v}_{C_{\gamma C}}(t)\\ \mathbf{i}_{L_{\lambda L}}(t)\end{pmatrix} - \dot{\mathbf{u}}_x(t)\end{aligned} \tag{7.23b}$$

$$\begin{aligned}\mathbf{M}_y\begin{pmatrix}\dot{\mathbf{v}}_{C_{\gamma G}}(t)\\ \frac{d}{dt}\mathbf{i}_{L_{\lambda R}}(t)\\ \dot{\mathbf{v}}_{C_{\gamma C}}(t)\\ \frac{d}{dt}\mathbf{i}_{L_{\lambda L}}(t)\end{pmatrix} &= -\mathbf{H}_{21}\begin{pmatrix}\mathbf{v}_{C_{\gamma M}}(t)\\ \mathbf{i}_{L_{\lambda M}}(t)\end{pmatrix}\\ &\quad - \mathbf{H}_{22}\begin{pmatrix}\mathbf{v}_{C_{\gamma G}}(t)\\ \mathbf{i}_{L_{\lambda R}}(t)\\ \mathbf{v}_{C_{\gamma C}}(t)\\ \mathbf{i}_{L_{\lambda L}}(t)\end{pmatrix} - \dot{\mathbf{u}}_y(t).\end{aligned} \tag{7.23c}$$

This is a system of $n_x + n = n_M + n_C + n_L$ SEs in the state variables $\mathbf{w}(t)$ in the (v, i)-domain given by

$$\begin{pmatrix} \boldsymbol{\varphi}_{M_{\gamma_M}}(t) \\ \mathbf{q}_{M_{\lambda_M}}(t) \\ \mathbf{v}_{C_{\gamma_M}}(t) \\ \mathbf{i}_{L_{\lambda_M}}(t) \\ \mathbf{v}_{C_{\gamma_G}}(t) \\ \mathbf{i}_{L_{\lambda_R}}(t) \\ \mathbf{v}_{C_{\gamma_C}}(t) \\ \mathbf{i}_{L_{\lambda_L}}(t) \end{pmatrix} = \begin{pmatrix} \mathbf{x}_M(t) \\ \dot{\mathbf{x}}(t) \\ \dot{\mathbf{y}}(t) \end{pmatrix} = \mathbf{w}(t). \tag{7.24}$$

The initial conditions at t_0 are

$$\mathbf{w}(t_0) = (\mathbf{x}_M(t_0), \dot{\mathbf{x}}(t_0), \dot{\mathbf{y}}(t_0)). \tag{7.25}$$

Proof It suffices to rewrite (7.22) by using (7.17), (7.18), and (7.19). ■

Finally, the following relationships hold between solutions of the SEs in the (φ, q)-domain and (v, i)-domain.

Property 7.1 If $(\mathbf{x}(t), \mathbf{y}(t))$ is the solution of the IVP (7.20) and (7.21) in the (φ, q)-domain, then $(\mathbf{x}(t) + \mathbf{x}_0, \dot{\mathbf{x}}(t), \dot{\mathbf{y}}(t))$ is the solution of the IVP (7.23)–(7.25) in the (v, i)-domain. Conversely, if $\mathbf{w}(t) = (\mathbf{x}_M(t), \dot{\mathbf{x}}(t), \dot{\mathbf{y}}(t))$ is the solution of the IVP (7.23)–(7.25) in the (v, i)-domain, then $(\int_{t_0}^{t} \dot{\mathbf{x}}(\tau)d\tau, \int_{t_0}^{t} \dot{\mathbf{y}}(\tau)d\tau)$ is the solution of the IVP (7.20) and (7.21) in the (φ, q)-domain.

Proof The verification is straightforward and is left to the reader. ■

7.5 Analysis of Manifolds

As shown in the previous section, under assumptions (A1)–(A3), the class of memristor circuits $N = N_D \bigcup N_R$ admits of the SE representation (7.20) in the (φ, q)-domain and the SE representation (7.22) in the (v, i)-domain. The aim of this section is to show that on the basis of these representations we are able to investigate the existence of invariant manifolds and the nonlinear dynamics on manifolds for such class of memristor circuits.

To this end, it is convenient to use the change of variables in (7.20)

$$\mathbf{X}(t) = \mathbf{x}(t) + \mathbf{x}_M(t_0) = \mathbf{x}_M(t) \tag{7.26a}$$

$$\mathbf{Y}(t) = \mathbf{y}(t) - \mathbf{H}_{22}^{-1}\left(\mathbf{H}_{21}\mathbf{x}_M(t_0) + \mathbf{M}_y\dot{\mathbf{y}}(t_0)\right) \tag{7.26b}$$

which is well defined under the next additional assumption (named *Assumption 4*, or (A4) for short; see also Appendix 1).

(A4) *Submatrix* $\mathbf{H}_{22}$ *is nonsingular, i.e.,* $\det \mathbf{H}_{22} \neq 0$.

It follows from (7.26b) that $\mathbf{Y}(t) - \mathbf{Y}(t_0) = \mathbf{y}(t) - \mathbf{y}(t_0)$ with $\mathbf{Y}(t_0) = -\mathbf{H}_{22}^{-1}\left(\mathbf{H}_{21}\mathbf{x}_M(t_0) + \mathbf{M}_y\dot{\mathbf{y}}(t_0)\right)$. Hence, the change of variables (7.26) is also equivalent to using the following compact expressions $\mathbf{X}(t) - \mathbf{x}_M(t_0) = \mathbf{x}(t)$ and $\mathbf{Y}(t) - \mathbf{Y}(t_0) = \mathbf{y}(t)$, that permits to identify the relationship with the circuit variables given in (7.17).

By (7.26) and (7.20) we can derive the following form of the SEs in the (φ, q)-domain such that manifolds and the dynamics of N with respect to manifolds can be grasped:

$$\begin{pmatrix} \mathbf{M}_x\dot{\mathbf{X}}(t) \\ \mathbf{M}_y\dot{\mathbf{Y}}(t) \end{pmatrix} = -\begin{pmatrix} \mathbf{H}_{11} & \mathbf{H}_{12} \\ \mathbf{H}_{21} & \mathbf{H}_{22} \end{pmatrix}\begin{pmatrix} \mathbf{X}(t) \\ \mathbf{Y}(t) \end{pmatrix} - \begin{pmatrix} \mathbf{F}(\mathbf{X}(t)) \\ \mathbf{0} \end{pmatrix} - \begin{pmatrix} \mathbf{u}_x(t) \\ \mathbf{u}_y(t) \end{pmatrix} + \begin{pmatrix} \mathbf{k}_0 \\ \mathbf{0} \end{pmatrix} \tag{7.27}$$

where we have let

$$\mathbf{k}_0 = \mathbf{S}_{22}\mathbf{x}_M(t_0) + \mathbf{F}(\mathbf{x}_M(t_0)) + \mathbf{M}_x\dot{\mathbf{x}}(t_0) - \mathbf{H}_{12}\mathbf{H}_{22}^{-1}\mathbf{M}_y\dot{\mathbf{y}}(t_0) \tag{7.28}$$

and

$$\mathbf{S}_{22} = \mathbf{H}/\mathbf{H}_{22} = \mathbf{H}_{11} - \mathbf{H}_{12}\mathbf{H}_{22}^{-1}\mathbf{H}_{21}$$

is the Schur complement of $\mathbf{H}_{22}$ in $\mathbf{H}$ [14].

Note that $\mathbf{k}_0 \in \mathbb{R}^{n_M}$ depends upon the ICs $(\mathbf{x}_M(t_0), \dot{\mathbf{x}}(t_0), \dot{\mathbf{y}}(t_0))$ for the state variables in the (v, i)-domain (see (7.24)). The SEs (7.27) make clear that $\mathbf{k}_0$ plays a crucial role in the nonlinear dynamics and bifurcation phenomena of N.[6]

In the following we investigate:

- *the geometric properties of manifolds*, i.e., how the state space in the (v, i)-domain can be foliated into manifolds with specific geometric properties;
- *the dynamic properties of manifolds*, i.e., conditions under which the manifolds are positively invariant for the dynamics of N in the (v, i)-domain described by the SEs (7.23). In the latter case, the goal is to study how solutions of (7.23) evolve in time through different manifolds.

[6]Note that $\mathbf{k}_0$ depends also on circuit parameters and memristor nonlinearities (through the vector $\mathbf{F}(\cdot)$ and the submatrices of $\mathbf{H}$ and $\mathbf{M}$). In this chapter, the chief interest is on the dependency on ICs in order to highlight the concept of bifurcation without parameters, whereas standard bifurcations due to the change of circuit parameters and memristor nonlinearities are not considered.

7.5.1 Geometric Properties

Using the expression of $\mathbf{k}_0$ in (7.28), let us introduce:

- the function $\mathbf{K}(\cdot) : \mathbb{R}^{(n_M+n_C+n_L)} \to \mathbb{R}^{n_M}$ of the state vector $\mathbf{w} = (\mathbf{x}_M, \dot{\mathbf{x}}, \dot{\mathbf{y}})$ in the (v, i)-domain (omitting dependence on t)

$$\mathbf{K}(\mathbf{x}_M, \dot{\mathbf{x}}, \dot{\mathbf{y}}) = \mathbf{S}_{22}\mathbf{x}_M + \mathbf{F}(\mathbf{x}_M) + \mathbf{M}_x\dot{\mathbf{x}} - \mathbf{H}_{12}\mathbf{H}_{22}^{-1}\mathbf{M}_y\dot{\mathbf{y}} \tag{7.29}$$

- for any vector $\mathbf{k} \in \mathbb{R}^{n_M}$, the level set $\mathcal{M}(\mathbf{k}) \subset \mathbb{R}^{(n_M+n_C+n_L)}$ of $\mathbf{K}(\mathbf{x}_M, \dot{\mathbf{x}}, \dot{\mathbf{y}})$, which is defined as

$$\mathcal{M}(\mathbf{k}) = \{\mathbf{w} = (\mathbf{x}_M, \dot{\mathbf{x}}, \dot{\mathbf{y}}) \in \mathbb{R}^{n_M+n_C+n_L} : \mathbf{K}(\mathbf{w}) = \mathbf{k}\}. \tag{7.30}$$

The next result illustrates some main geometric properties of sets $\mathcal{M}(\mathbf{k})$.

Theorem 7.3 *If (A1)–(A4) are satisfied by N, then the following geometric properties hold:*

1. *for any $\mathbf{k} \in \mathbb{R}^{n_M}$, $\mathcal{M}(\mathbf{k})$ defines a nonempty, nonplanar, $(n_C + n_L)$-dimensional manifold in the state space in the (v, i)-domain;*
2. *for any $\mathbf{k}_1 \neq \mathbf{k}_2 \in \mathbb{R}^{n_M}$, we have $\mathcal{M}(\mathbf{k}_2) = \mathcal{M}(\mathbf{k}_1) + \mathbf{M}_x^{-1}(\mathbf{k}_2 - \mathbf{k}_1)$, i.e., $\mathcal{M}(\mathbf{k}_2)$ is a rigid translation of $\mathcal{M}(\mathbf{k}_1)$, and conversely;*
3. *there are ∞^{n_M} nonintersecting manifolds, obtained by varying $\mathbf{k}$ in $\mathbb{R}^{n_M}$, which span the whole $(n_M + n_C + n_L)$-dimensional state space in the (v, i)-domain.*

Proof See Appendix 2. ■

The result confirms in particular the geometric properties of manifolds seen in specific low-order circuits with one memristor in Chap. 6.

7.5.2 Dynamic Properties

The dynamic properties of manifolds can be inferred from the time evolution of the solutions of the SEs (7.23) through the $(n_M + n_C + n_L)$-dimensional state space in the (v, i)-domain. Given $\mathbf{w}_0 \in \mathbb{R}^{(n_M+n_C+n_L)}$, let $\mathbf{w}(t; t_0, \mathbf{w}_0) = (\mathbf{x}_M(t), \dot{\mathbf{x}}(t), \dot{\mathbf{y}}(t))$ be the solution with ICs $\mathbf{w}_0$ at $t = t_0$ of the SEs (7.23) in the (v, i)-domain. Using (7.29) and (7.30), if we let

$$\mathbf{k}(t; t_0, \mathbf{w}_0) = \mathbf{K}(\mathbf{w}(t; t_0, \mathbf{w}_0)), \ \forall t \geq t_0 \tag{7.31}$$

then we have

$$\mathbf{w}(t; t_0, \mathbf{w}_0) \in \mathcal{M}(\mathbf{k}(t; t_0, \mathbf{w}_0)), \ \forall t \geq t_0. \tag{7.32}$$

In particular,

$$\mathbf{w}_0 = (\mathbf{x}_M(t_0), \dot{\mathbf{x}}(t_0), \dot{\mathbf{y}}(t_0)) \in \mathcal{M}(\mathbf{k}_0)$$

where $\mathbf{k}_0 = \mathbf{K}(\mathbf{x}_M(t_0), \dot{\mathbf{x}}(t_0), \dot{\mathbf{y}}(t_0))$ is given in (7.28).

The next property permits to study the link between the solution $\mathbf{w}(t; t_0, \mathbf{w}_0)$ and its associated manifold defined in (7.32) at any instant $t \geq t_0$.

Property 7.2 Suppose that (A1)–(A4) are satisfied by N. Then, the time derivative of $\mathbf{k}(t; t_0, \mathbf{w}_0)$ in (7.31) is given by

$$\dot{\mathbf{k}}(t; t_0, \mathbf{w}_0) = \mathbf{H}_{12}\mathbf{H}_{22}^{-1}\dot{\mathbf{u}}_y(t) - \dot{\mathbf{u}}_x(t) \tag{7.33}$$

for any $\mathbf{w}_0 \in \mathbb{R}^n$ and $t \geq t_0$.

Proof See Appendix 3. ■

The expression (7.33) can be rewritten as follows to highlight how we can *drive solutions through different manifolds by suitable independent voltage and/or current sources*

$$\dot{\mathbf{k}}(t; t_0, \mathbf{w}_0) = \mathbf{H}_{12}\mathbf{H}_{22}^{-1}(\mathbf{B}_{21}\mathbf{e}(t) + \mathbf{B}_{22}\mathbf{a}(t)) - (\mathbf{B}_{11}\mathbf{e}(t) + \mathbf{B}_{12}\mathbf{a}(t)) \tag{7.34}$$

from which, by integrating between t_0 and $t \geq t_0$, we have

$$\begin{aligned}\mathbf{k}(t; t_0, \mathbf{w}_0) = \mathbf{k}_0 + \mathbf{H}_{12}\mathbf{H}_{22}^{-1}&(\mathbf{B}_{21}\boldsymbol{\varphi}_e(t; t_0) + \mathbf{B}_{22}\mathbf{q}_a(t; t_0))\\ &-(\mathbf{B}_{11}\boldsymbol{\varphi}_e(t; t_0) + \mathbf{B}_{12}\mathbf{q}_a(t; t_0)).\end{aligned} \tag{7.35}$$

The next theorem summarizes the dynamic properties of manifolds proved so far.

Theorem 7.4 *Suppose that (A1)–(A4) are satisfied by N. Then, for any* $\mathbf{w}_0 \in \mathbb{R}^{(n_M+n_C+n_L)}$ *we have*

$$\mathbf{w}(t; t_0, \mathbf{w}_0) \in \mathcal{M}(\mathbf{k}(t; t_0, \mathbf{w}_0)), \ \forall t \geq t_0$$

where $\mathbf{k}(t; t_0, \mathbf{w}_0)$ *is a term depending on the independent sources in N given by*

$$\begin{aligned}\mathbf{k}(t; t_0, \mathbf{w}_0) = \mathbf{k}_0 + \mathbf{H}_{12}\mathbf{H}_{22}^{-1}&(\mathbf{B}_{21}\boldsymbol{\varphi}_e(t; t_0) + \mathbf{B}_{22}\mathbf{q}_a(t; t_0))\\ &-(\mathbf{B}_{11}\boldsymbol{\varphi}_e(t; t_0) + \mathbf{B}_{12}\mathbf{q}_a(t; t_0)))\end{aligned} \tag{7.36}$$

for any $t \geq t_0$*, and* $\mathbf{k}_0$ *is as in (7.28).*

Given any ICs $\mathbf{w}_0 \in \mathbb{R}^{(n_M+n_C+n_L)}$, Theorem 7.4 permits to find instant by instant the manifold $\mathcal{M}(\mathbf{k}(t; t_0, \mathbf{w}_0))$ containing the solution $\mathbf{w}(t; t_0, \mathbf{w}_0)$ as a *linear* function of the fluxes $\boldsymbol{\varphi}_e(t; t_0)$, charges $\mathbf{q}_a(t; t_0)$, and parameters of the hybrid representation of N_R.

In Sect. 7.6, the r.h.s. of (7.34) (or equivalently (7.35)) is specified by voltage and/or current sources exploited in practical applications of memristor circuits.

7.5.3 Manifolds in a Relevant Class of Memristor Circuits

The results derived so far can be rewritten in a simplified form for the relevant class of memristor circuits $N = N_D \bigcup N_R$ having a capacitor in parallel to any flux-controlled memristor and/or an inductor in series with any charge-controlled memristor. This means that N_D has only elements $\mathcal{D}_M^{\varphi}$ and $\mathcal{D}_M^{q}$, i.e., $\gamma_G = \gamma_C = \lambda_R = \lambda_L = \mu_F = \mu_Q = \rho_G = \rho_R = 0$. It follows that $n_y = 0$ and that (A2), (A3) are satisfied. Note that $\mathbf{H}$ coincides with $\mathbf{H}_{11}$, hence also (A4) is satisfied. In addition, $\mathbf{M}_x = \mathbf{M}$ and $\mathbf{u}_x(t) = \mathbf{u}(t) = \mathbf{B}_{11}\boldsymbol{\varphi}_e(t; t_0) + \mathbf{B}_{12}\mathbf{q}_a(t; t_0)$.

If (A1) is met, the SEs in the (φ, q)-domain (7.27) reduce to

$$\mathbf{M}\dot{\mathbf{X}}(t) = -\mathbf{H}\mathbf{X}(t) - \mathbf{F}(\mathbf{X}(t)) - \mathbf{u}(t) + \mathbf{k}_0 \tag{7.37a}$$

$$\mathbf{k}_0 = \mathbf{F}(\mathbf{x}_M(t_0)) + \mathbf{M}\dot{\mathbf{x}}(t_0) + \mathbf{H}\mathbf{x}_M(t_0) \tag{7.37b}$$

and those in the (v, i)-domain to

$$\dot{\mathbf{x}}_M(t) = \begin{pmatrix} \dot{\boldsymbol{\varphi}}_{\gamma_M}(t) \\ \dot{\mathbf{q}}_{\lambda_M}(t) \end{pmatrix} = \begin{pmatrix} \mathbf{v}_{C_{\gamma_M}}(t) \\ \mathbf{i}_{L_{\lambda_M}}(t) \end{pmatrix} \tag{7.38a}$$

$$\mathbf{M}\begin{pmatrix} \dot{\mathbf{v}}_{C_{\gamma_M}}(t) \\ \dot{\mathbf{i}}_{L_{\lambda_M}}(t) \end{pmatrix} = -(\mathbf{H} + \mathbf{J}_{\mathbf{F}}(\mathbf{x}_M(t)))\begin{pmatrix} \mathbf{v}_{C_{\gamma_M}}(t) \\ \mathbf{i}_{L_{\lambda_M}}(t) \end{pmatrix} - \dot{\mathbf{u}}(t) \tag{7.38b}$$

whereas (7.29) simplifies to

$$\mathbf{K}(\mathbf{x}_M, \dot{\mathbf{x}}) = \mathbf{H}\mathbf{x}_M + \mathbf{F}(\mathbf{x}_M) + \mathbf{M}\dot{\mathbf{x}} \tag{7.39}$$

and (7.33) in Property 7.2 becomes

$$\dot{\mathbf{k}}(t; t_0, \mathbf{w}_0) = -\dot{\mathbf{u}}(t) = -\left(\mathbf{B}_{11}\mathbf{e}(t) + \mathbf{B}_{12}\mathbf{a}(t)\right) \tag{7.40}$$

for any $t \geq t_0$. We have

$$\mathbf{k}(t; t_0, \mathbf{w}_0) = -\left(\mathbf{B}_{11}\boldsymbol{\varphi}_e(t; t_0) + \mathbf{B}_{12}\mathbf{q}_a(t; t_0)\right) + \mathbf{k}_0$$

for any $t \geq t_0$, yielding the manifolds $\mathcal{M}(\mathbf{k}(t; t_0, \mathbf{w}_0))$ as in (7.32). A result analogous to Theorem 7.4 can also be obtained.

7.6 Programming Memristor Circuits with Pulses and Time-Varying Inputs

Theorems 7.3 and 7.4 permit to investigate the dynamic properties of autonomous and non-autonomous memristor circuits $N = N_D \bigcup N_R$ by studying how the independent sources drive solutions of N through different manifolds in the $(n_M + n_C + n_L)$-dimensional state space in the (v, i)-domain. In practical applications, the following three main cases can be considered:

- autonomous memristor circuits N with *no external sources*, i.e., $\mathbf{e}(t) = \mathbf{a}(t) = 0$ for any $t \geq t_0$;
- non-autonomous memristor circuits N subject to external *pulses with finite time duration*, i.e., voltage $\mathbf{e}(t)$ and current $\mathbf{a}(t)$ sources vary over a finite time interval $[t_0, t_1]$ with $0 < t_1 - t_0 = \Delta < \infty$. The external sources are zero for any $t \geq t_1$. This class of pulses[7] defines $\boldsymbol{\varphi}_e(t; t_0)$ and $\mathbf{q}_a(t; t_0)$ as *sources with "constant momentum"* in the (φ, q)-domain for $t \geq t_1$ (following the nomenclature introduced in [15]) and it is widely used in the experimental characterization of memristor devices and the programming of memristor-based neuromorphic systems;
- non-autonomous memristor circuits N with voltage sources $\mathbf{e}(t)$ and/or current sources $\mathbf{a}(t)$ varying over the infinite time interval $[t_0, +\infty)$, i.e., N has *sources with "time-varying momentum"* in the (φ, q)-domain. Typical examples used in several applications are:
  - the class of sinusoidal voltage $\mathbf{e}(t)$ and/or current $\mathbf{a}(t)$ generators (usually considered in memristor-based filters), giving sources $\boldsymbol{\varphi}_e(t; t_0)$ and $\mathbf{q}_a(t; t_0)$ with sinusoidal momentum in the (φ, q)-domain;
  - the class of constant voltage $\mathbf{e}(t)$ and/or current $\mathbf{a}(t)$, providing sources $\boldsymbol{\varphi}_e(t; t_0)$ and $\mathbf{q}_a(t; t_0)$ with linear time-varying momentum in the (φ, q)-domain.

7.6.1 *Memristor Circuits with No External Sources*

Let us consider memristor circuits $N = N_D \bigcup N_R$ satisfying (A1)–(A4) with no voltage and/or current sources, that is N is an *autonomous memristor circuit* with $\mathbf{e}(t) = \mathbf{a}(t) = 0$ for any $t \geq t_0$. It turns out that $\boldsymbol{\varphi}_e(t; t_0) = 0$ and $\mathbf{q}_a(t; t_0) = 0$ in the (φ, q)-domain as well. Hence, $\dot{\mathbf{u}}_x(t; t_0) = 0$ and $\dot{\mathbf{u}}_y(t; t_0) = 0$ and (7.33) becomes

[7] For simplicity, impulsive voltage and current sources instantaneously applied at t_0 can be considered by including their effects in the ICs.

$$\dot{\mathbf{k}}(t; t_0, \mathbf{w}_0) = 0$$

for any $\mathbf{w}_0 \in \mathbb{R}^n$ and $t \geq t_0$, i.e., $\mathbf{K}(\mathbf{x}_M, \dot{\mathbf{x}}, \dot{\mathbf{y}})$ in (7.29) is an *invariant of motion* for N in the (v, i)-domain.

The following fundamental property summarizes the dynamics on manifolds of memristor circuits with no external sources.

Property 7.3 Let us consider an autonomous memristor circuit N with no external sources satisfying (A1)–(A4). Then, for any $\mathbf{w}_0 \in \mathbb{R}^{(n_M+n_C+n_L)}$ we have

$$\mathbf{w}(t; t_0, \mathbf{w}_0) \in \mathcal{M}(\mathbf{k}_0), \ \forall t \geq t_0$$

where $\mathbf{k}_0$ is given in (7.28).

Moreover, the reduced-order dynamics on $\mathcal{M}(\mathbf{k}_0)$ for any $t \geq t_0$ is described in the (φ, q)-domain by the SEs (7.27).

Finally, for any $\mathbf{k} \in \mathbb{R}^{n_M}$, manifold $\mathcal{M}(\mathbf{k})$ **defined in** (7.30) is positively invariant for the dynamics of N in the (v, i)-domain.

7.6.2 Memristor Circuits Subject to Pulses

Let us consider non-autonomous memristor circuits $N = N_D \bigcup N_R$ satisfying (A1)–(A4) and subject to sources with constant momentum in the (φ, q)-domain for $t \geq t_1$. We have

$$\boldsymbol{\varphi}_e(t; t_0) = \boldsymbol{\varphi}_e(t_1; t_0) = \int_{t_0}^{t_1} \mathbf{e}(\tau)d\tau = \bar{\mathbf{e}}\Delta, \quad \forall t \geq t_1 \tag{7.41}$$

$$\mathbf{q}_a(t; t_0) = \mathbf{q}_a(t_1; t_0) = \int_{t_0}^{t_1} \mathbf{a}(\tau)d\tau = \bar{\mathbf{a}}\Delta, \quad \forall t \geq t_1$$

where $\bar{\mathbf{e}}$ and $\bar{\mathbf{a}}$ are the mean values of $\mathbf{e}(t)$ and $\mathbf{a}(t)$ over $[t_0, t_1]$, respectively. Using these expressions, (7.35) becomes

$$\mathbf{k}(t_1; t_0, \mathbf{w}_0) = \mathbf{k}_1 = \mathbf{k}_0 + \mathbf{k}_u \tag{7.42}$$

where

$$\mathbf{k}_u = \mathbf{H}_{12}\mathbf{H}_{22}^{-1}(\mathbf{B}_{21}\bar{\mathbf{e}} + \mathbf{B}_{22}\bar{\mathbf{a}})\Delta - (\mathbf{B}_{11}\bar{\mathbf{e}} + \mathbf{B}_{12}\bar{\mathbf{a}})\Delta \tag{7.43}$$

takes into account the effect of the external pulses on the initial manifold $\mathcal{M}(\mathbf{k}_0)$.

Given the initial manifold $\mathcal{M}(\mathbf{k}_0)$, (7.42) and (7.43) permit to design the external pulses of $\mathbf{e}(t)$ and/or $\mathbf{a}(t)$ such that the dynamics of N evolves on an assigned manifold $\mathcal{M}(\mathbf{k}_1) = \mathcal{M}(\mathbf{k}_0 + \mathbf{k}_u)$ for any $t \geq t_1$. The dynamics of N is governed by the non-autonomous SEs (7.27) with

$$\mathbf{u}_x(t) = \mathbf{B}_{11} \int_{t_0}^{t} \mathbf{e}(\tau)d\tau + \mathbf{B}_{12} \int_{t_0}^{t} \mathbf{a}(\tau)d\tau \tag{7.44a}$$

$$\mathbf{u}_y(t) = \mathbf{B}_{21} \int_{t_0}^{t} \mathbf{e}(\tau)d\tau + \mathbf{B}_{22} \int_{t_0}^{t} \mathbf{a}(\tau)d\tau \tag{7.44b}$$

for any $t \in [t_0, t_1)$. Point (3) in Theorem 7.3 ensures that the effect of $\mathbf{u}_x(t)$ and $\mathbf{u}_y(t)$ is to drive the solution $\mathbf{w}(t; t_0, \mathbf{w}_0)$ from manifold $\mathcal{M}(\mathbf{k}_0)$ to $\mathcal{M}(\mathbf{k}_1)$, i.e., for any $t \in [t_0, t_1)$ the dynamics of N is continuously embedded in $\mathcal{M}(\mathbf{k}(t, t_0, \mathbf{w}_0))$—see (7.31)—where $\mathcal{M}(\mathbf{k}_0)$ and $\mathcal{M}(\mathbf{k}_1)$ represent the "initial" and "final" manifolds, respectively.

The following property summarizes the dynamics on manifolds for memristor circuits subject to pulses with constant momentum for $t \geq t_1$.

Property 7.4 Let us consider a non-autonomous memristor circuit N satisfying (A1)–(A4) and with constant momentum sources (7.6.2), in the (φ, q)-domain, for $t \geq t_1$. Then, for any $\mathbf{w}_0 \in \mathbb{R}^{(n_M+n_C+n_L)}$ we have

$$\mathbf{w}(t; t_0, \mathbf{w}_0) \in \mathcal{M}(\mathbf{k}_1), \ \forall t \geq t_1$$

where $\mathbf{k}_1 = \mathbf{k}_0 + \mathbf{k}_u$, $\mathbf{k}_0$ is given in (7.28) and $\mathbf{k}_u$ in (7.43) describes the effect of constant momentum sources.

The reduced-order dynamics on $\mathcal{M}(\mathbf{k}_1)$ are described in the (φ, q)-domain by the SEs (7.27) with $\mathbf{k}_0$ replaced by $\mathbf{k}_1$, $\mathbf{u}_x(t)$ by $(\mathbf{B}_{11}\bar{\mathbf{e}} + \mathbf{B}_{12}\bar{\mathbf{a}})\Delta$ and $\mathbf{u}_y(t)$ by $(\mathbf{B}_{21}\bar{\mathbf{e}} + \mathbf{B}_{22}\bar{\mathbf{a}})\Delta$, for any $t \geq t_1$.

Remark 7.2 We stress that voltage and current sources with constant momentum for $t \geq t_1$ as in (7.6.2) can be obtained by means of voltage and/or current sources with *different pulse duration and shape*. This means that *only the area (i.e., the momentum)* of the waveforms $\mathbf{e}(t)$ and/or $\mathbf{a}(t)$ over different finite time intervals is important to set the manifold $\mathcal{M}(\mathbf{k}_1)$ on which the dynamics of N takes place once all the pulses are over. This result agrees with the experimental results available in literature that show how memristors can be programmed (almost) in the same way by using for instance triangular or squared pulses or finite impulse trains.

7.6.3 *Memristor Circuits with Time-Varying Sources*

Let us consider memristor circuits N with voltage sources $\mathbf{e}(t)$ and/or current sources $\mathbf{a}(t)$ varying over the infinite time interval $[t_0, +\infty)$, i.e., N has *sources with "time-varying momentum"* in the (φ, q)-domain. In this case the dynamics of

N for any $t \geq t_0$ is governed by the non-autonomous SEs (7.27) with input $\mathbf{u}_x(t)$ and $\mathbf{u}_y(t)$ given by (7.44). Hence, any solution $\mathbf{w}(t; t_0, \mathbf{w}_0)$ of N is embedded into the (continuous) family of manifolds $\mathcal{M}(\mathbf{k}_0 + \mathbf{H}_{12}\mathbf{H}_{22}^{-1}\mathbf{u}_y(t) - \mathbf{u}_x(t))$ parametrized by time $t \in [t_0, +\infty)$ (see (7.35) and Theorem 7.4).

Although the solution of N is instant by instant on a known manifold, complex dynamic behavior and bifurcation phenomena can emerge due to the effect of the external sources with time-varying momentum in the (φ, q)-domain.

The next section presents selected examples for a thorough illustration of the explicit form of the SEs of memristor circuits $N = N_D \bigcup N_R$ and their programming by tuning invariant manifolds via suitable external sources.

7.7 Examples on Memristor Circuit Programming

In this section we discuss some fundamental examples of memristor circuits in $\mathcal{LM}$ where we apply the technique developed in the chapter for programming via pulses different manifolds and reduced-order dynamics. First, we reconsider the Memristor Chaotic Circuit (MCC) studied in Chap. 6, then we study a circuit with a flux- and a charge-controlled memristor, and, finally, a class of memristor neural networks with a star topology.

7.7.1 Memristor Chaotic Circuit

In Sect. 6.3 of Chap. 6 we have considered a chaotic circuit named MCC obtained by replacing the nonlinear resistor of a Chua's oscillator with a flux-controlled memristor. We have seen that for the MCC there coexist different regimes, i.e., convergent, periodic, and chaotic regimes for the same set of circuit parameters and fixed nonlinearities. Here, for programming purposes we add a charge source $q_a(t; t_0)$ and a flux source $\varphi_e(t; t_0)$ in the MCC as shown in Fig. 7.7.

For MCC we have $\gamma_M = 1$, $\gamma_C = 1$ and $\lambda_L = 1$, hence $n_C = 2$, $n_L = 1$ and then $n_M = n_x = 1$, $n_y = 2$. MCC satisfies (A2) and (A3); moreover, by inspection it is seen that it also satisfies (A1).

A simple circuit analysis based on Fig. 7.7 permits to derive $\mathbf{H}$ and $\mathbf{B}$ for the hybrid representation in (7.16) (being $\mathbf{G} = \mathbf{0}$ since there are no negative resistors, and $n_E = n_A = 1$). Namely, we have

$$\mathbf{H}_{11} = \frac{1}{R}, \quad \mathbf{H}_{12} = \left(-\tfrac{1}{R} \; 0\right) = \mathbf{H}_{21}^T \tag{7.45a}$$

$$\mathbf{H}_{21} = \begin{pmatrix} -\frac{1}{R} \\ 0 \end{pmatrix}, \quad \mathbf{H}_{22} = \begin{pmatrix} \frac{1}{R} & -1 \\ 1 & r \end{pmatrix} \tag{7.45b}$$

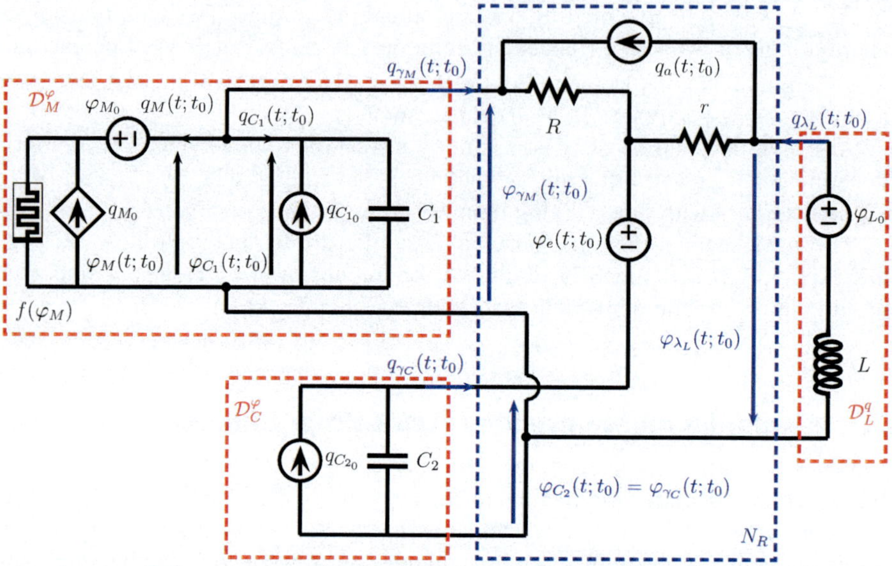

Fig. 7.7 Memristor Chaotic Circuit (MCC) with sources $q_a(t;t_0)$ and $\varphi_e(t;t_0)$ introduced for controlling the dynamics on manifolds. Matrix $\mathbf{H}$ corresponds to the hybrid representation of the three-port linear network N_R connected to the elements $\mathcal{D}_M^{\varphi}$, $\mathcal{D}_C^{\varphi}$ and $\mathcal{D}_L^{q}$

$$\mathbf{B}_{11} = -\frac{1}{R}, \quad \mathbf{B}_{12} = -1 \tag{7.45c}$$

$$\mathbf{B}_{21} = \begin{pmatrix} \frac{1}{R} \\ 1 \end{pmatrix}, \quad \mathbf{B}_{22} = \begin{pmatrix} 1 \\ -r \end{pmatrix}. \tag{7.45d}$$

Since $\det \mathbf{H}_{22} \neq 0$, (A4) holds and MCC is described by the SEs (7.23) in the (φ, q)-domain where (see Fig. 7.7) $\mathbf{x}_M(t) = \varphi_{M_{\gamma_M}}(t) = \varphi_{\gamma_M}(t;t_0) + \varphi_{M_{\gamma_M}}(t_0) = \varphi_M(t)$ (with $\varphi_{M_{\gamma_M}}(t_0) = \varphi_{M_0}$), $\mathbf{x}(t) = \varphi_{\gamma_M}(t;t_0) = \varphi_{C_1}(t;t_0)$, $\mathbf{y}(t) = (\varphi_{\gamma_C}(t;t_0), q_{\lambda_L}(t;t_0))^T = (\varphi_{C_2}(t;t_0), q_L(t;t_0))^T$,

$$\begin{aligned} u_x(t) &= -\frac{1}{R}\varphi_e(t;t_0) - q_a(t;t_0) \\ &= \mathbf{B}_{11}\varphi_e(t;t_0) + \mathbf{B}_{12}q_a(t;t_0) \end{aligned} \tag{7.46a}$$

$$\begin{aligned} \mathbf{u_y}(t) &= \begin{pmatrix} \frac{1}{R} \\ 1 \end{pmatrix} \varphi_e(t;t_0) + \begin{pmatrix} 1 \\ -r \end{pmatrix} q_a(t;t_0) \\ &= \mathbf{B}_{21}\varphi_e(t;t_0) + \mathbf{B}_{22}q_a(t;t_0). \end{aligned} \tag{7.46b}$$

Vector $\mathbf{w} = (\mathbf{x}_M, \dot{\mathbf{x}}, \dot{\mathbf{y}}) = (\varphi_M, v_{C_1}, v_{C_2}, i_L)^T$ is the vector of state variables in the (v, i)-domain and function $\mathbf{K}(\varphi_M, v_{C_1}, v_{C_2}, i_L)$ and the associated manifolds

$\mathcal{M}(k)$ are obtained from (7.29) and (7.30) as

$$\mathbf{K}(\varphi_M, v_{C_1}, v_{C_2}, i_L) = \frac{1}{r+R}\varphi_M + f(\varphi_M) + C_1 v_{C_1} + \frac{rC_2}{r+R} v_{C_2} + \frac{L}{r+R} i_L \quad (7.47a)$$

$$\mathcal{M}(k) = \{\mathbf{w} \in \mathbb{R}^4 : \mathbf{K}(\varphi_M, v_{C_1}, v_{C_2}, i_L) = k\} \quad (7.47b)$$

where $k \in \mathbb{R}$. Note that $\mathcal{M}(k)$ coincides with the expression obtained in Sect. 6.3 of Chap. 6. The four-dimensional state space $(\varphi_M, v_{C_1}, v_{C_2}, i_L)$ in the (v, i)-domain is completely spanned by the ∞^1 three-dimensional manifolds $\mathcal{M}(k)$ by varying k in $\mathbb{R}$.

Since $\mathbf{H}_{12}\mathbf{H}_{22}^{-1}\mathbf{B}_{21} = \mathbf{B}_{11} = -1/R$ and $\mathbf{H}_{12}\mathbf{H}_{22}^{-1}\mathbf{B}_{22} = 0$, a simple calculation based on Property 7.2 and (7.34) yields

$$\dot{\mathbf{k}}(t; t_0, \mathbf{w}_0) = a(t).$$

Then, the solution of the SEs (7.23) in the (v, i)-domain, with $\mathbf{H}$ and $\mathbf{B}$ in (7.45), and ICs $\mathbf{w}_0 = (\varphi_M(t_0), v_{C_1}(t_0), v_{C_2}(t_0), i_L(t_0))^T$

- evolves on the invariant manifold $\mathcal{M}(k_0)$ for any $t \geq t_0$, being $k_0 = \mathbf{K}(\mathbf{w}_0)$, when $q_a(t; t_0) = 0$ for any $t \geq t_0$ (cf. Property 7.3)
- evolves on the invariant manifold $\mathcal{M}(k_1)$ for $t \geq T_1$, being $k_1 = k_0 + \bar{a}\Delta$, when $q_a(t; t_0)$ is a constant momentum source for $t \geq T_1$ as in (7.6.2) (e.g., $q_a(t; t_0)$ in Fig. 7.8)
- explores the manifolds $\mathcal{M}(k_0 + q_a(t; t_0))$ when $q_a(t; t_0)$ is a time-varying momentum source.

The same analysis makes clear that the external source $\varphi_e(t; t_0)$ does not **influence the switching of solutions between different manifolds.**

The nonlinear dynamic behavior of MCC in Fig. 7.7 has been simulated for $t \geq t_0 = 0$ by using the SEs in the (φ, q)-domain (7.46). Note that $\mathbf{M}_x = C_1$, $\mathbf{M}_y = \mathrm{diag}(C_2, L)$ and introduce the dimensionless variables $z_1(\tau) = X(\tau)$, $z_2(\tau) = Y_1(\tau)$, $z_3(\tau) = RY_2(\tau)$ via (7.26) (see also Sect. 6.3 in Chap. 6). Then, the following normalized SEs are derived:

$$\frac{d z_1(\tau)}{d\tau} = \alpha[-z_1(\tau) + z_2(\tau) - n(z_1(\tau)) - w_1(\tau) + k_0] \quad (7.48a)$$

$$\frac{d z_2(\tau)}{d\tau} = z_1(\tau) - z_2(\tau) + z_3(\tau) - w_2(\tau) \quad (7.48b)$$

$$\frac{d z_3(\tau)}{d\tau} = -\beta z_2(\tau) - w_3(\tau) \quad (7.48c)$$

$$z_1(0) = \varphi_M(0)$$

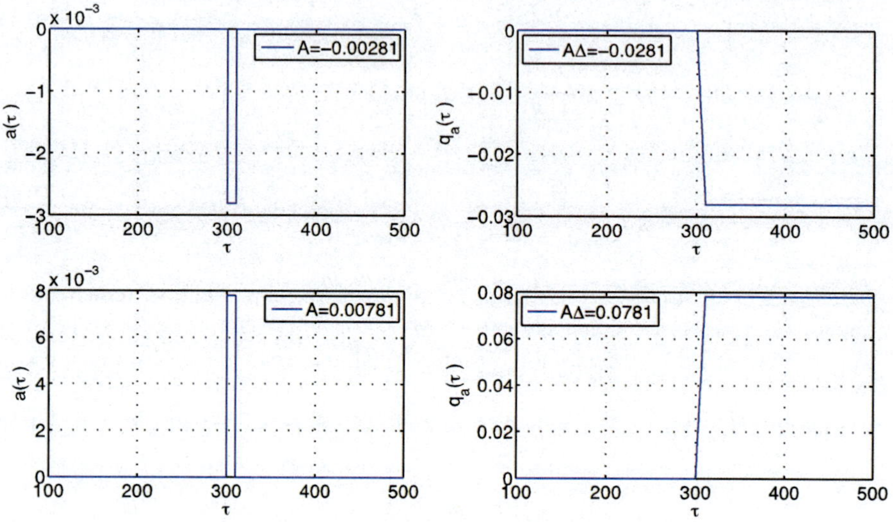

Fig. 7.8 The current source $a(t)$ is a rectangular pulse with finite time duration such that $q_a(t; t_0)$ in Fig. 7.7 represents a constant momentum source for $t \geq T_1$. The pulses have amplitude $A \neq 0$ only between $T_0 = 300$ and $T_1 = 310$ (i.e., $\Delta = 10$ in normalized time units); we have $A = -0.00281$ (upper part) and $A = 0.00781$ (bottom part)

$$z_2(0) = -Li_L(0)$$

$$z_3(0) = -\varphi_M(0) + C_2 v_{C_2}(0) - Li_L(0)$$

where we have assumed $R = 1\ \mathrm{k}\Omega$, $r = 0$, and let $n(\cdot) = Rf(\cdot)$. Moreover, $w_1(\tau) = -\varphi_e(\tau; \tau_0) - q_a(\tau; \tau_0) = -w_2(\tau)$, $w_3(\tau) = \varphi_e(\tau; \tau_0)$, $\mathbf{Y}(0) = -\mathbf{H}_{22}^{-1}\left(\mathbf{H}_{21}\varphi_M(0) + \mathbf{M}_y\dot{\mathbf{y}}(0)\right)$, the CR of the memristor is

$$n(z_1) = -m_0 z_1 + m_1 z_1^3$$

with $m_0 = 8/7$ and $m_1 = 4/63$, $\tau = t/(RC_2)$ is the normalized time (in [μs]), $\alpha = C_2/C_1$ and $\beta = (R^2 C_2)/L = C_2/L$.

Suppose $\varphi_e(\tau; \tau_0) = 0$ (i.e., $w_3(\tau) = 0$ for $\tau \geq 0$) and consider the constant momentum source $u_x(\tau) = -q_a(\tau; 0) = w_1(\tau)$ (cf. (7.46)) shown in Fig. 7.8, which is defined by rectangular pulses $a(\tau)$ with finite time duration $\Delta = 10$ and amplitude $A \neq 0$ between the instants $T_0 = 300$ and $T_1 = T_0 + \Delta = 310$ (normalized time units).

Given the ICs $\mathbf{w}_0 = (\varphi_M(0), v_{C_1}(0), v_{C_2}(0), i_L(0))^T$ in the (v, i)-domain, the SEs (7.48) describe the dynamics of MCC on the manifold $\mathcal{M}(k_0)$ with $k_0 = \varphi_M(0) + n(\varphi_M(0)) + C_1 v_{C_1}(0) + Li_L(0)$ (see (7.28) with the normalized values $R = 1$ and $r = 0$) as long as the constant momentum source $q_a(\tau; 0)$ is zero, i.e., for all $\tau \in [0, T_0]$. As we have seen in Fig. 6.29a of Chap. 6, MCC exhibits a double-scroll attractor when the ICs $\mathbf{w}_0$ are such that $k_0 = 0$.

Let us assume $k_0 = 0$ and show that the nonlinear dynamics of MCC can be programmed by means of suitable pulses $a(\cdot)$ with finite time duration as those in Fig. 7.8. Due to the pulse, the dynamics evolves from the manifold $\mathcal{M}(k_0)$ onto a different manifold $\mathcal{M}(k_1)$ for $\tau > T_1$, where k_1 is given by (7.42), i.e., $k_1 = k_u = A\Delta$.

Suppose we first choose $A = -0.00281$. In this case, as seen in Fig. 6.29c of Chap. 6, the dynamics for $\tau > T_1$ on the manifold $\mathcal{M}(-0.0281)$ is characterized by a spiral chaotic attractor. Figure 7.9 shows how the pulse with $A = -0.00281$ in Fig. 7.8 induces a bifurcation of the chaotic attractor in MCC with ***fixed circuit parameters*** $\alpha = 9.5$ and $\beta = 15$. Such results are also shown in Fig. 7.9b where the whole waveforms of $X(\tau)$, $Y_1(\tau)$ and $Y_2(\tau)$ are reported for $\tau \in [0, 1000]$. It is seen that the double-scroll chaotic attractor (blue curve) for $k_0 = 0$ (i.e., for $0 \le \tau \le T_0$) turns into a spiral chaotic attractor (green curve) for $k_1 = -0.0281$ (i.e., for $\tau \ge T_1$). When the pulse in Fig. 7.8 is applied (i.e., $T_0 \le \tau \le T_1$) the dynamics (red curve in Fig. 7.9a) evolves from the manifold $\mathcal{M}(k_0 = 0)$ to $\mathcal{M}(k_1 = 0.0281)$, i.e., from the double-scroll to the spiral chaotic attractor. Similar results have been obtained if the rectangular pulses in Fig. 7.8 are replaced by triangular pulses or pulses with arbitrary waveforms, provided their momentum is equal to -0.0281.

Figure 7.10 shows how the pulse with $A = 0.00781$ in Fig. 7.8 induces a different bifurcation. With reference to Fig. 7.10, the blue curve is the dynamics for $0 \le \tau \le 300$, (i.e., when the pulse has not yet been applied), the red curve is the dynamics for $300 \le \tau \le 310$ (i.e., during the pulse application) and the green curve is the dynamics for $\tau \ge 310$ (i.e., when the pulse is over). In this case the double-scroll chaotic attractor turns into a period-three limit cycle (black curve in Fig. 7.10) due to the pulse.

Remark 7.3 It is worth noting that a *single impulsive source* $a(t)$ **is sufficient to set** the desired manifolds and reduced-order dynamics on manifolds in the MCC. This is true although the MCC has four state variables in the (v, i)-domain.

Remark 7.4 In the MCC there coexists convergent, periodic, and complex chaotic dynamics for the same set of circuit parameters and nonlinearity. We have seen that via pulses we can easily control and set the desired regime. According to the viewpoint in [16], MCC can be thought of as being a potential source of controllable complex dynamics to be used in future neuromorphic architectures.

Remark 7.5 Several other *pulse-induced bifurcations* have been observed in other memristor circuits $N = N_R \bigcup N_D$. This allows us to draw the conclusion that the **nonlinear dynamics of memristor circuits can be programmed via pulse with finite** time duration (i.e., constant momentum sources in the (φ, q)-domain). The *dynamic analysis through manifolds in the* (φ, q)*-domain permits to effectively design the required pulse parameters* (i.e., duration and amplitude).

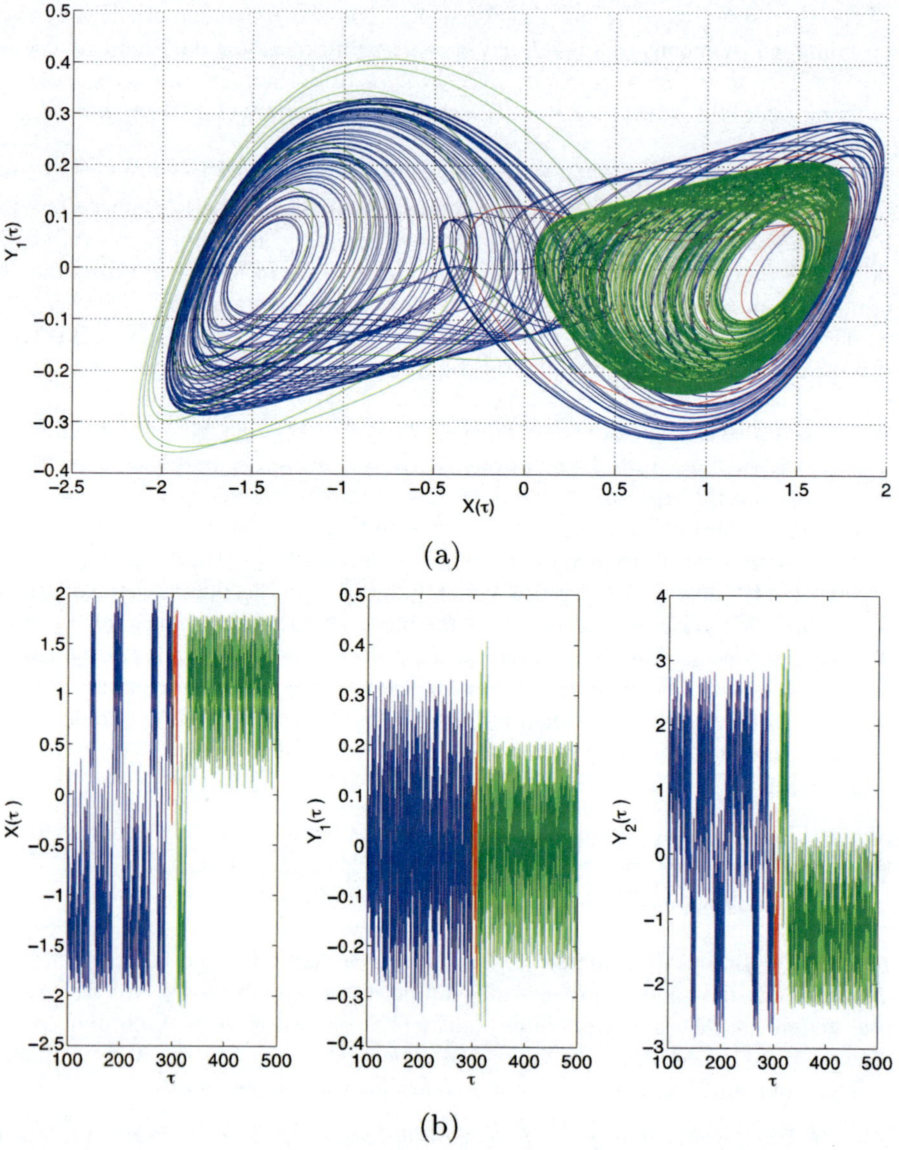

Fig. 7.9 Pulse-induced bifurcations of the chaotic attractor in MCC of Fig. 7.7 with fixed circuit parameters $\alpha = 9.5$ and $\beta = 15$. The double-scroll chaotic attractor (blue curve) for $k_0 = 0$ (i.e., for $0 \leq \tau \leq T_0$) turns into a spiral chaotic attractor (green curve) for $k_1 = -0.0281$ (i.e., for $\tau \geq T_1$). During application of the pulse in Fig. 7.8 (i.e., $T_0 \leq \tau \leq T_1$) the dynamics (red curve) evolves from manifold $\mathcal{M}(k_0 = 0)$ to manifold $\mathcal{M}(k_1 = 0.0281)$, i.e., from the manifold with a double-scroll chaotic attractor to a manifold with a spiral chaotic attractor

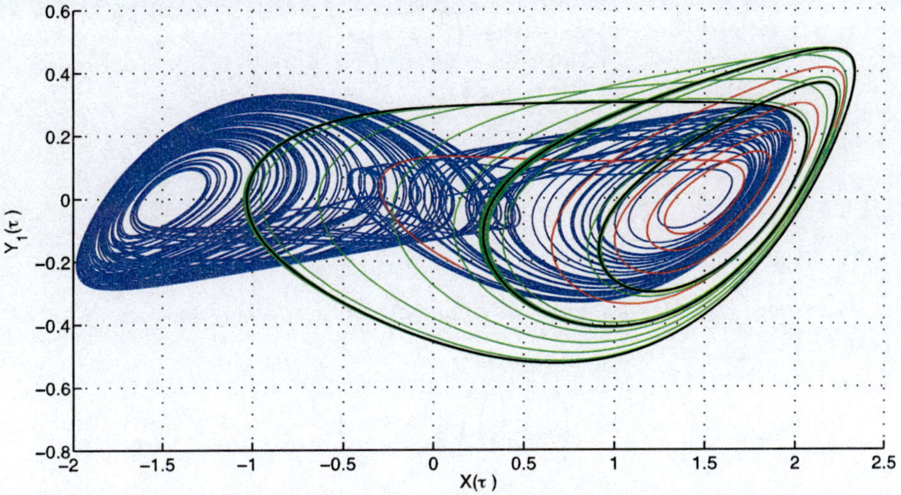

Fig. 7.10 Bifurcation induced by the pulse in Fig. 7.8 with $A = 0.00781$. The double-scroll chaotic attractor turns into a period-three limit cycle. The black curve represents the period-three limit cycle when $\tau \in [900, 1000]$

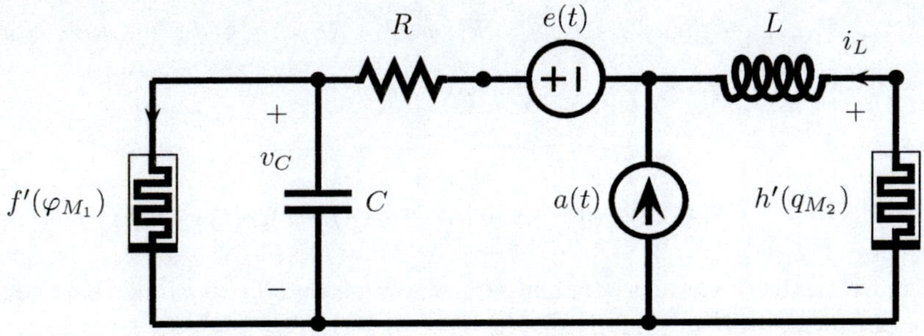

Fig. 7.11 A circuit with two memristors

7.7.2 *Circuit with Two Memristors*

Consider as in Fig. 7.11 a circuit with a flux-controlled memristor $q_{M_1} = f(\varphi_{M_1})$, a charge-controlled memristor $\varphi_{M_2} = h(q_{M_2})$, a current source $a(t)$, and a voltage source $e(t)$ introduced from programming purposes. It is easily seen that the circuit satisfies assumptions (A1)–(A3) for the existence of the SE representation in the (φ, q)-domain. We have $n_x = 2$, $n = n_C + n_L = 2$, hence $n_y = 0$ so that we are in the case considered in Sect. 7.5.3. Then, $\mathbf{H} = \mathbf{H}_{11}$, whereas matrices $\mathbf{H}_{12}$, $\mathbf{H}_{21}$ and $\mathbf{H}_{22}$ are not present. Similarly, $\mathbf{u}(t) = \mathbf{u}_x(t)$, whereas $\mathbf{u}_y(t)$ is not present. A simple circuit analysis yields the hybrid representation with

$$\mathbf{H} = \begin{pmatrix} 0 & -1 \\ 1 & R \end{pmatrix}$$

and

$$\mathbf{u}(t) = \begin{pmatrix} q_a(t; t_0) \\ -\varphi_e(t; t_0) + Rq_a(t; t_0) \end{pmatrix}$$

where $q_a(t; t_0) = \int_{t_0}^{t} a(\tau)d\tau$ and $\varphi_e(t; t_0) = \int_{t_0}^{t} e(\tau)d\tau$.

The state variables in the (v, i)-domain are $(\varphi_{M_1}, q_{M_2}, v_C, i_L)$. We have from (7.39)

$$\mathbf{K}(\varphi_{M_1}, q_{M_2}, v_C, i_L) = \begin{pmatrix} 0 & -1 \\ 1 & R \end{pmatrix}\begin{pmatrix} \varphi_{M_1} \\ q_{M_2} \end{pmatrix} + \begin{pmatrix} f(\varphi_{M_1}) \\ h(q_{M_2}) \end{pmatrix} + \begin{pmatrix} Cv_C(t) \\ L\varphi_L(t) \end{pmatrix}$$

and there are ∞^2 two-dimensional manifolds given by (cf. Sect. 7.5.3)

$$\begin{aligned} \mathcal{M}(\mathbf{k}) = \mathcal{M}\begin{pmatrix} k_1 \\ k_2 \end{pmatrix} &= \{(\varphi_{M_1}, q_{M_2}, v_C, i_L)^T \in \mathbb{R}^4 : \\ &-q_{M_2} + f(\varphi_{M_1}) + Cv_C = k_1; \varphi_{M_1} - Rq_{M_2} + h(q_{M_2}) + L\varphi_L = k_2\} \end{aligned}$$

for any $k_1, k_2 \in \mathbb{R}$. Moreover, due to (7.40)

$$\dot{\mathbf{k}}(t) = -\dot{\mathbf{u}}(t) = \begin{pmatrix} -\dot{q}_a(t; t_0) \\ \dot{\varphi}_e(t; t_0) - R\dot{q}_a(t; t_0) \end{pmatrix} = \begin{pmatrix} -a(t) \\ e(t) - Ra(t) \end{pmatrix}$$

so that, by means of sources $a(t)$ and $e(t)$, we can move the dynamics between any of the ∞^2 different manifolds.

The reduced-order dynamics on a given manifold is given by the second-order SEs

$$\begin{aligned} C\frac{d\varphi_C(t; t_0)}{dt} &= q_L(t; t_0) - f(\varphi_C(t; t_0) + \varphi_{M_{1_0}}) - q_a(t; t_0) \\ &\quad + f(\varphi_{M_{1_0}}) + Cv_{C_0} \\ L\frac{dq_L(t; t_0)}{dt} &= -\varphi_C(t; t_0) - Rq_L(t; t_0) - h(q_L(t; t_0) + q_{M_{2_0}}) \\ &\quad + \varphi_e(t; t_0) - Rq_a(t; t_0) + h(q_{M_{2_0}}) + L\varphi_{L_0}. \end{aligned}$$

Note that there coexist ∞^2 different second-order dynamics, one for each manifold. Also note that in this case we weed two different sources to set the desired second-order dynamics.

7.7.3 *Memristor Star-CNNs*

This short section discusses an application of the pulse programming of nonlinear dynamics on manifolds in large memristor circuits with a neural architecture.

Consider a neural network with a star topology and a (possibly) large number N of cells, where each cell is represented by an element $\mathcal{D}^{\varphi}$ with a capacitor and a flux-controlled memristor in parallel. Neural networks with a star topology are also referred to as Star-CNNs [17, 18]. Hereinafter, the studied Star-CNNs with memristors are named Memristor Star-CNNs (MS-CNNs).

MS-CNNs fall into the class of memristor circuits in Sect. 7.5.3. To simplify the notation, time is omitted, we let $t_0 = 0$ and denote $\varphi_{\gamma M}^{(i)} = \varphi_i$, $q_{\gamma M}^{(i)} = q_i$, $\varphi_M^{(i)}(0) = \varphi_{M0_i}$, and $G = \sum_{j=1}^{N} G_j$ (see Fig. 7.12). The vectors of fluxes $\boldsymbol{\varphi} = (\varphi_1, \ldots, \varphi_N)^T$, charges $\mathbf{q} = (q_1, \ldots, q_N)^T$, ICs $\boldsymbol{\varphi}_{M_0} = (\varphi_{M0_1}, \ldots, \varphi_{M0_N})^T$ and sources $\boldsymbol{\varphi}_e = (\varphi_{e_1}, \ldots, \varphi_{e_N})^T$ permit to write via Millmann's theorem the SEs of MS-CNNs in the (φ, q)-domain in the compact form (see also (7.37))

$$\mathbf{M}\dot{\boldsymbol{\varphi}} = -\mathbf{H}\boldsymbol{\varphi} - \mathbf{F}(\boldsymbol{\varphi} + \boldsymbol{\varphi}_{M_0}) + \mathbf{k}_0 - \mathbf{u} \tag{7.49}$$

$$\mathbf{k}_0 = \mathbf{F}(\boldsymbol{\varphi}_{M_0}) + \mathbf{M}\dot{\boldsymbol{\varphi}}_{M_0} + \mathbf{H}\boldsymbol{\varphi}_{M_0} \tag{7.50}$$

$$\boldsymbol{\varphi}(0) = \boldsymbol{\varphi}_{M_0} \tag{7.51}$$

where $\mathbf{u} = \mathbf{H}\boldsymbol{\varphi}_e + \mathbf{B}_{12}q_a$, the entry (i, j) of $\mathbf{B}_{11}$ (for $1 \leq i \leq N$ and $1 \leq j \leq N$) is $[\mathbf{B}_{11}]_{ij} = -G_i G_j / G$, $\mathbf{H} = \text{diag}(G_1, \ldots, G_N) + \mathbf{B}_{11}$, $\mathbf{B}_{12} = (-G_1/G, \ldots, -G_N/G)^T$ and $\mathbf{M} = \text{diag}(C_1, \ldots, C_N)$.

If $\mathbf{u}$ is constant, the *number and stability properties of EPs* in the MS-CNNs described by (7.51) depend on $\mathbf{k}_0$ and $\mathbf{u}$. Clearly, bifurcations without parameters of such equilibria can be induced by applying pulses with finite time duration by means of $\boldsymbol{\varphi}_e$ and q_a, which cause a change of $\mathbf{k}_0$ into $\mathbf{k}_0 + \mathbf{k}_u$ with $\mathbf{k}_u$ given in (7.43).

7.8 Further Examples

In the previous examples we considered memristor circuits that belong to the class $\mathcal{LM}$ and satisfy assumptions (A1)–(A3) guaranteeing the existence of the hybrid representation and the SE description. Of course, (A1)–(A3) are only sufficient conditions for the existence of the hybrid representation. It can be seen that if those conditions fail, but the hybrid representation exists, then we can still use the developed procedure for writing the SEs and studying the invariant manifolds. In this section, we provide some examples of memristor circuits containing active elements as controlled sources and thus not belonging to the class $\mathcal{LM}$. Due to the presence of active elements, it is not easy to give conditions that *a priori* guarantee the existence of the hybrid representation. Nevertheless, in the examples we directly

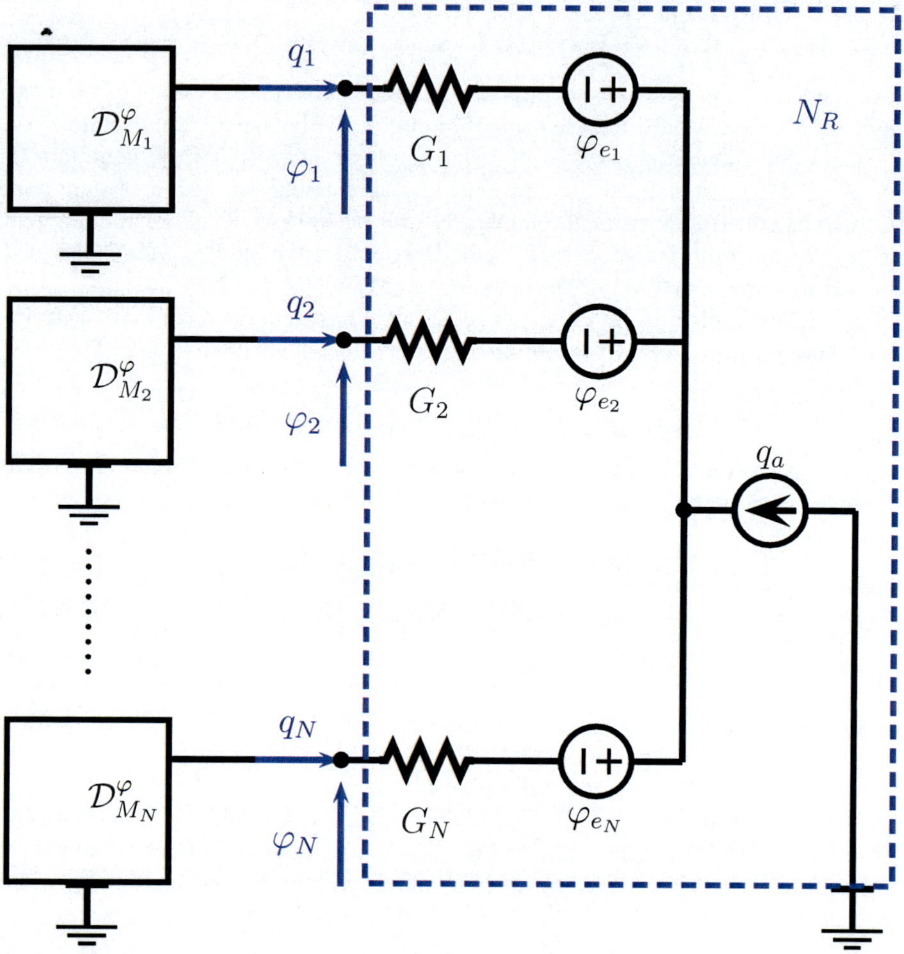

Fig. 7.12 Memristor Star-CNNs. Each cell is represented by an element $\mathcal{D}^{\varphi}_{M}$ with a capacitor and a flux-controlled memristor in parallel. The matrix $\mathbf{H}$ corresponds to the hybrid representation of the N-port linear network N_R

verify the existence of the hybrid representation and show that the procedure developed in the chapter for finding the SEs and studying invariant manifolds can indeed be used also in this more general situation.

Example 7.6 Consider a circuit with a flux-controlled memristor, an inductor and a current-controlled voltage-source as shown in Fig. 7.13. The hybrid representation of N_R is such that

$$\mathbf{H} = \begin{pmatrix} \frac{1}{R} & -1 \\ 1 + \frac{\rho}{R} & 2R \end{pmatrix}$$

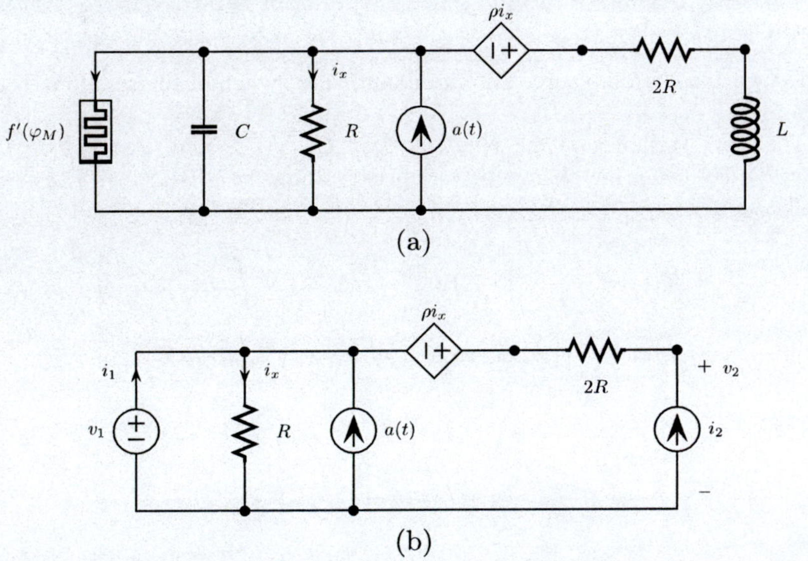

Fig. 7.13 (**a**) Memristor circuit with a current-controlled voltage-source and (**b**) resistive circuit for finding the hybrid representation

and

$$\mathbf{u}(t) = \begin{pmatrix} -q_a(t;t_0) \\ 0 \end{pmatrix}$$

with $u_x(t) = -q_a(t;t_0) = -\int_{t_0}^{t} \boldsymbol{a(\tau)d\tau}$ and $u_y(t) = 0$. Note that $H_{22} = 2R$ is nonsingular and $[H_{22}]^{-1} = 1/2R$. The representation always exists, even if $R < 0$.

The state variables in the (v, i)-domain are given by $(\varphi_M(t), v_C(t), i_L(t))$. We have $\mathbf{x}_M(t) = \varphi_M(t)$, $\mathbf{M}_x\dot{\mathbf{x}}(t) = Cv_C(t)$, and $\mathbf{M}_y\dot{\mathbf{y}}(t) = Li_L(t)$.

From (7.29), we have

$$\mathbf{K}(\varphi_M(t), v_C(t), i_L(t)) = \left(\frac{3}{2R} + \frac{\rho}{2R^2}\right)\varphi_M(t) + f(\varphi_M(t))$$
$$+Cv_C(t) + \frac{Li_L(t)}{2R}$$

so that on the basis of (7.30) the manifolds are given as

$$\mathcal{M}(\bar{k}) = \{(\varphi_M(t), v_C(t), i_L(t))^T \in \mathbb{R}^3 : \mathbf{K}(\varphi_M(t), v_C(t), i_L(t)) = \bar{k}\}$$

where $\bar{k} \in \mathbb{R}$. Moreover, from (7.34)

$$\dot{\mathbf{k}}(\varphi_M(t), v_C(t), i_L(t)) = a(t)$$

hence via the current-source we can control the switching of solutions between manifolds.

The state variables in the (φ, q)-domain are $x(t) = \varphi_C(t; t_0)$ and $y(t) = q_L(t; t_0)$. By using the change of variables (7.26b), the SEs in the (φ, q)-domain can be written as

$$C\dot{X}(t) = -\frac{1}{R}X(t) + Y(t) - f(X(t)) + \int_{t_0}^{t} a(\tau)d\tau + k_0$$

$$L\dot{Y}(t) = -(1 + \frac{\rho}{R})X(t) - 2RY(t)$$

where

$$k_0 = k(\varphi_{M_0}, v_{C_0}, i_{L_0}).$$

Example 7.7 Consider a circuit with a flux-controlled memristor, two capacitors, an inductor, and a current-controlled current-source as shown in Fig. 7.14a. To find the hybrid representation, consider Fig. 7.14b from which we have

$$\mathbf{H} = \begin{pmatrix} \frac{1}{R} & -\frac{1}{R} & -1 \\ -\frac{1+\gamma}{R} & \frac{1+\gamma}{R} & -\gamma \\ 1 & 0 & R \end{pmatrix}$$

and so

$$H_{11} = \frac{1}{R}, \mathbf{H}_{12} = \begin{pmatrix} -\frac{1}{R} & -1 \end{pmatrix}, \mathbf{H}_{21} = \begin{pmatrix} -\frac{1+\gamma}{R} \\ 1 \end{pmatrix}, \mathbf{H}_{22} = \begin{pmatrix} \frac{1+\gamma}{R} & -\gamma \\ 0 & R \end{pmatrix}.$$

Furthermore

$$\mathbf{u}(t) = \begin{pmatrix} 0 \\ 0 \\ -\varphi_e(t; t_0) \end{pmatrix}$$

with

$$u_x(t) = 0, \mathbf{u}_y(t) = \begin{pmatrix} 0 \\ -\varphi_e(t; t_0) \end{pmatrix}$$

and $\varphi_e(t; t_0) = \int_{t_0}^{t} e(\tau)d\tau$. The representation always exists, even if $R < 0$. Note that H_{22} is nonsingular if and only if $\gamma \neq -1$.

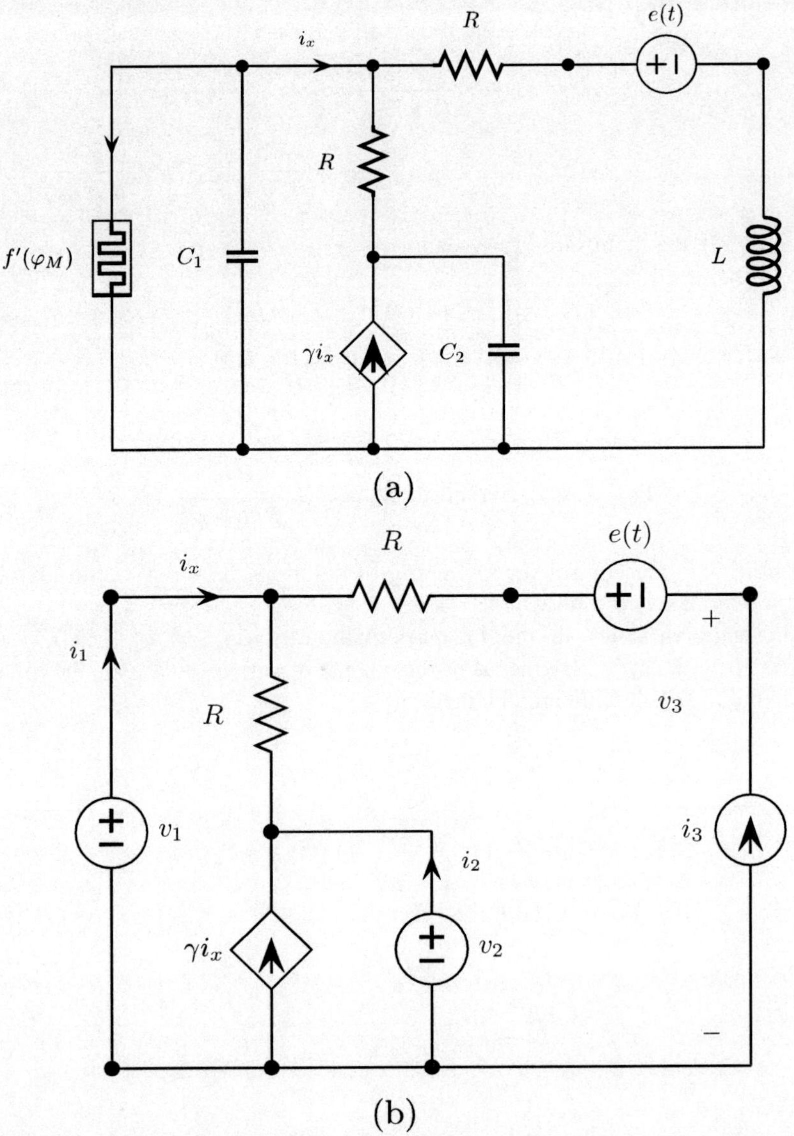

Fig. 7.14 (**a**) Memristor circuit with a current-controlled current source and (**b**) resistive circuit for finding the hybrid representation

The state variables in the (v, i)-domain are $(\varphi_M(t), v_{C_1}(t), v_{C_2}(t), i_L(t))$. We have $\mathbf{x}_M(t) = \varphi_M(t)$, $\mathbf{M}_x\dot{\mathbf{x}}(t) = C_1 v_{C_1}(t)$, $\mathbf{M}_y\dot{\mathbf{y}}(t) = (C_2 v_{C_2}(t), L i_L(t))^T$.

Equation (7.29) yields

$$\mathbf{K}(\varphi_M(t), v_{C_1}(t), v_{C_2}(t), i_L(t)) = \frac{2\gamma+1}{R(\gamma+1)}\varphi_M(t) + f(\varphi_M(t))$$
$$+C_1 v_{C_1}(t) + \frac{C_2}{1+\gamma} v_{C_2}(t) + \frac{2\gamma+1}{R(1+\gamma)} L i_L(t).$$

Using (7.30), the manifolds are given as

$$\mathcal{M}(\bar{k}) = \{(\varphi_M(t), v_{C_1}(t), v_{C_2}(t), i_L(t))^T \in \mathbb{R}^4 :$$
$$\mathbf{K}(\varphi_M(t), v_{C_1}(t), v_{C_2}(t), i_L(t)) = \bar{k}\}.$$

Moreover, from (7.34)

$$\dot{k}(\varphi_M(t), v_{C_1}(t), v_{C_2}(t), i_L(t)) = \frac{2\gamma+1}{R(1+\gamma)} e(t).$$

Therefore, we can control the switching of solutions between manifolds via the voltage source $e(t)$ provided $\gamma \neq -1/2$.

The state variables in the (φ, q)-domain are $x(t) = \varphi_{C_1}(t; t_0)$, $\mathbf{y}(t) = (\varphi_{C_2}(t; t_0), q_L(t; t_0))^T$. By means of the change of variables (7.26b), the following SEs in the (φ, q)-domain are obtained:

$$C_1 \dot{X}(t) = -\frac{1}{R} X(t) + \frac{1}{R} Y_1(t) + Y_2(t) - f(X(t)) + k_0$$
$$C_2 \dot{Y}_1(t) = \frac{1+\gamma}{R} X(t) - \frac{1+\gamma}{R} Y_1(t) + \gamma Y_2(t)$$
$$L \dot{Y}_2(t) = -X(t) + R Y_2(t)$$

where

$$k_0 = k(\varphi_{M0}, v_{C_{10}}, v_{C_{20}}, i_{L0}).$$

7.9 Discussion

This chapter has provided an explicit general form for the SEs of a broad class of memristor circuits with an arbitrary number of inductors, capacitors, and flux- or charge-controlled memristors. Conditions for the existence of the SEs are couched in graph-theoretic terms and as such they can be checked by inspection on the circuit topology. On the basis of the SEs, it is shown that the state space in the (v, i)-domain of a memristor circuit can be foliated in a continuum of manifolds and that, in the autonomous case, each manifold is positively invariant. Different regimes, reduced-order dynamics, and attractors therefore coexist in a memristor circuit for a fixed set of parameters and nonlinearities. The chief results in the chapter concern the nonautonomous case where time-varying independent sources are present. In such a case, general explicit formulas are obtained for effectively designing impulsive sources with finite time-duration in order to programme the different regimes and nonlinear dynamics that can be displayed by the memristor circuits, i.e., to drive at one will solutions between different manifolds, reduced-order dynamics and attractors. This gives a sound theoretical foundation to a common practice for changing the dynamics in memristor circuits that has been so far mainly based on experimental trials and/or heuristic means.

Appendix 1: Nonsingularity of $\mathbf{H}_{22}$

We provide some brief considerations about assumption (A4). Consider for simplicity a network N satisfying (A1)–(A3) with only one flux-controlled memristor M in parallel to C_1. Also suppose that N has only positive resistors, i.e., $\gamma_G = \lambda_R = \rho_G = \rho_R = 0$. In this case, we can show that (A4) is satisfied, i.e., $\det \mathbf{H}_{22} \neq 0$, if and only if the next topological assumption holds.

(A5) *The network N has no cut-sets made of capacitors $C_2, \ldots, C_{n_C}$ and/or charge sources and no loops made of M, inductors $L_1, \ldots, L_{n_L}$ and/or flux sources.*

The proof follows by an argument for testing the nonsingularity of the hybrid representation of N_R as in Corollary 3 in [13].[8] Furthermore, if (A4) fails, then the state space in the (v, i)-domain can be foliated in ∞^1 *planar* manifolds, as illustrated in the next example.

Example 7.8 Consider the circuit in Fig. 7.15 containing a loop formed by a flux-controlled memristor, an inductor, and an independent voltage source. For the circuit (A5) (and thus (A4)) fails.

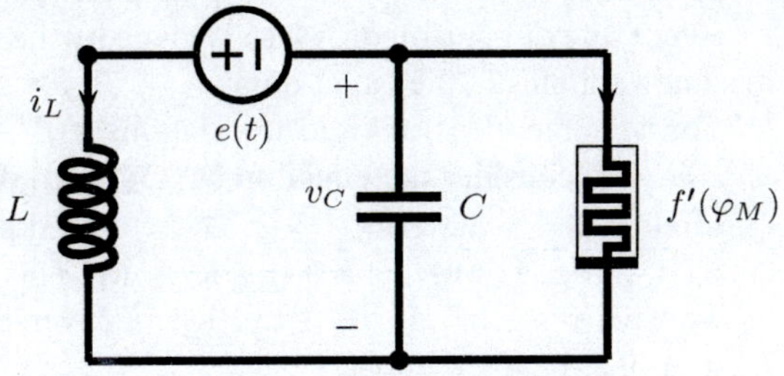

Fig. 7.15 Memristor circuit with planar invariant manifolds

[8] For example, it can be verified that the memristor chaotic circuit in Fig. 7.7 satisfies (A5), and indeed we have explicitly seen that $\det \mathbf{H}_{22} \neq 0$.

A simple analysis shows that in this case we have

$$\mathbf{H} = \begin{pmatrix} 0 & -1 \\ 1 & 0 \end{pmatrix}, \mathbf{u}(t) = \begin{pmatrix} 0 \\ e(t) \end{pmatrix}$$

hence, as expected, $H_{22} = 0$ is singular.

The state variables in the (v, i)-domain are (φ_M, v_C, i_L). By KVL we obtain

$$L\frac{di_L(t)}{dt} = e(t) + \frac{d\varphi_M(t)}{dt}.$$

Let

$$\mathbf{K}(\varphi_M(t), v_C(t), i_L(t)) = Li_L(t) - \varphi_M(t).$$

If $e(t) = 0$ there are *planar* invariant manifolds

$$\mathcal{M}(k) = \{(\varphi_M(t), v_C(t), i_L(t)) \in \mathbb{R}^3 : \mathbf{K}(\varphi_M(t), v_C(t), i_L(t)) = k\}$$

for any $k \in \mathbb{R}$. If $e(t) \neq 0$ we have

$$\dot{\mathbf{k}}(\varphi_M(t), i_L(t)) = e(t)$$

and so via the voltage source we are able to programme the different reduced-order dynamics on manifolds.

Appendix 2: Proof of Theorem 7.3

1. Given any $\mathbf{k} \in \mathbb{R}^{n_M}$, it can be checked that $\mathcal{M}(\mathbf{k})$ contains at least point $\mathbf{w} = (\mathbf{0}, \mathbf{M}_x^{-1}(\mathbf{k} - \mathbf{F}(\mathbf{0})), \mathbf{0})$, hence $\mathcal{M}(\mathbf{k}) \neq \emptyset$. The manifold $\mathcal{M}(\mathbf{k})$ is nonplanar due to the nonlinear function $\mathbf{F}(\cdot)$. Finally, function $\mathbf{K}(\mathbf{w}) = \mathbf{k}$ specifies n_M constraints in the $(n_M + n_C + n_L)$-dimensional state space in the (v, i)-domain. Hence, the dimension of $\mathcal{M}(\mathbf{k})$ is given by the number $(n_C + n_L)$ of independent state variables in the (v, i)-domain.
2. Consider the class of rigid translations $\mathbf{x}_M \to \mathbf{x}_M$, $\dot{\mathbf{x}} \to \dot{\mathbf{x}} + \Delta_{\dot{\mathbf{x}}}$, $\dot{\mathbf{y}} \to \dot{\mathbf{y}} + \Delta_{\dot{\mathbf{y}}}$. Also consider the equation $\mathbf{M}_x \Delta_{\dot{\mathbf{x}}} - \mathbf{H}_{12}\mathbf{H}_{22}^{-1}\mathbf{M}_y \Delta_{\dot{\mathbf{y}}} = \mathbf{k}_2 - \mathbf{k}_1$, that has a solution $\Delta_{\dot{\mathbf{x}}} = \mathbf{M}_x^{-1}(\mathbf{k}_2 - \mathbf{k}_1)$, $\Delta_{\dot{\mathbf{y}}} = \mathbf{0}$ (and possibly other solutions). We have $\mathbf{K}(\mathbf{w} + (\mathbf{0}, \ \Delta_{\dot{\mathbf{x}}}, \ \mathbf{0})) = \mathbf{K}(\mathbf{w}) + \mathbf{k}_2 - \mathbf{k}_1$. This means that, for any $\mathbf{w} \in \mathcal{M}(\mathbf{k}_1)$, we have $\mathbf{w} + (\mathbf{0}, \ \Delta_{\dot{\mathbf{x}}}, \mathbf{0}) \in \mathcal{M}(\mathbf{k}_2)$. Conversely, for any $\mathbf{w} \in \mathcal{M}(\mathbf{k}_2)$ we have $\mathbf{w} - (\mathbf{0}, \ \Delta_{\dot{\mathbf{x}}}, \ \mathbf{0}) \in \mathcal{M}(\mathbf{k}_1)$.
3. Manifolds are nonintersecting because if there exist $\mathbf{w} \in \mathcal{M}(\mathbf{k}_1) \bigcap \mathcal{M}(\mathbf{k}_2)$, with $\mathbf{k}_1 \neq \mathbf{k}_2$, then $\mathbf{K}(\mathbf{w}) = \mathbf{k}_1$ and $\mathbf{K}(\mathbf{w}) = \mathbf{k}_2$, i.e., $\mathbf{w}$ is mapped by $\mathbf{K}(\cdot)$ in two distinct vectors $\mathbf{k}_1$ and $\mathbf{k}_2$, while $\mathbf{K}(\cdot)$ in (7.29) is a (single-valued) function.

Since manifolds are parametrized by the n_M-dimensional vector $\mathbf{k}$, we conclude that there are ∞^{n_M} nonintersecting manifolds. To see that they fill the whole state space, it is enough to note that, given any point $\mathbf{w}$, we have $\mathbf{w} \in \mathcal{M}(\mathbf{K}(\mathbf{w}))$.

Appendix 3: Proof of Property 7.2

Due to (7.29), considering that $\dot{\mathbf{x}}_M(t) = \dot{\mathbf{x}}(t)$, we have $\dot{\mathbf{k}}(t; t_0, \mathbf{w}_0) = \mathbf{S}_{22}\dot{\mathbf{x}}(t) + \mathbf{J}_\mathbf{F}(\mathbf{x}_M(t))\dot{\mathbf{x}}(t) + \mathbf{M}_x\ddot{\mathbf{x}}(t) - \mathbf{H}_{12}\mathbf{H}_{22}^{-1}\mathbf{M}_y\ddot{\mathbf{y}}(t)$. Using such equation, the expression in (7.33) is easily obtained by the following relationships derived from (7.23):

$$\begin{aligned}\mathbf{J}_\mathbf{F}(\mathbf{x}_M(t))\dot{\mathbf{x}}(t) + \mathbf{M}_x\ddot{\mathbf{x}}(t) &= -\mathbf{H}_{11}\dot{\mathbf{x}}(t) - \mathbf{H}_{12}\dot{\mathbf{y}}(t) - \dot{\mathbf{u}}_x(t)\\ \mathbf{M}_y\ddot{\mathbf{y}}(t) &= -\mathbf{H}_{21}\dot{\mathbf{x}}(t) - \mathbf{H}_{22}\dot{\mathbf{y}}(t) - \dot{\mathbf{u}}_y(t).\end{aligned}$$

References

1. M. Itoh, L.O. Chua, Memristor oscillators. Int. J. Bifurc. Chaos **18**(11), 3183–3206 (2008)
2. R. Riaza, C. Tischendorf, Semistate models of electrical circuits including memristors. Int. J. Circuit Theory Appl. **39**(6), 607–627 (2011)
3. A. Ascoli, F. Corinto, R. Tetzlaff, Generalized boundary condition memristor model. Int. J. Circuit Theory Appl. **44**(1), 60–84 (2016)
4. J. Ma, F. Wu, G. Ren, J. Tang, A class of initials-dependent dynamical systems. Appl. Math. Comput. **298**, 65–76 (2017)
5. B. Bao, T. Jiang, Q. Xu, M. Chen, H. Wu, Y. Hu, Coexisting infinitely many attractors in active band-pass filter-based memristive circuit. Nonlinear Dyn. **86**(3), 1711–1723 (2016)
6. A. Buscarino, C. Corradino, L. Fortuna, M. Frasca, L.O. Chua, Turing patterns in memristive cellular nonlinear networks. IEEE Trans. Circuits Syst. I Regul. Pap. **63**(8), 1222–1230 (2016)
7. V.-T. Pham, S. Vaidyanathan, C.K. Volos, S. Jafari, N.V. Kuznetsov, T.M. Hoang, A novel memristive time-delay chaotic system without equilibrium points. Eur. Phys. J. Spec. Top. **225**(1), 127–136 (2016)
8. R. Riaza, Manifolds of equilibria and bifurcations without parameters in memristive circuits. SIAM J. Appl. Math. **72**(3), 877–896 (2012)
9. B. Bao, Z. Ma, J. Xu, Z. Liu, Q. Xu, A simple memristor chaotic circuit with complex dynamics. Int. J. Bifurc. Chaos **21**(9), 2629–2645 (2011)
10. Q. Li, S. Hu, S. Tang, G. Zeng, Hyperchaos and horseshoe in a 4D memristive system with a line of equilibria and its implementation. Int. J. Circuit Theory Appl. **42**(11), 1172–1188 (2014)
11. F. Corinto, A. Ascoli, M. Gilli, Analysis of current–voltage characteristics for memristive elements in pattern recognition systems. Int. J. Circuit Theory Appl. **40**(12), 1277–1320 (2012)
12. L.O. Chua, Dynamic nonlinear networks: state-of-the-art. IEEE Trans. Circuits Syst. **27**(11), 1059–1087 (1980)
13. H.C. So, On the hybrid description of a linear n–port resulting from the extraction of arbitrarily specified elements. IEEE Trans. Circuit Theory **CT-12**(3), 381–387 (1965)
14. F. Zhang, *The Schur Complement and Its Applications*, vol. 4 (Springer, Berlin, 2006)
15. F. Corinto, P.P. Civalleri, L.O. Chua, A theoretical approach to memristor devices. IEEE J. Emerg. Sel. Topics Circuits Syst. **5**(2), 123–132 (2015)

16. S. Kumar, J.P. Strachan, R. Stanley Williams, Chaotic dynamics in nanoscale NbO_2 Mott memristors for analogue computing. Nature **548**(7667), 318 (2017)
17. M. Itoh , L.O. Chua, Star cellular neural networks for associative and dynamic memories. Int. J. Bifurc. Chaos **14**(5), 1725–1772 (2004)
18. F. Corinto, M. Gilli, T. Roska, On full-connectivity properties of locally connected oscillatory networks. IEEE Trans. Circuits Syst. I: Regul. Pap. **58**(5), 1063–1075 (2011)

Part III
Applications and Extension of the FCAM

Chapter 8
Complex Dynamics and Synchronization Phenomena in Arrays of Memristor Oscillators

8.1 Introduction

A great deal of efforts have been traditionally devoted in circuit theory to analyze nonstationary steady-state behaviors in networks obtained by locally coupled arrays of simple dynamic circuits (also named cells, oscillators, units, etc.) [1]. These arrays can be thought of as a bio-inspired circuit model of complex nonlinear phenomena observable in nature with potential applications in signal processing and computing systems. In fact, on one hand, they are a mean for reproducing, analyzing, and understanding spatiotemporal nonlinear phenomena displayed by spatially extended networks found in such diverse fields as electrical engineering, computer science, biology, and physics. On the other hand, complex spatiotemporal dynamics including chaos are potentially useful for developing future analogue computing systems. Recent studies have shown for instance that chaos can play a crucial role in searching for the global solution of combinatorial optimization problems and chaotic relaxation oscillators with memristors have been used to boot efficiency and accuracy of Hopfield-like computing networks [2].

One of the most relevant aspects concerns bifurcations and synchronization phenomena [3, 4], i.e., a scenario where all the oscillators adjust their dynamic behavior so that the whole array acts in unison and spatiotemporal patterns emerge. Synchronization phenomena in one-dimensional (1D) and bi-dimensional (2D) dynamic arrays using Chua's oscillators as building block, with various types of uni-directional, bi-directional, static, and dynamic diffusive interactions, have been considered and analyzed by numerical simulations and analytic tools [5–7]. The influence of the network topology properties and their link with complex dynamic periodic/chaotic attractors in large biological and artificial systems have been studied in detail in several works [8–13].

Memristors provide an accurate and power efficient emulator of neural synapses and biological neural codes [14–16]. Recent works have shown the use of memristors as adaptive couplings for connecting simple Chua's oscillators in a crossbar

F. Corinto et al., *Nonlinear Circuits and Systems with Memristors*,
https://doi.org/10.1007/978-3-030-55651-8_8

architectures that is well suited for implementation in nanotechnology [17–19]. The articles [20–23] explore the dynamics of arrays of diffusively coupled Chua's oscillators where the nonlinear resistor in each oscillator is replaced by a memristor. By simulations and experiments, it is demonstrated that, due to the nonlinear dynamics of memristors, such arrays are endowed with a new and intriguing variety of complex spatiotemporal phenomena [24, 25]. Interestingly, several form of synchronization can be observed by modifying not only circuit parameters and coupling strengths, but also initial conditions of dynamic elements and, especially, of memristors. Considering that locally connected regular architectures are especially well suited for nanoscale implementation, and that emerging dynamic phenomena due to the presence of memristors are observed, it is crucial to develop analytical and numerical tools to investigate complex dynamics including synchronization phenomena in bio-inspired networks of memristor-based oscillatory cells.

This chapter considers a 1D array of N diffusively coupled memristor chaotic circuits (MCCs), where each MCC is obtained by replacing the nonlinear resistor of a Chua's oscillator with a flux-controlled memristor (cf. Chap. 6). We name NMCC the considered array of N interconnected MCCs. The goal is to analyze the complex dynamics, bifurcations, and synchronization phenomena in NMCC by using FCAM.

The main contributions in this chapter can be summarized as follows:

- it is shown that the state space in the voltage-current (v, i)-domain can be foliated in a continuum of manifolds that are invariant for the dynamics. Moreover, it is possible to explicitly find the state equations (SEs) describing the reduced-order dynamics on each invariant manifold. Compared to the result in Chap. 7, this step is accomplished via a newly developed circuit technique based on writing suitable sets of Kirchhoff laws for the NMCC in the (φ, q)-domain;
- via the concept of invariant manifolds, it is analytically shown that for the NMCC there coexist infinitely many different complex attractors and dynamics, and that bifurcations without parameters, i.e., bifurcations due to changes of initial conditions for a fixed set of circuit parameters, occur. Such results give an analytic explanation of initial-condition dependent nonlinear phenomena experimentally observed and reported in several publications [20, 21, 24, 25];
- it is shown how the explicit knowledge of invariant manifolds, and the reduced-order dynamics on each manifold, enable to exploit results available in the literature for analyzing some relevant features of the complex nonlinear dynamics in the NMCC. In particular, the theoretic results in the chapter permit to choose initial conditions such that the nonlinear dynamics in the NMCC take place on a selected manifold with some special properties, named "zero-manifold," see Sect. 8.4 for details. By relying on previous analytic results in [26–28], various types of synchronization phenomena on the "zero-manifold" are investigated, including in-phase and/or anti-phase synchronization of periodic/chaotic attractors.

Although the chapter focuses on a 1D array of N diffusively coupled memristor oscillators, all the obtained results can be extended, *mutatis mutandis*, to any network of MCCs with diffusive couplings arranged in 2D or 3D structures.

8.2 Memristor-Based Chaotic Circuit (MCC)

Consider the memristor chaotic circuit (MCC) in Fig. 8.1, obtained from Chua's oscillator (Chap. 4), by letting $r = 0$ and replacing the nonlinear locally active resistor (Chua's diode) with an ideal locally active flux-controlled memristor $q_M(t) = f(\varphi_M(t))$, where $\varphi_M(t)$ (resp., $q_M(t)$) is the memristor flux (resp., charge) and $f : \mathbb{R} \to \mathbb{R}$ is a smooth function which will be defined later. The ideal two-terminal elements C_1, C_2, L, and R are assumed to be passive. Such an MCC has been frequently considered in literature as a prototypical circuit for studying the nonlinear dynamics, bifurcations, and complex oscillatory/chaotic phenomena emerging in memristor-based bioinspired networks, see, e.g., [29–32], and references therein.

The nonlinear dynamics of the MCC has been studied in Sect. 6.3 of Chap. 6. Next, we briefly recall some chief properties of the dynamics and we introduce a novel circuit technique for finding the invariant manifolds of the MCC, that is effective also to study the invariant manifolds in arrays of diffusively coupled MCCs (see Sect. 8.4).

Let us introduce the vector of state variables of the four memory elements in the (v, i)-domain, i.e., C_1, C_2, L and M, given by

$$\mathbf{w}(t) = (v_{C_1}(t), v_{C_2}(t), i_L(t), \varphi_M(t))^T \in \mathbb{R}^4.$$

Denote by $-\infty < t_0 < +\infty$ a given finite instant and let $v_{C_1}(t_0)$, $v_{C_2}(t_0)$, $i_L(t_0)$, $\varphi_M(t_0)$ be the initial conditions (ICs) at t_0 for the state variables. Moreover, let $q_{C_1}(t_0) = C_1 v_{C_1}(t_0)$, $q_{C_2}(t_0) = C_2 v_{C_2}(t_0)$, $\varphi_L(t_0) = L i_L(t_0)$, and $q_M(t_0) = f(\varphi_M(t_0))$.

Given the ICs in the (v, i)-domain, FCAM permits to associate the MCC in Fig. 8.1 with the reduced-order, i.e., third-order circuit in the (φ, q)-domain reported in Fig. 8.2, where each two-terminal element is represented by its equivalent circuit in the (φ, q)-domain.[1] The circuit can be analyzed by Kirchhoff flux law (KφL) and

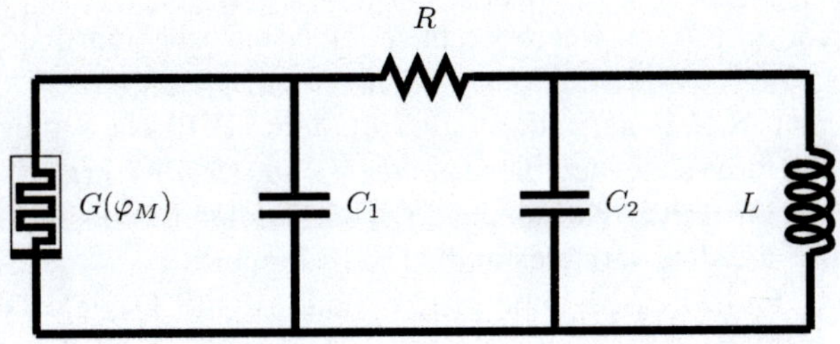

Fig. 8.1 MCC obtained from Chua's oscillator once the nonlinear locally active resistor is replaced by an ideal locally active flux-controlled memristor (we have let $G(\varphi_M) = f'(\varphi_M)$)

[1] For convenience in writing the dynamic equations of coupled MCCs, the reference direction of the current (charge) and voltage (flux) on the inductor has been reversed with respect to the MCC studied in Chap. 6.

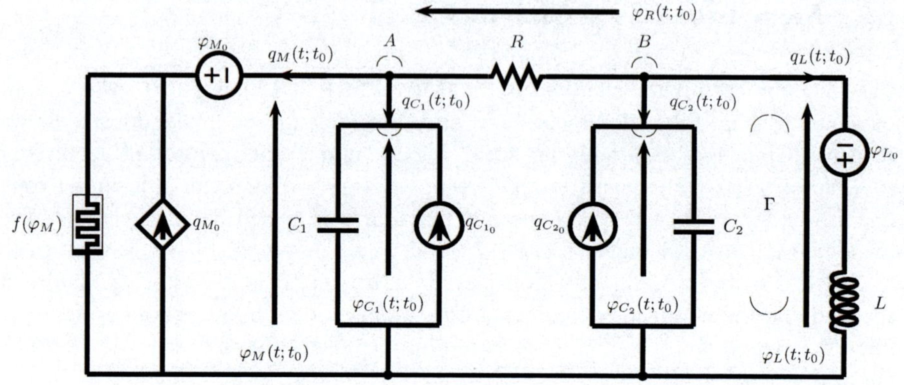

Fig. 8.2 Equivalent circuit of MCC in the (φ, q)-domain. We have let $q_{C_{1_0}} = q_{C_1}(t_0)$, $q_{C_{2_0}} = q_{C_2}(t_0)$, $\varphi_{L_0} = \varphi_L(t_0)$ and $\varphi_{M_0} = \varphi_M(t_0)$

Kirchhoff charge law (KqL) for incremental charges and fluxes. The reduction of order for the associated circuit follows from the fact that in the (φ, q)-domain the ideal memristor results to be a memoryless nonlinear element.

By using KφLs and KqLs, the following SEs describing the MCC in the (φ, q)-domain are obtained:

$$C_1 \frac{d\varphi_{C_1}(t;t_0)}{dt} = -\frac{1}{R}(\varphi_{C_1}(t;t_0) - \varphi_{C_2}(t;t_0)) + q_{C_1}(t_0) - f(\varphi_{C_1}(t;t_0) + \varphi_M(t_0)) + f(\varphi_M(t_0)) \tag{8.1a}$$

$$C_2 \frac{d\varphi_{C_2}(t;t_0)}{dt} = -\frac{1}{R}(\varphi_{C_2}(t;t_0) - \varphi_{C_1}(t;t_0)) + q_{C_2}(t_0) - q_L(t;t_0) \tag{8.1b}$$

$$L \frac{dq_L(t;t_0)}{dt} = \varphi_{C_2}(t;t_0) + \varphi_L(t_0) \tag{8.1c}$$

for all $t \geq t_0$, where we have taken into account that $\varphi_{C_1}(t;t_0) = \varphi_M(t;t_0)$. The initial conditions are $\varphi_{C_1}(t_0;t_0) = 0$, $\varphi_{C_2}(t_0;t_0) = 0$ and $q_L(t_0;t_0) = 0$. Note that (8.1) is an initial value problem (IVP) for a third-order system of ODEs in the three state variables $(\varphi_{C_1}(t;t_0), \varphi_{C_2}(t;t_0), q_L(t;t_0))$ in the (φ, q)-domain. Also note that the right-hand side of (8.1) contains constant terms depending on the ICs for the state variables in the (v, i)-domain.

From FCAM, time differentiation of (8.1) yields the SEs of the MCC in the (v, i)-domain

$$C_1 \frac{dv_{C_1}(t)}{dt} = -\frac{1}{R}(v_{C_1}(t) - v_{C_2}(t)) - f'(\varphi_M(t))v_{C_1}(t) \tag{8.2a}$$

$$C_2 \frac{dv_{C_2}(t)}{dt} = -\frac{1}{R}(v_{C_2}(t) - v_{C_1}(t)) - i_L(t) \tag{8.2b}$$

$$L\frac{di_L(t)}{dt} = v_{C_2}(t) \tag{8.2c}$$

$$\frac{d\varphi_M(t)}{dt} = v_{C_1}(t) \tag{8.2d}$$

for all $t \geq t_0$, with ICs $\mathbf{w}(t_0) = (v_{C_1}(t_0), v_{C_2}(t_0), i_L(t_0), \varphi_M(t_0))^T$. Here, we have taken into account that $\varphi_{C_1}(t;t_0) + \varphi_M(t_0) = \varphi_M(t;t_0) + \varphi_M(t_0) = \varphi_M(t)$. The SEs (8.2) are an IVP for a fourth-order system of ODEs in the state variables $\mathbf{w}(t)$.

For a better understanding of the order reduction of the MCC dynamics in the (φ, q)-domain, with respect to the (v, i)-domain, let us verify that the state space in the (v, i)-domain can be foliated in a continuum of invariant manifolds. Consider the function $Q : \mathbb{R}^4 \to \mathbb{R}$ of the state variables in the (v, i)-domain

$$Q(\mathbf{w}) = f(\varphi_M) + \frac{1}{R}\varphi_M + C_1 v_{C_1} - \frac{L}{R} i_L \tag{8.3}$$

and, for any $Q_0 \in \mathbb{R}$, let

$$\mathcal{M}(Q_0) = \{\mathbf{w} \in \mathbb{R}^4 : Q(\mathbf{w}) = Q_0\} \tag{8.4}$$

which is a three-dimensional manifold in $\mathbb{R}^4$ that coincides with the Q_0-level set of function $Q(\cdot)$. Note that, for any $\mathbf{w} \in \mathbb{R}^4$, we have $\mathbf{w} \in \mathcal{M}(Q(\mathbf{w}))$.

Property 8.1 The state space $\mathbb{R}^4$ of the MCC in the (v, i)-domain can be foliated in ∞^1 3D manifolds $\mathcal{M}(Q_0)$ by varying $Q_0 \in \mathbb{R}$. Manifolds are nonintersecting and they span the whole state space $\mathbb{R}^4$. Each manifold is positively invariant for the dynamics of the MCC in the (v, i)-domain, i.e., if the ICs $\mathbf{w}(t_0) \in \mathcal{M}(Q_0)$, where $Q_0 = Q(\mathbf{w}(t_0))$, then the solution $(v_{C_1}(t), v_{C_2}(t), i_L(t), \varphi_M(t))$ of the IVP (8.2) belongs to $\mathcal{M}(Q_0)$ for any $t \geq t_0$. On each manifold $\mathcal{M}(Q_0)$ the dynamics of the MCC is described in the (φ, q)-domain by the third-order system of ODEs (8.1).

Proof Suppose for purposes of contradiction that there exists $\mathbf{x} \in \mathbb{R}^4$ such that $\mathbf{w} \in \mathcal{M}(Q_1) \cap \mathcal{M}(Q_2)$, with $Q_1 \neq Q_2$. This implies $Q(\mathbf{w}) = Q_1$ and $Q(\mathbf{w}) = Q_2$, which is a contradiction. Then, manifolds are nonintersecting. To see that they cover the whole space $\mathbb{R}^4$ it is enough to recall that, for any $\mathbf{w} \in \mathbb{R}^4$, we have $\mathbf{w} \in \mathcal{M}(Q(\mathbf{w}))$.

Let us now show that each manifold is positively invariant by analyzing the associated circuit of the MCC in the (φ, q)-domain. We have from KqL at the cut-set made of nodes (A) and (B) (see Fig. 8.2)

$$q_M(t;t_0) + q_{C_1}(t;t_0) + q_{C_2}(t;t_0) + q_L(t;t_0) = 0$$

for any $t \geq t_0$. From KqL at node (B)

$$q_R(t;t_0) + q_{C_2}(t;t_0) + q_L(t;t_0) = 0$$

and from KφL at the loop Γ (see again Fig. 8.2)

$$\varphi_M(t; t_0) = \varphi_R(t; t_0) + \varphi_L(t; t_0).$$

These two last equations yield

$$q_L(t; t_0) = \frac{\varphi_M(t; t_0) - \varphi_L(t; t_0)}{R} - q_{C_2}(t; t_0)$$

and substituting in the first equation we obtain

$$q_M(t; t_0) + q_{C_1}(t; t_0) + \frac{\varphi_M(t; t_0) - \varphi_L(t; t_0)}{R} = 0.$$

Since $q_M(t) = f(\varphi_M(t))$, we have

$$f(\varphi_M) + \frac{1}{R}\varphi_M + C_1 v_{C_1} - \frac{L}{R} i_L = f(\varphi_M(t_0)) + \frac{1}{R}\varphi_M(t_0) + C_1 v_{C_1}(t_0) - \frac{L}{R} i_L(t_0)$$

for any $t \geq t_0$. ■

Remark 8.1 The technique here used for proving invariance of $\mathcal{M}(Q_0)$ in Property 8.1 is based on KφLs and KqLs of the associated circuit and, as such, it differs from that given in Chap. 6, that is instead based on algebraic manipulations of SEs in the (φ, q)-domain. This new proof, in addition to being simpler and based on circuit theoretic ideas, has also the advantage of lending itself to an extension to arrays of coupled MCCs (cf. Property 8.2 in Sect. 8.3). Of course, the invariance of manifold $\mathcal{M}(Q_0)$ is equivalent to saying that $Q(\cdot)$ as in (8.3) is an *invariant of motion* for the dynamics of the MCC in the (v, i)-domain.

8.2.1 Complex Dynamics in the Single MCC

To analyze the dynamics of the MCC, it is convenient to recast the SEs (8.1) in a more compact normal form. Let us define parameters (cf. Table 6.1 in Chap. 6)

$$\alpha = \frac{C_1}{C_2}, \quad \beta = \frac{R^2 C_2}{L} \tag{8.5}$$

introduce the normalized time $t \to t/(RC_2)$ and consider the change of variables

$$x(t) = \varphi_{C_1}(t; t_0) + \varphi_M(t_0) = \varphi_M(t) \tag{8.6a}$$

$$y(t) = \varphi_{C_2}(t; t_0) + \varphi_L(t_0) \tag{8.6b}$$

$$z(t) = -Rq_L(t; t_0) + \varphi_L(t_0) - \varphi_M(t_0) + Rq_{C_2}(t_0). \tag{8.6c}$$

The following SEs in adimensional form for $t \geq t_0$ are obtained

$$\frac{dx(t)}{dt} = \alpha(-x(t) + y(t) - n(x(t))) + X_0 \tag{8.7a}$$

$$\frac{dy(t)}{dt} = x(t) - y(t) + z(t) \tag{8.7b}$$

$$\frac{dz(t)}{dt} = -\beta y(t) \tag{8.7c}$$

where we let $n(x(t)) = Rf(x(t))$ and

$$X_0 = \alpha R Q(\mathbf{w}(t_0)) = \alpha(n(\varphi_M(t_0)) + \varphi_M(t_0) + RC_1 v_{C_1}(t_0) - Li_L(t_0)) \tag{8.8}$$

which is a term depending on the ICs for the state variables in the (v, i)-domain. The ICs are $x(t_0) = \varphi_M(t_0)$, $y(t_0) = \varphi_L(t_0) = Li_L(t_0)$, $z(t_0) = \varphi_L(t_0) - \varphi_M(t_0) + Rq_{C_2}(t_0) = Li_L(t_0) - \varphi_M(t_0) + RC_2 v_{C_2}(t_0)$.

We stress that, when the ICs $v_{C_1}(t_0)$, $i_L(t_0)$ and $\varphi_M(t_0)$ are such that $X_0 = 0$, i.e., $Q(\mathbf{w}(t_0)) = 0$, and the nonlinear dynamics of the MCC is on the invariant *zero-manifold* $\mathcal{M}(0)$, the SEs (8.7) formally *coincide with those of the classical Chua's oscillator* (Chap. 4). Therefore, the MCC can undergo *standard bifurcations* on the ***fixed manifold*** $\mathcal{M}(0)$ by varying the circuit parameters α and β. On the other hand, due to the term X_0 at the right-hand side of (8.7), MCC exhibits a rich variety of dynamic behaviors and different coexisting attractors. The dynamic behavior of a nonlinear system (number and stability of EPs and limit cycles, etc.) is indeed known to be heavily dependent also on constant terms in the vector field [33, 34]. We can then envisage an alternative mechanism to induce bifurcations, i.e., bifurcations due to the change of ICs and X_0 for ***fixed circuit parameters*** (bifurcations without parameters).

Bifurcations in the MCC have been investigated in Sect. 6.3 of Chap. 6, where, in particular, the memristor characteristic is chosen as

$$f(\varphi_M) = -\frac{8}{7}\varphi_M + \frac{4}{63}\varphi_M^3 \tag{8.9}$$

which results to be a good smooth approximation of the nonlinear Chua's diode characteristic [35]. For the sake of completeness, and to make the chapter more self-consistent, next we briefly summarize the Hopf and period-doubling bifurcations without parameters leading to a scenario where complex dynamics depending on the choice of ICs and invariant manifold are observed. Some selected numerical simulations of (8.7) by varying Q_0 (and, hence, X_0), for suitable fixed sets of circuit parameters (L, C_1, C_2, R), are reported.

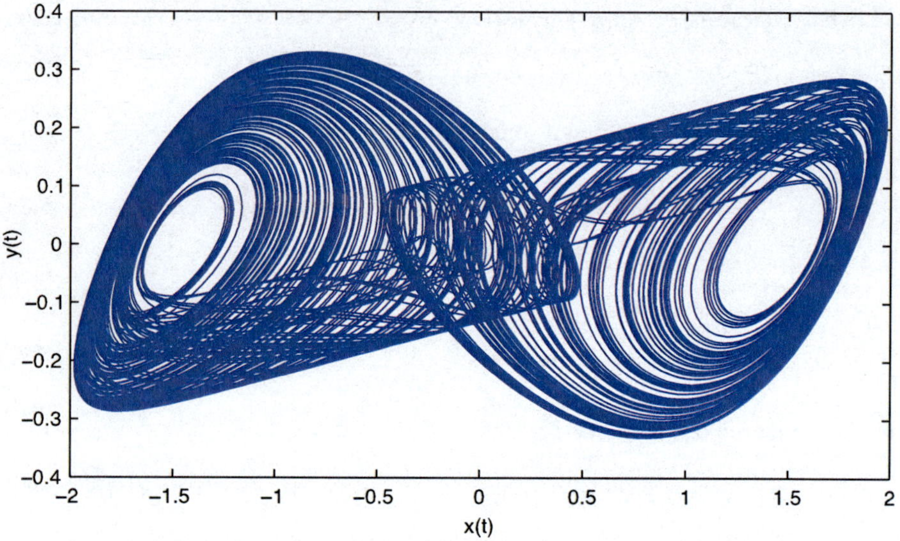

Fig. 8.3 Double-scroll chaotic attractor of the third-order memristor-based oscillator in Fig. 8.2 with $Q_0 = 0$

Figures 8.3 and 8.4 depict simulations of (8.7) with the different sets of ICs $\varphi_M(t_0) \in \{0, 0.25, 0.5\}$, $q_{C_1}(t_0) = \varphi_L(t_0) = 0$, $q_{C_2}(t_0) = 1$ and fixed parameters $C_1 = 1$, $R = 1$, $\alpha = 9.5$, and $\beta = 15$. Accordingly, we have

- $Q_0 = 0$ when $\varphi_M(t_0) = 0$
- $Q_0 = -0.0347$ when $\varphi_M(t_0) = 0.25$
- $Q_0 = -0.0635$ when $\varphi_M(t_0) = 0.5$.

The projection on the (x, y) plane of the chaotic attractor obtained for $Q_0 = 0$ (resp., $Q_0 = -0.0347$) is depicted in Fig. 8.3 (resp., Fig. 8.4). The period-4 attractor obtained for $Q_0 = -0.0635$ is in Fig. 8.5. Clearly, the MCC undergoes a sequence of period-doubling bifurcations originating a change from the periodic attractor to the single-scroll chaotic attractor and subsequently to the double-scroll chaotic attractor. Such bifurcations are induced by changing the ICs (i.e., the constant Q_0) whereas the circuit parameters (L, C_1, C_2, R) are held fixed, i.e., we are dealing with a bifurcation without parameters scenario with coexistence of different attractors for the same set of parameters.

8.3 One-Dimensional Arrays of MCCs

Several categories of biological and physical systems are described as a collection of interacting cells (e.g., neurons, oscillators, etc.) and the emergence of a common complex dynamical behavior, which might differ significantly from those of

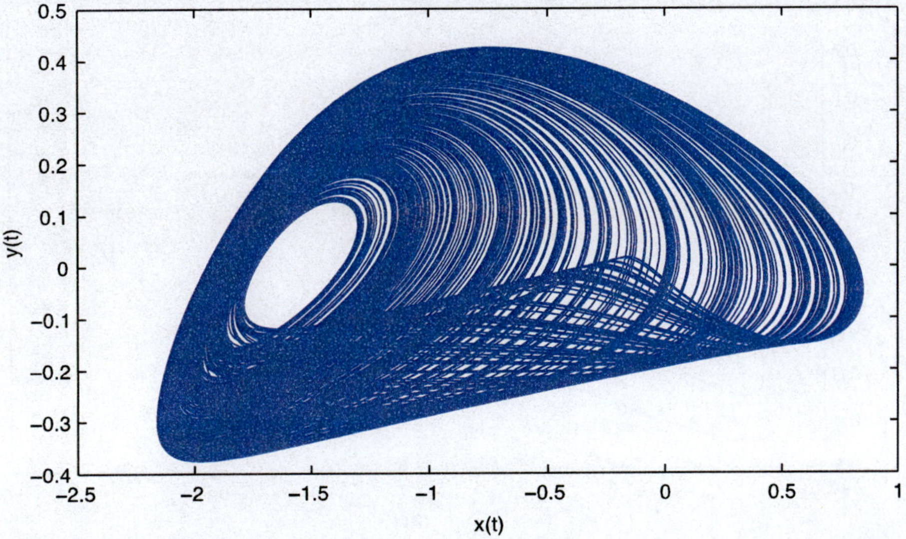

Fig. 8.4 Single-scroll chaotic attractor of the third-order memristor-based oscillator in Fig. 8.2 with $Q_0 = -0.0347$

each individual subsystem, is due to the interactions. Synchronization phenomena represent one of the chief aspects of dynamical processes in nontrivial complex network topologies. Roughly speaking, synchronization can be considered as the "adjustment of rhythms of self-sustained periodic/chaotic oscillators due to their weak interaction (coupling)" and it is thought of as being one of the best way to explore the collective behavior of interconnected networks.

In order to study synchronization phenomena in a 1D array of N diffusively coupled identical MCCs (hereinafter denoted by NMCC), let us consider their description in the flux-charge domain given by the following SEs:

$$\begin{aligned} C_1 \frac{d\varphi_{C_{1,i}}(t;t_0)}{dt} &= -\frac{1}{R}(\varphi_{C_{1,i}}(t;t_0) - \varphi_{C_{2,i}}(t;t_0)) \\ &\quad - f(\varphi_{C_{1,i}}(t;t_0) + \varphi_{M_{0,i}}) + f(\varphi_{M_{0,i}}) + q_{C_{1,i}}(t_0) \\ &\quad + \sum_{k \in \mathcal{N}_i} \frac{1}{R_{ik}}(\varphi_{C_{1,k}}(t;t_0) - \varphi_{C_{1,i}}(t;t_0)) \end{aligned} \quad (8.10a)$$

$$\begin{aligned} C_2 \frac{d\varphi_{C_{2,i}}(t;t_0)}{dt} &= -\frac{1}{R}(\varphi_{C_{2,i}}(t;t_0) - \varphi_{C_{1,i}}(t;t_0)) \\ &\quad - q_{L,i}(t;t_0) + q_{C_{2,i}}(t_0) \end{aligned} \quad (8.10b)$$

$$L \frac{dq_{L,i}(t;t_0)}{dt} = \varphi_{C_{2,i}}(t;t_0) + \varphi_{L,i}(t_0) \quad (8.10c)$$

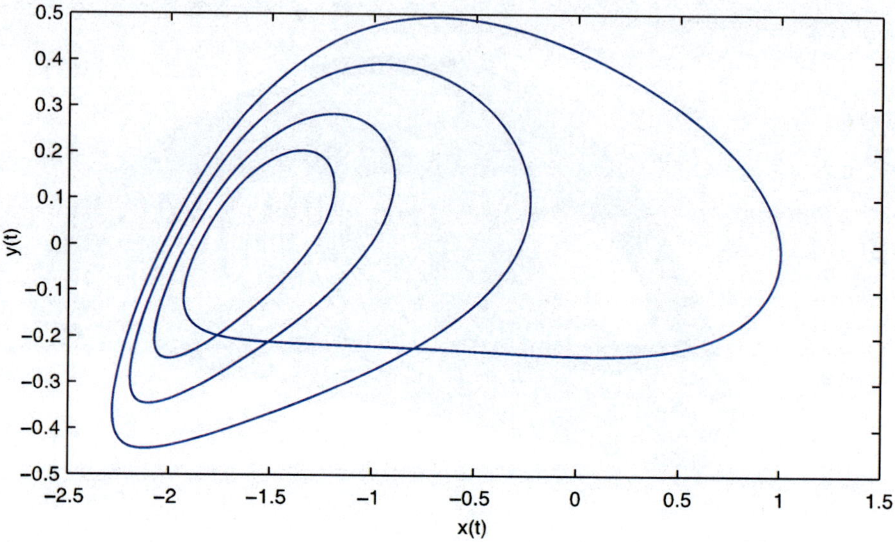

Fig. 8.5 Periodic attractor (cycle of period 4) of the third-order memristor-based oscillator in Fig. 8.2 with $Q_0 = -0.0635$

with ICs $\varphi_{C_{1,i}}(t_0; t_0) = 0$, $\varphi_{C_{2,i}}(t_0; t_0) = 0$ and $q_{L,i}(t_0; t_0) = 0$, where $i = 1, \ldots, N$ identifies the ith MCC, whereas k assumes values in the set $\mathcal{N}_i = \{i - r, \ldots, i, \ldots, i + r\}$ and specifies the $2r + 1$ $(r \geq 1)$ MCCs in the neighbor of the ith MCC and connected to it. Denote by R_{ik} the resistor connecting the ith and k^{th} MCCs through $\varphi_{C_{1,i}}(t; t_0)$ and $\varphi_{C_{1,k}}(t; t_0)$. Hereinafter, we also suppose that the boundary conditions be of Dirichlet type, i.e., we have $\varphi_{C_{1,0}}(t; t_0) = \varphi_{C_{1,N+1}}(t; t_0) = 0$. Such SEs define an IVP for a system of $3N$ coupled first-order ODEs. The structure of the NMCC is depicted in Fig. 8.6.

The SEs in the (v, i)-domain of the NMCC are readily obtained by time differentiation of (8.10)

$$C_1 \frac{dv_{C_{1,i}}(t)}{dt} = -\frac{1}{R}(v_{C_{1,i}}(t) - v_{C_{2,i}}(t)) - f'(\varphi_{M,i}(t))v_{C_{1,i}}(t) + \sum_{k \in \mathcal{N}_i} \frac{1}{R_{ik}}(v_{C_{1,k}}(t) - v_{C_{1,i}}(t)) \tag{8.11a}$$

$$C_2 \frac{dv_{C_{2,i}}(t)}{dt} = -\frac{1}{R}(v_{C_{2,i}}(t) - v_{C_{1,i}}(t)) - i_{L,i}(t) \tag{8.11b}$$

$$L \frac{di_{L,i}(t)}{dt} = v_{C_{2,i}}(t) \tag{8.11c}$$

$$\frac{d\varphi_{M,i}(t)}{dt} = v_{C_{1,i}}(t) \tag{8.11d}$$

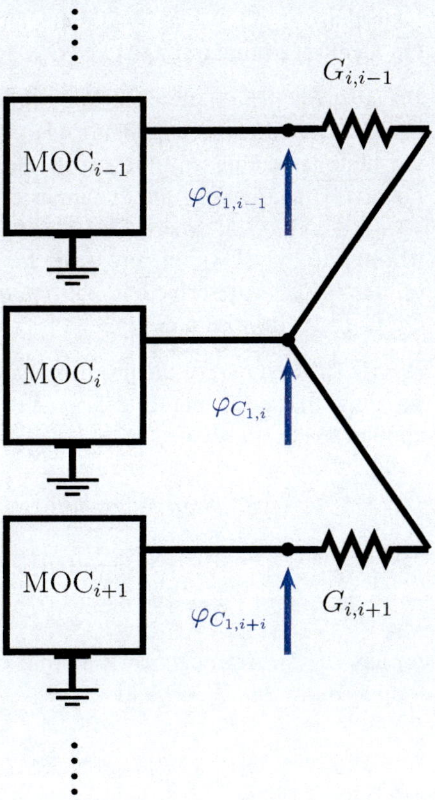

Fig. 8.6 One-dimensional array of N diffusively coupled identical MCCs

with ICs $v_{C_{1,i}}(t_0) = v_{C_{1,i}}(t_0)$, $v_{C_{2,i}}(t_0) = v_{C_{2,i}}(t_0)$, $i_{L,i}(t_0) = i_{L,i}(t_0)$, $\varphi_{M,i}(t_0) = \varphi_{M,i}(t_0)$ and $i = 1, \ldots, N$. The SEs in the (v, i)-domain are an IVP for a system of $4N$ coupled first-order ODEs in the state variables

$$\mathbf{w}_c(t) = (v_{C_{1,1}}, v_{C_{2,1}}, i_{L,1}, \varphi_{M,1}, \ldots, v_{C_{1,N}}, v_{C_{2,N}}, i_{L,N}, \varphi_{M,N})^T \in \mathbb{R}^{4N}.$$

Invariant manifolds in the state space $\mathbb{R}^{4N}$ in the (v, i)-domain of (8.11) can be identified as follows. Define functions $Q_i : \mathbb{R}^{4N} \to \mathbb{R}$, $i = 1, 2, \ldots, N$ as

$$Q_i(\mathbf{w}_c) = f(\varphi_{M,i}) + \frac{1}{R}\varphi_{M,i} + C_1 v_{C_{1,i}} - \frac{L}{R} i_{L,i} - \sum_{k \in \mathcal{N}_i} \frac{1}{R_{ik}}(\varphi_{M,k} - \varphi_{M,i}) \qquad (8.12)$$

and, for any $\mathbf{Q}_0 = (Q_{0,1}, \ldots, Q_{0,N})^T \in \mathbb{R}^N$, let

$$\mathcal{M}_c(\mathbf{Q}_0) = \{\mathbf{w}_c \in \mathbb{R}^{4N} : Q_i(\mathbf{w}_c) = Q_{0,i}, i = 1, 2, \ldots, N\}.$$

Note that $\mathcal{M}_c(\mathbf{Q}_0)$ is a $3N$-dimensional manifold in $\mathbb{R}^{4N}$ that coincides with the $\mathbf{Q}_0$-level set of function $\mathbf{Q}(\cdot) = (Q_1(\cdot), Q_2(\cdot), \ldots, Q_n(\cdot))^T$.

Property 8.2 The state space $\mathbb{R}^{4N}$ of the NMCC in the (v, i)-domain can be foliated in ∞^N $3N$-dimensional manifolds $\mathcal{M}_c(\mathbf{Q}_0)$ by varying $\mathbf{Q}_0 \in \mathbb{R}^N$. Manifolds are nonintersecting and they span the whole state space $\mathbb{R}^{4N}$. Each manifold is positively invariant for the dynamics of NMCC in the (v, i)-domain, i.e., if the ICs $\mathbf{w}(t_0) \in \mathcal{M}_c(\mathbf{Q}_0)$, where $\mathbf{Q}_0 = \mathbf{Q}(\mathbf{w}_c(t_0))$, then the solution of the IVP (8.11) belongs to $\mathcal{M}_c(\mathbf{Q}_0)$ for any $t \geq t_0$. On each manifold $\mathcal{M}_c(\mathbf{Q}_0)$ the dynamics of the MCC is described in the (φ, q)-domain by the reduced-order system of ODEs (8.10), whose order is $3N$.

Proof The first part of the proof is analogous to that of Property 8.1 and is omitted. To show that each manifold is positively invariant, note that we have from KqL applied to the ith MCC

$$q_{M,i}(t;t_0) + q_{C_{1,i}}(t;t_0) + q_{C_{2,i}}(t;t_0) + q_{L,i}(t;t_0)$$
$$-\sum_{k\in\mathcal{N}_i} \frac{1}{R_{ik}}(\varphi_{M,k}(t;t_0) - \varphi_{M,i}(t;t_0)) = 0$$

for any $t \geq t_0$. Arguing as in the proof of Property 8.1, KqL at node Ⓑ and KφL at loop Γ of the ith MCC yield

$$q_{L,i}(t;t_0) = \frac{\varphi_{M,i}(t;t_0) - \varphi_{L,i}(t;t_0)}{R} - q_{C_{2,i}}(t;t_0)$$

and substituting in the first equation we obtain

$$q_{M,i}(t;t_0) + q_{C_{1,i}}(t;t_0) + \frac{\varphi_{M,i}(t;t_0) - \varphi_{L,i}(t;t_0)}{R}$$
$$-\sum_{k\in\mathcal{N}_i} \frac{1}{R_{ik}}(\varphi_{M,k}(t;t_0) - \varphi_{M,i}(t;t_0)) = 0.$$

Then, we obtain

$$f(\varphi_{M,i}(t)) + \frac{1}{R}\varphi_{M,i}(t) + C_1 v_{C_{1,i}}(t)$$
$$-\frac{L}{R} i_{L,i}(t) + \sum_{k\in\mathcal{N}_i} \frac{1}{R_{ik}}(\varphi_{M,k}(t) - \varphi_{M,i}(t))$$
$$= f(\varphi_{M,i}(t_0)) + \frac{1}{R}\varphi_{M,i}(t_0) + C_1 v_{C_{1,i}}(t_0)$$

$$-\frac{L}{R}i_{L,i}(t_0) - \sum_{k\in\mathcal{N}_i} \frac{1}{R_{ik}}(\varphi_{M0,k} - \varphi_{M0,i})$$

for any $t \geq t_0$ and $i = 1, 2, \ldots, N$. ■

Remark 8.2 From the proof of Property 8.2 it is seen that $Q_i(\mathbf{w}_c)$ $(i = 1, 2, \ldots, N)$ provide N invariants of motion for the dynamics of the NMCC in the (v, i)-domain.

8.3.1 Adimensional Normal Form of State Equations in the Flux-Charge Domain

By using parameters defined in (8.5), and the change of variables in (8.6) for each MCC, the adimensional SEs of the whole NMCC is obtained as (for all $i = 1, \ldots, N$)

$$\begin{aligned}\frac{dx_i(t)}{dt} &= \alpha(-x_i(t) + y_i(t) - n(x_i(t))) + X_{0,i}\\ &\quad + \Big(\sum_{k\in\mathcal{N}_i} d_{ik}(x_k(t) - x_i(t))\\ &\quad - \sum_{k\in\mathcal{N}_i} d_{ik}(\varphi_{M,k}(t_0) - \varphi_{M,i}(t_0))\Big) \qquad (8.13a)\end{aligned}$$

$$\frac{dy_i(t)}{dt} = x_i(t) - y_i(t) + z_i(t) \qquad (8.13b)$$

$$\frac{dz_i(t)}{dt} = -\beta y_i(t) \qquad (8.13c)$$

with ICs $x_i(t_0) = \varphi_{M,i}(t_0)$, $y_i(t_0) = \varphi_{L,i}(t_0) = Li_{L,i}(t_0)$, $z_i(t_0) = \varphi_{L,i}(t_0) - \varphi_{M,i}(t_0) + Rq_{C_{2,i}}(t_0) = Li_{L,i}(t_0) - \varphi_{M,i}(t_0) + RC_2v_{C_{2,i}}(t_0)$. We also have $d_{ik} = \alpha R/R_{ik}$ and

$$\begin{aligned}X_{0,i} &= \alpha RQ(v_{C_{1,i}}(t_0), v_{C_{2,i}}(t_0), i_{L,i}(t_0), \varphi_{M,i}(t_0))\\ &= \alpha(n(\varphi_{M,i}(t_0)) + \varphi_{M,i}(t_0) + RC_1v_{C_{1,i}}(t_0) - Li_{L,i}(t_0)) \qquad (8.14)\end{aligned}$$

where $Q(\cdot)$ is as in (8.3).

Let us introduce the parameters

$$X_{c0,i} = X_{0,i} - \sum_{k\in\mathcal{N}_i} d_{ik}(\varphi_{M,k}(t_0) - \varphi_{M,i}(t_0)). \qquad (8.15)$$

When $r = 1$ and $\mathcal{N}_i = \{i-1, i, i+1\}$, the SEs (8.13) can be written in the form ($i = 1, \ldots, N$)

$$\frac{dx_i(t)}{dt} = \alpha(-x_i(t) + y_i(t) - n(x_i(t))) + \sum_{k\in\mathcal{N}_i} d_{ik}(x_k(t) - x_i(t)) + X_{c0,i} \tag{8.16a}$$

$$\frac{dy_i(t)}{dt} = x_i(t) - y_i(t) + z_i(t) \tag{8.16b}$$

$$\frac{dz_i(t)}{dt} = -\beta y_i(t). \tag{8.16c}$$

This shows that *the NMCC is analogous to an array of diffusively coupled Chua's oscillators* [1]. A relevant difference is however due to the additional constants terms $X_{c0,i}$ at the right-hand side of (8.16), depending on the ICs for the state variables in the (v, i)-domain, which need to be carefully taken into account in the investigation of synchronization phenomena.

8.4 Synchronization Phenomena in the NMCC

The analytical results on invariant manifolds for the single MCC and the whole NMCC are fundamental to analyze synchronization phenomena and to exploit the theory of weakly connected oscillatory networks developed in the literature [36]. In particular, in order to demonstrate the role of invariant manifolds on the occurrence of synchronization in the NMCC, let us introduce the synchronization manifold of the whole NMCC in the (φ, q)-domain

$$\mathcal{M}_S = \{(x_1, y_1, z_1, \ldots, x_N, y_N, z_N) \in \mathbb{R}^{3N} : x_1 = \cdots = x_N; \; y_1 = \cdots = y_N; z_1 = \cdots = z_N\} \subset \mathbb{R}^{3N}$$

and define the synchronization errors

$$\mathbf{e}_{ij} = \begin{pmatrix} x_i - x_j \\ y_i - y_j \\ z_i - z_j \end{pmatrix}$$

for $i, j = 1, 2 \ldots, N$.

Manifold $\mathcal{M}_S$ is positively invariant for the dynamics of (8.16) if and only if $\mathbf{w}_c(t) \in \mathcal{M}_S$ implies $d\mathbf{e}_{ij}(t)/dt = 0$, $t \geq t_0$. Since for $\mathbf{w}_c(t) \in \mathcal{M}_S$

$$\frac{d\mathbf{e}_{ij}(t)}{dt} = \begin{pmatrix} X_{c0,i} - X_{c0,j} \\ 0 \\ 0 \end{pmatrix}$$

it follows that $\mathcal{M}_S$ is positively invariant for the dynamics of (8.16) if and only if

$$X_{c0,i} = X_{c0} \tag{8.17}$$

for any $i = 1, 2, \ldots, N$, and this means that the uncoupled MCCs need to be identical (same set of Eqs. (8.7)). Note that (8.17) is equivalent to

$$X_{0,i} - \sum_{k \in \mathcal{N}_i} d_{ik}(\varphi_{M,k}(t_0) - \varphi_{M,i}(t_0)) = X_{c0} \tag{8.18}$$

for any $i = 1, 2, \ldots, N$.

If the NMCC reach the state of complete synchronization (CS), that is

$$\lim_{t \to +\infty} \mathbf{e}_{ij}(t) = 0$$

for any $i, j = 1, 2, \ldots, N$, it can be seen that we have

$$\lim_{t \to +\infty} Q(v_{C_{1,i}}(t), v_{C_{2,i}}(t), i_{L,i}(t), \varphi_{M,i}(t)) = \frac{X_{c0}}{\alpha R}$$

for any $i = 1, 2, \ldots, N$. In other words, under the condition of CS each uncoupled MCC asymptotically evolves on the same manifold defined by

$$\begin{pmatrix} x_i(t) \\ y_i(t) \\ z_i(t) \end{pmatrix} \to \mathcal{M}\left(\frac{X_{c0}}{\alpha R}\right) \tag{8.19}$$

as $t \to +\infty$, for any $i = 1, 2, \ldots, N$, where $\mathcal{M}(\cdot)$ is as in (8.4).

To sketch the proof of this fact, consider that $Q_i(\cdot)$ as in (8.12) are invariants of motion for NMCC in the (v, i)-domain (cf. Property 8.2), hence

$$\alpha R Q_i(\mathbf{w}_c(t)) = \alpha R Q_i(\mathbf{w}_c(t_0)) = X_{c0}$$

for any $t \geq t_0$ and $i = 1, \ldots, N$. From (8.12) we obtain

$$\alpha R\Big[Q(v_{C_{1,i}}(t), v_{C_{2,i}}(t), i_{L,i}(t), \varphi_{M,i}(t)) - \sum_{k \in \mathcal{N}_i} \frac{1}{R_{ik}}(\varphi_{M,k}(t) - \varphi_{M,i}(t))\Big] = X_{c0}$$

for $t \geq t_0$. The result follows by taking the limit as $t \to +\infty$ in the previous expression and considering that $x_i(t) = \varphi_{M,i}(t)$ and that in the case of CS we have $\sum_{k \in \mathcal{N}_i} d_{ik}(\varphi_{M,k}(t) - \varphi_{M,i}(t)) \to 0$ as $t \to +\infty$.

Remark 8.3 Since $x_i(t) = \varphi_M(t)$, it can be immediately checked that, in the case of CS of (8.16) in the (φ, q)-domain, the state variables $\mathbf{w}_c(t)$ in the (v, i)-domain also achieve CS.

8.4.1 Numerical Simulations

Here, numerical simulations of 1D arrays with $N = 4$ identical MCCs are provided. For the sake of simplicity, assume homogeneous ICs on the fluxes of memristors, i.e., $\varphi_{M,i}(t_0) = \varphi_0$, and $v_{C_{1,i}}(t_0) = i_{L,i}(t_0) = 0$ for $i = 1, \ldots, 4$. It follows that condition (8.18) is satisfied and simplifies to (see also (8.14))

$$X_{c0} = X_{0,i} = \alpha(n(\varphi_0) + \varphi_0) \tag{8.20}$$

for $i = 1, \ldots, 4$. Moreover, it is assumed that φ_0 is such that $X_{c0} = 0$, that is, the values $\varphi_0 \in \{-1.5, 0, 1.5\}$ are derived by using the expression (8.9). Under such assumptions, each uncoupled MCC evolves on the zero-manifold $\mathcal{M}(0)$ on which its nonlinear dynamics is the same as that of the classical Chua's oscillator.

The dynamics of NMCC is described by (8.16) with ICs

$$x_i(t_0) = \varphi_0 \tag{8.21a}$$

$$y_i(t_0) = 0 \tag{8.21b}$$

$$z_i(t_0) = -\varphi_0 + \bar{\varphi}_{0,i} \tag{8.21c}$$

where[2] $R = 1$ and $RC_2 v_{C_{2,i}}(t_0) = \bar{\varphi}_{0,i}$. Finally, it is considered a uniform space-invariant weak coupling among the MCCs, i.e., $d_{ik} = d$ for all $i = 1, \ldots, 4$. Then, the dynamics of the whole NMCC takes place on the $3N$-dimensional zero-manifold $\mathcal{M}(\mathbf{0})$ and the same periodic/chaotic attractors and bifurcations occurring in locally connected networks of Chua's oscillators can be observed.

A summary of the nonlinear dynamics in cellular nonlinear networks of Chua's oscillators [26] is described by the following scenario:

- for the single uncoupled Chua's oscillator (or equivalently for the uncoupled MCC on $\mathcal{M}(0)$) with

$$\alpha = 8, \beta = 15$$

[2] See also the normalization values in Table 6.1 of Chap. 6.

there are:

- two stable asymmetric limit cycles A^+ and A^- surrounding the unstable equilibria $(+1.5, 0, -1.5)$ and $(-1.5, 0, +1.5)$, respectively
- one stable symmetric limit cycle S^s surrounding the unstable EP $(0, 0, 0)$
- one unstable symmetric limit cycle S^u

- for a network composed by an arbitrary number (N) of Chua's oscillators and small coupling d it is seen that (see [26, p. 953]):

  (a) there are 2^N stable asymmetric limit cycles with all the phase shifts equal to π;
  (b) the other $2^N \times (2^{N-1} - 1)$ asymmetric limit cycles are unstable
  (c) there is only one stable symmetric limit cycle with all the phase shifts equal to zero
  (d) the other $(2^N - 1)$ symmetric limit cycles are unstable.

Cases (a) and (c), referred to as phase-locking in oscillatory arrays, are reported in the next subsection, whereas the final subsection shows how the four MCC arrays split into periodic and chaotic clusters.

8.4.1.1 Phase-Locking in Oscillatory NMCC

Consider a simple array made of four MCCs ($N = 4$) with uniform space-invariant weak coupling $d = 0.05$ and ICs as in (8.21) with $\varphi_0 = 0$, $\bar{\varphi}_{0,1} = 10$ and $\bar{\varphi}_{0,2} = \bar{\varphi}_{0,3} = \bar{\varphi}_{0,4} = -10$. We have $X_{c0} = 0$, $x_i(t_0) = 0$ and $y_i(t_0) = 0$ for all $i = 1, \ldots, 4$, whereas $-z_1(t_0) = z_2(t_0) = z_3(t_0) = z_4(t_0) = 10$.

The previous analysis permits to conclude that for each MCC parameter $Q_0 = Q(\mathbf{w}(t_0)) = 0$ and then $\mathbf{Q}_0 = 0$ for the NMCC in (8.16). Then, the nonlinear dynamics of the NMCC takes place on the manifold $\mathcal{M}_c(\mathbf{0})$ for $t \geq t_0$. In addition, the necessary condition for CS (in-phase synchronization), i.e., $X_{c0,i}$ are equal for all i (cf. (8.18)), is satisfied. The numerical simulation shown in Fig. 8.7 confirms the analysis and the scenario reported in (c), that is each MCC oscillates according to the symmetric limit cycle (only the variables $x_i(t)$ are reported in Fig. 8.7). In particular, Fig. 8.8 presents the waveforms $x_i(t)$ over the time intervals $[0, 25]$ and $[60, 100]$. Note that, after a transient lasting almost until $t = 60$, an in-phase oscillatory synchronized state emerges due to the weak coupling.

We may change ICs in such a way that each uncoupled MCC would evolve on an asymmetric limit cycle. In particular the ICs in (8.21) with $\varphi_0 = 0$, $\bar{\varphi}_{0,1} = 1.1$, and $\bar{\varphi}_{0,2} = \bar{\varphi}_{0,3} = \bar{\varphi}_{0,4} = -1$ have been used (all the other parameters are the same as in the previous case) in the numerical simulation reported in Fig. 8.9. In such a case, referred to as scenario (a) in the summary above, the weak coupling $d = 0.005$ gives rise to a global periodic oscillation in the NMCC such that there exist π phase shifts among the asymmetric limit cycles of each MCC (note the oscillations of $x_i(t)$, for $i = 1, \ldots, 4$, around -1.5). Such anti-phase synchronized state is shown in the

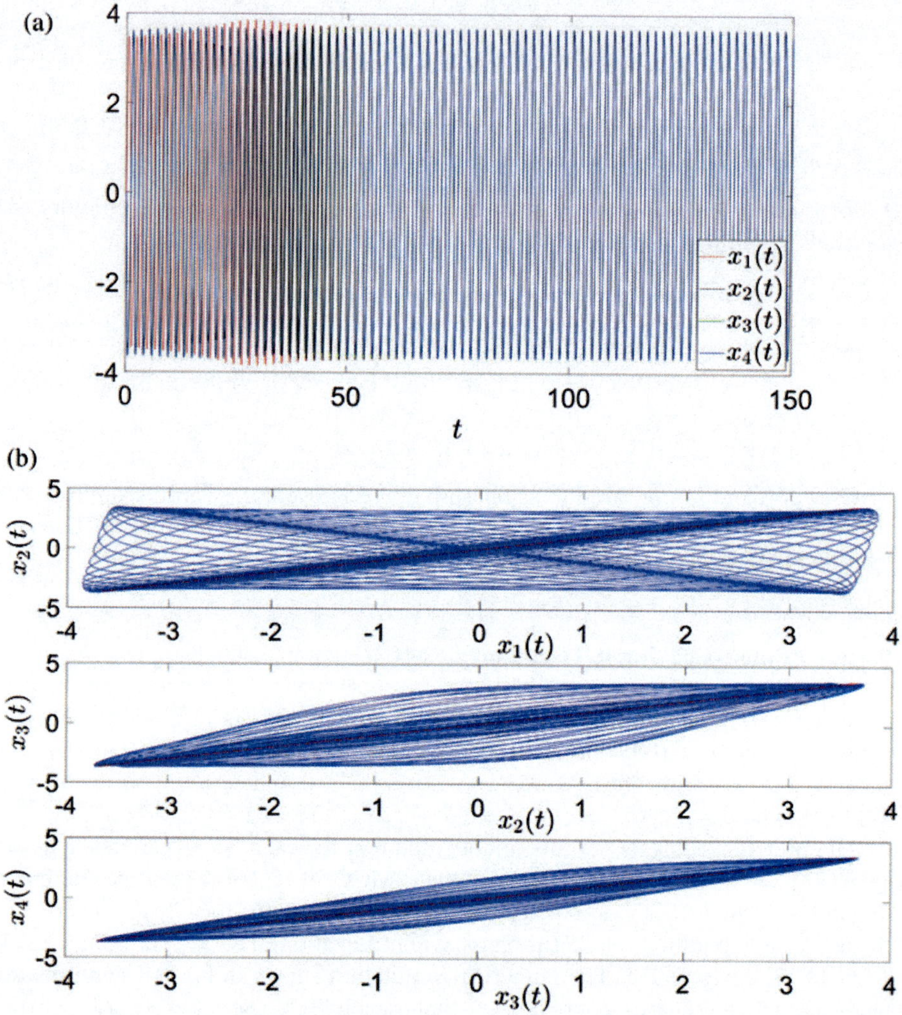

Fig. 8.7 In-phase synchronization among four MCCs with coupling strength $d = 0.05$. Waveforms over the time period [0, 200] are shown in the upper part, whereas the phase shifts equal to zero (among the variables $x_i(t)$ for $t \to \infty$) are represented by the red curves in the bottom part of the figure

bottom part of Fig. 8.9 over a time interval [450, 500], whereas the middle part of Fig. 8.9 presents the initial transient for $t \in [0, 50]$.

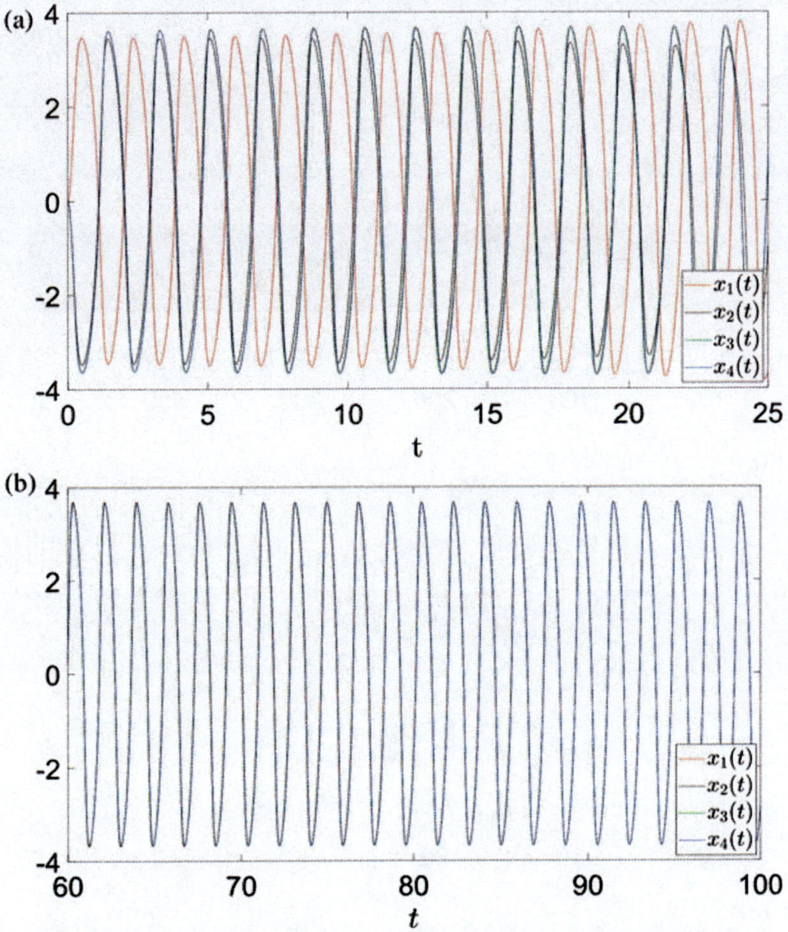

Fig. 8.8 Zoom of the waveforms in Fig. 8.7 over time intervals [0, 25] (upper part) and [60, 100] (bottom part)

8.4.1.2 Periodic and Chaotic Clusters in NMCC

Suppose to keep ICs values as in the previous case (i.e., $\varphi_0 = 0$, $\bar{\varphi}_{0,1} = 1.1$, and $\bar{\varphi}_{0,2} = \bar{\varphi}_{0,3} = \bar{\varphi}_{0,4} = -1$), but $\alpha = 8$ changes over to $\alpha = 9.5$, thus each uncoupled MCC exhibits on the zero-manifold $\mathcal{M}(0)$ a double-scroll chaotic attractor (see Fig. 8.3 in Sect. 8.2.1). In such a case the change of parameter α induces a (standard) bifurcation and the complex dynamic scenario is reported in Figs. 8.10 and 8.11 for $d = 0.005$ and Fig. 8.12 for $d = 0.05$. From Fig. 8.11 it is seen that when $d = 0.05$ the whole NMCC exhibits a chaotic attractor such that the external 1st and 4th MCCs have a periodic synchronized behavior whereas the internal 2nd and

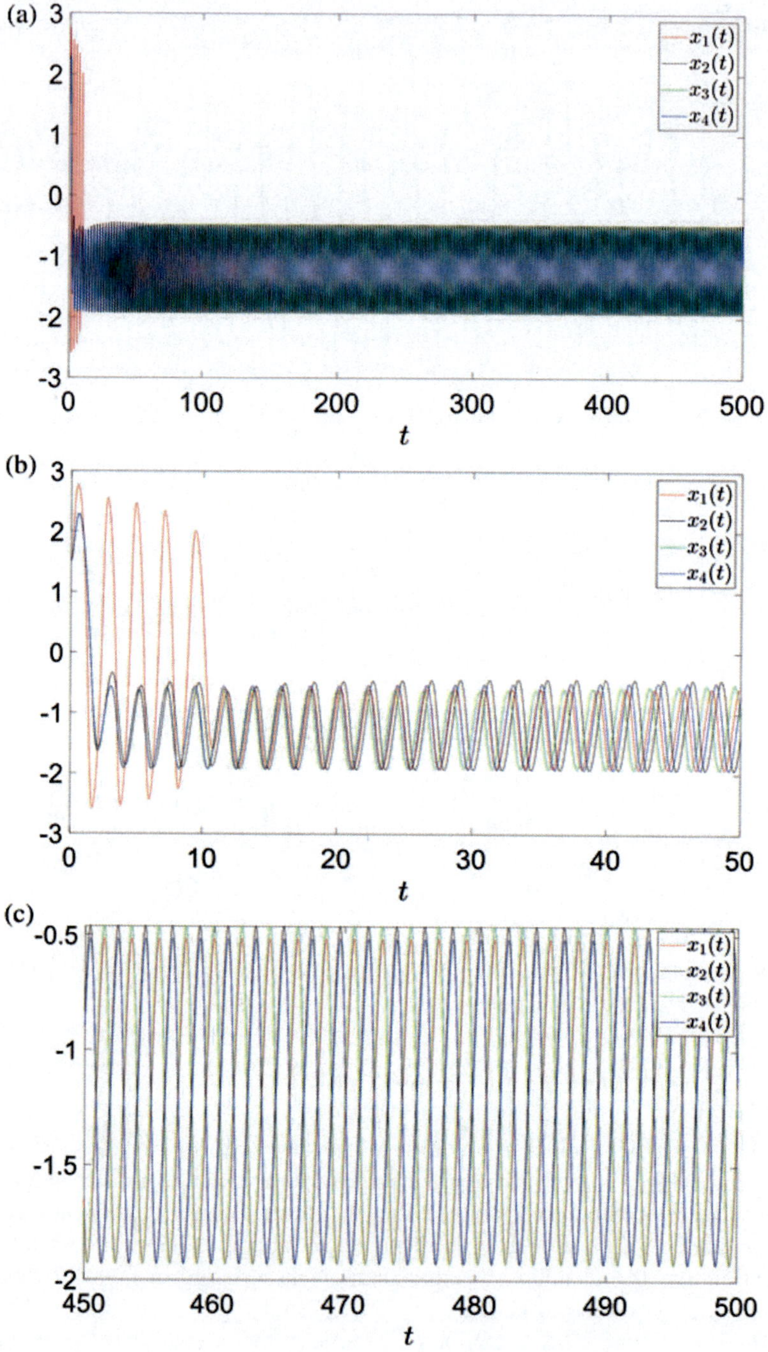

Fig. 8.9 Anti-phase synchronized state in the NMCC such that there exists a π phase shift among the periodic oscillation of each MCC. The middle part is the initial transient for $t \in [0, 50]$. In the bottom part we observe a π phase shift among $x_i(t)$ $(i = 1, \ldots, 4)$ over the time interval $[450, 500]$

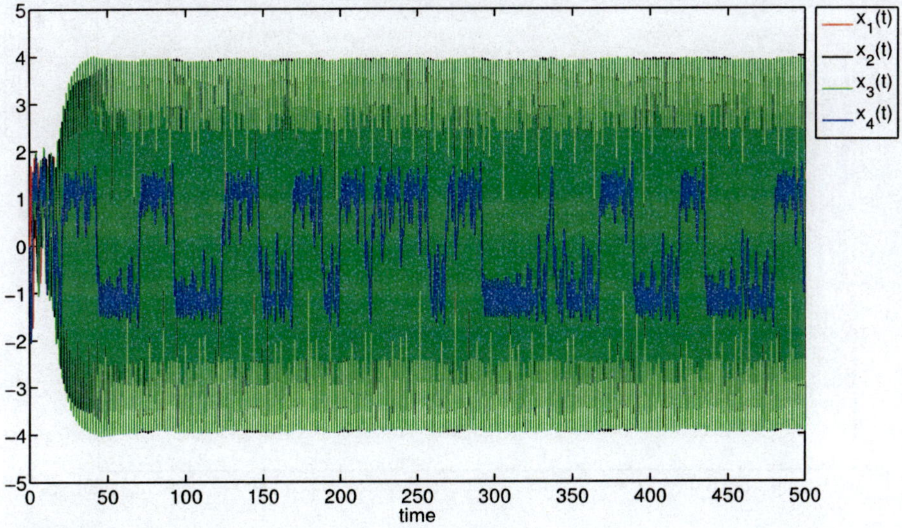

Fig. 8.10 Chaotic behavior in the NMCC with $d = 0.05$

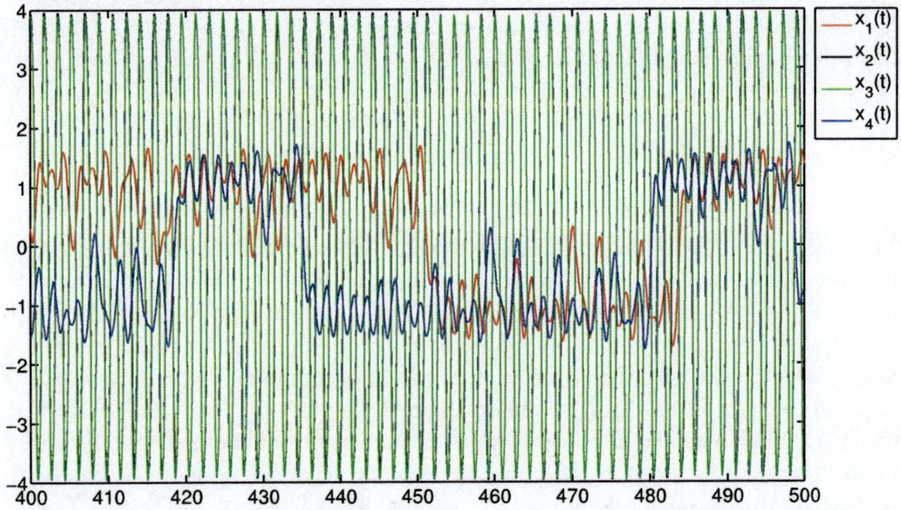

Fig. 8.11 Zoom of the waveforms in Fig. 8.10

3rd MCCs are chaotic. If the coupling strength d is reduced to 0.005 the whole NMCC presents a chaotic behavior (see Fig. 8.12).

Extensive numerical simulations have shown that a wide range of complex dynamical phenomena and bifurcations (standard and without parameters) occur on the zero-manifold of such NMCC. The invariant manifold analysis presented in

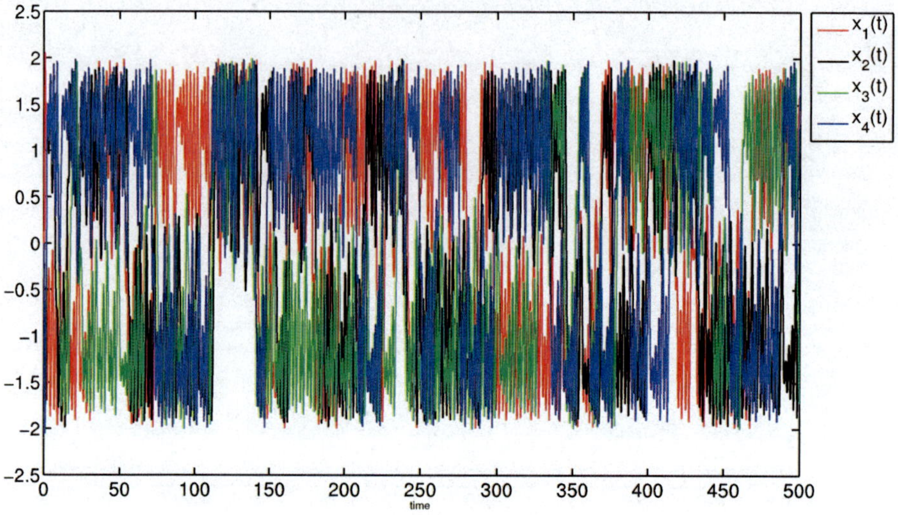

Fig. 8.12 Chaotic behavior in the NMCC with $d = 0.005$

Sects. 8.2 and 8.4 permits to control the manifolds, on which the nonlinear dynamics takes place, via a suitable choice of ICs.

8.5 Discussion

This chapter has analyzed complex dynamics, bifurcations, and synchronization phenomena in 1D arrays of diffusively coupled MCCs. FCAM has been used to describe the NMCC nonlinear dynamics via an order-reduced dynamical circuit, namely, the state space in the (v, i)-domain has been foliated in invariant manifolds where NMCC obeys a reduced-order dynamics, depending on the manifold, that is explicitly known in the (φ, q)-domain. In particular, ICs for the state variables in the (v, i)-domain appear as constant inputs in the vector field defining the dynamics in the (φ, q)-domain on each manifold (see the terms $X_{c0,i}$ in the SEs (8.16) of NMCC in the (φ, q)-domain). The explicit knowledge of the terms $X_{c0,i}$ is crucial for addressing synchronization of NMCC. Indeed, in order to have complete synchronization it is needed that all $X_{c0,i}$ are the same, which can be guaranteed by choosing uniformly distributed ICs for memristor fluxes in the uncoupled MCCs. We stress that this necessary condition for synchronization of NMCC is clearly identifiable via FCAM in the (φ, q)-domain, but it would be very difficult to identify by a standard analysis of NMCC in the (v, i)-domain or a daunting task by means of a numerical approach. In addition to complete synchronization, the chapter has also focused on other types of synchronization, such as anti-phase synchronization between limit cycles and complex chaotic behavior in NMCC. Overall, the results in the chapter demonstrate the effectiveness of FCAM to study in the (φ, q)-domain the global nonlinear dynamics and collective behavior of large arrays of interconnected cells.

References

1. L.O. Chua (Ed.), Special issue on nonlinear waves, patterns and spatio-temporal chaos in dynamic arrays. IEEE Trans. Circuits Syst. I Fundam. Theory Appl. **42**(10), 557–823 (1995)
2. S. Kumar, J.P. Strachan, R. Stanley Williams, Chaotic dynamics in nanoscale NbO_2 Mott memristors for analogue computing. Nature **548**(7667), 318 (2017)
3. A. Pikovsky, M. Rosenblum, J. Kurths, *Synchronization: A Universal Concept in Nonlinear Sciences*, vol. 12 (Cambridge University Press, Cambridge, 2003)
4. A.-L. Barabási, *Linked: The New Science of Networks* (American Association of Physics Teachers, College Park, 2003)
5. G.V. Osipov, V.D. Shalfeev, The evolution of spatio-temporal disorder in a chain of unidirectionally-coupled Chua's circuits. IEEE Trans. Circ. Syst. I Fund. Theory Appl. **42**(10), 687–692 (1995)
6. M.J. Ogorzalek, Z. Galias, A.M. Dabrowski, W.R. Dabrowski, Chaotic waves and spatio-temporal patterns in large arrays of doubly-coupled Chua's circuits. IEEE Trans. Circ. Syst. I Fund. Theory Appl. **42**(10), 706–714 (1995)
7. F. Kavaslar, C. Guzelis, A computer-assisted investigation of a 2-D array of Chua's circuits. IEEE Trans. Circ. Syst. I Fund. Theory Appl. **42**(10), 721–735 (1995)
8. E. Sánchez, M.A. Matías, V. Pérez-Muñuzuri, Chaotic synchronization in small assemblies of driven Chua's circuits. IEEE Trans. Circ. Syst. I Fund. Theory Appl. **47**(5), 644–654 (2000)
9. M. de Magistris, M. di Bernardo, E. Di Tucci, S. Manfredi, Synchronization of networks of non-identical Chua's circuits: analysis and experiments. IEEE Trans. Circ. Syst. I Regul. Pap. **59**(5), 1029–1041 (2012)
10. A. Bogojeska, M. Mirchev, I. Mishkovski, L. Kocarev, Synchronization and consensus in state-dependent networks. IEEE Trans. Circ. Syst. I Regul. Pap. **61**(2), 522–529 (2014)
11. H. Liu, M. Cao, C.W. Wu, Coupling strength allocation for synchronization in complex networks using spectral graph theory. IEEE Trans. Circ. Syst. I Regul. Pap. **61**(5), 1520–1530 (2014)
12. W.K. Wong, W. Zhang, Y. Tang, X. Wu, Stochastic synchronization of complex networks with mixed impulses. IEEE Trans. Circ. Syst. I Regul. Pap. **60**(10), 2657–2667 (2013)
13. D. Kim, M. Jin, P.H. Chang, Control and synchronization of the generalized Lorenz system with mismatched uncertainties using backstepping technique and time-delay estimation. Int. J. Circuit Theory Appl. (2017). https://doi.org/10.1002/cta.2353
14. L.O. Chua, Memristor-The missing circuit element. IEEE Trans. Circuit Theory **18**(5), 507–519 (1971)
15. A. Ascoli, V. Lanza, F. Corinto, R. Tetzlaff, Synchronization conditions in simple memristor neural networks. J. Franklin Inst. **352**(8), 3196–3220 (2015)
16. V. Erokhin, T. Berzina, P. Camorani, A. Smerieri, D. Vavoulis, J. Feng, M.P. Fontana, Material memristive device circuits with synaptic plasticity: learning and memory. BioNanoScience **1**(1–2), 24–30 (2011)
17. Z. Wang, S. Ambrogio, S. Balatti, D. Ielmini, A 2-transistor/1-resistor artificial synapse capable of communication and stochastic learning in neuromorphic systems. Front. Neurosci. **8**, (2014)
18. A. Buscarino, C. Corradino, L. Fortuna, M. Frasca, L.O. Chua, Turing patterns in memristive cellular nonlinear networks. IEEE Trans. Circuits Syst. I Regul. Pap. **63**(8), 1222–1230 (2016)
19. L.V. Gambuzza, A. Buscarino, L. Fortuna, M. Frasca, Memristor-based adaptive coupling for consensus and synchronization. IEEE Trans. Circ. Syst. I Regul. Pap. **62**(4), 1175–1184 (2015)

20. E. Bilotta, F. Chiaravalloti, P. Pantano, Spontaneous synchronization in two mutually coupled memristor-based Chua's circuits: numerical investigations. Math. Problems Eng. **2014**, 594962 (2014)
21. E. Bilotta, F. Chiaravalloti, P. Pantano, Synchronization and waves in a ring of diffusively coupled memristor-based Chua's circuits. Acta Appl. Math. **132**(1), 83–94 (2014)
22. V.-T. Pham, S. Vaidyanathan, C.K. Volos, S. Jafari, N.V. Kuznetsov, T.M. Hoang, A novel memristive time-delay chaotic system without equilibrium points. Eur. Phys. J. Spec. Top **225**(1), 127–136 (2016)
23. J. Ma, F. Wu, G. Ren, J. Tang, A class of initials-dependent dynamical systems. Appl. Math. Comput. **298**, 65–76 (2017)
24. Q. Xu, Y. Lin, B. Bao, M. Chen, Multiple attractors in a non-ideal active voltage-controlled memristor based Chua's circuit. Chaos Solitons Fractals **83**, 186–200 (2016)
25. B. Bao, T. Jiang, G. Wang, P. Jin, H. Bao, M. Chen, Two-memristor-based Chua's hyperchaotic circuit with plane equilibrium and its extreme multistability. Nonlinear Dyn. 1–15 (2017)
26. M. Gilli, F. Corinto, P. Checco, Periodic oscillations and bifurcations in cellular nonlinear networks. IEEE Trans. Circ. Syst. I Regul. Pap. **51**(5), 948–962 (2004)
27. W. Lu, T. Chen, New approach to synchronization analysis of linearly coupled ordinary differential systems. Phys. D Nonlinear Phenom. **213**(2), 214–230 (2006)
28. C.W. Wu, L.O. Chua, Synchronization in an array of linearly coupled dynamical systems. IEEE Trans. Circuits Syst. I Fund. Theory Appl. **42**(8), 430–447 (1995)
29. M. Itoh, L.O. Chua, Memristor oscillators. Int. J. Bifurc. Chaos **18**(11), 3183–3206 (2008)
30. B.-C. Bao, Q. Xu, H. Bao, M. Chen, Extreme multistability in a memristive circuit. Electron. Lett. **52**(12), 1008–1010 (2016)
31. B. Bao, Z. Ma, J. Xu, Z. Liu, Q. Xu, A simple memristor chaotic circuit with complex dynamics. Int. J. Bifurc. Chaos **21**(9), 2629–2645 (2011)
32. M.E. Yalcin, Dynamic behavior of 1-D array of the memristively-coupled Chua's circuits, in *2013 8th International Conference on Electrical and Electronics Engineering (ELECO)* (IEEE, Piscataway, 2013), pp. 13–16
33. L. Ponta, V. Lanza, M. Bonnin, F. Corinto, Emerging dynamics in neuronal networks of diffusively coupled hard oscillators. Neural Netw. **24**(5), 466–475 (2011)
34. V. Lanza, L. Ponta, M. Bonnin, F. Corinto, Multiple attractors and bifurcations in hard oscillators driven by constant inputs. Int. J. Bifurc. Chaos **22**(11), 1250267 (2012)
35. R. Genesio, A. Tesi, Distortion control of chaotic systems: the Chua's circuit. J. Circuits Syst. Comput. **3**(1), 151–171 (1993)
36. F.C. Hoppensteadt, E.M. Izhikevich, *Weakly Connected Neural Networks*, vol. 126 (Springer, Berlin, 2012)

Chapter 9
Memristor Cellular Neural Networks Computing in the Flux-charge Domain

9.1 Introduction

A memristor is a nonlinear device obeying Ohm's law but, unlike a resistor, the memristor resistance, also called memristance, depends upon the history of the voltage applied or the current flowing through it. A memristor is then both a *nonlinear and a memory element* in the (v, i)-domain. Another unique property is *nonvolatility*, namely, when current (or voltage) is turned off, the memristor can keep in memory the final value of charge, flux, or memristance, thereafter (see Chap. 2).

It is widely believed that memristors are potentially useful and will play a major role in the design of neural networks (NNs), smart computers, and future brain-like machines for efficient applications in edge computing and the Internet of Things (IoT) era [1–7]. Memristors are indeed expected to provide various advantages, such as scalability, small on-chip area, low power dissipation in the synapse implementation, efficiency, and better adaptation capability with respect to their CMOS counterparts.

So far the literature has mainly investigated the use of memristors to implement adaptive synapses in neuromorphic architectures [2, 3]. In these applications, use is made of the fine-resolution programming of the memristance, tuned by the input amplitude, pulsewidth, and frequency, in memristor acting as a nonvolatile memory. Memristors are subject to low voltages during their operation as analog circuit elements and (relatively) high voltages to program their memristance, i.e., memristors are exploited as pulse-programmable resistances.

Goal of the chapter is to take a different viewpoint and exploit the nonlinear dynamic features in the (v, i)-domain and nonvolatility of memristors to implement a class of NNs that features some potential advantages over the standard cellular neural networks (SCNNs) proposed by Chua and Yang [8]. We refer to the networks as memristors SCNNs ($\mathcal{M}$-SCNNs). In a $\mathcal{M}$-SCNN, nonlinear memristors are used within the cells in place of linear (memoryless) resistors of SCNNs. Via a suitable

F. Corinto et al., *Nonlinear Circuits and Systems with Memristors*,
https://doi.org/10.1007/978-3-030-55651-8_9

design procedure based on FCAM, and the use of a prototypical memristor as that proposed by HP (cf. Chap. 2), we obtain an *$\mathcal{M}$-SCNN model that is the analogous in the (φ, q)-domain of that describing the dynamics in the (v, i)-domain of a SCNN*. However, due to the use of memristors during the analog computation, $\mathcal{M}$-SCNNs display peculiar and basically different properties with respect to SCNNs, as remarked next.

- One salient feature is that the analog processing of $\mathcal{M}$-SCNNs takes place in the *(φ, q)-domain*, instead of the typical *(v, i)-domain*, as it happens for SCNNs. In the case of charge-controlled memristors, the inputs are provided via the initial values of memristor charges $q_{M_i}(0)$, the processing is accomplished during the time evolutions of $q_{M_i}(t)$, and the result of processing, for convergent $\mathcal{M}$-SCNNs, is given by the asymptotic values of charges $q_{M_i}(\infty)$ (or asymptotic values of memristances).
- There are *potential advantages in terms of power consumption for NNs operating in the (φ, q)-domain*. Indeed, when a steady state is reached, i.e., the memristor charges reach a constant value, the memristor currents and voltages, as well as the capacitors and all other voltages and currents in the $\mathcal{M}$-SCNNs, vanish. Said another way, *at a steady state a $\mathcal{M}$-SCNN turns off* and so the dissipated power is null. Yet, in steady state the memristors act as *nonvolatile devices keeping in memory the processing result*, i.e., the asymptotic values of charges $q_{M_i}(\infty)$ (or the corresponding memristances). We stress that this is different from SCNNs, where voltages, currents, and power do not vanish when a steady state is reached, and batteries are needed to hold in memory the processing result.
- In a $\mathcal{M}$-SCNN the role played by memristors is twofold. They participate in the nonlinear dynamics used for real-time signal processing and they also store the result of computation in the asymptotic values of the charge. Namely, in a $\mathcal{M}$-SCNN, *processing and storing of information are at the same physical location*, according to the principle of in-memory computing. This is a potential advantage also with respect to traditional Von Neumann computing machines, where one main bottleneck is due to the fact that processing and storing are at different physical locations (e.g., the CPU and the RAM).

Since $\mathcal{M}$-SCNNs are analogous from a mathematical viewpoint to SCNNs, it is *possible to exploit the bulk of results already available in the literature for studying the dynamics* of SCNNs in the (φ, q)-domain and to design them in order to accomplish a large variety of signal processing tasks [9].

The chapter also gives a foundation to the nonlinear dynamics in the (φ, q)-domain of the $\mathcal{M}$-SCNN model and addresses convergence of solutions in the case of symmetric interconnections between cells. Applications to the solution of some simple image processing tasks in real time are also discussed to confirm that the introduced model has processing capabilities analogous to those of a SCNN.

9.2 Memristor Neural Network Model

The equations describing in the (v, i)-domain the dynamics of the SCNN model [8] can be written in matrix-vector form as follows:

$$C\frac{d\mathbf{v}(t)}{dt} = -\frac{\mathbf{v}(t)}{R} + \mathbf{G}\mathbf{S}(\mathbf{v}(t)) + \mathbf{I} \tag{9.1}$$

where C and R are the cell capacitor and resistor, respectively, $\mathbf{v}(t) \in \mathbb{R}^n$ is the vector of capacitor voltages, $\mathbf{G} \in \mathbb{R}^{n\times n}$ is the interconnection matrix, and $\mathbf{I} \in \mathbb{R}^n$ is a biasing input. Moreover, $\mathbf{S}(\mathbf{v}(t)) = (s(v_1(t)), s(v_2(t)), \ldots, s(v_n(t)))^T : \mathbb{R}^n \to \mathbb{R}^n$, where

$$s(\rho) = \frac{1}{2}(|\rho + 1| - |\rho - 1|) \tag{9.2}$$

is the typical piecewise linear function with unity gain in the linear region and flat saturation levels ± 1 shown in Fig. 9.1.

Our goal is to design, via FCAM, a NN with memristors whose dynamics in the (φ, q)-domain is described by a set of differential equations analogous to those in (9.1) describing a SCNN in the (v, i)-domain. To this end, first we tackle the problem of synthesizing in the (φ, q)-domain a saturation nonlinearity that closely approximates that of a SCNN (Sect. 9.2.1). Then, we address the design of the interconnecting structure so that the network implements in the (φ, q)-domain the multiplicative term given by the interconnection matrix times the saturation nonlinearity (Sect. 9.2.2).

Recall that a charge-controlled memristor $\varphi_M(t) = h(q_M(t))$ has a CR in the (φ, q)-domain in terms of incremental flux and charge

$$\varphi_M(t; t_0) = -h(q_{M0}) + h(q_M(t; t_0) + q_{M0}) \doteq h^s(q_M(t; t_0); q_{M0}) \tag{9.3}$$

for $t \geq t_0$, where $q_{M0} = q_M(t_0)$ is the initial memristor charge. The equivalent circuit in the (φ, q)-domain is in Fig. 9.2, where the initial condition at t_0 for the

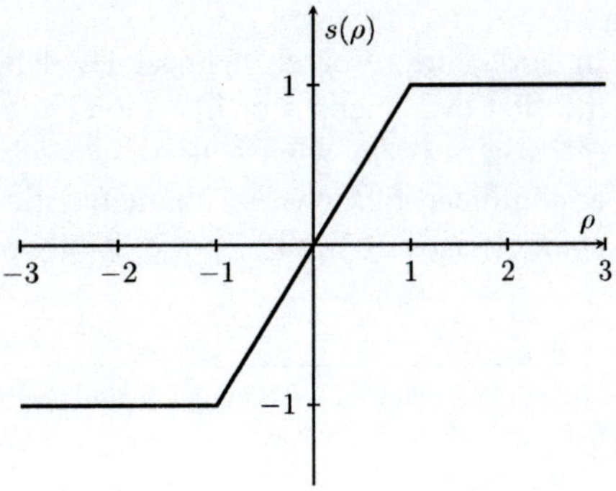

Fig. 9.1 Nonlinearity $s(\cdot)$ of a SCNN

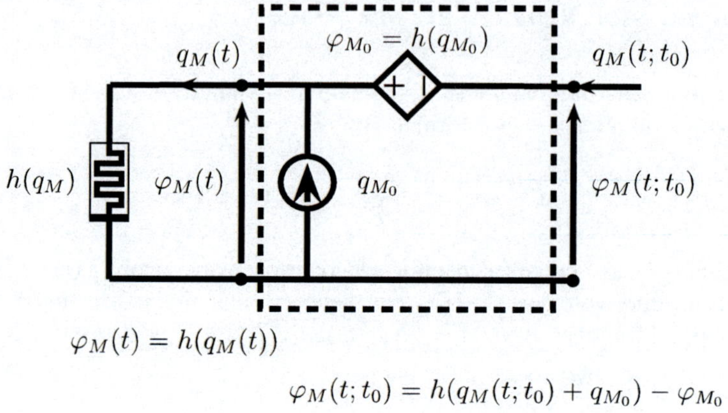

Fig. 9.2 Equivalent circuit in the (φ, q)-domain of a charge-controlled memristor

state variable in the (v, i)-domain, i.e., q_{M_0}, explicitly appears as a constant input. We stress that a memristor has an algebraic CR in the (φ, q)-domain given by a nonlinear relation between $q_M(t; t_0)$ and $\varphi_M(t; t_0)$.[1] It is known that the initial condition q_{M_0} has a remarkable effect on the nonlinear memristor characteristic $h^s(q_M(t; t_0); q_{M_0})$ which results to be shifted with respect to the characteristic $\varphi_M(t) = h(q_M(t))$ (see also Sect. 5.7 in Chap. 5). This initial condition dependent property is clearly highlighted by FCAM and needs to be carefully accounted for in the NN design (cf. Sect. 9.2.1).

Finally, we recall that for an ideal operational amplifier obeying $v_1(t) = 0$, $i_1(t) = 0$ the CRs in the (φ, q)-domain are given as (see Fig. 9.3)

$$\varphi_1(t; t_0) = 0; \quad q_1(t; t_0) = 0 \tag{9.4}$$

for $t \geq t_0$.

9.2.1 HP Memristors in Antiparallel

In 2008, the research team at HP laboratories guided by S. Williams developed the first ever memristor in nanotechnology based on a $Pt/TiO_2/Pt$ device [10] (cf. Chap. 2). In what follows we discuss how we can obtain a memristor with a nonlinear flux-charge characteristic approximating the ideal piecewise linear characteristic $s(\cdot)$ in (9.2) of a SCNN by using two HP memristors in antiparallel.

[1] Recall that in the (v, i)-domain a memristor has instead a CR in differential form. For example, in the (v, i)-domain a charge-controlled memristor obeys the relations $v_M(t) = h'(q_M(t))i_M(t)$ and $dq_M(t)/dt = i_M(t)$.

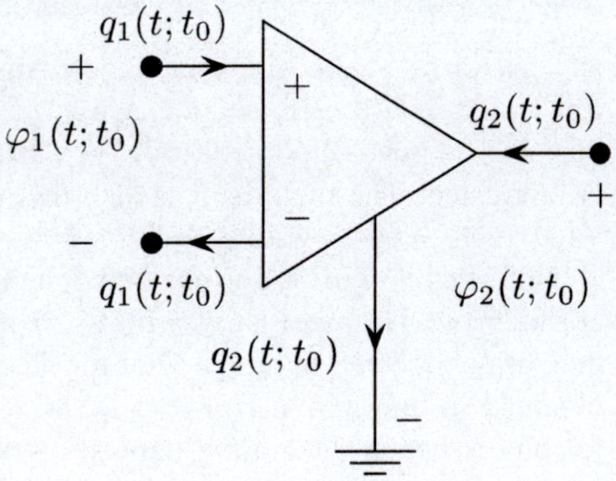

Fig. 9.3 Equivalent circuit in the (φ, q)-domain of an ideal operational amplifier

Let us consider the HP memristor model

$$v_M(t) = [R_{\text{on}}x(t) + R_{\text{off}}(1 - x(t))]i_M(t)$$
$$\frac{dx(t)}{dt} = \alpha i_M(t)F(x(t), p) \tag{9.5}$$

where α is a quantity with dimension Coulomb^{-1}, $x(t) \in [0, 1]$ is the dimensionless length of the conductive layer, $R_{\text{on}}x(t) + R_{\text{off}}(1 - x(t))$ is the device memristance in Ohm, taking values between R_{on} (when $x(t) = 1$ and the memristor is in its fully conductive state) and R_{off} (when $x(t) = 0$ and the memristor is in its fully insulating state). We suppose in the chapter that $R_{\text{off}}/R_{\text{on}} \gg 1$.[2] Function $F(x(t), p)$ is a window generally used to enforce state constraints and account for nonlinearity of ion transport (Chap. 2). More specifically, $F(x(t), p)$ satisfies $F(0, p) = F(1, p) = 0$ and $F(x(t), p) > 0$ for any $x(t) \in (0, 1)$, thus guaranteeing zero speed at boundary values and ensuring that if $x(t_0) \in (0, 1)$, then $x(t) \in (0, 1)$, $t \geq t_0$.

Henceforth we consider the classic Joglekar window [11]

$$F(x, p) = 1 - (2x - 1)^{2p} \tag{9.6}$$

with $p = 1$, but similar results would be obtained for other values of p and with other typical windows as those proposed in [12].

With these windows the HP memristor (9.5) is equivalent to an ideal memristor,[3] which is both charge- and flux-controlled, and whose characteristic can be obtained by the procedure described for instance in [13, 14]. It is shown in Appendix that in the case of the Joglekar window with $p = 1$ the equivalent ideal memristor has a flux-charge relation $\varphi_M(t) = h(q_M(t)) : \mathbb{R} \to \mathbb{R}$ with the following

[2]Some values reported in [10] are $R_{\text{off}}/R_{\text{on}} = 160$ and $R_{\text{off}}/R_{\text{on}} = 380$.

[3]Actually, as discussed in Chap. 2, it is a sibling of an ideal memristor.

properties. Function h is *analytic* in $\mathbb{R}$.[4] We have $h(0) = 0$, $h''(q_M(t)) < 0$ for any $q_M(t) \in \mathbb{R}$, $h(\cdot)$ is strictly increasing ($h'(q_M(t)) > 0$ for any $q_M(t) \in \mathbb{R}$), $\lim_{q_M(t)\to-\infty} h'(q_M(t)) = R_{\text{off}}$, $\lim_{q_M(t)\to+\infty} h'(q_M(t)) = R_{\text{on}}$. Figure 9.4 shows that $h(\cdot)$ depends upon $x(-\infty)$, i.e., the initial value of x when the memristor is fabricated. The memristor is also flux-controlled with a characteristic $q_M(t) = f(\varphi_M(t)) : \mathbb{R} \to \mathbb{R}$ with $f = h^{-1}$.

Note that h and f are not odd functions. To obtain an odd characteristic we consider two HP memristors with CR $q_{M_j}(t) = f(\varphi_{M_j}(t))$, $j = 1, 2$, and suppose they are connected in antiparallel for all t (Fig. 9.5). For the two-terminal element obtained in this way $v_M(t) = v_{M_1}(t) = -v_{M_2}(t)$ and $i_M(t) = i_{M_1}(t) - i_{M_2}(t)$ for any t, hence integrating between $-\infty$ and t, $q_M(t) = q_{M_1}(t) - q_{M_2}(t)$ and $\varphi_M(t) = \varphi_{M_1}(t) = -\varphi_{M_2}(t)$. Then

$$q_M(t) = f(\varphi_M(t)) - f(-\varphi_M(t)) \doteq f_{\text{ap}}(\varphi_M(t)).$$

This shows that the two-terminal element is equivalent to a flux-controlled memristor with an odd CR $f_{\text{ap}} : \mathbb{R} \to \mathbb{R}$. It is also charge-controlled with relation $h_{\text{ap}} = f_{\text{ap}}^{-1}$.

If we consider an HP memristor with a Joglekar window and $p = 1$, then h_{ap} is odd, $h_{\text{ap}}(0) = 0$, h_{ap} is analytic in $\mathbb{R}$, strictly increasing ($h'_{\text{ap}}(q_M(t)) > 0$) and $h''_{\text{ap}}(q_M(t)) < 0$ for $q_M(t) > 0$, $h''_{\text{ap}}(q_M(t)) > 0$ for $q_M(t) < 0$. We have $\lim_{|q_M(t)|\to\infty} h'_{\text{ap}}(q_M(t)) = R_{\text{on}} \| R_{\text{off}} \simeq R_{\text{on}}$. As shown in Fig. 9.6, h_{ap} depends upon $x(-\infty)$. In particular, if $x(-\infty) \ll 1$, the maximum slope $h'_{\text{ap}}(0) \simeq R_{\text{off}}/2$.

9.2.1.1 An Approximating Memristor Characteristic

Next we show that under the assumption $x(-\infty) \ll 1$ we can perform a suitable change of variables in order that the nonlinearity $h_{\text{ap}}(\cdot)$ is brought back to a nonlinearity closely approximating the nonlinearity $s(\cdot)$ of a SCNN for not too large values of the charge.[5]

Suppose to choose $R_{\text{on}} = 100\,\Omega$, $R_{\text{off}} = 16\,\text{k}\Omega$, $\alpha = 10^4\,\text{C}^{-1}$, as in the HP memristor model in [10], and also let $x(-\infty) = 0.001$. Consider the change of variable $q_M(t) = k_q q(t)$ and the nonlinearity $\hat{\varphi} : \mathbb{R} \to \mathbb{R}$ given by

$$\hat{\varphi}(q(t)) = \frac{2h_{\text{ap}}(k_q q(t))}{R_{\text{off}} k_q} \tag{9.7}$$

where parameter $k_q > 0$.

[4] This means that $h(\cdot)$ has derivatives of all orders and it is the sum of its Taylor series in some neighborhood of any $q_M \in \mathbb{R}$.

[5] For simplicity we refer to a set of values for parameters R_{on}, R_{off} as those in [10]. However, similar conclusions hold for other sets of parameters provided the standing assumption $R_{\text{on}} \ll R_{\text{off}}$ is satisfied.

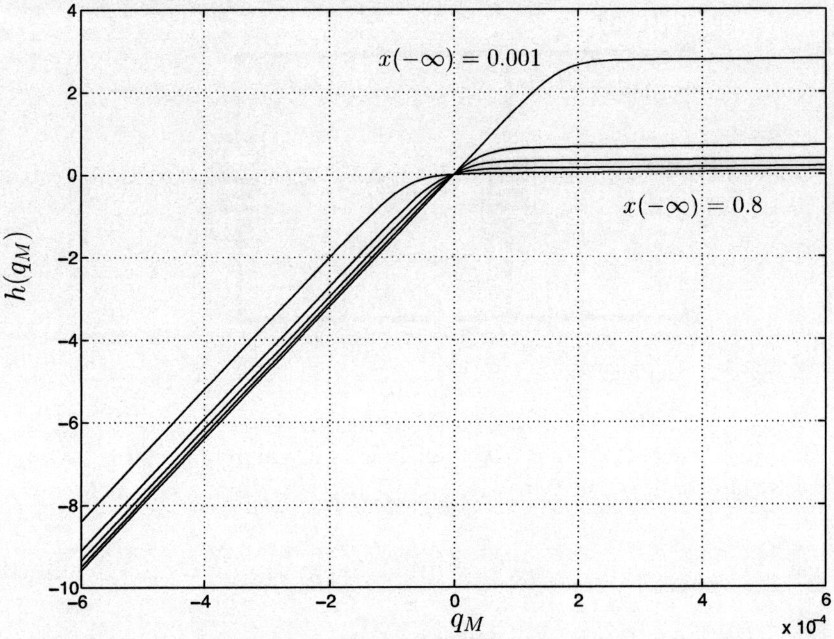

(a)

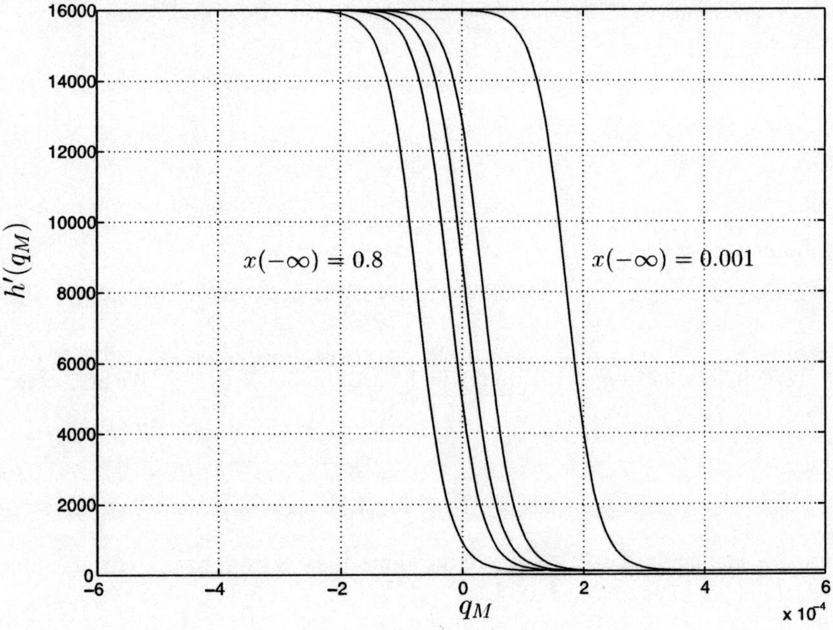

(b)

Fig. 9.4 (**a**) Flux-charge characteristics $h(\cdot)$ and (**b**) memristance $h'(\cdot)$ of an HP memristor modeled with Joglekar window with $p = 1$ for different choices of initial values $x(-\infty) \in \{0.001, 0.05, 0.3, 0.55, 0.8\}$

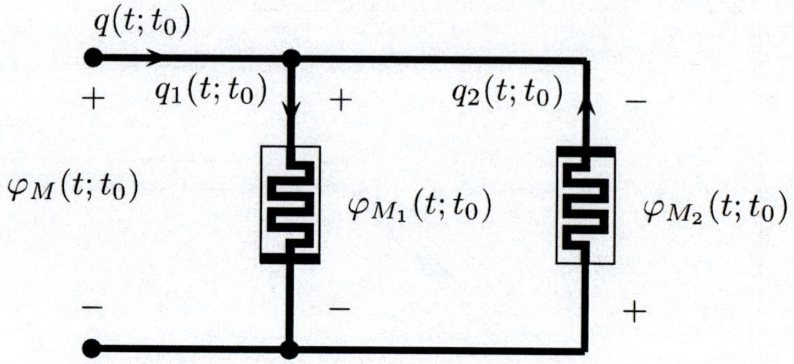

Fig. 9.5 Memristors in antiparallel

Clearly, $\hat{\varphi}(\cdot)$ is odd, $\hat{\varphi}(0) = 0$, $\hat{\varphi}$ is analytic in $\mathbb{R}$, strictly increasing ($\hat{\varphi}'(q(t)) > 0$) and we have $\hat{\varphi}''(q(t)) < 0$ for $q(t) > 0$, $\hat{\varphi}''(q(t)) > 0$ for $q(t) < 0$. Moreover,

$$\lim_{|q(t)|\to\infty} \hat{\varphi}'(q(t)) \simeq \frac{2R_{\rm on}}{R_{\rm off}}. \tag{9.8}$$

Given a point $\bar{q}_M > 1$, if k_q is such that the following condition is satisfied:

$$\frac{2\hat{\varphi}(k_q\bar{q}_M)}{R_{\rm off}k_q} = 1$$

then the maximum slope of $\hat{\varphi}(\cdot)$ is

$$\hat{\varphi}'(0) = 1 \tag{9.9}$$

and moreover we have

$$\hat{\varphi}(\bar{q}_M) = 1, \hat{\varphi}(-\bar{q}_M) = -1. \tag{9.10}$$

A convenient choice is $\bar{q}_M = 1.6$, in which case $k_q = 3.48 \cdot 10^{-4}$. We also note that for $|q(t)| > \bar{q}_M$ we have

$$\hat{\varphi}'(q(t)) \leq \hat{\varphi}'(\bar{q}_M) = \frac{2R_{\rm on}}{R_{\rm off}}(1 + \Delta R) \tag{9.11}$$

where ΔR is the relative difference of the slope at $\bar{q}_M$ with respect to the asymptotic slope. Here, we have $\Delta R = 0.034$.

It is seen from Fig. 9.7 that the obtained nonlinearity $\hat{\varphi}(\cdot)$ is a good approximation for $|q|$ not too large of the nonlinearity $s(\cdot)$ of a SCNN.

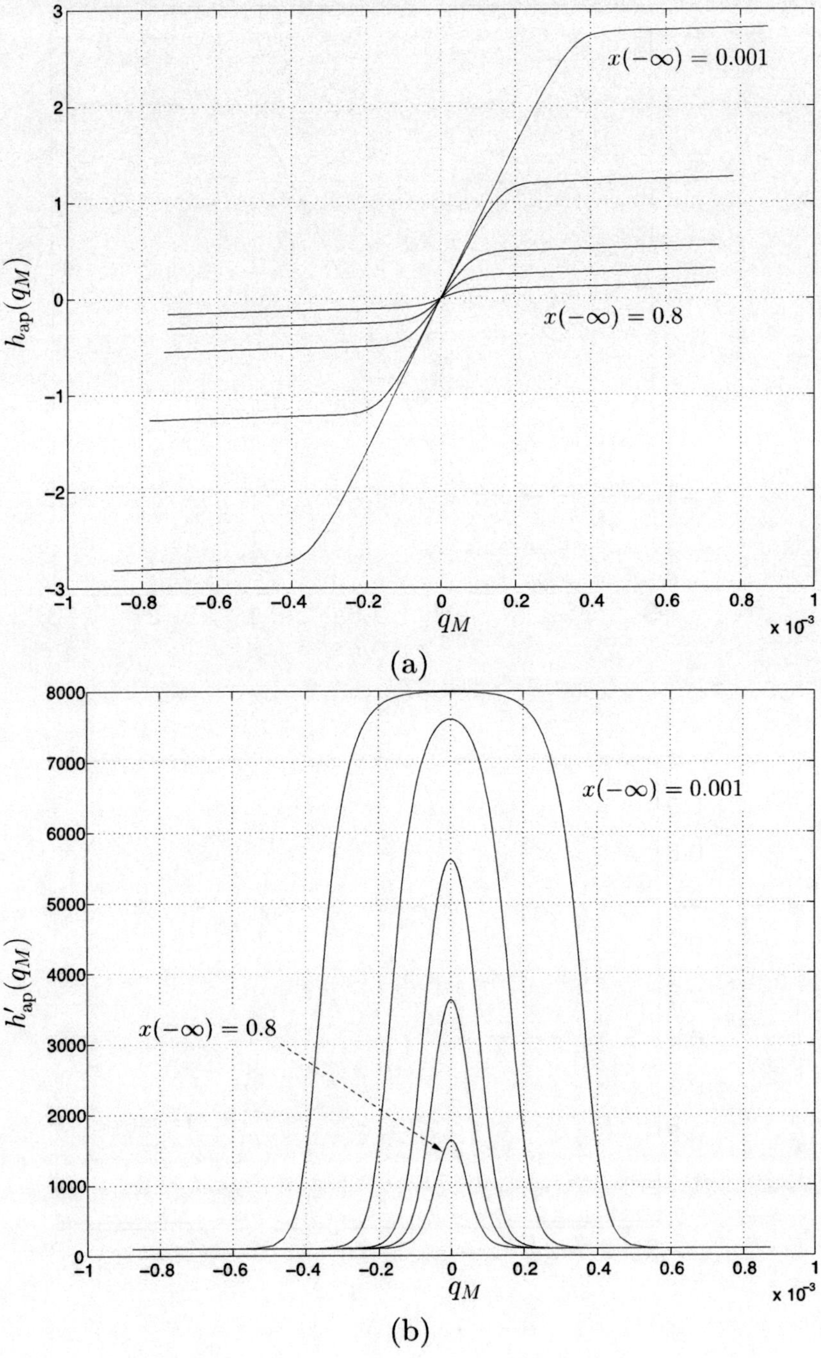

Fig. 9.6 (**a**) Flux-charge characteristics $h_{\mathrm{ap}}(\cdot)$ and (**b**) memristance $h'_{\mathrm{ap}}(\cdot)$ of two HP memristors in antiparallel for different choices of initial values $x(-\infty) \in \{0.001, 0.05, 0.3, 0.55, 0.8\}$

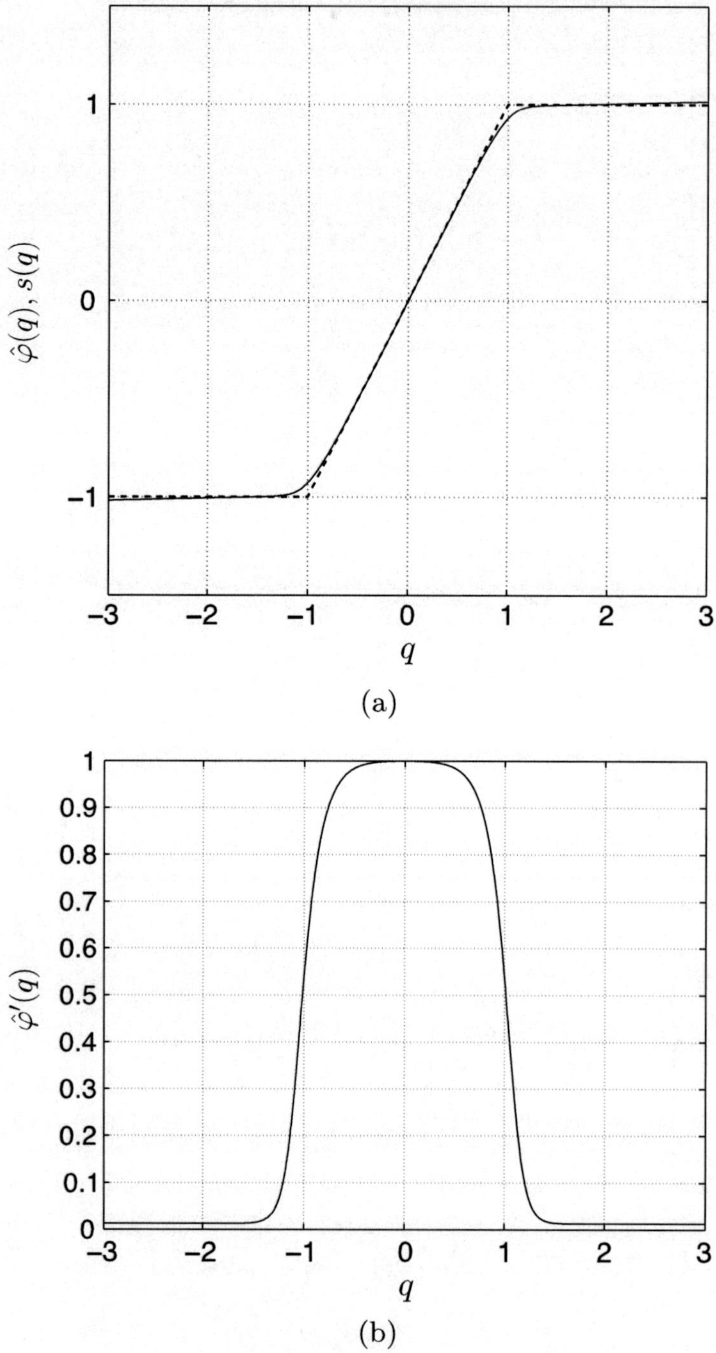

Fig. 9.7 (**a**) Normalized flux-charge characteristics $\hat{\varphi}(\cdot)$ of two HP memristors in antiparallel (solid) and piecewise linear characteristic $s(\cdot)$ of a SCNN (dashed) (**b**) Memristance $\hat{\varphi}'(\cdot)$

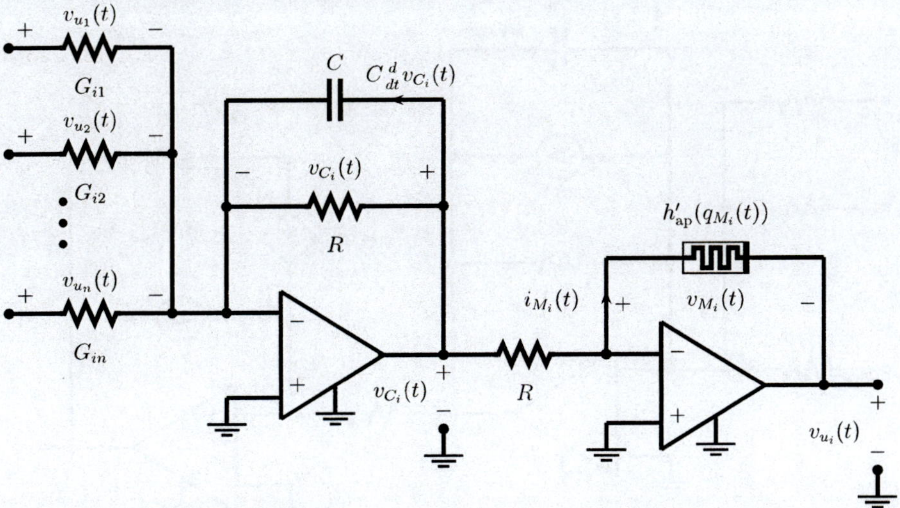

Fig. 9.8 Circuit implementation of the i-th cell in the (v, i)-domain

9.2.2 Cell and Interconnecting Structure

Let us consider a 1D array of n cells. The i-th cell is represented in Fig. 9.8. Conductances G_{ij}, $i, j = 1, 2, \ldots, n$, represent the interconnections with other cells. Henceforth we denote by $\mathbf{G} = (G_{ij}) \in \mathbb{R}^{n\times n}$ the cell interconnection matrix. Moreover, R is an ideal resistor and C is an ideal capacitor. Each operational amplifier is assumed to be ideal. The operational amplifier at the right has a charge-controlled memristor in feedback, implemented by 2 HP memristors in antiparallel with flux-charge characteristic as that described in Sect. 9.2.1, and satisfying (9.8)–(9.11). We denoted by $v_{u,i}(t)$ the output voltage of the operational amplifier at the right-hand side.

Figure 9.9 depicts the equivalent circuit in the (φ, q)-domain as obtained with FCAM. By applying KqL to the operational amplifier at the left we obtain

$$C\frac{d}{dt}\varphi_{C_i}(t; t_0) = -\frac{\varphi_{C_i}(t; t_0)}{R} - \sum_{j=1}^{n} G_{ij}\varphi_{u_j}(t; t_0) + q_{Ci0}$$

where $\varphi_{uj}(t; t_0) = \int_{t_0}^{t} v_{uj}(\sigma)d\sigma$, $j = 1, 2, \ldots, n$.

For the operational amplifier at the right we can write the KqL

$$\frac{\varphi_{C_i}(t; t_0)}{R} = q_{M_i}(t; t_0) \tag{9.12}$$

and KφL

$$\varphi_{u_i}(t; t_0) = -\varphi_{M_i}(t; t_0).$$

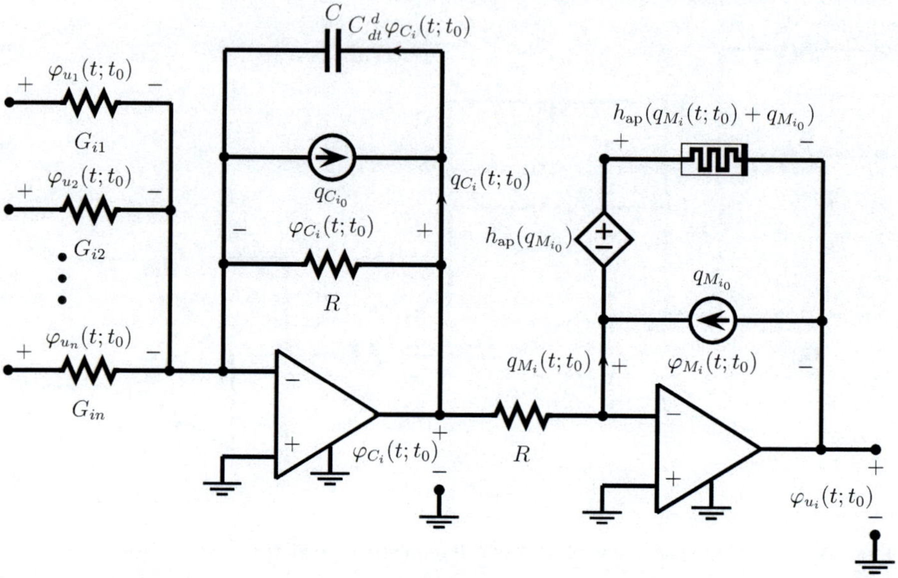

Fig. 9.9 Equivalent circuit of the i-th cell in the (φ, q)-domain as obtained via FCAM

The CR of the charge-controlled memristor in the (φ, q)-domain is given by (cf. (9.3))

$$\varphi_{M_i}(t; t_0) = -h_{\text{ap}}(q_{Mi0}) + h_{\text{ap}}(q_{M_i}(t; t_0) + q_{Mi0}).$$

By substitution we obtain for the 1D NN the following system of n differential equations in the incremental memristor charges

$$\tau \frac{d}{dt} q_{M_i}(t; t_0) = -q_{M_i}(t; t_0) + \sum_{j=1}^{n} G_{ij} h_{\text{ap}}(q_{M_j}(t; t_0) + q_{Mj0})$$
$$- \sum_{j=1}^{n} G_{ij} h_{\text{ap}}(q_{M_{j0}}) + q_{Ci0}$$

for $i = 1, 2, \ldots, n$ and $t \geq t_0$, where $\tau = RC$ is the cell time constant. In terms of charge we have

$$\tau \frac{d}{dt} q_{M_i}(t) = -q_{M_i}(t) + \sum_{j=1}^{n} G_{ij} h_{\text{ap}}(q_{M_j}(t)) + q_{Mi0} + q_{Ci0}$$
$$- \sum_{j=1}^{n} G_{ij} h_{\text{ap}}(q_{M_{j0}})$$

for $i = 1, 2, \ldots, n$ and $t \geq t_0$.

As in Sect. 9.2.1, consider the change of variables $q_{M_i}(t) = k_q q_i(t)$, $i = 1, 2, \ldots, n$, and nonlinearity (9.7), which is reported below for convenience

$$\hat{\varphi}(q_i(t)) = \frac{2h_{\text{ap}}(k_q q_i(t))}{R_{\text{off}} k_q}.$$

Omitting dependence on t, and assuming $t_0 = 0$, we obtain in vector notation that the memristor NN satisfies in the (φ, q)-domain the following system of n differential equations

$$\tau \frac{d\mathbf{q}}{dt} = -\mathbf{q} + \mathcal{G}\hat{\boldsymbol{\Phi}}(\mathbf{q}) + \mathcal{Q}_0 \tag{9.13}$$

for $t \geq 0$, where $\mathbf{q} = (q_1, q_2, \ldots, q_n)^T$, $\hat{\boldsymbol{\Phi}}(\mathbf{q}) = (\hat{\varphi}(q_1), \hat{\varphi}(q_2), \ldots, \hat{\varphi}(q_n))^T$,

$$\mathcal{G} = \frac{R_{\text{off}}\mathbf{G}}{2} \tag{9.14}$$

and

$$\mathcal{Q}_0 = \frac{C\mathbf{v}_{C0}}{k_q} + \mathbf{q}_0 - \frac{\mathbf{G}R_{\text{off}}}{2}\hat{\boldsymbol{\Phi}}(\mathbf{q}_0). \tag{9.15}$$

Here we have let $\mathbf{v}_{C0} = (v_{C_1}(0), v_{C_2}(0), \ldots, v_{C_n}(0))^T$ and $\mathbf{q}_0 = \mathbf{q}(0) = \mathbf{q}_{M0}/k_q$, where $\mathbf{q}_{M0} = (q_{M_1}(0), q_{M_2}(0), \ldots, q_{M_n}(0))^T$. In view of the applications, it is important to remark that the constant term $\mathcal{Q}_0$ depends upon the initial conditions $C\mathbf{v}_{C0}$ and $\mathbf{q}_{M0}$ for the state variables in the (v, i)-domain.

Equations (9.13) is the SE representation of a memristor NN in terms of the n state variables given by the memristor incremental charges. As discussed in Sect. 9.2.1, if $R_{\text{on}} \ll R_{\text{off}}$ and $x(-\infty) \ll 1$, the nonlinearity $\hat{\varphi}(\cdot)$ enjoys properties (9.8)–(9.11) and is a good approximation for not too large $|q|$ of the nonlinearity $s(\cdot)$ of a SCNN. As a consequence, model (9.13) closely approximates and is analogous to the SCNN model (9.1). Henceforth, we will use the acronym $\mathcal{M}$-SCNN for model (9.13).

9.3 Convergence Results

In this section, we give a mathematical foundation to the $\mathcal{M}$-SCNN model (9.13) by studying the existence, uniqueness, boundedness, and prolongability of solutions. Thereafter, we give conditions ensuring that $\mathcal{M}$-SCNN is convergent, i.e., any solution converges toward an EP.[6] The typical scenario in a convergent NN can be described as follows. The NN possesses a huge number of asymptotically stable EPs; moreover, due to convergence, the state space can be subdivided in attraction basins of the asymptotically stable EPs, where the attraction basin of an EP is defined as the set of initial conditions for which the corresponding solution converges to the considered EP. The analog transient displayed by the network during convergence to the EP can be suitably exploited for real-time computational purposes. For example, in a content addressable memory (CAM) the target patterns are stored in the network as asymptotically stable EPs. Given a partial information (e.g., a pattern corrupted with noise), the NN is designed so as to retrieve the uncorrupted pattern during the transient toward the asymptotically stable EP. Convergence is one of the most important global dynamical properties of NNs in view of the application to the solution of signal processing tasks in real time. In fact, convergent NNs are potentially useful not only to implement CAMs but also in the field of pattern formation, to solve image processing tasks and combinatorial optimization problems [8, 9, 15, 16]. In Sect. 9.4, we will discuss some simple applications of convergent $\mathcal{M}$-SCNNs to the solution of image processing tasks in real time.

The vector field $-\mathbf{q} + \mathcal{G}\hat{\boldsymbol{\Phi}}(\mathbf{q}) + \mathcal{Q}_0$ defining the $\mathcal{M}$-SCNN (9.13) is smooth, indeed, it is an analytic function of $\mathbf{q}$. From standard results on ordinary differential equations it follows that the $\mathcal{M}$-SCNN model (9.13) enjoys the property of local existence and uniqueness of the solution with respect to the initial conditions [17]. Moreover, it can be easily checked that the norm of the vector field increases at most linearly with the norm of $\mathbf{q}$, thus ensuring that any solution is defined on the whole time interval $[0, +\infty)$ [17]. However, since the nonlinearity $\hat{\varphi}(\cdot)$ is unbounded, it is not guaranteed *a priori* that solutions are bounded. The next results show that, under a suitable constraint on the norm of $\mathcal{G}$, we can guarantee boundedness of solutions and also obtain a useful estimate of the norm of solutions starting with prescribed initial conditions. In the statements we use the notations introduced in (9.8)–(9.11).

Property 9.1 Suppose that

$$\|\mathcal{G}\|_1 \doteq \max_i \sum_{j=1}^{n} |G_{ij}| < \frac{R_{\text{off}}}{2R_{\text{on}}(1+\Delta R)} \tag{9.16}$$

and let

$$\rho_{\min} = \max\left\{\bar{q}_M, \frac{\|\mathcal{G}\|_1 + \|\mathcal{Q}_0\|_\infty}{1 - \|\mathcal{G}\|_1\left(\frac{2R_{\text{on}}}{R_{\text{off}}}\right)(1+\Delta R)}\right\} \tag{9.17}$$

[6]Such a dynamic property is also referred to in the literature as complete stability.

where $\|\mathcal{Q}_0\|_\infty = \max_i\{|Q_{0i}|\}$. Then, for any $\rho \geq \rho_{\min}$ the hypercube

$$\mathbf{K}_\rho = \{\mathbf{q} \in \mathbb{R}^n : \|\mathbf{q}\|_\infty \leq \rho\}$$

is a positively invariant set for the dynamics of (9.13), i.e., if $\mathbf{q}(0) \in \mathbf{K}_\rho$, we have $\mathbf{q}(t) \in \mathbf{K}_\rho$ for any $t \geq 0$, where $\mathbf{q}(\cdot)$ is the solution of (9.13) with initial condition $\mathbf{q}(0)$.

Proof As it is intuitively clear, it is enough to show that the vector field at the right-hand side of (9.13) points inward on the boundary of the hypercube $\mathbf{K}_\rho$. This ensures that any solution starting in $\mathbf{K}_\rho$ cannot leave the same hypercube. In mathematical terms we should have

$$\begin{aligned} -\rho + \sum_{i=1}^{n} \mathcal{G}_{ik}\hat{\varphi}(q_i) + \mathcal{Q}_{0k} &\leq 0 \\ \rho + \sum_{i=1}^{n} \mathcal{G}_{ik}\hat{\varphi}(q_i) + \mathcal{Q}_{0k} &\geq 0 \end{aligned} \tag{9.18}$$

for any $k = 1, 2, \ldots, n$ and $\rho \geq \rho_{\min}$. These are satisfied if $-\rho + |\sum_{i=1}^{n} \mathcal{G}_{ik}\hat{\varphi}(q_i)| + |\mathcal{Q}_{0k}| \leq -\rho + \sum_{i=1}^{n} |\mathcal{G}_{ik}|\,|\hat{\varphi}(q_i)| + |\mathcal{Q}_{0k}| \leq -\rho + \left(\sum_{i=1}^{n} |\mathcal{G}_{ik}|\right) |\hat{\varphi}(\rho)| + |\mathcal{Q}_{0k}| \leq 0$.

Since $\rho \geq \bar{q}_M$, and $\hat{\varphi}(\cdot)$ is odd, we have $0 < \hat{\varphi}(\rho) \leq 1 + \left(\frac{2R_{\text{on}}}{R_{\text{off}}}\right)(1 + \Delta R)(\rho - \bar{q}_M)$ so that inequalities (9.18) hold if

$$-\rho + \left(\sum_{i=1}^{n} |\mathcal{G}_{ik}|\right)\left[1 + \left(\frac{2R_{\text{on}}}{R_{\text{off}}}\right)(1 + \Delta R)(\rho - \bar{q}_M)\right] + |\mathcal{Q}_{0k}| \leq 0.$$

Note that $\sum_{i=1}^{n} |\mathcal{G}_{ik}| \leq \|\mathcal{G}\|_1$, and $|\mathcal{Q}_{0k}| \leq \|\mathcal{Q}_0\|_\infty$, for any $k = 1, 2, \ldots, n$, so that by (9.16) we conclude that (9.18) are satisfied if

$$\rho \geq \frac{\|\mathcal{G}\|_1\left[1 - \bar{q}_M\left(\frac{2R_{\text{on}}}{R_{\text{off}}}\right)(1 + \Delta R)\right] + \|\mathcal{Q}_0\|_\infty}{1 - \|\mathcal{G}\|_1\left(\frac{2R_{\text{on}}}{R_{\text{off}}}\right)(1 + \Delta R)}.$$

■

Property 9.2 Suppose that (9.16) is satisfied. Then, the following hold.

(i) Any solution $\mathbf{q}(\cdot)$ of (9.13) is bounded, i.e., $\|\mathbf{q}(t)\| \leq \max\{\|\mathbf{q}(0)\|_\infty, \rho_{\min}\}$ for $t \geq 0$.
(ii) There exists at least an EP of (9.13).

Proof

(i) Immediately follows from Property 9.1.
(ii) We have seen in Property 9.1 that for sufficiently large $\rho > 0$ the hypercube K_ρ is positively invariant for the dynamics of (9.13). By using a standard result based on a fixed point theorem (see, e.g., [18, Theorem 8.2, p. 49]) we obtain that (9.13) has at least an EP within $\mathbf{K}_\rho$.

∎

Remark 9.1 On the basis of Property 9.2 we can clarify what we mean by saying that the nonlinearity $\hat{\varphi}(\cdot)$ is a good approximation of $s(\cdot)$ (cf. Sect. 9.2.1). If we choose $\|\mathbf{q}(0)\|_\infty \le \rho_{\min}$, by i) of Property 9.2 we have $\|\mathbf{q}(t)\|_\infty \le \rho_{\min}$ for any $t \ge 0$, i.e., $\mathbf{q}(t)$ evolves within $\mathbf{K}_{\rho_{\min}}$ for $t \ge 0$. So, actually it is enough that $\hat{\varphi}(q_i)$ is a good approximation of $s(\cdot)$ in $\mathcal{I}_{\rho_{\min}} = \{q_i \in \mathbb{R} : |q_i| \le \rho_{\min}\}$. We have $|\hat{\varphi}(q_i) - s(q_i)| < 0.072$ in $\mathcal{I}_{\rho_{\min}}$ up to $\rho_{\min} = 7.3$, $|\hat{\varphi}(q_i) - s(q_i)| < 0.1$ in $\mathcal{I}_{\rho_{\min}}$ up to $\rho_{\min} = 9.7$ and $|\hat{\varphi}(q_i) - s(q_i)| < 0.13$ in $\mathcal{I}_{\rho_{\min}}$ up to $\rho_{\min} = 12$. Note that typical values of $\rho_{\min}$ in the applications are of a few units (cf. examples in Sect. 9.4). It is also worth remarking that the approximation improves if the memristor ratio $R_{\text{off}}/R_{\text{on}}$ increases. In the chapter we considered the HP memristor model where this ratio equals 160. However, in the literature memristors are reported with values of this ratio as large as 1000 and well above [1].

Let us now address convergence of $\mathcal{M}$-SCNNs. The main result is as follows.

Theorem 9.1 *Suppose that (9.16) is satisfied and the interconnection matrix $\mathcal{G}$ is symmetric. Then, (9.13) is convergent, i.e., any solution of (9.13) converges to an EP as $t \to +\infty$.*

Proof We have seen in Property 9.2 that any solution of (9.13) is bounded for $t \ge 0$. For symmetric interconnections and a strictly increasing activation function it is a standard technique to prove that any solution converges to the set of EPs via a Lyapunov approach. For the reader convenience we report the details of the verification. Suppose without loss of generality $\tau = 1$. Let $\mathbf{y} = \hat{\boldsymbol{\Phi}}(\boldsymbol{q})$ and consider as in the case of a symmetric SCNN [8] the candidate Lyapunov function

$$V(\mathbf{y}) = \sum_{i=1}^{n} \int_0^{y_i} \hat{\varphi}_i^{-1}(\rho)d\rho - \mathbf{y}^T \mathcal{G}\mathbf{y} - \mathbf{y}^T \mathcal{Q}_0.$$

By exploiting the symmetry of $\mathcal{G}$ it can be easily checked that

$$\nabla V(\mathbf{y}) = \left(\frac{\partial V(\mathbf{y})}{\partial y_1}, \ldots, \frac{\partial V(\mathbf{y})}{\partial y_n}\right)^T = \mathbf{q} - \mathcal{G}\hat{\boldsymbol{\Phi}}(\mathbf{q}) - \mathcal{Q}_0$$

hence the $\mathcal{M}$-SCNN equations can be rewritten as the gradient-type system

$$\frac{d\mathbf{q}}{dt} = -\nabla V(\mathbf{y}). \tag{9.19}$$

Computing the time derivative of V along the solutions of (9.19) we obtain

$$\begin{aligned}\dot{V}(\mathbf{y}) &= [\nabla V(\mathbf{y})]^T \frac{d\mathbf{y}}{dt} \\ &= [\nabla V(\mathbf{y})]^T \operatorname{diag}\left(\frac{dy_i}{dq_1}, \ldots, \frac{dy_n}{dq_n}\right) \frac{d\mathbf{q}}{dt} \\ &= -[\nabla V(\mathbf{y})]^T \operatorname{diag}(\hat{\varphi}'(q_1), \ldots, \hat{\varphi}'(q_n)) \nabla V(\mathbf{y}) \\ &= -\sum_{i=1}^{n} \hat{\varphi}'(q_i) \left(\frac{\partial V(\mathbf{y})}{\partial y_i}\right)^2 \\ &= -\sum_{i=1}^{n} \hat{\varphi}'(q_i) \left(\frac{dq_i}{dt}\right)^2.\end{aligned}$$

Since $\hat{\varphi}'(q_i) > 0$ for any $q_i \in \mathbb{R}$, we have $\dot{V}(\cdot) \leq 0$ for any $\mathbf{q}$ and $\dot{V}(\cdot) < 0$ when $\mathbf{q}$ is not an EP, i.e., $V(\cdot)$ is strictly decreasing along nonstationary solutions of (9.13). This implies by LaSalle's invariance principle that any solution of (9.13) converges to the set of EPs of (9.13) as $t \to +\infty$ [16].

Now, recall that $\hat{\varphi}(\cdot)$ is analytic in $\mathbb{R}$ (Sect. 9.2.1). Then, we can use an argument based on Łojasiewicz inequality, and the principle of trajectories with finite length, as in the proof of Theorem 1 in [19], to show that any solution of (9.13) converges to a singleton, which is necessarily an EP of (9.13), as $t \to +\infty$. The argument holds independently of the geometric structure of the set of EPs, hence also in the case when (9.13) has infinitely many nonisolated EPs. ■

Remark 9.2 The convergence result in Theorem 9.1 holds even when (9.13) has infinitely many nonisolated EPs. However, it is worth to remark that vector fields as those defining (9.13) enjoy the generic property of possessing isolated EPs. Indeed, given the interconnection matrix $\mathcal{G}$ and the nonlinearity $\hat{\boldsymbol{\varphi}}(\cdot)$, it can be shown via a technique based on Sard's lemma as that used in the proof of Property 2 in [19] that, for almost all $\mathcal{Q}_0 \in \mathbb{R}^n$, in the sense of the Lebesgue measure, there exists a finite number of isolated EPs of (9.13).

9.3.1 Voltage-Current Domain

The equations describing the dynamics of a $\mathcal{M}$-SCNN in the standard (v, i)-domain can be obtained according to FCAM (Chap. 5) by differentiating with respect to time the Eqs. (9.13) in the (φ, q)-domain. We have

$$\frac{dq_i(t)}{dt} = \frac{1}{k_q} \frac{dq_{M_i}(t; t_0)}{dt} = \frac{1}{k_q} i_{M_i}(t; t_0) = \frac{1}{k_q R} v_{C_i}(t) \tag{9.20}$$

for $i = 1, 2, \ldots, n$, where $v_{C_i}(t) = d\varphi_{C_i}(t)/dt$ and we have taken into account (9.12). Let $\mathbf{v}_C(t) = (v_{C_1}(t), v_{C_2}(t), \ldots, v_{C_n}(t))^T$, where $v_{C_i}(t) = d\varphi_{C_i}(t)/dt$, $i = 1, 2, \ldots, n$, is the vector of capacitor voltages for the circuit in Fig. 9.8.

Then, we obtain a system of $2n$ SEs describing the dynamics in the (v, i)-domain in the normalized state variables $\mathbf{q}(t) = \mathbf{q}_M(t)/k_q$, $\mathbf{v}(t) = \mathbf{v}_C(t)/k_q$ (we omit dependence on t)

$$\begin{cases} \tau \dfrac{d\mathbf{v}}{dt} = -\mathbf{v} + \boldsymbol{\mathcal{G}}\hat{\boldsymbol{\Phi}}'(\boldsymbol{q})\boldsymbol{v} \\ \\ \dfrac{d\mathbf{q}}{dt} = \frac{1}{R}\mathbf{v} \end{cases} \tag{9.21}$$

for $t \geq 0$. The initial conditions are $\mathbf{v}(0) = \mathbf{v}_C(0)/k_q = \mathbf{v}_{C0}/k_q$ and $\mathbf{q}(0) = \mathbf{q}_0 = \mathbf{q}_{M0}/k_q$. From the standard theory of ordinary differential equations we have that (9.21) enjoys the properties of the existence, uniqueness with respect to initial conditions and prolongability of solutions to $+\infty$.

Remark 9.3 It is worth noting that the SEs (9.21) can also be derived by a direct analysis in the (v, i)-domain. Indeed, let us consider the cell implementation in Fig. 9.8. By applying the Kirchhoff Current Law (KCL) to the operational amplifier at the left we obtain (omitting dependence on t)

$$C_i \frac{d}{dt} v_{C_i} = -\frac{v_{C_i}}{R} - \sum_{j=1}^{n} G_{ij} v_{uj}.$$

The KCL for the operational amplifier at the right leads us to

$$v_{ui} = -h'_{\mathrm{ap}}(q_{M_i}) \frac{v_{C_i}}{R}.$$

Taking into account that $v_{M_i} = -v_{uj}$ and $v_{C_i}/R = i_{M_i} = dq_{M_i}/dt$, and letting $\mathbf{H}'_{\mathrm{ap}}(\mathbf{q}_M) = (h'_{\mathrm{ap}}(q_{M_1}), \ldots, h'_{\mathrm{ap}}(q_{M_n}))^T$ and $\mathbf{q}_M = (q_{M_1}, q_{M_2}, \ldots, q_{M_n})^T$, the above equations become

$$\begin{cases} \tau \dfrac{d\mathbf{v}_C}{dt} = -\mathbf{v}_C + \mathbf{G}\mathbf{H}'_{\mathrm{ap}}(\mathbf{q}_M)\mathbf{v}_C \\ \\ \dfrac{d\mathbf{q}_M}{dt} = \frac{1}{R}\mathbf{v}_C. \end{cases}$$

The change of variables presented in Sect. 9.2.1 yields (9.21).

By letting $d\mathbf{v}/dt = 0$ and $d\mathbf{q}/dt = 0$ it is seen that (9.21) has an n-dimensional manifold of EPs

$$\{(\bar{\mathbf{v}}, \bar{\mathbf{q}}) \in \mathbb{R}^{2n} : \bar{\mathbf{v}} = \mathbf{0}, \bar{\mathbf{q}} \in \mathbb{R}^n\}.$$

It is then a structural property of the SEs describing the dynamics of a $\mathcal{M}$-SCNN in the (v, i)-domain to possess a manifold (a continuum) of nonisolated EPs. Note that at any EP the capacitor voltages vanish.

Theorem 9.2 *Suppose that (9.16) is satisfied and $\mathcal{G}$ is symmetric. Then, any solution $(\mathbf{v}_C(t), \mathbf{q}(t))$ of (9.21) is bounded for $t \geq 0$ and converges to an EP of (9.21) as $t \to +\infty$, i.e., (9.21) is convergent. In particular, $\mathbf{v}_C(t) \to 0$ as $t \to +\infty$, i.e., the capacitor voltages tend to 0.*

Proof Let $(\mathbf{v}(t), \mathbf{q}(t))$, $t \geq 0$, be the solution of (9.21) with initial condition $(\mathbf{v}(0), \boldsymbol{\varphi}(0)) = (\mathbf{v}_0, \mathbf{q}_0)$. On the basis of FCAM, if we choose $\mathcal{Q}_0$ as in (9.15), then $\mathbf{q}(t)$, $t \geq 0$, is the unique solution of (9.13) with initial condition $\mathbf{q}(0) = \mathbf{q}_0$. By Theorem 9.1 we have that $\mathbf{q}(t) \to \bar{\mathbf{q}}$ and $d\mathbf{q}(t)/dt \to 0$ as $t \to +\infty$, where $\bar{\mathbf{q}}$ is an EP of (9.13). The proof is concluded by noting that $\mathbf{v}_C(t) = k_q R(d\mathbf{q}(t)/dt)$. ■

Remark 9.4 We point out that the results in Theorems 9.1, 9.2 are consistent with each other. In fact, the latter theorem shows that capacitor voltages tend to 0 as $t \to +\infty$, hence, due to (9.20), also memristor currents vanish as $t \to +\infty$. This is in agreement with Theorem 9.1, according to which memristor charges, i.e., the integral of memristor currents, tend to a constant value as $t \to +\infty$. It is not difficult to verify from Fig. 9.8 that for a convergent $\mathcal{M}$-SCNN all voltages and currents in the network vanish when a steady state is reached. This means that voltages and currents are not useful for processing purposes. However, their integrals, i.e., fluxes and charges, tend to finite constants in steady state and in particular the memristor charges can be used for processing purposes, as discussed via the examples in Sect. 9.4.

9.4 Applications to Image Processing

We have seen in Sect. 9.2.2 that the SEs (9.13) describing the $\mathcal{M}$-SCNN model are a good approximation, in the (φ, q)-domain, of the SEs describing the SCNN model (9.1) in the (v, i)-domain, i.e., the two models are formally analogous. In this section we show by numerical means that the two models indeed display similar processing capabilities when applied to the solution of some image processing tasks in real time.

9.4.1 Horizontal Line Detection

In order to illustrate how the dynamics in the (φ, q)-domain of a convergent $\mathcal{M}$-SCNN can be used for processing purposes, let us consider a specific application where a 2D $\mathcal{M}$-SCNN array is used for extracting the horizontal lines of an image. After reordering cells row-wise the $\mathcal{M}$-SCNN can be described in the (φ, q)-domain by the set of differential equations (9.13) and in the (v, i)-domain by (9.21).

According to Theorem 9.2, in a convergent $\mathcal{M}$-SCNN all voltages and currents vanish in steady state, so that the network cannot compute in the standard (v, i)-domain. However, it can perform a processing in the (φ, q)-domain via the transient dynamics of system (9.13). More precisely, first we design the interconnections of a $\mathcal{M}$-SCNN in order that the assumptions of Theorem 9.1 are satisfied, in particular, we choose a symmetric matrix $\mathcal{G}$. The input image to be processed is provided to the $\mathcal{M}$-SCNN (9.13) via the initial conditions $\mathbf{q}(0) = \mathbf{q}_0$, the processing is performed during the time evolution of the charges $\mathbf{q}(\cdot)$ in (9.13) and the processing result is given by the asymptotic values of the charges $\mathbf{q}(\infty) = \lim_{t\to+\infty} \mathbf{q}(t)$.

For the horizontal line detection suppose to choose the same 3×3 feedback cloning template as in [15, Sect. III]

$$\mathcal{A} = \begin{pmatrix} 0 & 0 & 0 \\ 1 & 2 & 1 \\ 0 & 0 & 0 \end{pmatrix} \tag{9.22}$$

from which we obtain a symmetric interconnection matrix $\mathcal{G}$ by following a procedure as that described, e.g., in [19, App. C]. Moreover, let the $\mathcal{M}$-SCNN input $\mathcal{Q}_0$ be equal to 0. Note that, given the initial image $\mathbf{q}(0) = \mathbf{q}_0$, in order to have $\mathcal{Q}_0 = 0$, we need to choose for system (9.21) the following initial conditions for the state variables in the (v, i)-domain (cf. (9.15))

$$\mathbf{q}_M(0) = k_q \mathbf{q}_0$$

and

$$\mathbf{v}_C(0) = \frac{k_q}{C}[-\mathbf{q}_0 + \mathcal{G}\hat{\boldsymbol{\Phi}}(\mathbf{q}_0)].$$

Consider again the memristor nonlinearity $\hat{\varphi}(\cdot)$ as in Sect. 9.2.1, which is relative to the case $R_{\text{off}} = 16\ \text{k}\Omega$, $R_{\text{on}} = 100\ \Omega$. Recall that $\Delta R = 0.034$. We have $\|\mathcal{G}\|_1 = 4$, i.e., $\mathcal{G}$ satisfies condition (9.16), hence the hypotheses of Property 9.2 and Theorems 9.1, 9.2 are satisfied. Moreover, we have from (9.17) $\rho_{\min} = 4.22$. We simulated the $\mathcal{M}$-SCNN for this task using MATLAB. As an example, Fig. 9.10a reports the initial 20×20 image to be processed, and Fig. 9.10b the final result of processing, i.e., the asymptotic values of charges. The figure shows that the $\mathcal{M}$-SCNN performs a correct horizontal line detection as it happens for a SCNN as that

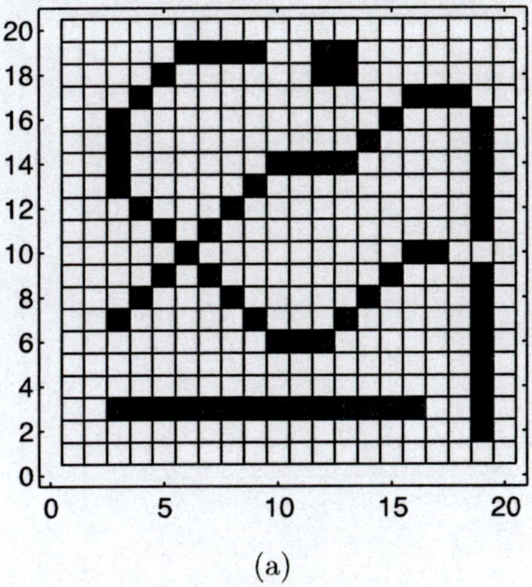

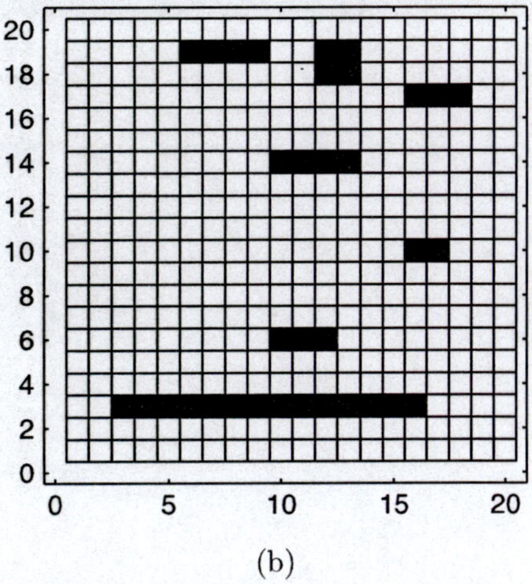

Fig. 9.10 (**a**) Initial image and (**b**) result of processing for a $\mathcal{M}$-SCNN used for horizontal line detection. White (resp., black) pixels represent values of charges ≥ 1 (resp., ≤ -1)

considered in [8, Sect. IV]. In Fig. 9.11a we have depicted the time evolution of charges $q_i(\cdot)$ for cells in the 7th row, whereas Fig. 9.11b reports the corresponding time evolution of voltages $v_i(\cdot)$. The latter figure shows that, as predicted by Theorem 9.2, capacitor voltages vanish when the $\mathcal{M}$-SCNN transient is settled down. A simple inspection of the circuit in Fig. 9.8 shows that all voltages and currents, as well as power, vanish when a steady state is reached.

For comparison, we simulated the behavior of a SCNN as in (9.1), with the same template $\boldsymbol{\mathcal{A}}$ as in (9.22), for solving the horizontal line detection problem (cf. [8, Sect. IV]). The simulations confirmed that $\mathcal{M}$-SCNNs and SCNNs provide the same processing result for the same input images. Also the transient behavior is similar. As an example, Fig. 9.12 reports the time evolution of capacitor voltages, for cells in the 7th row of the SCNN (9.1), when the input image in Fig. 9.10a is supplied. By comparing Fig. 9.11a with Fig. 9.12 it is seen that the two transient evolutions are very similar.

We have performed experiments by adding a uniformly distributed noise taking values in the interval $[-\theta, \theta]$ to each input image pixel in order to investigate the noise tolerance capabilities of the $\mathcal{M}$-SCNN. We found that the $\mathcal{M}$-SCNN is able to correctly extract horizontal lines up to a noise threshold of about $\theta_M = 0.6$. Beyond this threshold, a number of errors are frequently detected in experiments. As an example, Fig. 9.13 reports the input image corrupted with noise and the final result of computation showing three erroneous black pixels when $\theta = 0.75$.

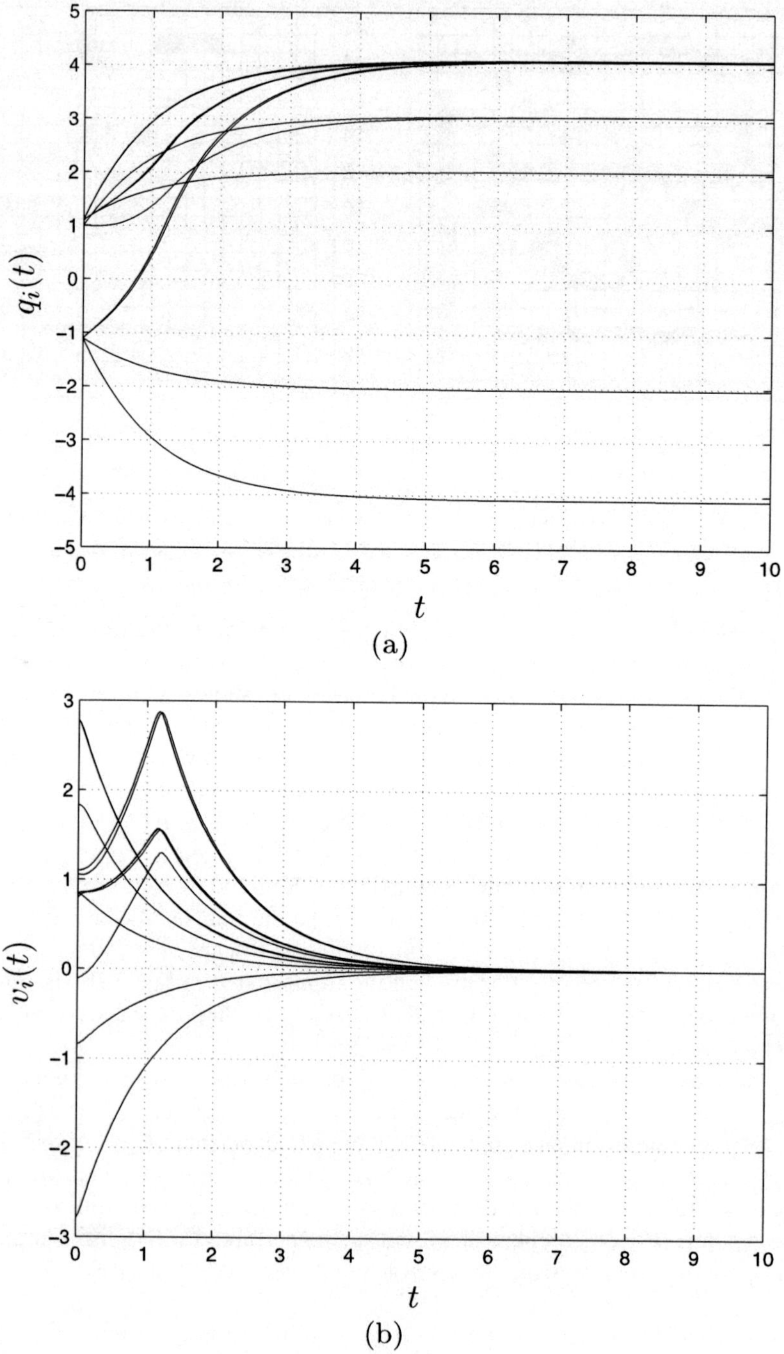

Fig. 9.11 (**a**) Time-domain evolution of charges $q_i(\cdot)$ for cells in the 7th row and (**b**) corresponding time evolution of voltages $v_i(\cdot)$

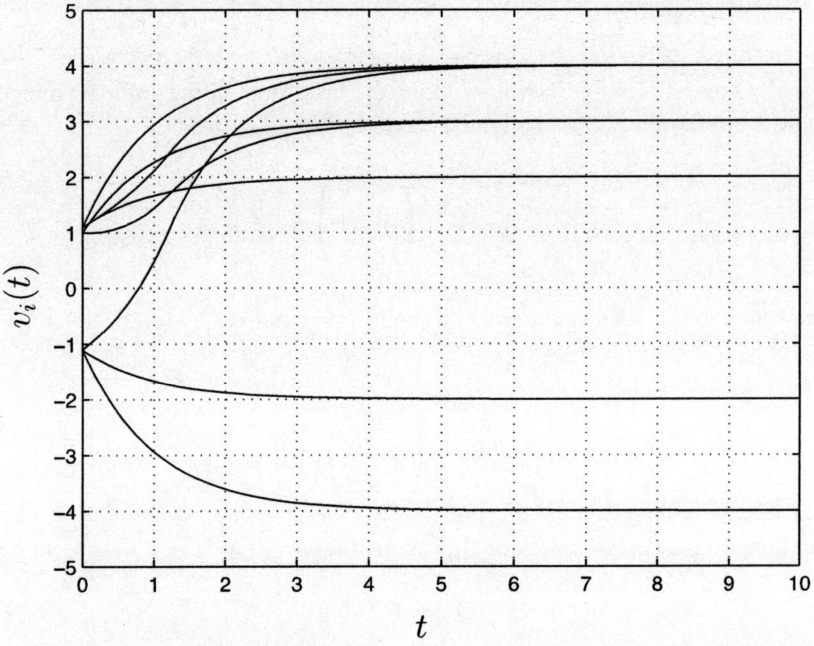

Fig. 9.12 Time-domain evolution of capacitor voltages $v_i(\cdot)$ of a SCNN for cells in the 7th row

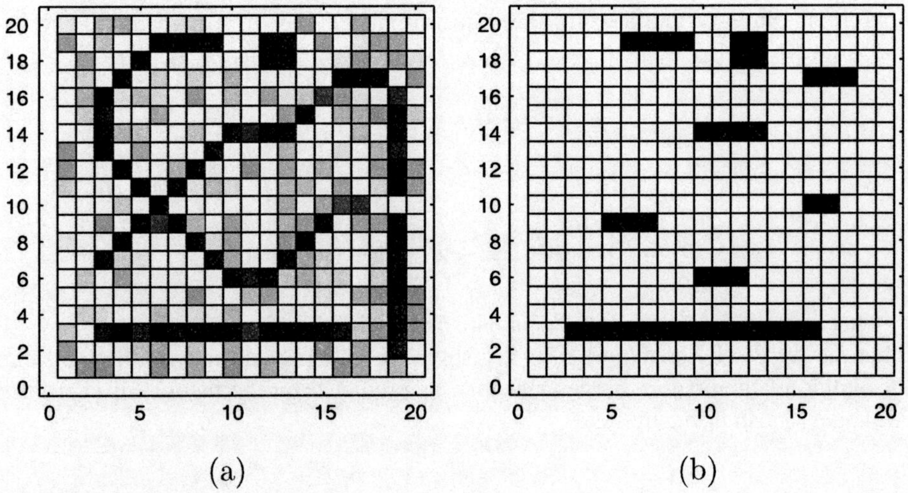

Fig. 9.13 (**a**) Initial image corrupted with noise and (**b**) result of processing for a $\mathcal{M}$-SCNN used for horizontal line detection

9.4.2 Hole Filling

Consider now a 2D $\mathcal{M}$-SCNN for solving the task "hole filling" of a 2D image. Let us choose the 3×3 feedback cloning template as in [20, p. 44]

$$\mathcal{A} = \begin{pmatrix} 0 & 1 & 0 \\ 1 & 3 & 1 \\ 0 & 1 & 0 \end{pmatrix}$$

from which we obtain a symmetric interconnection matrix $\mathcal{G}$. Moreover, let the 3×3 control template be given by

$$\mathcal{B} = \begin{pmatrix} 0 & 0 & 0 \\ 0 & 4 & 0 \\ 0 & 0 & 0 \end{pmatrix}.$$

According to the design procedure in [20, p. 44] let

$$\mathcal{Q}_0 = \mathcal{B}\mathcal{I} + \mathbf{z}$$

where $\mathcal{I}$ is the initial image and $\mathbf{z}$ is a biasing vector with all elements equal to 1. Moreover, choose the initial conditions $q_i(0) = q_{i_0} = -1.1$ for all $i = 1, 2, \ldots, n$. Note that in this case the image to be processed is provided via $\mathcal{Q}_0$. As discussed before, the initial conditions for the state variables in the (v, i)-domain are chosen according to (9.15) as

$$\mathbf{q}_M(0) = k_q \mathbf{q}_0$$

and

$$\mathbf{v}_C(0) = \frac{k_q}{C}[\mathcal{B}\mathcal{I} + \mathbf{z} - \mathbf{q}_0 + \mathcal{G}\hat{\boldsymbol{\Phi}}(\mathbf{q}_0)].$$

It can be checked that (9.16) is satisfied. It is seen from Figs. 9.14 and 9.15 that the $\mathcal{M}$-SCNN is able to correctly solve the task in real time via the transient evolution of memristor charges and that, as predicted by the theory, all capacitor voltages vanish in steady state.

9.5 Discussion

A number of remarks on the $\mathcal{M}$-SCNNs introduced in this chapter are in order.

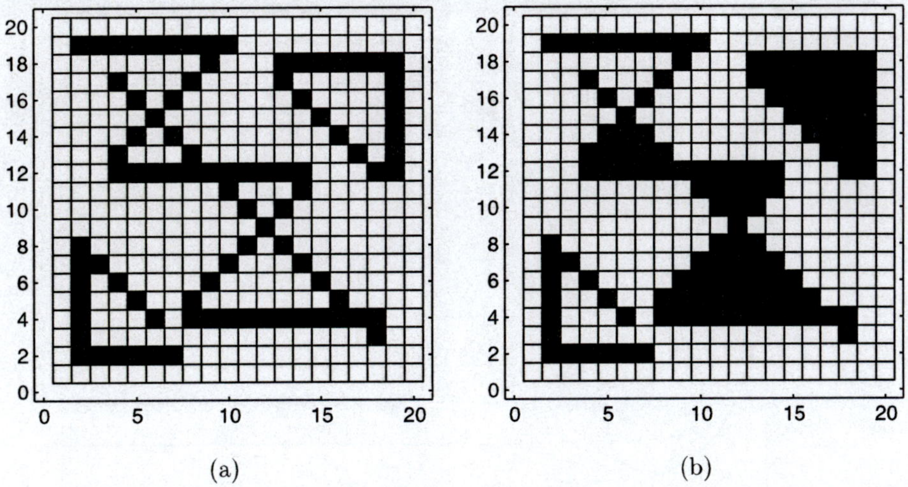

Fig. 9.14 (**a**) Initial image and (**b**) result of processing for a $\mathcal{M}$-SCNN used for solving the task "hole filling"

Remark 9.5 The examples discussed in Sect. 9.4 demonstrate that the class of $\mathcal{M}$-SCNNs has processing capabilities, in the (φ, q)-domain, similar to those displayed by SCNNs [8] in the typical (v, i)-domain. This was confirmed by several other numerical experiments where $\mathcal{M}$-SCNNs have been applied to other kinds of processing tasks. On this basis, we may conclude that it is potentially possible to use the bulk of results already available in the literature for SCNNs (see, e.g., [9]) in order to design $\mathcal{M}$-SCNNs for solving specific signal processing tasks.

Remark 9.6 Although from an abstract mathematical viewpoint the way a $\mathcal{M}$-SCNN processes signals is similar to that of a SCNN, since both NNs obey analogous systems of differential equations, in practice there are some fundamental differences. Indeed, *the processing of a $\mathcal{M}$-SCNN takes place in the (φ, q)-domain so that voltages, currents and power drop off when a $\mathcal{M}$-SCNN reaches a steady state. This is an advantage in terms of power consumption with respect to SCNNs [8] operating in the (v, i)-domain, where capacitor voltages, currents and power do not vanish in steady state (as an example, see Fig. 9.12).* We also remark that there are basic differences in the way the result of computation is memorized. Indeed, although a $\mathcal{M}$-SCNN drops off in steady state, the memristors act as nonvolatile memories keeping the result of computation, i.e., the asymptotic values of charges. No battery is needed in a $\mathcal{M}$-SCNN to hold the processing result. Quite on the contrary, batteries are needed to hold in memory the processing result in a SCNN.

Remark 9.7 One main bottleneck of the classical Von Neumann architecture is that processing and storing of information occur on physically distinct locations, such as in the CPU and in the RAM. This limits the rate at which information can be transferred and processed. A possible and promising way to overcome this issue

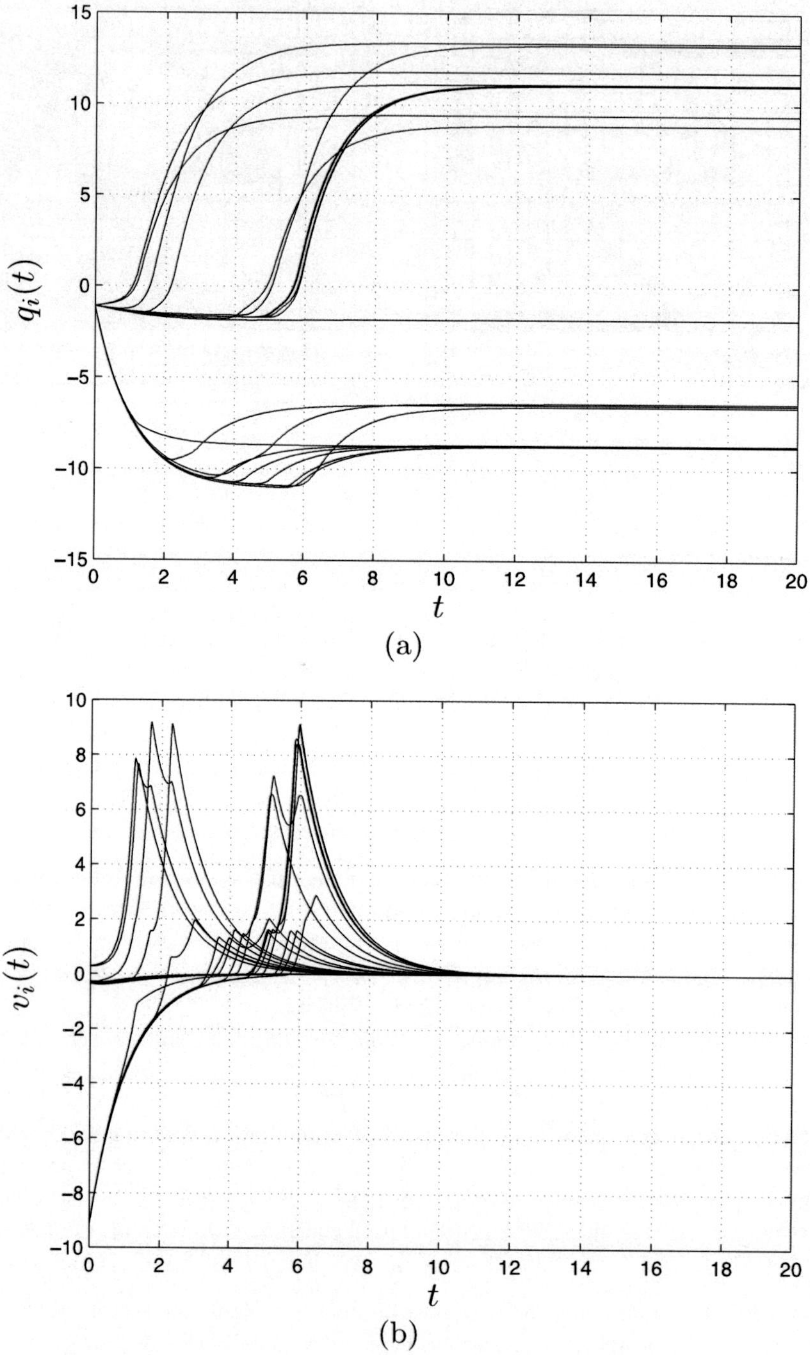

Fig. 9.15 (**a**) Time-domain evolution of charges $q_i(\cdot)$ for cells in the 15th row and (**b**) corresponding time evolution of voltages $v_i(\cdot)$

is to use a parallel computational approach, as that offered by a neuromorphic architecture, together with unconventional electric devices as the memristors, that are able to process and store information on the same physical device according to the principle of *in-memory computing*. The design of the $\mathcal{M}$-SCNNs here introduced goes along this direction. In fact, the role played by the memristors in the behavior of $\mathcal{M}$-SCNNs is twofold. First, the nonlinear dynamics of memristors is exploited during the transient analog computation; moreover, the same memristors are also used to store the result of the processing in steady state, i.e., in a $\mathcal{M}$-SCNN processing and storing of information occur in the same physical location and are performed by the same device.

Remark 9.8 Recall that in the (φ, q)-domain a memristor has an algebraic CR (cf. (9.3)), i.e., it establishes a static nonlinear relationship between flux and charge. The cell model here considered for a $\mathcal{M}$-SCNN (cf. Fig. 9.8) differs from that of a SCNN due to the use of memristors as the nonlinear elements in the (φ, q)-domain. In a SCNN the nonlinearity is instead implemented in the (v, i)-domain by using a saturated operational amplifier [8].

Remark 9.9 The use of memristors permits to obtain for a $\mathcal{M}$-SCNN analogous equations in the (φ, q)-domains as those describing SCNNs in the (v, i)-domain. It is known however that a memristor has a CR in differential form in the (v, i)-domain. In fact we have seen in Sect. 9.3.1 that a $\mathcal{M}$-SCNN obeys in the (v, i)-domain a system of $2n$ differential equations for an array of n cells. Clearly, there is not an analogy between $\mathcal{M}$-SCNNs and SCNNs when considering the (v, i)-domain.

Remark 9.10 Other related articles in the literature have dealt with the design of memristor NNs whose mathematical model in the (φ, q)-domain is analogous to the Full-Range model of cellular NNs [19]. Convergence in the presence of multiple EPs in the (φ, q)-domain has been studied for symmetric and also for some classes of nonsymmetric (e.g., cooperative) cell interconnections [21]. Moreover, the issue of global stability of the unique EP in the (φ, q)-domain has been addressed in the general case of nonsymmetric interconnections [22] and also in the case where there are delays in the interconnections [23]. In the case of convergence, or global stability, those memristor NNs feature the same advantages as $\mathcal{M}$-SCNNs in terms of reduction of power consumption at steady state and in terms of the possibility to implement the processing and storing phase at the same physical location.

Remark 9.11 The theoretical foundation for a class of memristor CNNs that is conceptually similar to $\mathcal{M}$-CNNs is given in [24–26]. Such NNs also adopt a nonvolatile memristor in place of the linear resistor in the circuit of each processing cell. A generic memristor model is considered (Chap. 2) inspired to the mathematical description presented in [27] to capture the learning process of a unicellular organism. The resulting cells are second order.

Two main cases are studied. In the first case, parameters are chosen such that there are only abrupt transitions between lowest and highest resistive memristor states, so that the dynamic analysis can be performed via a dynamic route map for a first-order neuron. In the second case, an analog behavior with smooth memristance changes is considered resulting in a true second-order dynamics. The behavior is studied via an extended form of dynamic route map that can be applied to second-order nonlinear dynamical systems.

Such papers stress that the unique capability of a nonvolatile resistance switching memory to store or process data in the same physical nano-scale volume may be leveraged in such CNNs to improve the performance of sensor-processor systems, especially the spatial resolution of state-of-the-art visual microprocessors consisting of Cellular Nonlinear Network computing machines [28] integrated on top of CMOS image sensor arrays. The maximum number of smart sensors in these hardware components is currently limited by the excessive integrated circuit area consumed by each processing element, due to the need to accommodate large memory units allowing to reprogram its circuit parameters as well as to store computation results or to retrieve them at a later stage for further data processing. Relevant advantages are potentially obtained, allowing to resolve the spatial resolution issue of modern sensor-processor arrays, thanks to the capability to store and retrieve data into and from the resistance switching memories without the necessity to reserve additional cell circuit area for memory banks.

Appendix: HP Memristor with Joglekar Window

By integrating (9.5) with Joglekar window (9.6), in the case $p = 1$, we obtain

$$q_M(t) = \frac{1}{\alpha} \int_{x(-\infty)}^{x(t)} \frac{1}{1 - (2\sigma - 1)^2} d\sigma$$

where we considered that $q_M(-\infty) = 0$, i.e., the overall charge of the memristor is null at its fabrication. Then

$$q_M(t) = F(x(t)) = \frac{1}{4\alpha} \log\left(\frac{x(t)}{1 - x(t)}\right) + \tilde{C} \tag{9.23}$$

where $\tilde{C}$ is a constant depending on the state initial value $x(-\infty) \in (0, 1)$. Since $x(t) \in (0, 1)$ for any $t \geq 0$, it turns out that $F : (0, 1) \to \mathbb{R}$ is an analytic function. Additionally, $F'(x) > 0$ for any $x \in (0, 1)$, hence $F(\cdot)$ is globally invertible, and their inverse function $x = \hat{x}(q_M) : \mathbb{R} \to (-1, 1)$ is analytic in $\mathbb{R}$.

We have

$$x(t) = \hat{x}(q_M(t) - \tilde{C}) = \frac{e^{4\alpha(q_M(t) - \tilde{C})}}{1 + e^{4\alpha(q_M(t) - \tilde{C})}}$$

so that (9.5) can be rewritten as

$$v_M(t) = \left[\frac{R_{\text{off}} + R_{\text{on}} e^{4\alpha(q_M(t)-\tilde{C})}}{1 + e^{4\alpha(q_M(t)-\tilde{C})}}\right] i_M(t).$$

Recalling that $v_M(t) = d\varphi_M(t)/dt$ and $i_M(t) = dq_M(t)/dt$, by integrating between $-\infty$ and t we obtain

$$\varphi_M(t) = \int_{q_M(-\infty)}^{q_M(t)} \left[\frac{R_{\text{off}} + R_{\text{on}} e^{4\alpha(q-\tilde{C})}}{1 + e^{4\alpha(q-\tilde{C})}}\right] dq$$

where we considered that $\varphi(-\infty) = 0$.

Then, we obtain the following flux-charge relation for the HP memristor

$$\varphi_M = h(q_M) = \int_0^{q_M} \left[\frac{R_{\text{off}} + R_{\text{on}} e^{4\alpha(q-\tilde{C})}}{1 + e^{4\alpha(q-\tilde{C})}}\right] dq. \tag{9.24}$$

From the previous discussion we have that $h(\cdot) : \mathbb{R} \to \mathbb{R}$ is analytic in $\mathbb{R}$. Moreover, it is easily verified that $h(0) = 0$,

$$h'(q_M) = \left[\frac{R_{\text{off}} + R_{\text{on}} e^{4\alpha(q_M-\tilde{C})}}{1 + e^{4\alpha(q_M-\tilde{C})}}\right] > 0$$

and

$$h''(q_M) = -\frac{4\alpha(R_{\text{off}} - R_{\text{on}}) e^{4\alpha(q_M-\tilde{C})}}{\left(1 + e^{4\alpha(q_M-\tilde{C})}\right)^2} < 0$$

for any $q_M \in \mathbb{R}$.

References

1. L. Chua, G. Sirakoulis, A. Adamatzky (eds.), *Handbook of Memristor Networks*, vol. 1 and 2 (Springer, New York, 2019)
2. H. Kim, M. Sah, C. Yang, T. Roska, L.O. Chua, Memristor bridge synapses. Proc. IEEE **100**(6), 2061–2070 (2012)
3. H. Kim, M. Sah, C. Yang, T. Roska, L.O. Chua, Neural synaptic weighting with a pulse-based memristor circuit. IEEE Trans. Circuits Syst. I. Regul. Pap. **59**(1), 148–158 (2012)
4. M. Sah, H. Kim, L.O. Chua, Brains are made of memristors. IEEE Circuits Syst. Mag. **14**(1), 12–36 (2014)
5. R. Stanley Williams, What's next? [The end of Moore's law]. Comput. Sci. Eng. **19**(2), 7–13 (2017)
6. M.A. Zidan, J.P. Strachan, W.D. Lu, The future of electronics based on memristive systems. Nat. Electron. **1**(1), 22 (2018)

7. O. Krestinskaya, A.P. James, L.O. Chua, Neuromemristive circuits for edge computing: a review. IEEE Trans. Neural Netw. Learn. Syst. **31**(1), 4–23 (2020). https://doi.org/10.1109/TNNLS.2019.2899262
8. L.O. Chua, L. Yang, Cellular neural networks: theory. IEEE Trans. Circuits Syst. **35**(10), 1257–1272 (1988)
9. L.O. Chua, T. Roska, *Cellular Neural Networks and Visual Computing: Foundation and Applications* (Cambridge University Press, Cambridge, 2005)
10. D.B. Strukov, G.S. Snider, D.R. Stewart, R.S. Williams, The missing memristor found. Nature **453**(7191), 80–83 (2008)
11. Y.N. Joglekar, S.J. Wolf, The elusive memristor: properties of basic electrical circuits. Eur. J. Phys. **30**(4), 661 (2009)
12. R. Stanley Williams, How we found the missing memristor. IEEE Spectr. **45**(12), 28–35 (2008)
13. F. Corinto, A. Ascoli, A boundary condition-based approach to the modeling of memristor nanostructures. IEEE Trans. Circuits Syst. I. Regul. Pap. **59**(11), 2713–2726 (2012)
14. D. Biolek, Z. Biolek, V. Biolkova, Z. Kolka, Reliable modeling of ideal generic memristors via state-space transformation. Radioengineering **24**(2), 393–407 (2015)
15. L.O. Chua, L. Yang, Cellular neural networks: applications. IEEE Trans. Circuits Syst. **35**(10), 1273–1290 (1988)
16. M. Hirsch, Convergent activation dynamics in continuous time networks. Neural Netw. **2**, 331–349 (1989)
17. H.K. Khalil, *Nonlinear Systems* (Prentice Hall, Upper Saddle River, 2002)
18. J.K. Hale, *Ordinary Differential Equations* (Wiley Interscience, New York, 1969)
19. M. Di Marco, M. Forti, L. Pancioni, Complete stability of feedback CNNs with dynamic memristors and second-order cells. Int. J. Circuit Theory Appl. **44**, 1959–1981 (2016)
20. L. Kék, K. Karacs, T. Roska, Cellular wave computing library (templates, algorithms, and programs). Version 2.1, in *Research report of Cellular Sensory and Wave Computing Laboratory CSW-1-2007* (2007)
21. M. Di Marco, M. Forti, L. Pancioni, Convergence and multistability of nonsymmetric cellular neural networks with memristors. IEEE Trans. Cybern. **47**(10), 2970–2983 (2017)
22. M. Di Marco, M. Forti, L. Pancioni, New conditions for global asymptotic stability of memristor neural networks. IEEE Trans. Neural Netw. Learn. Syst. **29**, 1822–1834 (2018)
23. M. Di Marco, M. Forti, L. Pancioni, Stability of memristor neural networks with delays operating in the flux-charge domain. J. Franklin Inst. **355**(12), 5135–5162 (2018)
24. R. Tetzlaff, A. Ascoli, I. Messaris, L.O. Chua, Theoretical foundations of memristor cellular nonlinear networks: memcomputing with bistable-like memristors. IEEE Trans. Circuits Syst. I. Regul. Pap. **67**(2), 502–515 (2020)
25. A. Ascoli, I. Messaris, R. Tetzlaff, L.O. Chua, Theoretical foundations of memristor cellular nonlinear networks: stability analysis with dynamic memristors. IEEE Trans. Circuits Syst. I: Regul. Pap. **67**(4), 1389–1401 (2020)
26. A. Ascoli, R. Tetzlaff, S.-M. Kang, L.O. Chua, Theoretical foundations of memristor cellular nonlinear networks: a DRM2-based method to design memcomputers with dynamic memristors. IEEE Trans. Circuits Syst. I: Regul. Pap. **67**(8), 2753–2766 (2020). https://doi.org/10.1109/TCSI.2020.2978460
27. Y.V. Pershin, S. La Fontaine, M. Di Ventra, Memristive model of amoeba learning. Phys. Rev. E **80**(2), 021926 (2009)
28. T. Roska, L.O. Chua, The CNN universal machine: an analogic array computer. IEEE Trans. Circuits Syst. II. Analog Digit. Signal Process. **40**(3):163–173 (1993)

Chapter 10
Extended Memristor Devices

Memristor devices embedded in pioneering future computing platforms are believed to represent one of the most promising key-enabling technologies for the treatment of massive amount of data. The ideal memristor, a two-terminal circuit element described by a nonlinear relationship between the time integrals of current and voltage (i.e., the current and voltage momenta, a.k.a. charge and flux), was theoretically envisioned by Chua back in 1971 in the seminal paper [1]. In 2008, some main peculiar features of the memristor were identified in a nano-scale film based on titanium dioxide by a team of Hewlett Packard researchers headed by S. Williams [2] (Chap. 2). After this discovery, the scientific community has focused the interest on the experimental observation of some aspects of memristor behavior in other nano-structures (Chap. 2). Memristor nano-devices typically adopt a metal/insulator/metal (MIM) structure. The electrode materials and the switching layer are carefully designed so as to obtain desirable programming voltage, memristance on/off ratio, power consumption, and device variation that enable fast, low-power yet reliable data processing.

Real memristor devices may exhibit a number of peculiar features that are in general not displayed by ideal memristors (cf. Chap. 2). For instance, different from an ideal memristor, a real memristor device may be volatile, it may have voltage thresholds and asymmetries in the hysteresis loops in the (v, i)-domain, it may display boundary effects due to the interface between the electrodes and the oxide, or it may show a dependence of memductance on different (nonelectrical) physical state variables (e.g., temperature, ion concentration, etc.) [3–6]. Roughly speaking, memristor devices can be modeled by a generic nonlinear dynamical input–output system just satisfying the fundamental *zero-crossing property*, i.e., the output (e.g., the current) vanishes when the input (e.g., the voltage) is zero. As a consequence, pinched hysteresis loops in the (v, i)-domain are observed when memristor devices are subject to a (zero-mean bipolar) periodic input. Actually, such systems satisfying a zero-crossing property fall into the class of the so-called memristive devices and systems introduced by Chua and Kang in [7], that have been discussed in Sect. 2.3

F. Corinto et al., *Nonlinear Circuits and Systems with Memristors*,
https://doi.org/10.1007/978-3-030-55651-8_10

of Chap. 2. A complete classification of memristor devices has been introduced in Sect. 2.4 of the same chapter (see also the articles [8, 9]) to provide a systematic tool to circuit designers. According to this classification, three main classes of memristor device models can be identified: *extended memristors*, *generic memristors* and *ideal memristors*.

An examination of the literature shows an evident gap between the experimental characterization of memristor devices and their systematic description and modeling in terms of fundamental circuit elements introduced in Chap. 1. In particular, the following question naturally arises:

> Which is the set of extended memristors that can be described by a composite circuit made of the interconnection of linear/nonlinear circuit elements and ideal memristors?

This chapter provides a study on composite circuits obtained by interconnection of nonlinear circuit elements and ideal memristors for describing and modeling extended memristors. The results rely upon the concept of two-terminal algebraic (α, β)-elements introduced in Chap. 1, i.e., elements whose constitutive relation is an algebraic link between $v^{(\alpha)}(t)$ (the α-derivative of $v(t)$ for positive α, or the α-integral of $v(t)$ for negative α, see Chap. 1 for the exact definition), and $i^{(\beta)}(t)$, where α, β are integers. Accordingly, it is possible to classify the four basic algebraic circuit elements, i.e., (possibly) nonlinear resistors, inductors, capacitors and memristors, and also higher-order algebraic circuit elements, by arranging them into a periodic table. For example, a nonlinear resistor corresponds to an $(\alpha, \beta) = (0, 0)$ element, whereas an ideal memristor corresponds to an $(\alpha, \beta) = (-1, -1)$ element.

In the chapter, it is shown that we can identify a significant class of extended memristors (i.e., memristor devices that cannot be reduced to just ideal memristors) that admits of the description in terms of a combination of algebraic circuit elements. To this end, it is preliminary observed that: (a) an ideal memristor with a linear relationship between charge and flux is equivalent to an ideal resistor (Chap. 1); (b) due to the Element Closure Property (Chap. 1), an arbitrary interconnection of (α, β)-elements is equivalent to a (α, β) element. Then, in order to implement an extended memristor, it is necessary to consider the interconnection of (α, β) elements with at least two different pairs of (α, β). One of the main results in the chapter is that an extended memristor can be realized by a combination of ideal memristors ($(\alpha, \beta) = (-1, -1)$ elements) and nonlinear resistors ($(\alpha, \beta) = (0, 0)$ elements). For brevity, we focus on the simplest class $\mathcal{D}_{\text{ext}}$ of extended memristors with one ideal memristor and one nonlinear (possibly locally active) resistor. By massaging the characteristics of the constitutive elements, such class $\mathcal{D}_{\text{ext}}$ of extended memristors is shown to be able to approximate rectifying effects and asymmetric pinched hysteresis loops of real nonvolatile switching memristor devices. Finally, it is shown that if the nonlinear resistors are described in terms of piecewise linear characteristics, then FCAM can still be employed to effectively analyze the dynamics of nonlinear circuits with extended memristors in $\mathcal{D}_{\text{ext}}$.

10.1 A Class of Extended Memristors

Consider the class of voltage-controlled extended memristor devices described by

$$i_M = G(\mathbf{x}, v_M)v_M \tag{10.1}$$

$$\frac{d\mathbf{x}}{dt} = \mathbf{h}(\mathbf{x}, v_M) \tag{10.2}$$

where i_M and v_M are the current through and the voltage across the memristor, $G(\mathbf{x}, v_M)$ is its memductance, whereas $\mathbf{x} \in \mathbf{R}^n$ is a vector of n internal state variables (possibly) including the memristor flux φ_M. According to the definition in Sect. 2.4.4 of Chap. 2, the zero-crossing property of the i_M–v_M curves, i.e.,

$$\lim_{v_M \to 0} i_M = \lim_{v_M \to 0} G(\mathbf{x}, v_M)v_M = 0 \tag{10.3}$$

requires that, for any $\mathbf{x} \in \mathbb{R}^n$, the memductance $G(\mathbf{x}, v_M) : \mathbf{R}^{n+1} \to \mathbb{R}$ is a *bounded* function in the neighbor of $(\mathbf{x}, 0)$. Then, (10.1)–(10.2) satisfying (10.3) describe a generic dynamical system satisfying the distinctive zero-crossing property [7].

In this chapter, the focus is on the class of memristor devices $\mathcal{D}_{\text{ext}}$, represented in Fig. 10.1, given by the parallel connection of an ideal flux-controlled memristor and a voltage-controlled nonlinear resistor. The ideal memristor is an algebraic $(\alpha, \beta) = (-1, -1)$ element characterized by the nonlinear relationship $q_M = f(\varphi_M)$ between the flux φ_M and charge q_M, where f is a *locally Lipschitz function.*[1] Since $v_M = v$, the CR of the ideal memristor in the (v, i)-domain is

$$i_M = W(\varphi_M)v \tag{10.4}$$

$$\frac{d\varphi_M}{dt} = v \tag{10.5}$$

where $W(\varphi_M) = f'(\varphi_M)$ is the memductance. The nonlinear resistor is an algebraic $(\alpha, \beta) = (0, 0)$ element defined by (being $v_R = v$)

$$i_R = F_R(v) \tag{10.6}$$

where $F_R(\cdot)$ is *locally Lipschitz* and $F_R(0) = 0$.

In the (v, i)-domain, $\mathcal{D}_{\text{ext}}$ is described by the CR

$$i = W(\varphi_M)v + F_R(v) \tag{10.7}$$

[1] This means that, for any φ_M, there exists a neighborhood U of φ_M and $k_f(\varphi_M) > 0$ such that $|f(\varphi_{M,1}) - f(\varphi_{M,2})| \leq k_f(\varphi_M)|\varphi_{M,1} - \varphi_{M,2}|$ for any $\varphi_{M,1}, \varphi_{M,2} \in U$, where $k_f(\varphi_M)$ is the Lipschitz constant of f at φ_M. This constant depends on φ_M and U. If f is locally Lipschitz, then the derivative f' is defined almost everywhere [10].

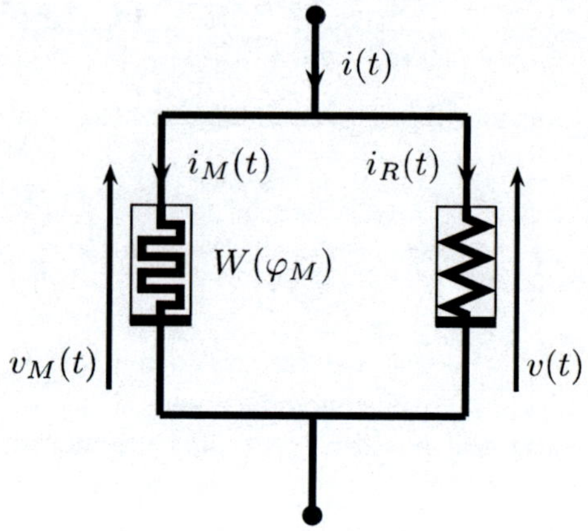

Fig. 10.1 Two-terminal element made of the parallel connection of an ideal flux-controlled memristor and a voltage-controlled nonlinear resistor

$$\frac{d\varphi_M}{dt} = v. \tag{10.8}$$

Note that the state variable of $\mathcal{D}_{\text{ext}}$ is the memristor flux φ_M. Also note that the class $\mathcal{D}_{\text{ext}}$ of extended memristors describes *nonvolatile memory devices*. Indeed, by turning off power, i.e., letting $v = 0$, we obtain that any φ_M is a stable EP of

$$\frac{d\varphi_M}{dt} = 0.$$

This property is analogous to that of an ideal memristor discussed in Chap. 2.

Equation (10.7) can be written as follows for $v \neq 0$

$$i = \left(\frac{F_R(v)}{v} + W(\varphi_M) \right) v = G(\varphi_M, v)v \tag{10.9}$$

where

$$G(\varphi_M, v) = \frac{F_R(v)}{v} + W(\varphi_M). \tag{10.10}$$

Since f and F_R are locally Lipschitz, it is not difficult to show that, for any $\varphi_M \in \mathbb{R}$, the memductance $G(\varphi_M, v)$ in (10.10) is bounded in a neighborhood of $(\varphi_M, 0)$ and then the zero-crossing property (10.3) is satisfied. In fact, we have $|F_R(v)| \leq k_{F_R}(0)|v|$ in a neighborhood of $v = 0$ and $W(\varphi_M) = f'(\varphi_M) \leq k_f(0)$ in a neighborhood of $\varphi_M = 0$. Then, $|G(\varphi_M, v)| \leq k_{F_R}(0) + k_f(0)$ in a neighborhood of $(0, 0)$. As a consequence, $\mathcal{D}_{\text{ext}}$ in Fig. 10.1 corresponds to an extended memristor as defined in Sect. 2.4.4 of Chap. 2.

Remark 10.1 It can be easily seen that $\mathcal{D}_{\text{ext}}$ is *not* an ideal memristor, nor it can be brought back to an ideal memristor via a transformation of variables, i.e., $\mathcal{D}_{\text{ext}}$ is not an ideal memristor sibling (Chap. 2), when F_R is a nonlinear function.

Remark 10.2 It is worth remarking that the assumption F_R is Lipschitz near 0 does not imply that there exists the $\lim_{v\to 0} F_R(v)/v$. As a counterexample, if $F_R(v) = G_1 v$ when $v \geq 0$ and $F_R(v) = G_2 v$ when $v < 0$, with $G_1 \neq G_2$, then $\lim_{v\to 0+} F_R(v)/v = G_1 \neq \lim_{v\to 0-} F_R(v)/v = G_2$.

Remark 10.3 If F_R is not locally Lipschitz, then $F_R(v)/v$ may be unbounded near $v = 0$ (for example, $F_R(v) = v^{1/2}$) and in consequence the zero-crossing property (10.3) might fail. It is worth to observe that we are interested in using extended memristors in dynamic nonlinear networks and then it is crucial to assume that the nonlinear elements are described by Lipschitz functions, otherwise the uniqueness of the solution for the dynamic equations would be in general not guaranteed.

Remark 10.4 If the ideal memristor (10.4) and (10.5) and the nonlinear resistor (10.6) are passive, then the extended memristor $\mathcal{D}_{\text{ext}}$ is passive as well. In fact, the memductance $G(\mathbf{x}, v) = G(\varphi_M, v) \geq 0$ for all (φ_M, v) and then the electric power always enters the memristor.

Remark 10.5 As noticed before, the state variable of $\mathcal{D}_{\text{ext}}$ is the memristor flux φ_M. Several real memristor devices are actually modeled by extended memristors where nonelectrical physical variables play an important role, e.g., the length of the doped part of the oxide in MIM structures, the radius of the conductive filament in resistive memories, or the temperature in phase-change memory devices and thermistors (see Chap. 2 and also the article [11] for a review).

If the nonelectrical state variables are in a one-to-one correspondence with the memristor flux φ_M (see [9, Theorem 2]), as in the case of the linear drift model (with or without window function) of the HP memristor [12], and of phase change memories [13], then the memristor device is basically described by an ideal memristor. As a consequence, such memristor devices fall in the class $\mathcal{D}_{\text{ext}}$. On the other hand, if the state variables cannot be put in correspondence with φ_M, then the memristor device may not be modeled via the class $\mathcal{D}_{\text{ext}}$. This is true for instance for the thermistor (Chap. 2), which is a volatile generic memristor, and as such does not belong to $\mathcal{D}_{\text{ext}}$.

Nevertheless, the considered class $\mathcal{D}_{\text{ext}}$ of extended memristors is of interest and practical value for many fabricated memristor devices. A relevant case in point is the real memristors in [14], that have been in fact modeled by the parallel connection of an ideal memristor and a passive nonlinear resistor (a diode) taking into account the rectifying effects experimentally observed due to the Schottky barrier at the junction between platinum and oxide (see also [15]).

Example 10.1 As a specific example, consider for the extended memristor in Fig. 10.1 the case where the nonlinear resistor is a Shockley diode obeying

$$F_R(v) = I_S \left(\exp(\frac{v}{\eta V_T}) - 1 \right) \tag{10.11}$$

where I_S is the reverse bias saturation current, V_T is the thermal voltage, and η is the ideality factor. By considering a cubic ideal memristor $q_M = f(\varphi_M) = a\varphi_M + \frac{1}{3}b\varphi_M^3$, the memconductance

$$G(\varphi_M, v) = \frac{I_S \left(\exp(\frac{v}{\eta V_T}) - 1 \right)}{v} + a + b\varphi_M^2$$

as in (10.10) is obtained. By De l'Hôpital rule, for any $\varphi_M \in \mathbb{R}$, we have

$$\lim_{v \to 0} G(\varphi_M, v) = \frac{I_S}{\eta V_T} + a + b\varphi_M^2$$

hence the memductance is bounded in a neighborhood of $(\varphi_M, v) = (\varphi_M, 0)$. The numerical simulation in Fig. 10.2 illustrates the rectifying effect of the diode in the pinched hysteresis loop of the extended memristor. Note that the hysteresis loop is nonsymmetric about the origin, which differs from what would be observed in an ideal memristor (Chap. 2).

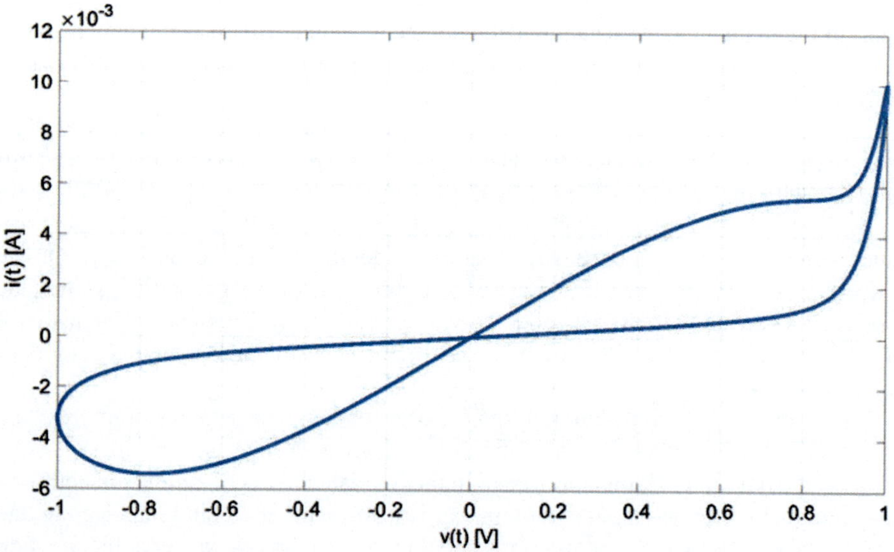

Fig. 10.2 Pinched i–v curve in response to a sinusoidal voltage signal $v(t) = \sin(2t)$ V applied to an extended memristor as in Fig. 10.1 made of the parallel connection of an ideal flux-controlled cubic memristor with $W(\varphi_M) = a + b\varphi_M^2$ ($a = 10^{-3}$, $b = 3 \cdot 10^{-3}$) and a Shockley diode with $F_R(v)$ as in (10.11). We have $I_S = 10^{-12}$ A, $V_T = 26 \cdot 10^{-3}$ V and $\eta = 1.7$. We have also let $\varphi_M(0) = 0$

10.1.1 Examples of Extended Memristors

This section reports some selected examples of an extended memristor $\mathcal{D}_{\text{ext}}$ as in Fig. 10.1 obtained by combining an ideal memristor with a circuit implementing a (locally active) nonlinear resistor. The goal is to show how the parameters of the nonlinear resistor can be massaged to capture the features of the experimental i–v curves observed during the characterization of real memristor devices. Simulations are carried out in the LTSPICE simulator to emulate the properties of real components and the ideal memristor is described by the LTSPICE code reported in [16, pag. 969]. For the sake of completeness, the code of the ideal memristor and the locally active nonlinear resistor are reported in Fig. 10.3.

The circuit in Fig. 10.3 is implemented by connecting in parallel (between node $N001$ and ground):

- a negative resistor implemented by the two-terminal element B_{NR}.[2] A detailed analysis of negative resistors can be found in [17]. The resistance R_6 is in

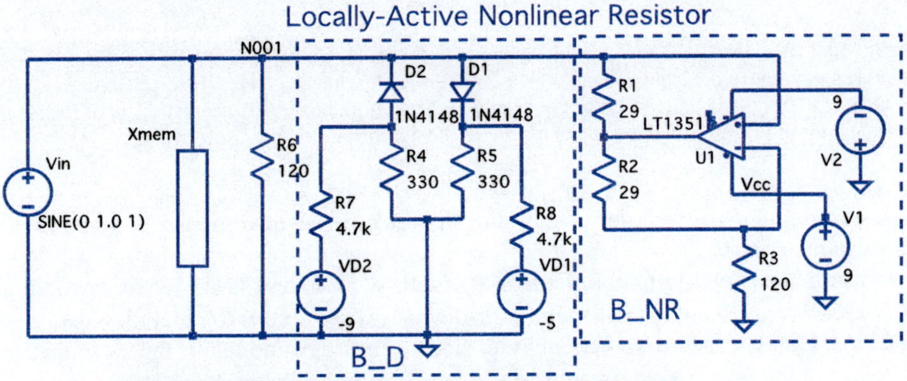


```
.subckt memristorR1 plus minus params: Ron=200 Roff=20k Rini=10k
.param uv=10f D=10n k={uv*Ron/D**2} a={(Rini-Ron)/(Roff-Rini)}
* model of memristive port
Roff plus aux {Roff}
Eres aux minus value={(Ron-Roff)/(1+a*exp(-4*k*V(q)))*I(Eres)}
*end of the model of memristive port
*integrator model
Gx 0 Q value={i(Eres)}
Cint Q 0 1
Raux Q 0 100meg
*end of integrator model
.ends memristorR1
.options method=gear ;use only for LTSpice
Xmem N001 0 memristorR1
```

```
.tran 0 1 0 0.1m
```

Fig. 10.3 Two-terminal element made of the parallel connection of an ideal flux-controlled memristor and a (locally active) nonlinear resistor. The inset in the figure reports the LTSPICE code of the ideal memristor and the locally active nonlinear resistor

[2] A similar circuit is also reported in http://www.chaotic-circuits.com/.

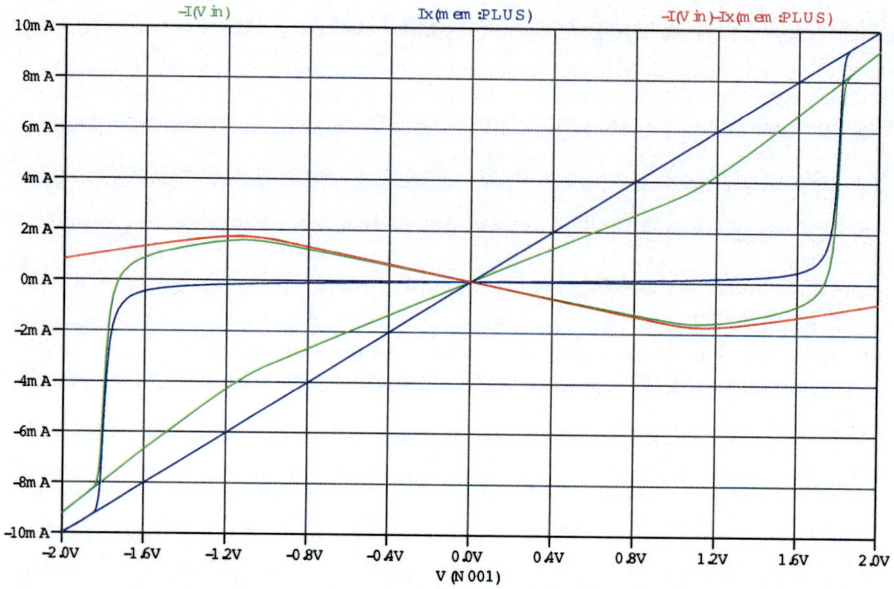

Fig. 10.4 Results for the two-terminal element in Fig. 10.3, where $R_6 = 150\,\Omega$ and $VD1 = 9$ V, subject to a sinusoidal input voltage $v(t) = 2\sin(2\pi ft)$ V with $f = 1$ Hz. The green, blue, and red curves are the i–v characteristic of the extended memristor, the i_M–v_M characteristic of the ideal memristor in Fig. 10.3, and the characteristic of the nonlinear resistor, respectively

parallel to B_{NR} in order to tune the slope of the negative part of the nonlinear characteristic;

- a nonlinear two-terminal element B_D with two-diodes (1N4148) for modeling rectifying effects in real memristor devices. The circuit B_D includes batteries $VD1$ and $VD2$ to switch off the diodes and then modulate the asymmetric switching features of the memristor devices. Suitable values (e.g., $VD1 = -9$ V) can be selected to obtain pinched i–v curves off the origin (i.e., the nano-battery effect reported in [18]).

Figures 10.4 and 10.5 show the simulation of the circuit in Fig. 10.3 subject to a sinusoidal input. Figure 10.4 depicts the results for a locally active extended memristor (i.e., green curve) due to the locally active nonlinear resistor (i.e., red curve) obtained by the parallel connection of $R_6 = 150\ \Omega$, B_D, and B_{NR} (the i_M–v_M curve of the passive ideal memristor is in blue). In this figure we have set $v(t) = 2\sin(2\pi ft)$ with $f = 1$ Hz and $VD1 = 9$ V.

If the amplitude of the sinusoidal input is reduced (e.g., $v(t) = \sin(2\pi ft)$ with $f = 1$ Hz), $VD1$ is set to -5 V to deactivate $D1$ and $R_6 = 120\,\Omega$ to null B_{NR}, then the nonlinear resistor acts like a diode (see the red curve in Fig. 10.5). As a consequence, the whole extended memristor exhibits i–v green curve in Fig. 10.5, that is asymmetric with respect to the origin. The i_M–v_M curve of the passive ideal memristor is in blue.

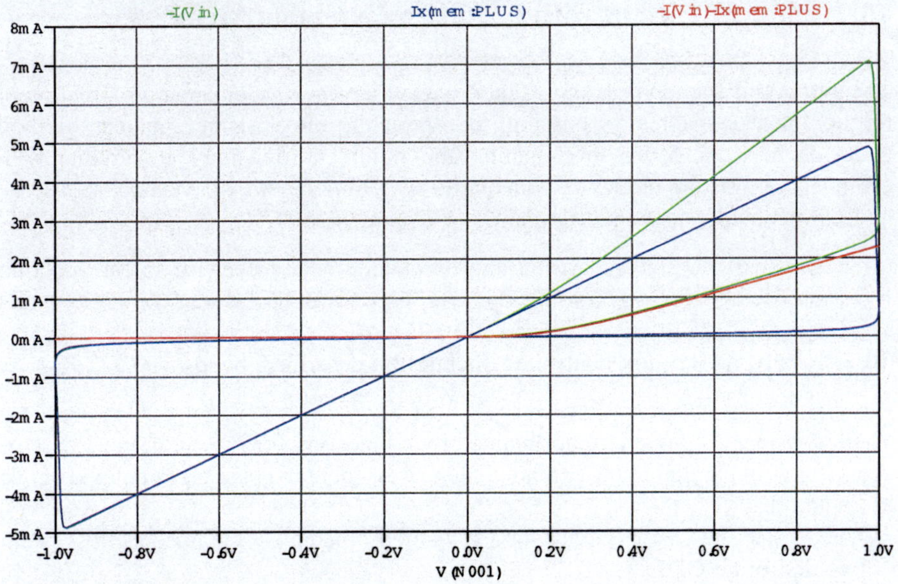

Fig. 10.5 Results for the two-terminal element as in Fig. 10.3. The green, blue, and red curves are the i–v characteristic of the extended memristor, the i_M–v_M characteristic of the ideal memristor in Fig. 10.3, and the characteristic of the nonlinear resistor, respectively

Remark 10.6 Chapter 1 provides a discussion on the two chief approaches to circuit device modeling, i.e., the physical approach and the black-box approach (refer also to the fundamental article [19]). In practice, a trade-off between the two approaches is desirable in order to derive the simpler and more accurate circuit model embedding the physics of the device. Such circuit modeling approach is used in this chapter to identify the class $\mathcal{D}_{\text{ext}}$ of extended memristor devices. In fact, it can be argued that an extended memristor in $\mathcal{D}_{\text{ext}}$ contains a part corresponding to an ideal memristor, modeling the physical phenomenon according to which the memductance depends upon φ_M, and additional circuit elements that take into account physical phenomena observed in the experimental characterization of real devices and not captured by ideal memristors, such as, for example, rectifying effects and asymmetric pinched i–v curves [14].

Remark 10.7 The technique in this chapter for modeling the class $\mathcal{D}_{\text{ext}}$ is quite different from common approaches reported in the literature. In fact, physical and mathematical memristor models available in the literature [20] are not oriented toward nonlinear network synthesis [19], namely, an approach that aims to model the device as the interconnection of fundamental two-terminal algebraic (α, β)-elements (cf. Chap. 1). It is also noted that the approach in this chapter to model $\mathcal{D}_{\text{ext}}$ can be easily extended to include circuit elements accounting for the presence of parasitic effects in real memristors (cf. [21]).

10.2 Simple Circuit with C and an Extended Memristor

The aim of this section is to show that if we use a piecewise linear approximation of the nonlinear resistor, it is possible to exploit the Flux-Charge Analysis Method (FCAM) in Chap. 5 for investigating the nonlinear dynamics in circuits with extended memristors modeling nonvolatile switching memory devices.

Let us consider an extended memristor $\mathcal{D}_{ext}$ made of:

(a) a (passive) nonlinear resistor approximated by a piecewise linear function $F_R(v) = G_1 v$ for $v \geq 0$ while $F_R(v) = G_2 v$ for $v < 0$, where $G_1, G_2 \geq 0$ and $G_1 \neq G_2$;
(b) a locally active flux-controlled memristor described by $q_M = f(\varphi_M) = -a\varphi_M + b\varphi_M^3$ with $a, b > 0$, hence $W(\varphi_M) = -a + 3b\varphi_M^2$.

Then, consider a simple circuit obtained by connecting for $t \geq t_0$ an ideal passive capacitor C in parallel to the extended memristor (denoted by eM) as defined above in (a) and (b).

The SEs of the eM–C circuit for $t \geq t_0$ are immediately obtained as the system of two first-order ODEs

$$C\frac{dv_C}{dt} = -W(\varphi_M)v_C - F_R(v_C) \tag{10.12}$$

$$\frac{d\varphi_M}{dt} = v_C \tag{10.13}$$

for any $t \geq t_0$. The state variables in the (v, i)-domain are v_C and φ_M and the initial conditions $v_C(t_0) = v_0$, $\varphi_M(t_0) = \varphi_0$. Since f and F_R are locally Lipschitz, given any initial condition $(v_0, \varphi_0)^T \in \mathbb{R}^2$, there exists a unique solution (v_C, φ_M) of (10.12)–(10.13), that is defined on a maximal interval of existence $[t_0, t_0 + T)$, where $T > 0$ is possibly $+\infty$ [10].

It can be easily checked that the following properties hold.

P_1: there exists a continuum of EPs $\{(\bar{v}, \bar{\varphi}) \in \mathbb{R}^2 : \bar{v} = 0\}$
P_2: if the initial condition $v_0 = 0$, then $(v_C(t), \varphi_M(t)) = (0, \varphi_0)$ for any $t \geq t_0$.

Properties P_1 and P_2 imply that the whole state-space $(v_C, \varphi_M)^T \in \mathbb{R}^2$ in the (v, i)-domain can be split into two regions $\mathcal{R}_+$ and $\mathcal{R}_-$ separated by the line of EPs, i.e., $\mathcal{R}_+ = \{(v_C, \varphi_M) \in \mathbb{R}^2 : v_C > 0\}$ whereas $\mathcal{R}_- = \{(v_C, \varphi_M)^T \in \mathbb{R}^2 : v_C < 0\}$. Hence, for any initial condition in $\mathcal{R}_+$, that is $v_0 > 0$, the dynamic route analysis of (10.12) and (10.13) and the uniqueness of the solution imply that $v_C(t) > 0$ for all $t \geq t_0$. The r.h.s. of (10.12) is $W_+(\varphi_M) = -(a + G_1) + 3b\varphi_M^2$ for any $t \in [t_0, t_0 + T)$. It follows that the eM–C circuit is equivalent to a capacitor in parallel to an ideal memristor defined by $f_+(\varphi_M) = -(a + G_1)\varphi_M + b\varphi_M^3$ and thus FCAM can be used in order to analyze the dynamic behavior in $\mathcal{R}_+$. Similar considerations hold in $\mathcal{R}_-$ and permit to reduce the analysis of the eM–C circuit to that of a memristor-circuit with a capacitor in parallel to an ideal memristor given by $f_{-}(\varphi_M) = -(a + G_2)\varphi_M + b\varphi_M^3$.

Let us focus on the nonlinear dynamics in $\mathcal{R}_+$. FCAM can be applied to reduce (10.12) and (10.13) to the first-order ODE in the (φ, q)-domain ($\forall t \geq t_0$)

$$C\frac{d\varphi_M(t;t_0)}{dt} = -f_{+}(\varphi_M(t;t_0) + \varphi_0) + Q_0^+ \tag{10.14}$$

where $\varphi_M(t;t_0) = \varphi_M(t) - \varphi_0$ is the incremental flux across the memristor and $Q_0^+ = Cv_0 + f_{+}(\varphi_0)$ is a constant quantity depending on the initial conditions $(\varphi_0, v_0)^T \in \mathcal{R}_+$ for the state variables in the (v, i)-domain.

It can be easily seen that (10.14) has a unique solution $\varphi_M(t;t_0)$ which is bounded and hence defined for any $t \geq t_0$. In addition, since (10.14) is an autonomous first-order ODE, $\varphi_M(t;t_0) \to \bar{\varphi}$ and its time derivative $\dot{\varphi}_M(t;t_0) \to 0$ as $t \to +\infty$ (Chap. 2). According to FCAM, the solution of (10.12) and (10.13) is obtained as $(v_C(t), \varphi_M(t)) = (\dot{\varphi}_M(t), \varphi_M(t) + \varphi_0)$ and then any solution of (10.12) and (10.13) is bounded and defined for $t \geq t_0$. Moreover $(v_C(t), \varphi_M(t)) \to (0, \bar{\varphi})$ as $t \to +\infty$.

To wrap up, the following result has been proved.

Proposition 10.1 *Any solution (v_C, φ_M) of the SEs describing the eM–C circuit in the (v, i)-domain is defined and bounded for any $t \geq t_0$. Furthermore, the solution converges to an EP $(0, \bar{\varphi})$ as $t \to +\infty$.*

Such a property is analogous to that of the circuit with a capacitor and an ideal memristor in Chap. 6. The investigation of invariant manifolds for the eM–C circuit can follow a discussion analogous to that in Example 5.15 of Chap. 5. In brief, it can be observed that when $v_0 > 0$, then Q_0^+ is an invariant of motion for (10.12) and (10.13). Therefore, we can define infinitely many positively invariant manifolds for (10.12) and (10.13) given by $\mathcal{M}(Q_0^+) = \{(v_C, \varphi_M)^T \in \mathbb{R}^2 : v_C > 0, Cv_C + f_{+}(\varphi_M) = Q_0^+\}$, where $Q_0^+ \in \mathbb{R}$, and on each manifold the dynamics is described in the (φ, q)-domain by the first-order ODE (10.14). The manifolds are nonintersecting and span the whole semi-plane $v_C > 0$ of the state-space in the (v, i)-domain. Moreover, we can prove the existence of bifurcations without parameters of EPs, i.e., bifurcations due to a change in the initial conditions and Q_0^+ for a fixed set of parameters, for the reduced-order dynamic equation in the (φ, q)-domain.

10.3 An $L - C$ Circuit with an Extended Memristor

Consider the $L - C$ circuit with an extended memristor in the class $\mathcal{D}_{\text{ext}}$ obtained via the parallel connection of an ideal flux-controlled memristor with memductance $W(\varphi_M)$ and a nonlinear resistor $i = F_R(v)$ as shown in Fig. 10.6.

In the (v, i)-domain the circuit satisfies the third-order SE

$$\begin{aligned} C\frac{dv_C}{dt} &= -i_L - W(\varphi_M)v_C - F_R(v_C) \\ L\frac{di_L}{dt} &= v_C \\ \frac{d\varphi_M}{dt} &= v_C \end{aligned}$$

where v_C is the voltage on the capacitor and i_L is the current through the inductor.

It can be immediately checked from the SEs that

$$\frac{d}{dt}(Li_L(t) - \varphi_M(t)) = 0$$

for any $t \geq t_0$. Therefore, the dynamics in the (v, i)-domain admits planar invariant manifolds given by

$$\mathcal{M}(Q_0) = \{(v_C, i_L, \varphi_M)^T \in \mathbb{R}^3 : Li_L - \varphi_M = \Phi_0\}$$

where $\Phi_0 \in \mathbb{R}$. We conclude that the circuit is intrinsically second-order, hence, being autonomous, it cannot display complex dynamics.

By substitution, the dynamics on an invariant manifold satisfies in the (v, i)-domain the second-order SE

$$C\frac{dv_C}{dt} = -i_L - W(Li_L - \Phi_0)v_C - F_R(v_C)$$

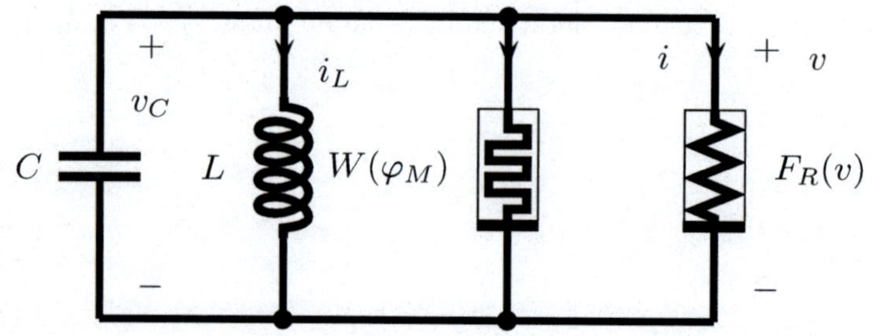

Fig. 10.6 An $L - C$ circuit with an extended memristor made of the parallel connection of an ideal flux-controlled memristor and a voltage-controlled nonlinear resistor

$$L\frac{di_L}{dt} = v_C.$$

Such a system is similar to that describing a planar Van der Pol oscillator (Chap. 4). Then, we may take advantage of the theoretic results on planar Van der Pol oscillators to study the dynamics of the considerer circuit. In particular, we can expect that there are bifurcations without parameters when varying Φ_0 via the initial conditions $i_L(t_0)$ and $\varphi_M(t_0)$ in the (v, i)-domain.

10.4 Discussion

This chapter has introduced a class $\mathcal{D}_{\text{ext}}$ of extended memristors obtained by interconnecting basic algebraic circuit elements given by ideal memristors and nonlinear resistors. The proposed approach for modeling such a class is in the spirit of classical techniques for analyzing and synthesizing nonlinear circuits. The class $\mathcal{D}_{\text{ext}}$ can be used for instance to yield a more accurate characterization and modeling of real memristor devices with rectifying effects and complex switching features. Finally, it has been shown that simple dynamic circuits containing extended memristors in $\mathcal{D}_{\text{ext}}$ with piecewise-linear resistors can be effectively investigated via FCAM.

References

1. L.O. Chua, Memristor-the missing circuit element. IEEE Trans. Circuit Theory **18**(5), 507–519 (1971)
2. D.B. Strukov, G.S. Snider, D.R. Stewart, R.S. Williams, The missing memristor found. Nature **453**(7191), 80–83 (2008)
3. S.H. Jo, T. Chang, I. Ebong, B.B. Bhadviya, P. Mazumder, W. Lu, Nanoscale memristor device as synapse in neuromorphic systems. Nano Lett. **10**(4), 1297–1301 (2010)
4. A. Ascoli, S. Slesazeck, H. Mähne, R. Tetzlaff, T. Mikolajick, Nonlinear dynamics of a locally-active memristor. IEEE Trans. Circuits Syst. I. Regul. Pap. **62**(4), 1165–1174 (2015)
5. V. Erokhin, T. Berzina, P. Camorani, A. Smerieri, D. Vavoulis, J. Feng, M.P. Fontana, Material memristive device circuits with synaptic plasticity: learning and memory. BioNanoScience **1**(1–2), 24–30 (2011)
6. S. Kumar, J.P. Strachan, R. Stanley Williams, Chaotic dynamics in nanoscale NbO_2 Mott memristors for analogue computing. Nature **548**(7667), 318 (2017)
7. L.O. Chua, S.M. Kang, Memristive devices and systems. Proc. IEEE **64**(2), 209–223 (1976)
8. L. Chua, Everything you wish to know about memristors but are afraid to ask. Radioengineering **24**(2), 319–368 (2015)
9. F. Corinto, P.P. Civalleri, L.O. Chua, A theoretical approach to memristor devices. IEEE J. Emerg. Select. Top. Circuits Syst. **5**(2), 123–132 (2015)
10. H.K. Khalil, *Nonlinear Systems* (Prentice Hall, Upper Saddle River, 2002)
11. D. Panda, P.P. Sahu, T. Y. Tseng, A collective study on modeling and simulation of resistive random access memory. Nanoscale Res. Lett. **13**(1), 8 (2018)

12. F. Corinto, A. Ascoli, A boundary condition-based approach to the modeling of memristor nanostructures. IEEE Trans. Circuits Syst. I. Regul. Pap. **59**(11), 2713–2726 (2012)
13. J. Secco, F. Corinto, A. Sebastian, Flux-charge memristor model for phase change memory. IEEE Trans. Circuits Syst. II. Expr. Briefs **65**(1), 111–114 (2018)
14. J. Joshua Yang, M.D. Pickett, X. Li, D.A.A. Ohlberg, D.R. Stewart, R. Stanley Williams, Memristive switching mechanism for metal/oxide/metal nanodevices. Nat. Nanotechnol. **3**(7), 429 (2008)
15. S.G. Hu, S.Y. Wu, W.W. Jia, Q. Yu, L.J. Deng, Y.Q. Fu, Y. Liu, T.P. Chen, Review of nanostructured resistive switching memristor and its applications. Nanosci. Nanotechnol. Lett. **6**(9), 729–757 (2014)
16. D. Biolek, M. Di Ventra, Y.V. Pershin, Reliable SPICE simulations of memristors, memcapacitors and meminductors. Radioengineering **22**(4), 945 (2013)
17. L.O. Chua, C.A. Desoer, E.S. Kuh, *Linear and Nonlinear Circuits* (McGraw-Hill, New York, 1987)
18. I. Valov, E. Linn, S. Tappertzhofen, S. Schmelzer, J. Van den Hurk, F. Lentz, R. Waser, Nanobatteries in redox-based resistive switches require extension of memristor theory. Nat. Commun. **4**, 1771 (2013)
19. L. Chua, Device modeling via basic nonlinear circuit elements. IEEE Trans. Circuits Syst. **27**(11), 1014–1044 (1980)
20. A. Ascoli, F. Corinto, V. Senger, R. Tetzlaff, Memristor model comparison. IEEE Circuits Syst. Mag. **13**(2), 89–105 (2013)
21. M.P. Sah, C. Yang, H. Kim, B. Muthuswamy, J. Jevtic, L. Chua, A generic model of memristors with parasitic components. IEEE Trans. Circuits Syst. I. Regul. Pap. **62**(3), 891–898 (2015)

Chapter 11
Nonlinear Dynamics of Circuits with Mem-Elements

In recent years, the use of memristors as nonlinear dynamical elements for real-time analog signal processing has been a topic of ever increasing interest. Memristors are widely employed in neuromorphic architectures and cellular neural networks, where they behave as nonlinear dynamic devices within neurons (Chap. 9) or they implement synaptic connections in nanotechnology with adaptation capabilities and reduced area consumption and power dissipation, see, e.g., [1–13], and references therein. The article [14] stresses that scalable electronic devices, as Mott memristors, implementing a source of controllable chaotic behavior, and that can be incorporated into a neuromorphic network, are expected to be an essential component of future computational systems. Memristors have also been found effective for implementing rich nonlinear dynamics at the core of reservoir computing [15]. A new paradigm has been developed in [16], where memcomputing machines use memristors both as dynamical nonlinear devices for analog computing and as devices that memorize the result of computation in the same physical location, with the goal to overcome some bottlenecks of Von Neumann machines where the computing and memory phases are performed at different locations.

Memcapacitors and meminductors are also currently receiving an increasing attention. The recent article [17] reports on a voltage-controlled memcapacitor where capacitive memory arises from reversible and hysteretic geometrical changes in a lipid bilayer mimicking the composition and structure of biomembranes (cf. Example 2.28) in Chap. 2. It is pointed out that such mem-elements are capable of co-locating signal processing and memory via history-dependent reconfigurability at the nanoscale and, as such, they are expected to be vital for next generation computing materials striving to match the brain's efficiency and flexible cognitive capabilities. This should in turn pave the way to develop low-energy, biomolecular neuromorphic mem-elements as models to study capacitive memory and signal processing in neuronal membranes. From an application viewpoint, memcapacitive properties observed in certain metal-oxides nanostructures (e.g., VO_2) have suggested their use for frequency tuning of some metamaterials [18].

F. Corinto et al., *Nonlinear Circuits and Systems with Memristors*,
https://doi.org/10.1007/978-3-030-55651-8_11

In the paper [19], an adaptive reactive element is implemented and modeled as memcapacitor/meminductor and it is demonstrated that such component is ideal for creating adaptive metasurfaces for manipulating EM waves due to its ability to hold the electromagnetic state without external bias. The use of memcapacitors and meminductors for implementing nonlinear oscillators is increasingly explored and widely investigated theoretically in the literature, see [20–22], and references therein. More recently, the implementation of neural network synapses via memcapacitors has been suggested [23].

Memristors, memcapacitors, and meminductors are frequently used for modeling nanoscale devices in combination with nonlinear inductors or capacitors. As a relevant example, the classical circuit model for a Josephson junction consists of a parallel connection of a linear capacitor, a linear resistor, and a nonlinear flux-controlled inductor. However, a more rigorous quantum mechanical analysis of the Josephson junction dynamics reveals the presence of an additional current component due to interference among quasi-particle pairs, which can be modeled with the current flowing into a flux-controlled memristor (cf. Example 1.9 in Chap. 1).

These considerations show that a challenging topic is to develop ad hoc methods for investigating the peculiar nonlinear dynamical properties of circuits containing such emerging nanoscale elements with memory properties, a.k.a. *mem-elements*. This is believed to be a crucial step for further understanding stability, oscillatory and synchronization properties, and more generally the computational capabilities of neural architectures and reservoir systems with mem-elements, or memcomputing machines.

Chapter 5 introduced FCAM as an effective tool for analyzing the nonlinear dynamics of a class $\mathcal{LM}$ of nonlinear circuits containing ideal flux- or charge-controlled memristors and ideal (linear) R, L, C. Goal of the chapter is to show that FCAM can be naturally extended to a much larger class $\mathcal{N}_e$ of circuits containing, in addition to the elements of $\mathcal{LM}$, also memcapacitors, meminductors, and nonlinear capacitors and inductors. In the chapter we treat in a systematic way the extension to memcapacitors and meminductors, while discussing via selected examples the case of nonlinear capacitors and inductors.

The main results in the chapter are summarized as follows.

1. We obtain the CR and equivalent circuit in the (φ, q)-domain of each two-terminal element in $\mathcal{N}_e$ and then we identify wide and relevant subclasses of $\mathcal{N}_e$ for which it is possible to write an SE representation in the (φ, q)-domain and also in the traditional (v, i)-domain.
2. It is shown that there is a reduction of order for the SEs in the (φ, q)-domain, with respect to the (v, i)-domain, leading to advantages in the nonlinear dynamic analysis.
3. Via the extended FCAM it is shown, in the inputless (autonomous) case, that the state space can be foliated in a continuum of invariant manifolds and there coexist infinitely many different reduced-order dynamics and attractors for the same set of circuit parameters. This is the basis to show the existence of bifurcations due

to changing initial conditions (ICs) for fixed parameters, i.e., bifurcations without parameters.
4. In the non-autonomous case, formula for designing independent pulse current or voltage sources to drive trajectories through manifolds with different regimes and dynamics are obtained. These can be used to effectively design circuits with mem-elements that behave as controllable sources of complex dynamics to be used in future neuromorphic systems.

11.1 Nonlinear Circuits with Mem-Elements

For the reader's convenience, we first summarize some facts from Chap. 1 which are needed in this chapter. Let $v(t)$ and $i(t)$ be the voltage and current of a two-terminal circuit element $\mathcal{D}$. Define the higher-order derivatives and integrals of $v(t)$, denoted by $v^{(\alpha)}(t)$ and $i^{(\beta)}(t)$, respectively, where α, β are integers (i.e., $0, \pm 1, \pm 2, \ldots$). While $v^{(0)}(t) = v(t)$ and $i^{(0)}(t) = i(t)$, the following notation and nomenclature are also used:

$$v^{(-1)}(t) = \varphi(t) = \int_{-\infty}^{t} v(\tau) d\tau$$

is the flux (or voltage momentum) and

$$i^{(-1)}(t) = q(t) = \int_{-\infty}^{t} i(\tau) d\tau$$

is the charge (or current momentum). Note that, from a physical viewpoint, $\varphi(t)$ and $q(t)$ do not necessarily correspond to the flux of a magnetic field or the charge accumulated in $\mathcal{D}$. Let us also introduce the electrical variables

$$v^{(-2)}(t) = \rho(t) = \int_{-\infty}^{t} \left(\int_{-\infty}^{\tau} v(\tau_1) d\tau_1 \right) d\tau = \int_{-\infty}^{t} \varphi(\tau) d\tau$$

i.e., the time integral of flux (or flux momentum) and

$$i^{(-2)}(t) = \sigma(t) = \int_{-\infty}^{t} \left(\int_{-\infty}^{\tau} i(\tau_1) d\tau_1 \right) d\tau = \int_{-\infty}^{t} q(\tau) d\tau$$

i.e., the time integral of charge (or charge momentum).

According to the axiomatic approach in Chap. 1, a two-terminal element $\mathcal{D}$ is said to be algebraic if and only if its CR can be expressed by an algebraic relationship that involves at most two independent electric variables $v^{(\alpha)}(t)$ and $i^{(\beta)}(t)$. An algebraic element is also referred to as an (α, β)-element.

The four basic algebraic elements, i.e., the resistor, capacitor, inductor, and memristor, are defined axiomatically by an algebraic CR involving the electric pairs $(v^{(0)}, i^{(0)}) = (v, i)$, $(v^{(0)}, i^{(-1)}) = (v, q)$, $(v^{(-1)}, i^{(0)}) = (\varphi, i)$, and $(v^{(-1)}, i^{(-1)}) = (\varphi, q)$, respectively. Higher-order elements as a memcapacitor or a meminductor have been introduced axiomatically in a similar way in Chap. 1. Namely, a memcapacitor is defined by an algebraic CR involving the electric quantities $(v^{(-1)}, i^{(-2)}) = (\varphi, \sigma)$ and a meminductor by a CR involving $(v^{(-2)}, i^{(-1)}) = (\rho, q)$. The algebraic (α, β)-elements can be systematically organized in a "periodic table" where each element exhibits peculiar properties of the small-signal impedance at an operating point. The elements of the periodic table named memristors, memcapacitors, and meminductor are also designated as *mem-elements*, in contrast to the classical nonlinear resistors, capacitors, and inductors.

The next class $\mathcal{N}_e$ of nonlinear dynamical circuits containing mem-elements is the main subject of the chapter. Specifically, $\mathcal{N}_e$ includes the following types of two-terminal elements $\mathcal{D}$:

- flux- or charge-controlled memristors, denoted M_φ and M_q, resp.
- nonlinear charge- or voltage-controlled capacitors C_q and C_v, resp.
- nonlinear flux- or current-controlled inductors L_φ and L_i, resp.
- σ- or flux-controlled memcapacitors MC_σ and MC_φ, resp.
- ρ- or charge-controlled meminductors ML_ρ and ML_q, resp.
- linear resistors, capacitors, and inductors R, C, and L, resp.
- time-varying current or voltage sources, $a(t)$ and $e(t)$, resp.

Since a generic circuit in $\mathcal{N}_e$ is made of algebraic elements with different α and β (i.e., the trivial case of circuits obtained interconnecting elements of one single kind is excluded), the following question arises:

> Which is the domain (α, β), if any, where the dynamics of the circuit is described in the simplest form (e.g., the SEs contain the lowest number of differential equations)?

If a nonlinear dynamical circuit in $\mathcal{N}_e$ contains as mem-elements only memristors,[1] then the answer to the previous question, which is provided in Chap. 5, is that the most suitable domain is $(\alpha, \beta) = (-1, -1)$, i.e., the (φ, q)-domain. The goal of this chapter is to extend FCAM to circuits with not only memristors, but memcapacitors and meminductors as well, and to show that the analysis in the (φ, q)-domain presents several advantages over the traditional (v, i)-domain. In other words, the answer to the stated question continues to be that the most suitable domain is the (φ, q)-domain also for the extended class $\mathcal{N}_e$.

Suppose we wish to analyze the dynamics of a circuit in $\mathcal{N}_e$ for $t \geq t_0$, where t_0 is a finite initial instant, in the (φ, q)-domain. According to FCAM, the analysis has to exploit the *incremental flux* and *incremental charge*

[1] Such a class of nonlinear dynamical circuits is denoted by $\mathcal{LM}$ in Chap. 5. Note that $\mathcal{LM} \subset \mathcal{N}_e$.

$$\varphi(t;t_0)=\int_{t_0}^{t}\varphi(\tau)d\tau; \qquad q(t;t_0)=\int_{t_0}^{t}i(\tau)d\tau$$

at the terminals of any element $\mathcal{D}$, where $t \geq t_0$. Kirchhoff charge laws (KqLs) are expressed in the (φ, q)-domain as $\mathbf{Aq}(t;t_0) = \mathbf{0}$ while Kirchhoff flux laws (KφLs) read as $\mathbf{B}\boldsymbol{\varphi}(t;t_0) = \mathbf{0}$ for $t \geq t_0$. We denoted by $\mathbf{q}(t;t_0)$ and $\boldsymbol{\varphi}(t;t_0)$ the vectors of incremental charges and fluxes of the elements, whereas $\mathbf{A}$ and $\mathbf{B}$ are the reduced incidence matrix and fundamental loop matrix, respectively (Chap. 3).

Since KqLs and KφLs are written in incremental form, the CR of each element $\mathcal{D}$ has to be expressed as a link between the incremental flux and charge at its terminals. For instance, using the passive convention, a σ-controlled memcapacitor is defined by $\varphi_{MC}(t) = f_{MC}(\sigma_{MC}(t))$. In the (φ, q)-domain the CRs for $t \geq t_0$ are (see Sect. 11.4 for more details)

$$\varphi_{MC}(t;t_0) = f_{MC}(\sigma_{MC}(t;t_0)+\sigma_{MC_0}) - f_{MC}(\sigma_{MC_0})$$
$$q_{MC}(t;t_0) = \dot{\sigma}_{MC}(t;t_0) - q_{MC}(t_0).$$

For pedagogical reasons, we find it useful to first highlight key features and advantages of the analysis in the flux-charge domain by means of a simple circuit in $\mathcal{N}_e$ (Sect. 11.2). Then, we extend FCAM and develop a systematic approach for a comprehensive analysis of circuits in $\mathcal{N}_e$ in the (φ, q)-domain in the subsequent Sects. 11.3–11.5.

11.2 Motivating Example

11.2.1 Reduction of Order

Consider a nonlinear dynamical circuit in $\mathcal{N}_e$ with a σ-controlled memcapacitor MC$_\sigma$, a flux-controlled memristor M$_\varphi$, and a current source $a(t)$ (Fig. 11.1). For simplicity the circuit is named MC − M.

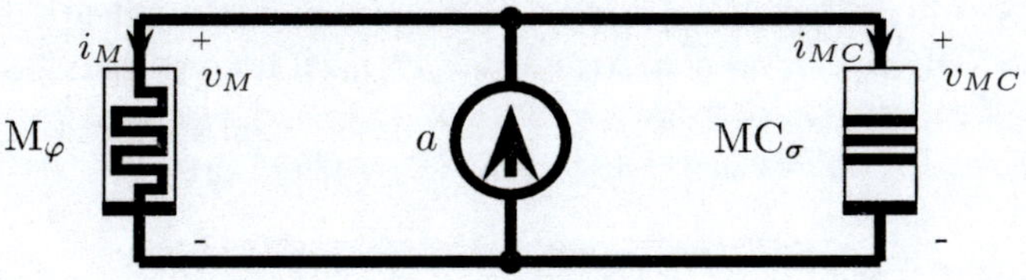

Fig. 11.1 The MC − M circuit in $\mathcal{N}_e$ that includes a σ-controlled memcapacitor, a flux-controlled memristor, and a current source

By using the classical approach in the (v, i)-domain based on:

- KCL: $i_{MC}(t) = -i_M(t) + a(t)$
- KVL: $v_{MC}(t) = v_M(t)$
- CR of MC_σ: $v_{MC}(t) = f'_{MC}(\sigma_{MC}(t))q_{MC}(t)$, with $\dot{\sigma}_{MC}(t) = q_{MC}(t)$ and $\dot{q}_{MC}(t) = i_{MC}(t)$
- CR of M_φ: $i_M(t) = f'_M(\varphi_M(t))v_M(t)$, with $\dot{\varphi}_M(t) = v_M(t)$

the SEs for $t \geq t_0$ result in the *third-order* system

$$\begin{aligned}\dot{q}_{MC}(t) &= -f'_M(\varphi_M(t))f'_{MC}(\sigma_{MC}(t))q_{MC}(t) + a(t)\\ \dot{\sigma}_{MC}(t) &= q_{MC}(t)\\ \dot{\varphi}_M(t) &= f'_{MC}(\sigma_{MC}(t))q_{MC}(t).\end{aligned}$$

The state variables for the SE description of the MC − M circuit in the (v, i)-domain are $(\sigma_{MC}(t), q_{MC}(t), \varphi_M(t))$ and the corresponding ICs are $\sigma_{MC}(t_0) = \sigma_{MC_0}$, $q_{MC}(t_0) = q_{MC_0}$ and $\varphi_M(t_0) = \varphi_{M_0}$.

In the (φ, q)-domain MC − M can be analyzed for $t \geq t_0$ via:

- KqL: $q_{MC}(t; t_0) = -q_M(t; t_0) + q_a(t; t_0)$
- KφL: $\varphi_{MC}(t; t_0) = \varphi_M(t; t_0)$
- CR of MC_σ: $\varphi_{MC}(t; t_0) = f_{MC}(\sigma_{MC}(t; t_0) + \sigma_{MC_0}) - f_{MC}(\sigma_{MC_0})$ and $q_{MC}(t; t_0) = \dot{\sigma}_{MC}(t; t_0) - q_{MC_0}$
- CR of memristor: $q_M(t; t_0) = f_M(\varphi_M(t; t_0) + \varphi_{M_0}) - f_M(\varphi_{M_0})$
- CR of current source: $q_a(t; t_0) = \int_{t_0}^{t} a(\tau)d\tau$ for any $\varphi_a(t; t_0)$.

By substitution, the following ***first-order*** SE in the (φ, q)-domain for $t \geq t_0$ is obtained:

$$\begin{aligned}\dot{\sigma}_{MC}(t; t_0) = &-f_M\big(f_{MC}(\sigma_{MC}(t; t_0) + \sigma_{MC_0}) - f_{MC}(\sigma_{MC_0}) + \varphi_{M_0}\big)\\ &+q_{MC_0} + f_M(\varphi_{M_0}) + q_a(t; t_0)\end{aligned} \tag{11.1}$$

where the only state variable is $\sigma_{MC}(t; t_0)$ and the IC is $\sigma_{MC}(t_0; t_0) = 0$.

Clearly, the SEs of the MC − M circuit exhibit a reduction of order equal to 2 passing from the (v, i)-domain to the (φ, q)-domain. Note also that the right-hand side of the SE (11.1) in the (φ, q)-domain depends upon ICs in the (v, i)-domain.

From the KqL $q_{MC}(t; t_0) + q_M(t; t_0) = q_a(t; t_0)$ it follows that

$$Q(t) \doteq q_{MC}(t) + f_M(\varphi_M(t)) = Q_0 + q_a(t; t_0)$$

for any $t \geq t_0$, where

$$Q_0 = Q_0(q_{MC_0}, \varphi_{M_0}) \doteq q_{MC_0} + f_M(\varphi_{M_0}).$$

In a similar way, the KφL $\varphi_{MC}(t; t_0) = \varphi_M(t; t_0)$ implies that

$$\Phi(t) \doteq f_{MC}(\sigma_{MC}(t)) - \varphi_M(t) = \Phi_0$$

for any $t \geq t_0$, where

$$\Phi_0 = \Phi_0(\sigma_{MC_0}) \doteq f_{MC}(\sigma_{MC_0}) - \varphi_{M_0}.$$

Using these notations, (11.1) can be conveniently rewritten as

$$\dot{\sigma}_{MC}(t) = -f_M\big(f_{MC}(\sigma_{MC}(t)) - \Phi_0\big) + Q_0 + q_a(t; t_0) \tag{11.2}$$

for $t \geq t_0$, with IC $\sigma_{MC}(t_0) = \sigma_{MC_0}$.

11.2.2 Invariant Manifolds, Coexisting Dynamics, and Bifurcations Without Parameters

Consider first the autonomous MC – M circuit where $a(t) = 0$, hence $q_a(t; t_0) = 0$, $t \geq t_0$. Then, the previous expressions of $Q(t)$ and $\Phi(t)$ simplify as follows (for $t \geq t_0$):

$$Q(t) = q_{MC}(t) + f_M(\varphi_M(t)) = Q_0$$
$$\Phi(t) = f_{MC}(\sigma_{MC}(t)) - \varphi_M(t) = \Phi_0$$

i.e., $Q(t)$ and $\Phi(t)$, which depend only on the state variables in the (v, i)-domain, are *invariants of motion* for the dynamics. Then, the three-dimensional state space $(\sigma_{MC}(t), q_{MC}(t), \varphi_M(t)) \in \mathbb{R}^3$ can be foliated in ∞^2 1D manifolds (each manifold is a curve in $\mathbb{R}^3$), which are (positively) invariant for the dynamics in the (v, i)-domain, given by

$$\mathcal{M}(Q_0, \Phi_0) = \{(\sigma_{MC}(t), q_{MC}(t), \varphi_M(t))^\tau \in \mathbb{R}^3 : q_{MC}(t) + f_M(\varphi_M(t)) = Q_0,$$
$$f_{MC}(\sigma_{MC}(t)) - \varphi_M(t) = \Phi_0\} \tag{11.3}$$

where $Q_0, \Phi_0 \in \mathbb{R}$.

On each manifold the MC – M circuit has a first-order dynamics described by the SE

$$\dot{\sigma}_{MC}(t) = -f_M\big(f_{MC}(\sigma_{MC}(t)) - \Phi_0\big) + Q_0. \tag{11.4}$$

From (11.4) it is clear that the nonlinear dynamics (number of equilibrium points (EPs) and their stability properties) is strongly dependent upon the terms Φ_0 and Q_0, which in turn depend upon the ICs in the (v, i)-domain. Hence, in the (v, i)-domain there coexist ∞^2 different first-order dynamics, one for each manifold $\mathcal{M}$ specified by Q_0 and Φ_0.

To clarify the implications of the presence of invariant manifolds $\mathcal{M}(Q_0, \Phi_0)$ in the state-space $(\sigma_{MC}(t), q_{MC}(t), \varphi_M(t))^T \in \mathbb{R}^3$, let us focus on a specific MC – M circuit with nonlinearities

$$f_{MC}(\sigma_{MC}) = \sigma_{MC} + \frac{1}{3}\sigma_{MC}^3$$

and

$$f_M(\varphi_M) = -\varphi_M + \frac{1}{3}\varphi_M^3.$$

By changing the ICs in the (v, i)-domain in such a way that Φ_0 and Q_0 are varied, different dynamical behaviors and bifurcations occur in the MC – M circuit described by (11.4), even if circuit parameters and $f_{MC}(\cdot)$ and $f_M(\cdot)$ are held fixed. For simplicity, assume $\Phi_0 = 0$, but Q_0 can change. In this case, each manifold $\mathcal{M}(Q_0) = \mathcal{M}(Q_0, 0)$ can be represented by a curve in the (σ_{MC}, q_{MC}) plane given by

$$\mathcal{M}(Q_0) = \{(\sigma_{MC}, \dot{\sigma}_{MC} = q_{MC})^\tau \in \mathbb{R}^2 : q_{MC} = -f_M(f_{MC}(\sigma_{MC})) + Q_0\}.$$

Three of these manifolds (for $Q_0 = 0$, $Q_0 = 0.3$, and $Q_0 = 0.7$) are represented in Fig. 11.2. The first-order dynamics on each manifold can be studied via the Dynamic Route Maps (DRMs) reported in the same figure (cf. Chap. 4), where the arrowheads denote the direction of motion on each invariant manifold.

The DRM analysis allows us to draw the following scenario:

- any solution of (11.4) is bounded and since (11.4) is a first-order autonomous system, any solution converges to an EP (cf. Chap. 4);
- the MC – M circuit exhibits a bistable behavior on the invariant manifold $\mathcal{M}(Q_0 = 0)$, named zero-manifold, due to one unstable EP at the origin and two asymptotically stable EPs;
- the MC – M circuit is bistable also on the invariant manifold $\mathcal{M}(Q_0 = 0.3)$, however, there is a different location of the three EPs;
- the MC – M circuit has a globally convergent dynamics, i.e., every solution of (11.4) approaches the unique EP, on the invariant manifold $\mathcal{M}(Q_0 = 0.7)$.

The results illustrated in Fig. 11.2 make clear the presence of infinitely many different dynamics for fixed circuit parameters, that can be tuned by the ICs in the (v, i)-domain. In particular, increasing (or decreasing) Q_0 causes the disappearance of a pair of EPs of (11.4) via a saddle-node bifurcation without parameters (Chap. 6).

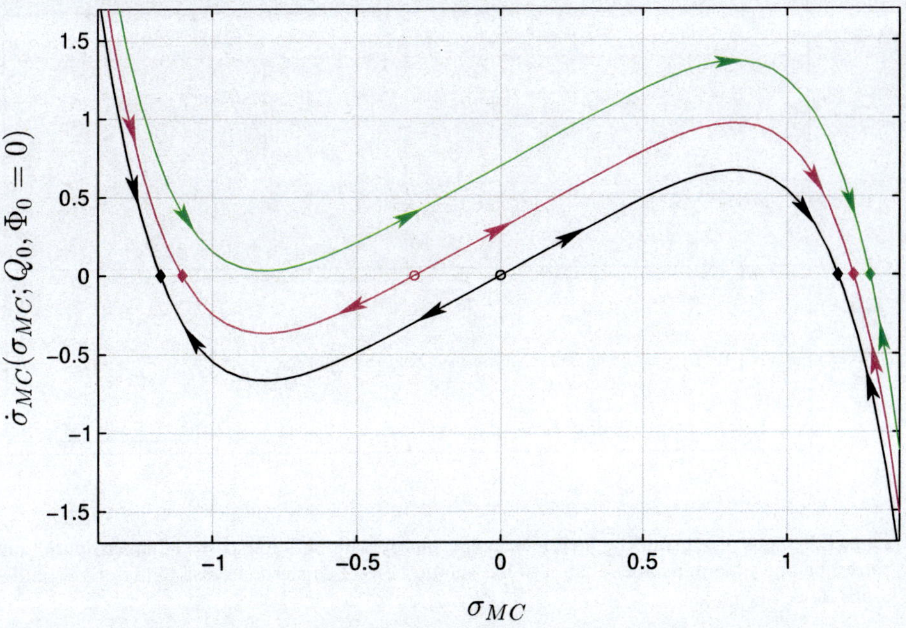

Fig. 11.2 Three different invariant manifolds corresponding to $Q_0 = 0$ (black), $Q_0 = 0.3$ (magenta), and $Q_0 = 0.7$ (green). Direction of motion along manifolds is shown via arrowheads. We have let $\Phi_0 = 0$

11.2.3 Programming the Circuit Dynamics via Input Pulses

The current source $a(t)$, or, equivalently, $q_a(t; t_0)$, can be used for driving the nonlinear dynamics of the MC – M circuit through different invariant manifolds and hence tuning different dynamics. Consider a solution of (11.2) starting in a neighborhood of the origin (red circle in Fig. 11.4) on the zero-manifold $\mathcal{M}(Q_0 = 0)$. Also, assume $a(t)$ is a sequence of three rectangular pulses, each with duration $\Delta = 1$ (see Fig. 11.3) so that the area under each pulse corresponds to its amplitude. The solution first moves along $\mathcal{M}(Q_0 = 0)$ approaching the stable EP marked by a black diamond in Fig. 11.4. Then, a pulse with area (charge) 0.3 is applied via the current source at $t = 15$. We have

$$Q(t) = Q(0) = 0$$

for $0 \le t \le 15$ while

$$Q(16) = Q(15) + \int_{15}^{16} a(\tau; 0)d\tau = 0.3$$

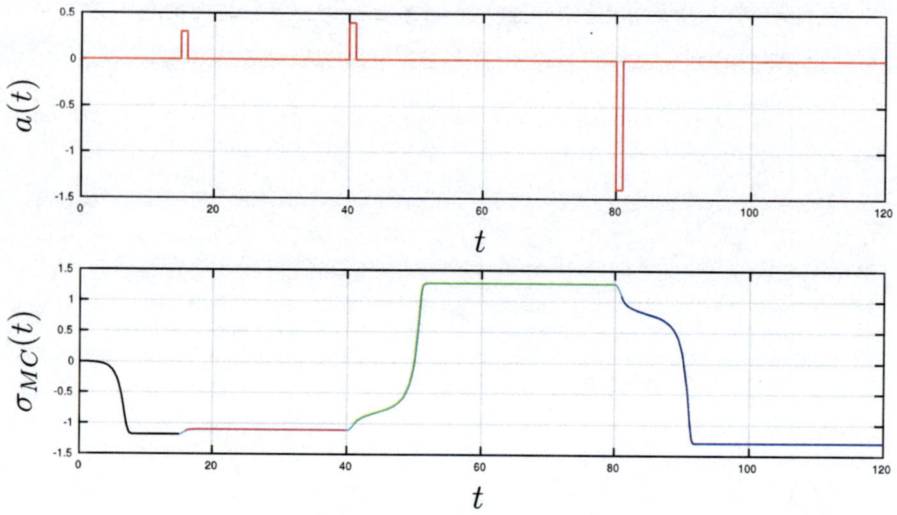

Fig. 11.3 Input $a(t)$ given by three rectangular pulses with duration $\Delta = 1$ (upper figure) and corresponding time-domain evolution of the solution of (11.2) starting close to the origin on the 0-manifold

i.e., the pulse forces the trajectory to pass from $\mathcal{M}(Q_0 = 0)$ to $\mathcal{M}(Q_0 = 0.3)$ in Δ (light blue curve emerging from the black diamond). Note that what really counts is the area of the pulse, not the shape. Any pulse (not necessarily rectangular) with the same area would cause the solution to pass from $\mathcal{M}(Q_0 = 0)$ to $\mathcal{M}(Q_0 = 0.3)$. For $t \geq 16$, the solution approaches a different EP (marked as a magenta diamond) along $\mathcal{M}(Q_0 = 0.3)$. A further pulse applied at $t = 40$ with area 0.4 causes a switch of the trajectory onto $\mathcal{M}(Q_0 = 0.7)$. The solution then slides on this invariant manifold $\mathcal{M}(Q_0 = 0.7)$ (green curve in Fig. 11.4) until it approaches the only EP (marked as green diamond). Finally, a pulse applied at $t = 80$ with area -1.4 drives the trajectory on the $\mathcal{M}(Q_0 = -0.7)$ manifold and the solution eventually approaches the EP indicated with a blue diamond in Fig. 11.4. The corresponding time-domain behavior of $\sigma_{MC}(t)$ is depicted in Fig. 11.3.

The investigation of the nonlinear dynamics in the MC – M circuit highlights the advantages of the analysis in the (φ, q)-domain due to the reduction of order for the SEs in the (φ, q)-domain with respect to the (v, i)-domain.[2] Such an approach is crucial to show the existence of invariant manifolds, the coexistence of infinitely many different dynamical behaviors, the phenomena of bifurcations without parameters, and is also crucial to design suitable current or voltage pulses to drive solutions through different manifolds.

[2]It is worth remarking that it would have been a much more complicated task to analyze the dynamics of the MC – M circuit via the third-order system in the (v, i)-domain.

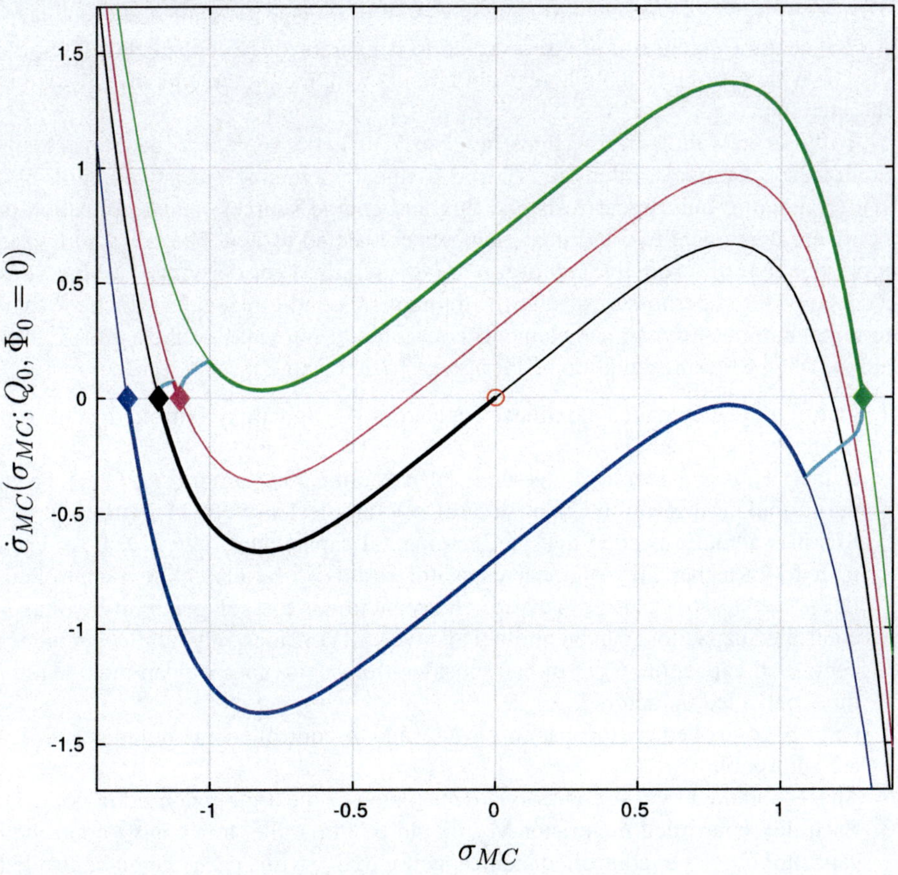

Fig. 11.4 Four different manifolds corresponding to $Q_0 = 0$ (black), $Q_0 = 0.3$ (magenta), $Q_0 = 0.7$ (green), and $Q_0 = -0.7$ (blue) and solution of (11.2) driven through different manifolds via the application of rectangular pulses. Switches between different manifolds during the application of pulses are drawn in light blue

Following the *modus operandi* presented for the MC − M circuit, a systematic and comprehensive method for a broad class of nonlinear dynamical circuits with mem-elements is developed in the remaining part of the chapter.

11.3 Classes $\mathcal{LME}$ and $\mathcal{LME}'$ of Nonlinear Circuits with Mem-Elements

We have discussed in Chap. 3 why an SE description is needed for a deep understanding of the qualitative behavior of the dynamics of nonlinear circuits.

Therefore, it is crucial to identify subclasses of nonlinear dynamical circuits in $\mathcal{N}_e$ for which the existence of an SE description is guaranteed. Such subclasses should also be wide enough to include relevant circuits with mem-elements investigated in the literature.

To this end, following the lines in Chap. 7, it is convenient to decompose any nonlinear dynamical circuit in $\mathcal{N}_e$ into a linear adynamic (resistive) multi-port N_R (containing only linear resistors, flux and charge sources) and a collection of nonlinear dynamical two-terminal elements connected to N_R. The standard hybrid representation of a resistive multi-port N_R is then used for deriving the SEs. Such a decomposition permits to specify the topological conditions under which the two-terminal nonlinear dynamical elements connected to N_R yield a subclass of $\mathcal{N}_e$ that admits of an SE representation (cf. Property 3.1 in Chap. 3):

1. Only charge-controlled nonlinear capacitors C_q and flux-controlled nonlinear inductors L_φ are admissible.
   In fact, let us consider a voltage-controlled nonlinear capacitor $q = f_C(v)$. From the formulation given in Chap. 3 (cf. also Example 1 in Sect. 11.2), to write the SEs it is actually needed that the incremental capacitance $f_C'(v) \neq 0$ for each v. It follows that the nonlinear capacitor results to be also charge-controlled, i.e., $v = f_C^{-1}(q)$ can be derived. Hence, without loss of generality, voltage-controlled capacitors can be omitted and we will consider only nonlinear charge-controlled capacitors C_q. Similar considerations show that we can include only flux-controlled inductors L_φ.
2. Only σ-controlled memcapacitors MC_σ and ρ-controlled meminductors ML_ρ are admissible.
   This is obtained by a straightforward extension of the argument used in point 1).
3. Each flux-controlled memristor M_φ should be in parallel to a charge-controlled capacitor C_q or a σ-controlled memcapacitor MC_σ, while each charge-controlled memristor M_q should be in series with a flux-controlled inductor L_φ or a ρ-controlled meminductor ML_ρ.

Note that linear capacitors C or inductors L can be considered as special cases of the corresponding nonlinear elements.

On the basis of the previous discussion and assumptions, the two following subclasses of nonlinear dynamical circuits with mem-elements in $\mathcal{N}_e$ can be identified:

1. $\mathcal{LME} \subset \mathcal{N}_e$ is obtained interconnecting N_R to the following nonlinear dynamical two-terminal elements $\mathcal{D}$:
   - $n_{\gamma\mu}$ flux-controlled memristors M_φ (each having in parallel a linear capacitor C)
   - $n_{\lambda\mu}$ charge-controlled memristors M_q (each having in series a linear inductor L)
   - n_γ linear capacitors C
   - n_λ linear inductors L
   - n_σ memcapacitors MC_σ

- n_ρ meminductors ML_ρ

2. $\mathcal{LME}' \subset \mathcal{N}_e$ is obtained interconnecting N_R to the following nonlinear dynamical two-terminal elements $\mathcal{D}'$:
   - $n_{\gamma\mu}$ flux-controlled memristors M_φ (each having in parallel a charge-controlled nonlinear capacitor C_q or a σ-controlled memcapacitor MC_σ)
   - $n_{\lambda\mu}$ charge-controlled memristors M_q (each having in series a flux-controlled nonlinear inductor L_φ or a ρ-controlled meminductor ML_ρ)
   - n_γ charge-controlled nonlinear capacitors C_q
   - n_λ flux-controlled nonlinear inductors L_φ
   - n_σ memcapacitors MC_σ
   - n_ρ meminductors ML_ρ.

The subclasses $\mathcal{LME}$ and $\mathcal{LME}'$ include several *mem-circuits*[3] studied in the literature. Henceforth, we focus on developing a systematic approach for the SEs in the (φ, q)-domain of $\mathcal{LME}$.[4] To this end, the CRs of any nonlinear dynamical two-terminal element $\mathcal{D} \in \{M_\varphi, M_q, C, L, MC_\sigma, ML_\rho\}$ in $\mathcal{LME}$ are derived in the next section and used to write the SEs in Sect. 11.5. We then discuss in Sect. 11.7 via a number of specific examples how we can write the SEs for circuits in the class $\mathcal{LME}'$.

11.4 Constitutive Relations of Two-Terminal Elements in $\mathcal{LME}$

In this section, the CRs are written both in the (φ, q)-domain and in the (v, i)-domain, with the aim of identifying the state variable(s) associated with $\mathcal{D}$ and the element order in both domains. The "element order" of $\mathcal{D}$ in a given (v^α, i^β)-domain (denoted by $O_{\mathcal{D}}^{(\alpha,\beta)}$) is defined as the number of differential/integral operators appearing in its CR. By comparing state variables and element order, it is possible to ascertain the reduction of $O_{\mathcal{D}}^{(\alpha,\beta)}$ in the CRs passing from the (v, i)-domain to the (φ, q)-domain. In the following, the CRs of M_φ and M_q are briefly summarized, whereas the CRs of MC_σ and ML_ρ and their possible equivalent circuit representations in the (φ, q)-domain are presented in detail. For the CRs and equivalent circuit in the (φ, q)-domain of linear resistors R, capacitors C, inductors L, and independent sources, we refer the reader to Chap. 5.

[3] Hereinafter, nonlinear dynamical circuits with mem-elements in $\mathcal{LME}$ are named mem-circuits.
[4] Note that $\mathcal{LM} \subset \mathcal{LME} \subset \mathcal{N}_e$.

11.4.1 *Memristor* $\mathbf{M}_\varphi$

A flux-controlled memristor M_φ is defined by the algebraic CR $q_M(t) = f_M(\varphi_M(t))$ in the (φ, q)-domain. CRs need to be expressed in terms of the incremental charge and flux, i.e., the CR of M_φ becomes

$$q_M(t; t_0) = f_M(\varphi_M(t; t_0) + \varphi_{M0}) - f_M(\varphi_{M0})$$

for $t \geq t_0$, where $\varphi_{M0} = \varphi_M(t_0)$. Hence, $O_{M_\varphi}^{(0,0)} = 0$ and then there is no state variable in view of the fact that a flux-controlled memristor acts as an algebraic element in the (φ, q)-domain (it is the analogous of a nonlinear resistor in the (v, i)-domain). It is worth noting that the CR depends upon φ_{M0}. The corresponding equivalent circuit in such domain is reported in Fig. 5.12 of Chap. 5. The time-derivative of the previous CR gives the state dependent Ohm's law describing M_φ in the (v, i)-domain, i.e.,

$$i_M(t) = f'_M(\varphi_M(t))v_M(t)$$

and

$$\dot{\varphi}_M(t) = v_M(t).$$

Hence, $O_{M_\varphi}^{(0,0)} = 1$ and the state variable is $\varphi_M(t)$. The IC is φ_{M0}.

11.4.2 *Memristor* $\mathbf{M}_\mathbf{q}$

Similar arguments hold for a charge-controlled memristor M_q defined by the algebraic CR $\varphi_M(t) = f_M(q_M(t))$, or equivalently by

$$\varphi_M(t; t_0) = f_M(q_M(t; t_0) + q_{M0}) - f_M(q_{M0})$$

for $t \geq t_0$, where $q_{M0} = q_M(t_0)$. Hence, $O_{M_q}^{(-1,-1)} = 0$ and there is no state variable in the (φ, q)-domain. The equivalent circuit is in Fig. 5.13 of Chap. 5. In the (v, i)-domain, M_q is described by

$$v_M(t) = f'_M(q_M(t))i(t)$$

and

$$\dot{q}_M(t) = i(t)$$

and then $O^{(0,0)}_{\mathrm{M}_q} = 1$ and the state variable is $q_M(t)$. The IC is q_{M0}.

11.4.3 *Memcapacitor* MC_σ

A σ-controlled memcapacitor MC_σ is defined by the algebraic CR $\varphi_{MC}(t) = f_{MC}(\sigma_{MC}(t))$ in the (φ, σ)-domain, i.e., $\alpha = -1$ and $\beta = -2$. The corresponding CR in the (φ, q)-domain is obtained by including the relationship between $\sigma_{MC}(t)$ and $q_{MC}(t)$, thus the CR results to be $\varphi_{MC}(t) = f_{MC}(\sigma_{MC}(t))$ and $\dot{\sigma}_{MC}(t) = q_{MC}(t)$, or equivalently ($\forall t \geq t_0$)

$$\varphi_{MC}(t; t_0) = f_{MC}(\sigma_{MC}(t; t_0) + \sigma_{MC0}) - f_{MC}(\sigma_{MC0})$$
$$\dot{\sigma}_{MC}(t; t_0) = q_{MC}(t; t_0) + q_{MC0}$$

where $\sigma_{MC0} = \sigma_{MC}(t_0)$ and $q_{MC0} = q_{MC}(t_0)$. Hence, $O^{(-1,-1)}_{\mathrm{MC}_\sigma} = 1$, the state variable is $\sigma_{MC}(t; t_0) = \sigma_{MC}(t) - \sigma_{MC0}$ and the IC is $\sigma_{MC}(t_0; t_0) = 0$. It is worth noting that the CR depends upon σ_{MC0} and q_{MC0}.

We can also write

$$\varphi_{MC}(t; t_0) = f_{MC}\Big(\sigma_{MC0} + \int_{t_0}^{t} (q_{MC}(\tau; t_0) + q_{MC0})d\tau\Big) - f_{MC}(\sigma_{MC0}) \tag{11.5}$$

which lends itself to an equivalent circuit representation in the (φ, q)-domain as in Fig. 11.5. Such equivalent circuit is synthesized by exploiting:

- a two-port network $\mathcal{B}_a$ that yields $\varphi_{MC}(t) = \varphi_{MC}(t; t_0) + \varphi_{MC0}$ and $q_{MC}(t) = q_{MC}(t; t_0) + q_{MC0}$ from the incremental variables $\varphi_{MC}(t; t_0)$ and $q_{MC}(t; t_0)$ at the input terminals of MC_σ
- a two-port network $\mathcal{B}_b$ with a (flux) source σ_{MC0}/C controlling the (flux) source $\varphi_{MC0} = f_{MC}(C\sigma_{MC0}/C)$ within $\mathcal{B}_a$

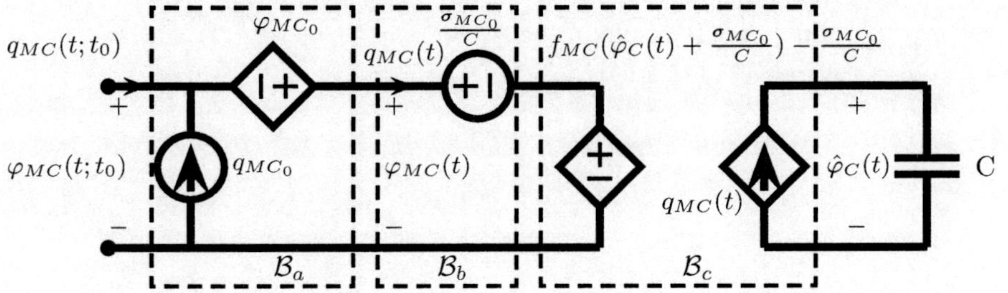

Fig. 11.5 Equivalent circuit in the (φ, q)-domain of a σ-controlled memcapacitor $\varphi_{MC}(t) = f_{MC}(\sigma_{MC}(t))$

- a two-port network $\mathcal{B}_c$ made of a linear charge-controlled charge-source $q_{MC}(t)$ and a nonlinear flux-controlled flux-source $f_{MC}(\hat{\varphi}_C(t) + \sigma_{MC_0}/C) - \sigma_{MC_0}/C$. The output port of $\mathcal{B}_c$ is connected to a linear capacitor C discharged at t_0.

By noting that the charge source $q_{MC}(t)$ supplying C yields a flux on C given by[5]

$$\hat{\varphi}_C(t) = \frac{1}{C}\int_{t_0}^{t} q_{MC}(\tau)d\tau = \frac{1}{C}\int_{t_0}^{t} (q_{MC}(\tau, t_0) + q_{MC_0})d\tau$$

then KφL at the input mesh yields (11.5) when $C = 1$.

It is remarkable that: (a) the equivalent circuit has two independent sources q_{MC_0} and σ_{MC_0}/C representing the ICs for the state variables of MC$_\sigma$ in the (v, i)-domain; (b) there is a unique dynamical element C in the equivalent circuit, according to the fact that MC$_\sigma$ has $O_{\mathrm{MC}_\sigma}^{(-1,-1)} = 1$, i.e., MC$_\sigma$ acts as a first-order memory element in the (φ, q)-domain.

The following description of a MC$_\sigma$ in the (v, i)-domain is obtained by differentiation in time:

$$v_{MC}(t) = f'_{MC}(\sigma_{MC}(t))q_{MC}(t)$$
$$\dot{\sigma}_{MC}(t) = q_{MC}(t)$$
$$\dot{q}_{MC}(t) = i_{MC}(t).$$

Hence, $O_{\mathrm{MC}_\sigma}^{(0,0)} = 2$, the state variables are $\sigma_{MC}(t)$ and $q_{MC}(t)$ with corresponding ICs σ_{MC_0} and q_{MC_0}.

11.4.4 Meminductors ML$_\rho$

A ρ-controlled meminductor ML$_\rho$ is defined by the algebraic CR $q_{ML}(t) = f_{ML}(\rho_{ML}(t))$ in the (ρ, q)-domain, i.e., $\alpha = -2$ and $\beta = -1$. The corresponding CR in the (φ, q)-domain is obtained by including the relationship between $\varphi_{ML}(t)$ and $\rho_{ML}(t)$, thus the CR results to be $q_{ML}(t) = f_{ML}(\rho_{ML}(t))$ and $\dot{\rho}_{ML}(t) = \varphi_{ML}(t)$, or equivalently ($\forall t \geq t_0$)

$$q_{ML}(t; t_0) = f_{ML}(\rho_{ML}(t; t_0) + \rho_{ML_0}) - f_{ML}(\rho_{ML_0})$$
$$\dot{\rho}_{ML}(t; t_0) = \varphi_{ML}(t; t_0) + \varphi_{ML_0}$$

[5]The "charge" $q_{MC}(t)$ and "flux" $\hat{\varphi}_C(t)$ in the linear capacitor C play the usual role of "current" and "voltage" due to the linearity of the element.

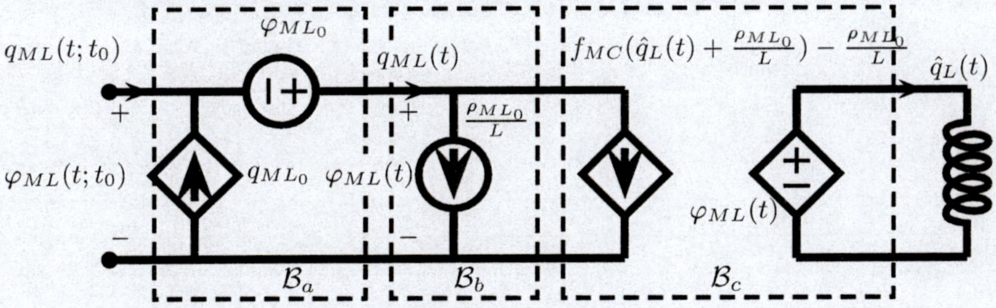

Fig. 11.6 Equivalent circuit in the (φ, q)-domain of a ρ-controlled meminductor $q_{ML}(t) = f_{ML}(\rho_{ML}(t))$

where $\rho_{ML_0} = \rho_{ML}(t_0)$ and $\varphi_{ML_0} = \varphi_{ML}(t_0)$. Hence, $O^{(-1,-1)}_{\text{ML}_\rho} = 1$, the state variable is $\rho_{ML}(t; t_0) = \rho_{ML}(t) - \rho_{ML_0}$ and the IC is $\rho_{MC}(t_0; t_0) = 0$.

The previous equations can be written in integral form as

$$q_{ML}(t; t_0) = f_{ML}\Big(\rho_{ML_0} + \int_{t_0}^{t} (\varphi_{ML}(\tau; t_0) + \varphi_{ML_0})d\tau\Big) - f_{ML}(\rho_{ML_0})$$

which yields, by a synthesis procedure similar to MC_σ, the equivalent circuit representation in Fig. 11.6.

The time-differentiation of the expressions above permits to derive the description of ML_ρ in the (v, i)-domain

$$i_{ML}(t) = f'_{ML}(\rho_{ML}(t))\varphi_{ML}(t)$$
$$\dot{\rho}_{ML}(t) = \varphi_{ML}(t)$$
$$\dot{\varphi}_{ML}(t) = v_{ML}(t).$$

Hence, $O^{(0,0)}_{\text{ML}_\rho} = 2$, the state variables are $\rho_{ML}(t)$ and $\varphi_{ML}(t)$ and the corresponding ICs are ρ_{ML_0} and φ_{ML_0}.

Remark 11.1 Table 11.1 wraps up the description of CRs for the elements $\mathcal{D}$ in the mem-circuits $\mathcal{LME}$. For each $\mathcal{D}$, the table includes the state variables, ICs, and the order $O^{(\alpha,\beta)}_{\mathcal{D}}$ in the (v, i)- and (φ, q)-domain. Moreover, the Reduction of Order (RO) in passing from the (v, i)-domain to the (φ, q)-domain is reported. It is apparent that *mem-elements (i.e., a memristor, a memcapacitor, and a meminductor) exhibit* RO $= 1$, *whereas linear elements* (R, C, L), *and independent sources have the same order in both domains.*

The CR in the (φ, q)-domain of $\mathcal{D}$ depends upon the ICs at t_0 for the state variables in the (v, i)-domain and such ICs are represented by independent charge or flux sources in the corresponding equivalent circuit in the (φ, q)-domain. Also note that the ICs of the state variables in the (φ, q)-domain are zero because in that domain the same state variables are given in incremental form.

Table 11.1 State variables and $O_{\mathcal{D}}^{(\alpha,\beta)}$

	(v, i)-domain			(φ, q)-domain			
$\mathcal{D}$	State variables	$O_{\mathcal{D}}^{(0,0)}$	ICs	State variables	$O_{\mathcal{D}}^{(-1,-1)}$	ICs	RO
R (lin.)	–	0	–	–	0	–	0
e	–	0	–	–	0	–	0
a	–	0	–	–	0	–	0
C (lin.)	$v_C(t)$	1	v_{C_0}	$\varphi_C(t; t_0)$	1	0	0
L (lin.)	$i_L(t)$	1	i_{L_0}	$q_L(t; t_0)$	1	0	0
C_q (nonlin.)	$q_C(t)$	1	q_{C_0}	$\varphi_C(t; t_0)$	1	0	0
L_φ (nonlin.)	$\varphi_L(t)$	1	φ_{L_0}	$q_L(t; t_0)$	1	0	0
M_φ	$\varphi_M(t)$	1	φ_{M_0}	–	0	–	1
M_q	$q_M(t)$	1	q_{M_0}	–	0	–	1
MC_σ	$\sigma_{MC}(t), q_{MC}(t)$	2	σ_{MC_0}, q_{MC_0}	$\sigma_{MC}(t; t_0)$	1	0	1
ML_ρ	$\rho_{ML}(t), \varphi_{ML}(t)$	2	$\rho_{ML_0}, \varphi_{ML_0}$	$\rho_{ML}(t; t_0)$	1	0	1

11.5 State Equations and Nonlinear Dynamics

This section presents a systematic methodology to derive the SE description for mem-circuits in $\mathcal{LME}$, in the (φ, q)- and (v, i)-domains, by means of the hybrid representation of N_R and the CRs in Table 11.1. Then, chief results that are useful for investigating invariant manifolds, coexisting attractors, and their bifurcations without parameters are derived.

Henceforth, assume that there exists the hybrid representation of N_R in terms of matrix **H**

$$\begin{pmatrix} \mathbf{q}^{\sigma}(t; t_0) \\ \boldsymbol{\varphi}^{\rho}(t; t_0) \\ \mathbf{q}^{\gamma\mu}(t; t_0) \\ \boldsymbol{\varphi}^{\lambda\mu}(t; t_0) \\ \mathbf{q}^{\gamma}(t; t_0) \\ \boldsymbol{\varphi}^{\lambda}(t; t_0) \end{pmatrix} = \mathbf{H} \begin{pmatrix} \boldsymbol{\varphi}^{\sigma}(t; t_0) \\ \mathbf{q}^{\rho}(t; t_0) \\ \boldsymbol{\varphi}^{\gamma\mu}(t; t_0) \\ \mathbf{q}^{\lambda\mu}(t; t_0) \\ \boldsymbol{\varphi}^{\gamma}(t; t_0) \\ \mathbf{q}^{\lambda}(t; t_0) \end{pmatrix} + \mathbf{U}(t; t_0) \tag{11.6}$$

where

$$\mathbf{U}(t; t_0) = \mathbf{B} \begin{pmatrix} \boldsymbol{\varphi}_e(t; t_0) \\ \mathbf{q}_a(t; t_0) \end{pmatrix} = \mathbf{B} \begin{pmatrix} \int_{t_0}^{t} \mathbf{e}(\tau)d\tau \\ \int_{t_0}^{t} \mathbf{a}(\tau)d\tau \end{pmatrix}$$

depends, via matrix **B**, on the flux and charge sources $\boldsymbol{\varphi}_e(t; t_0)$ and $\mathbf{q}_a(t; t_0)$ within N_R. This representation exists, for instance, when N_R contains only positive linear resistors and a condition analogous to (A1) in Chap. 7 is satisfied.

The elements $\mathcal{D}$ connected to N_R are (listed again for reader's convenience)[6]

- n_σ mem-capacitors MC_σ described in terms of $\mathbf{q}^\sigma(t; t_0)$ and $\boldsymbol{\varphi}^\sigma(t; t_0)$ in $\mathbb{R}^{n_\sigma}$ by

$$\boldsymbol{\varphi}^\sigma(t; t_0) = \mathbf{f}^\sigma(\boldsymbol{\sigma}^\sigma(t; t_0) + \boldsymbol{\sigma}_0^\sigma) - \mathbf{f}^\sigma(\boldsymbol{\sigma}_0^\sigma) \tag{11.7}$$

$$-\mathbf{q}^\sigma(t; t_0) = \dot{\boldsymbol{\sigma}}^\sigma(t; t_0) - \mathbf{q}_0^\sigma \tag{11.8}$$

  via state variables $\boldsymbol{\sigma}^\sigma(t; t_0)$
- n_ρ mem-inductors ML_ρ described in terms of $\boldsymbol{\varphi}^\rho(t; t_0)$ and $\mathbf{q}^\rho(t; t_0)$ in $\mathbb{R}^{n_\rho}$ by

$$\mathbf{q}^\rho(t; t_0) = \mathbf{f}^\rho(\boldsymbol{\rho}^\rho(t; t_0) + \boldsymbol{\rho}_0^\rho) - \mathbf{f}^\rho(\boldsymbol{\rho}_0^\rho) \tag{11.9}$$

$$-\boldsymbol{\varphi}^\rho(t; t_0) = \dot{\boldsymbol{\rho}}^\rho(t; t_0) - \boldsymbol{\varphi}_0^\rho \tag{11.10}$$

  via state variables $\boldsymbol{\rho}^\rho(t; t_0)$
- $n_{\gamma\mu}$ **flux-controlled memristors** $\mathbf{M}_\varphi$ (each having in parallel a capacitor C) described in terms of $\mathbf{q}^{\gamma\mu}(t; t_0)$ and $\boldsymbol{\varphi}^{\gamma\mu}(t; t_0)$ in $\mathbb{R}^{n_{\gamma\mu}}$ by (cf. Sect. 7.3.1 in Chap. 7)

$$\begin{aligned} -\mathbf{q}^{\gamma\mu}(t; t_0) = {} & \mathbf{C}^{\gamma\mu}\dot{\boldsymbol{\varphi}}^{\gamma\mu}(t; t_0) - \mathbf{q}_0^{\gamma\mu} \\ & + \mathbf{f}^{\gamma\mu}(\boldsymbol{\varphi}^{\gamma\mu}(t; t_0) + \boldsymbol{\varphi}_0^{\gamma\mu}) - \mathbf{f}^{\gamma\mu}(\boldsymbol{\varphi}_0^{\gamma\mu}) \end{aligned} \tag{11.11}$$

  where $\boldsymbol{\varphi}^{\gamma\mu}(t; t_0)$ are the state variables (i.e., fluxes of the capacitors C)
- $n_{\lambda\mu}$ charge-controlled memristors M_q (each having in series an inductor L) described in terms of $\boldsymbol{\varphi}^{\lambda\mu}(t; t_0)$ and $\mathbf{q}^{\lambda\mu}(t; t_0)$ in $\mathbb{R}^{n_{\lambda\mu}}$ by

$$\begin{aligned} -\boldsymbol{\varphi}^{\lambda\mu}(t; t_0) = {} & \mathbf{L}^{\lambda\mu}\dot{\mathbf{q}}^{\lambda\mu}(t; t_0) - \boldsymbol{\varphi}_0^{\lambda\mu} \\ & + \mathbf{f}^{\lambda\mu}(\mathbf{q}^{\lambda\mu}(t; t_0) + \mathbf{q}_0^{\lambda\mu}) - \mathbf{f}^{\lambda\mu}(\mathbf{q}_0^{\lambda\mu}) \end{aligned} \tag{11.12}$$

  where $\mathbf{q}^{\lambda\mu}(t; t_0)$ are the state variables (i.e., charges of the inductors L)
- n_γ linear capacitors C described in terms of $\mathbf{q}^\gamma(t; t_0)$ and $\boldsymbol{\varphi}^\gamma(t; t_0)$ in $\mathbb{R}^{n_\gamma}$ by

$$-\mathbf{q}^\gamma(t; t_0) = \mathbf{C}^\gamma\dot{\boldsymbol{\varphi}}^\gamma(t; t_0) - \mathbf{q}_0^\gamma \tag{11.13}$$

  with state variables $\boldsymbol{\varphi}^\gamma(t; t_0)$ (we let $\mathbf{q}_0^\gamma = \mathbf{C}^\gamma\mathbf{v}_0^\gamma$);
- n_λ linear inductors L described in terms of $\boldsymbol{\varphi}^\lambda(t; t_0)$ and $\mathbf{q}^\lambda(t; t_0)$ in $\mathbb{R}^{n_\lambda}$ by

$$-\boldsymbol{\varphi}^\lambda(t; t_0) = \mathbf{L}^\lambda\dot{\mathbf{q}}^\lambda(t; t_0) - \boldsymbol{\varphi}_0^\lambda \tag{11.14}$$

[6] The notation is simplified with respect to the previous part of the chapter by introducing the superscript that identifies the element $\mathcal{D} \in \{\mathrm{M}_\varphi, \mathrm{M}_q, \mathrm{C}, \mathrm{L}, \mathrm{MC}_\sigma, \mathrm{ML}_\rho\}$ in $\mathcal{LME}$ and the dimension of the corresponding vector of state variables.

with state variables $\mathbf{q}^{\lambda}(t;t_0)$ (we let $\boldsymbol{\varphi}_0^{\lambda} = \mathbf{L}^{\lambda}\mathbf{i}_0^{\gamma}$).

Substitution of (11.7)–(11.14) in (11.6) and simple algebraic manipulations yield the SEs in the (φ, q)-domain

$$
\mathbf{M}\begin{pmatrix} \dot{\boldsymbol{\sigma}}^{\sigma}(t;t_0) \\ \dot{\boldsymbol{\rho}}^{\rho}(t;t_0) \\ \dot{\boldsymbol{\varphi}}^{\gamma\mu}(t;t_0) \\ \dot{\mathbf{q}}^{\lambda\mu}(t;t_0) \\ \dot{\boldsymbol{\varphi}}^{\gamma}(t;t_0) \\ \dot{\mathbf{q}}^{\lambda}(t;t_0) \end{pmatrix} = -\mathbf{H}\begin{pmatrix} \mathbf{f}^{\sigma}(\boldsymbol{\sigma}^{\sigma}(t;t_0)+\boldsymbol{\sigma}_0^{\sigma}) \\ \mathbf{f}^{\rho}(\boldsymbol{\rho}^{\rho}(t;t_0)+\boldsymbol{\rho}_0^{\rho}) \\ \boldsymbol{\varphi}^{\gamma\mu}(t;t_0) \\ \mathbf{q}^{\lambda\mu}(t;t_0) \\ \boldsymbol{\varphi}^{\gamma}(t;t_0) \\ \mathbf{q}^{\lambda}(t;t_0) \end{pmatrix}
- \begin{pmatrix} \mathbf{0} \\ \mathbf{0} \\ \mathbf{f}^{\gamma\mu}(\boldsymbol{\varphi}^{\gamma\mu}(t;t_0)+\boldsymbol{\varphi}_0^{\gamma\mu}) \\ \mathbf{f}^{\lambda\mu}(\mathbf{q}^{\lambda\mu}(t;t_0)+\mathbf{q}_0^{\lambda\mu}) \\ \mathbf{0} \\ \mathbf{0} \end{pmatrix}
+ \mathbf{H}\begin{pmatrix} \mathbf{f}^{\sigma}(\boldsymbol{\sigma}_0^{\sigma}) \\ \mathbf{f}^{\lambda}(\boldsymbol{\rho}_0^{\rho}) \\ \mathbf{0} \\ \mathbf{0} \\ \mathbf{0} \\ \mathbf{0} \end{pmatrix} + \begin{pmatrix} \mathbf{q}_0^{\sigma} \\ \boldsymbol{\varphi}_0^{\rho} \\ \mathbf{q}_0^{\gamma\mu} \\ \boldsymbol{\varphi}_0^{\lambda\mu} \\ \mathbf{q}_0^{\gamma} \\ \boldsymbol{\varphi}_0^{\lambda} \end{pmatrix}
+ \begin{pmatrix} \mathbf{0} \\ \mathbf{0} \\ \mathbf{f}^{\gamma\mu}(\boldsymbol{\varphi}_0^{\gamma\mu}) \\ \mathbf{f}^{\lambda\mu}(\mathbf{q}_0^{\lambda\mu}) \\ \mathbf{0} \\ \mathbf{0} \end{pmatrix} - \mathbf{U}(t;t_0)
$$

where $\mathbf{M} = \mathrm{diag}(\mathbf{I}^{a}, \mathbf{I}^{b}, \mathbf{C}^{\gamma\mu}, \mathbf{L}^{\lambda\mu}, \mathbf{C}^{\gamma}, \mathbf{L}^{\lambda})$ and $\mathbf{I}^{a}$, $\mathbf{I}^{b}$ are identity matrices. Note that $\mathbf{M}$ is nonsingular.

The SEs can be written in compact form by means of the following vectors of state variables (see also Table 11.1):

- in the (φ, q)-domain

$$
\mathbf{x}(t) = \begin{pmatrix} \boldsymbol{\sigma}^{\sigma}(t;t_0) \\ \boldsymbol{\rho}^{\rho}(t;t_0) \end{pmatrix} \in \mathbb{R}^{n_{\sigma}+n_{\rho}} \tag{11.15}
$$

$$
\mathbf{y}(t) = \begin{pmatrix} \boldsymbol{\varphi}^{\gamma\mu}(t;t_0) \\ \mathbf{q}^{\lambda\mu}(t;t_0) \end{pmatrix} \in \mathbb{R}^{n_{\gamma\mu}+n_{\lambda\mu}} \tag{11.16}
$$

$$\mathbf{z}(t) = \begin{pmatrix} \boldsymbol{\varphi}^{\gamma}(t; t_0) \\ \mathbf{q}^{\lambda}(t; t_0) \end{pmatrix} \in \mathbb{R}^{n_\gamma + n_\lambda} \tag{11.17}$$

- in the (v, i)-domain[7]

$$\mathbf{X}(t) = \begin{pmatrix} \boldsymbol{\sigma}^{\sigma}(t) \\ \boldsymbol{\rho}^{\rho}(t) \end{pmatrix} \in \mathbb{R}^{n_\sigma + n_\rho} \tag{11.18}$$

$$\mathbf{W}(t) = \begin{pmatrix} \mathbf{q}^{\sigma}(t) \\ \boldsymbol{\varphi}^{\rho}(t) \end{pmatrix} \in \mathbb{R}^{n_\sigma + n_\rho} \tag{11.19}$$

for n_σ memcapacitors MC_σ and n_ρ meminductors ML_ρ

$$\mathbf{Y}(t) = \begin{pmatrix} \boldsymbol{\varphi}^{\gamma\mu}(t) \\ \mathbf{q}^{\lambda\mu}(t) \end{pmatrix} \in \mathbb{R}^{n_{\gamma\mu} + n_{\lambda\mu}} \tag{11.20}$$

$$\mathbf{T}(t) = \begin{pmatrix} \mathbf{v}^{\gamma\mu}(t) \\ \mathbf{i}^{\lambda\mu}(t) \end{pmatrix} \in \mathbb{R}^{n_{\gamma\mu} + n_{\lambda\mu}} \tag{11.21}$$

for $n_{\gamma\mu}$ memristors M_φ with in parallel C and $n_{\lambda\mu}$ memristors M_q with in series L

$$\mathbf{R}(t) = \begin{pmatrix} \mathbf{v}^{\gamma}(t) \\ \mathbf{i}^{\lambda}(t) \end{pmatrix} \in \mathbf{R}^{n_\gamma + n_\lambda} \tag{11.22}$$

for n_γ capacitors C and n_λ inductors L.

Finally, define the vectors of ICs

$$\mathbf{X_0} = \begin{pmatrix} \boldsymbol{\sigma}_0^{\sigma} \\ \boldsymbol{\rho}_0^{\rho} \end{pmatrix}; \mathbf{Y_0} = \begin{pmatrix} \boldsymbol{\varphi}_0^{\gamma\mu} \\ \mathbf{q}_0^{\lambda\mu} \end{pmatrix}; \mathbf{W}_0 = \begin{pmatrix} \mathbf{q}_0^{\sigma} \\ \boldsymbol{\varphi}_0^{\rho} \end{pmatrix}$$

and

$$\mathbf{T}_0 = \begin{pmatrix} \mathbf{v}_0^{\gamma\mu} \\ \mathbf{i}_0^{\lambda\mu} \end{pmatrix}; \mathbf{R}_0 = \begin{pmatrix} \mathbf{v}_0^{\gamma} \\ \mathbf{i}_0^{\lambda} \end{pmatrix}$$

and nonlinear functions

$$\mathbf{F}^a(\cdot) = \begin{pmatrix} \mathbf{f}^{\sigma}(\cdot) \\ \mathbf{f}^{\rho}(\cdot) \end{pmatrix} : \mathbb{R}^{n_\sigma + n_\rho} \to \mathbb{R}^{n_\sigma + n_\rho}$$

[7]It is worth to note that ICs permits to derive, by definition, incremental variables from the absolute ones. For instance, $\boldsymbol{\sigma}^{\sigma}(t; t_0) + \boldsymbol{\sigma}_0^{\sigma} = \boldsymbol{\sigma}^{\sigma}(t)$.

$$\mathbf{F}^b(\cdot) = \begin{pmatrix} \mathbf{f}^{\gamma\mu}(\cdot) \\ \mathbf{f}^{\lambda\mu}(\cdot) \end{pmatrix} : \mathbb{R}^{n_{\gamma\mu}+n_{\lambda\mu}} \to \mathbb{R}^{n_{\gamma\mu}+n_{\lambda\mu}}.$$

It follows that the SEs in the (φ, q)-domain result to be

$$\begin{aligned} \mathbf{M}\begin{pmatrix} \dot{\mathbf{x}}(t) \\ \dot{\mathbf{y}}(t) \\ \dot{\mathbf{z}}(t) \end{pmatrix} = -\mathbf{H}\begin{pmatrix} \mathbf{F}^a(\mathbf{x}(t)+\mathbf{X}_0) - \mathbf{F}^a(\mathbf{X}_0) \\ \mathbf{y}(t) \\ \mathbf{z}(t) \end{pmatrix} \\ -\begin{pmatrix} \mathbf{0} \\ \mathbf{F}^b(\mathbf{y}(t)+\mathbf{Y}_0) - \mathbf{F}^b(\mathbf{Y}_0) \\ \mathbf{0} \end{pmatrix} \\ +\mathbf{M}\begin{pmatrix} \mathbf{W}_0 \\ \mathbf{T}_0 \\ \mathbf{R}_0 \end{pmatrix} - \mathbf{U}(t; t_0) \end{aligned} \tag{11.23}$$

for $t \geq t_0$, with ICs

$$(\mathbf{x}(t_0), \mathbf{y}(t_0), \mathbf{z}(t_0)) = (\mathbf{0}, \mathbf{0}, \mathbf{0}). \tag{11.24}$$

The SEs (11.23) contain $n_{\varphi q} = n_\sigma + n_\rho + n_{\gamma\mu} + n_{\lambda\mu} + n_\gamma + n_\lambda$ ODEs, i.e., (11.23) defines a $n_{\varphi q}$-order Initial Value Problem (IVP) with $n_{\varphi q}$ unknowns corresponding to the state variables $\mathbf{x}(t)$, $\mathbf{y}(t)$, and $\mathbf{z}(t)$ in the (φ, q)-domain.

The time-differentiation of (11.23) provides the SEs in the (v, i)-domain

$$\begin{aligned} \mathbf{M}\begin{pmatrix} \dot{\mathbf{W}}(t) \\ \dot{\mathbf{T}}(t) \\ \dot{\mathbf{R}}(t) \end{pmatrix} = -\mathbf{H}\begin{pmatrix} (\mathbf{F}^a)'(\mathbf{X}(t))\mathbf{W}(t) \\ \mathbf{T}(t) \\ \mathbf{R}(t) \end{pmatrix} \\ -\begin{pmatrix} \mathbf{0} \\ (\mathbf{F}^b)'(\mathbf{Y}(t))\mathbf{T}(t) \\ \mathbf{0} \end{pmatrix} - \mathbf{B}\begin{pmatrix} \mathbf{e}(t) \\ \mathbf{a}(t) \end{pmatrix} \\ \dot{\mathbf{X}}(t) = \mathbf{W}(t) \\ \dot{\mathbf{Y}}(t) = \mathbf{T}(t) \end{aligned} \tag{11.25}$$

for $t \geq t_0$, with ICs

$$(\mathbf{X}(t_0), \mathbf{Y}(t_0), \mathbf{W}(t_0), \mathbf{T}(t_0), \mathbf{R}(t_0)) = (\mathbf{X}_0, \mathbf{Y}_0, \mathbf{W}_0, \mathbf{T}_0, \mathbf{R}_0). \tag{11.26}$$

The SEs (11.25) contain $n_{vi} = 2n_\sigma + 2n_\rho + 2n_{\gamma\mu} + 2n_{\lambda\mu} + n_\gamma + n_\lambda$ ODEs, i.e., (11.25) defines a n_{vi}-order IVP with n_{vi} unknowns corresponding to the state variables $\mathbf{W}(t)$, $\mathbf{T}(t)$, $\mathbf{R}(t)$, $\mathbf{X}(t)$ and $\mathbf{Y}(t)$ in the (v, i)-domain.

The next result provides the link between the solutions in the (φ, q)- and (v, i)-domains. The proof is straightforward and is left to the reader.

Property 11.1 If $(\mathbf{x}(t), \mathbf{y}(t), \mathbf{z}(t))$ is the solution of the IVP (11.23) and (11.24), then $(\mathbf{x}(t) + \mathbf{X}_0, \mathbf{y}(t) + \mathbf{Y}_0, \dot{\mathbf{x}}(t), \dot{\mathbf{y}}(t), \dot{\mathbf{z}}(t))$ is the solution of the IVP (11.25) and (11.26) for $t \geq t_0$. Conversely, if $(\mathbf{X}(t), \mathbf{Y}(t), \mathbf{W}(t), \mathbf{T}(t), \mathbf{R}(t))$ is the solution of the IVP (11.25) and (11.26), then $(\mathbf{X}(t) - \mathbf{X}_0, \mathbf{Y}(t) - \mathbf{Y}_0, \int_{t_0}^{t} \mathbf{R}(\tau)d\tau)$ is the solution of the IVP (11.23) and (11.24) for $t \geq t_0$.

Remark 11.2 The following chief result follows from the comparison between (11.23) and (11.25): in passing from the (v, i)-domain to the (φ, q)-domain we have an *order reduction* for the SEs

$$n = n_{vi} - n_{\varphi q} = n_\sigma + n_\rho + n_{\gamma\mu} + n_{\lambda\mu}$$

equal to the number of mem-elements in $\mathcal{LME}$. Note that the vector field defining the SEs (11.23) in the (φ, q)-domain depends upon the ICs for the state variables in the (v, i)-domain.

11.5.1 Invariant Manifolds

To identify the invariant manifolds of a mem-circuit in $\mathcal{LME}$, described by the SEs (11.23)–(11.24), let us introduce a suitable mapping $\mathbf{K}(\mathbf{\Gamma}) : \mathbb{R}^{n_{vi}} \to \mathbb{R}^n$

$$\mathbf{K}(\mathbf{\Gamma}) = \mathbf{M}_{11}\begin{pmatrix} \mathbf{W} \\ \mathbf{T} \end{pmatrix} + \mathbf{H}_{11}\begin{pmatrix} \mathbf{F}^a(\mathbf{X}) \\ \mathbf{Y} \end{pmatrix} + \begin{pmatrix} \mathbf{0} \\ \mathbf{F}^b(\mathbf{Y}) \end{pmatrix} + \mathbf{H}_{12}\mathbf{H}_{22}^{-1}\left[-\mathbf{M}_{22}\mathbf{R} - \mathbf{H}_{21}\begin{pmatrix} \mathbf{F}^a(\mathbf{X}) \\ \mathbf{Y} \end{pmatrix}\right] \quad (11.27)$$

where $\mathbf{\Gamma} = (\mathbf{X}, \mathbf{Y}, \mathbf{W}, \mathbf{T}, \mathbf{R}) \in \mathbb{R}^{n_{vi}}$ and the dimensions of the submatrices $\mathbf{M}_{11}$, $\mathbf{M}_{22}$, $\mathbf{H}_{11}$, $\mathbf{H}_{12}$, $\mathbf{H}_{21}$, and $\mathbf{H}_{22}$ are easily obtained by those of the corresponding vectors given above.

For any fixed vector $\bar{\mathbf{K}} \in \mathbb{R}^n$, consider the set

$$\mathcal{M}(\bar{\mathbf{K}}) = \{\mathbf{\Gamma} \in \mathbb{R}^{n_{vi}} : \mathbf{K}(\mathbf{\Gamma}) = \bar{\mathbf{K}}\}. \quad (11.28)$$

Note that $\mathcal{M}(\bar{\mathbf{K}})$ defines ∞^n manifolds in $\mathbb{R}^{n_{vi}}$ with geometrical properties analogous to those in Theorem 7.3 of Chap. 7. The dimension of each manifold is $n_{\varphi q}$.

The following holds.

Theorem 11.1 *Let us assume that* $\mathbf{H}_{22}$ *in* (11.27) *is nonsingular and let* $\mathbf{\Gamma}(t; t_0, \mathbf{\Gamma}_0)$, $t \geq t_0$, *with* $\mathbf{\Gamma}_0 = (\mathbf{X}_0, \mathbf{Y}_0, \mathbf{W}_0, \mathbf{T}_0, \mathbf{R}_0) \in \mathbb{R}^{n_{vi}}$, *be the solution of the IVP* (11.25)–(11.26) *for the SEs in the* (v, i)*-domain. Then, for any* $t \geq t_0$ *the* *solution* $\boldsymbol{\Gamma}(t; t_0, \boldsymbol{\Gamma}_0)$ *is such that*

$$\boldsymbol{\Gamma}(t; t_0, \boldsymbol{\Gamma}_0) \in \mathcal{M}(\mathbf{K}(\boldsymbol{\Gamma}(t; t_0, \boldsymbol{\Gamma}_0)))$$

where

$$\begin{aligned}\mathbf{K}(\boldsymbol{\Gamma}(t; t_0, \boldsymbol{\Gamma}_0)) &= \mathbf{K}(\boldsymbol{\Gamma}_0) + \mathbf{H}_{12}\mathbf{H}_{22}^{-1}\mathbf{U}_2(t; t_0) - \mathbf{U}_1(t; t_0) \\ &= \mathbf{K}(\boldsymbol{\Gamma}_0) + \mathbf{H}_{12}\mathbf{H}_{22}^{-1}(\mathbf{B}_{21}\boldsymbol{\varphi}_e(t; t_0) + \mathbf{B}_{22}\mathbf{q}_a(t; t_0)) \\ &\quad -(\mathbf{B}_{11}\boldsymbol{\varphi}_e(t; t_0) + \mathbf{B}_{12}\mathbf{q}_a(t; t_0)).\end{aligned} \tag{11.29}$$

Proof See Appendix 1. ■

This theorem permits to find at any instant $t \geq t_0$ the manifold $\mathcal{M}(\cdot)$ where the solution of the SEs in the (φ, q)-domain belongs to. Such a manifold $\mathcal{M}(\cdot)$ varies according to the variations of $\mathbf{K}(\boldsymbol{\Gamma}(t; t_0, \boldsymbol{\Gamma}_0))$ induced by the input sources $\boldsymbol{\varphi}_e(t; t_0)$ and $\mathbf{q}_a(t; t_0)$ (cf. (11.29)).

11.5.2 Coexisting Dynamics and Bifurcations Without Parameters

Suppose that a mem-circuit in $\mathcal{LME}$ has no input sources, thus $\boldsymbol{\varphi}_e(t; t_0) = \mathbf{0}$ and $\mathbf{q}_a(t; t_0) = \mathbf{0}$ (or equivalently $\mathbf{e}(t) = \mathbf{0}$, $\mathbf{a}(t) = \mathbf{0}$) for any $t \geq t_0$. Then, Theorem 11.1 can be simplified as follows.

Corollary 11.1 *If the assumptions of Theorem 11.1 hold, then*

$$\boldsymbol{\Gamma}(t; t_0, \boldsymbol{\Gamma}_0) \in \mathcal{M}(\mathbf{K}(\boldsymbol{\Gamma}_0))$$

for any $t \geq t_0$. *Moreover, the nonlinear dynamics is described for* $t \geq t_0$ *by the following reduced-order SEs:*

$$\mathbf{M}\begin{pmatrix}\dot{\mathbf{X}}(t)\\ \dot{\mathbf{Y}}(t)\\ \dot{\mathbf{Z}}(t)\end{pmatrix} = -\mathbf{H}\begin{pmatrix}\mathbf{F}^a(\mathbf{X}(t))\\ \mathbf{Y}(t)\\ \mathbf{Z}(t)\end{pmatrix} + \begin{pmatrix}\mathbf{0}\\ \mathbf{F}^b(\mathbf{Y}(t))\\ \mathbf{0}\end{pmatrix} + \begin{pmatrix}\mathbf{K}^a(\boldsymbol{\Gamma}_0)\\ \mathbf{K}^b(\boldsymbol{\Gamma}_0)\\ \mathbf{0}\end{pmatrix}$$

where

$$\mathbf{Z}(t) = \mathbf{z}(t) - \mathbf{H}_{22}^{-1}\mathbf{H}_{21}\begin{pmatrix}\mathbf{F}^a(\mathbf{X}_0)\\ \mathbf{Y}_0\end{pmatrix} - \mathbf{H}_{22}^{-1}\mathbf{M}_{22}\mathbf{R}_0. \tag{11.30}$$

Proof See Appendix 2. ■

The following remarks follow from Corollary 11.1 and extend the results in Chap. 7 to mem-circuits in $\mathcal{LME}$:

- if a mem-circuit has no input source, then any manifold (11.28) is positively invariant for the nonlinear dynamics in the (v, i)-domain;
- being $\mathbf{K}(\mathbf{\Gamma}_0)$ dependent on the vector $\mathbf{\Gamma}_0$ of ICs in the (v, i)-domain (see (11.27)), then a change in ICs results in general in a different invariant manifold on which the nonlinear dynamics take place. Two different invariant manifolds $\mathcal{M}(\mathbf{K}(\mathbf{\Gamma}_0'))$ and $\mathcal{M}(\mathbf{K}(\mathbf{\Gamma}_0''))$ are characterized by two different vectors $\mathbf{K}(\mathbf{\Gamma}_0')$ and $\mathbf{K}(\mathbf{\Gamma}_0'')$ associated with the ICs $\mathbf{\Gamma}_0'$ and $\mathbf{\Gamma}_0''$. Therefore, by (11.30), on each of the ∞^n different invariant manifolds the mem-circuit exhibits a different nonlinear dynamics. This explains theoretically the coexistence of infinitely many different reduced-order nonlinear dynamics. Moreover, in addition to standard bifurcations due to circuit parameter changes, also bifurcations due to changes in $\mathbf{\Gamma}_0$ for a fixed set of circuit parameters, i.e., bifurcations without parameters, can occur. Nonlinear dynamics can be also tuned by acting on $\mathbf{\Gamma}_0$ via suitable sequences of finite time duration voltage or current pulses, as illustrated by means of a specific example in the next section.

11.6 Chaotic Circuit with Memcapacitor and Memristor

Sustained oscillations and complex dynamics are observed in several mem-circuits belonging to $\mathcal{LME}$ (see [20–22, 24] and references therein). In order to illustrate the application of the methodology in this chapter, let us consider a variant of the mem-circuit with MC_σ and a M_φ studied in [24], where a current source $a(t)$ and a voltage source $e(t)$ are also introduced for programming purposes (see Fig. 11.7).

The hybrid representation of N_R is obtained via (11.6) with

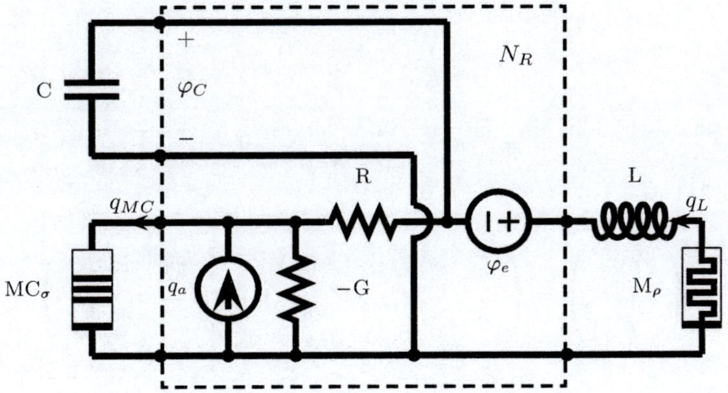

Fig. 11.7 Chaotic circuit with memcapacitor and memristor

$$\mathbf{H} = \begin{pmatrix} -G + \frac{1}{R} & 0 & -\frac{1}{R} \\ 0 & 0 & 1 \\ -\frac{1}{R} & -1 & \frac{1}{R} \end{pmatrix} \tag{11.31}$$

$$\mathbf{U}(t; t_0) = \begin{pmatrix} -q_a(t; t_0) \\ -\varphi_e(t; t_0) \\ 0 \end{pmatrix} \tag{11.32}$$

where $q_a(t; t_0) = \int_{t_0}^{t} a(\tau)d\tau$ and $\varphi_e(t; t_0) = \int_{t_0}^{t} e(\tau)d\tau$.

By defining

$$\mathbf{H}_{11} = \begin{pmatrix} -G + \frac{1}{R} & 0 \\ 0 & 0 \end{pmatrix}; \mathbf{H}_{12} = \begin{pmatrix} -\frac{1}{R} \\ 1 \end{pmatrix}$$

$$\mathbf{H}_{21} = \begin{pmatrix} -\frac{1}{R} & -1 \end{pmatrix}; H_{22} = \frac{1}{R}$$

where H_{22} is nonsingular and the state variables in the (φ, q)-domain $x(t) = \sigma^{\sigma}(t; t_0) = \sigma_{MC}(t; t_0)$, $y(t) = q^{\lambda\mu}(t; t_0) = q_L(t; t_0)$, and $z(t) = \varphi^{\gamma}(t; t_0) = \varphi_C(t; t_0)$, the following SEs (in the (φ, q)-domain) are obtained via (11.23)

$$\begin{aligned}
\dot{x}(t) &= (G - \frac{1}{R})[f_{MC}(x(t) + \sigma_{MC_0}) - f_{MC}(\sigma_{MC_0})] \\
&\quad + \frac{z(t)}{R} + q_{MC_0} + q_a(t; t_0) \\
L\dot{y}(t) &= -z(t) + f_M(y(t) + q_{M_0}) - f_M(q_{M_0}) \\
&\quad + Li_{L_0} + \varphi_e(t; t_0) \\
C\dot{z}(t) &= \frac{1}{R}[f_{MC}(x(t) + \sigma_{MC_0}) - f_{MC}(\sigma_{MC_0})] \qquad (11.33) \\
&\quad + y(t) - \frac{z(t)}{R} + Cv_{C_0}. \qquad (11.34)
\end{aligned}$$

The state variables in the (v, i)-domain are $X(t) = \sigma_{MC}(t), Y(t) = q_M(t), W(t) = q_{MC}(t), T(t) = i_L(t), R(t) = v_C(t)$ and the SEs are given from (11.25) as

$$\begin{aligned}
\begin{pmatrix} \dot{W}(t) \\ L\dot{T}(t) \\ C\dot{R}(t) \end{pmatrix} &= \begin{pmatrix} (G - \frac{1}{R})f'_{MC}(X(t))W(t) + \frac{Z(t)}{R} + a(t) \\ -Z(t) + f'_M(Y(t))T(t) + e(t) \\ \frac{1}{R}f'_{MC}(X(t))W(t) + Y(t) - \frac{Z(t)}{R} \end{pmatrix} \\
\dot{X}(t) &= W(t) \\
\dot{Y}(t) &= T(t). \qquad (11.35)
\end{aligned}$$

As predicted by the theoretical results in the previous section, the SEs have an order reduction $n_{vi} - n_{\varphi q} = 2$ in passing from the (v, i)-domain to the (φ, q)-domain, that is exactly the number of mem-elements in Fig. 11.7.

Let $\mathbf{\Gamma}(t) = (\sigma_{MC}(t), q_M(t), q_{MC}(t), i_L(t), v_C(t))^T$ be the vector of state variables in the (v, i)-domain, then (11.27) gives

$$\mathbf{K}(\mathbf{\Gamma}) = \begin{pmatrix} -Gf_{MC}(\sigma_{MC}) + q_{MC} + Cv_C - q_M \\ Li_L + f_M(q_M) - RCv_C + f_{MC}(\sigma_{MC}) + Rq_M \end{pmatrix}.$$

Moreover, by (11.32), $\mathbf{K}(\mathbf{\Gamma}(t; t_0, \mathbf{\Gamma}_0))$ along a solution $\mathbf{\Gamma}(t; t_0, \mathbf{\Gamma}_0)$ of the SEs in the (v, i)-domain (see (11.29)) is

$$\mathbf{K}(\mathbf{\Gamma}(t; t_0, \mathbf{\Gamma}_0)) = \mathbf{K}(\mathbf{\Gamma}_0) + \begin{pmatrix} q_a(t; t_0) \\ \varphi_e(t; t_0) \end{pmatrix} \tag{11.36}$$

for any $t \geq t_0$, where $\mathbf{\Gamma}_0 = (\sigma_{MC_0}, q_{M_0}, q_{MC_0}, i_{L_0}, v_{C_0})^T$ are the ICs for the state variables in the (v, i)-domain.

Let us focus first on the case in which the mem-circuit in Fig. 11.7 has zero input sources, i.e., $a(t) = 0$ and $e(t) = 0$, thus $q_a(t; t_0) = 0$ and $\varphi_e(t; t_0) = 0$, $t \geq t_0$. In this case, due to Theorem 11.1, there exist ∞^2 three-dimensional manifolds $\mathcal{M}(\bar{\mathbf{K}}) = \{\mathbf{\Gamma} \in \mathbb{R}^5 : \mathbf{K}(\mathbf{\Gamma}) = \bar{\mathbf{K}}\}$, for any $\bar{\mathbf{K}} \in \mathbb{R}^2$, that are positively invariant for the mem-circuit dynamics in the (v, i)-domain. On each manifold, due to (11.30), the reduced third-order nonlinear dynamics can be rewritten as

$$\begin{pmatrix} \dot{X}(t) \\ L\dot{Y}(t) \\ C\dot{Z}(t) \end{pmatrix} = \begin{pmatrix} (-G + \frac{1}{R})f_{MC}(X(t)) + \frac{Z(t)}{R} + \mathbf{K}_1(\mathbf{\Gamma}_0) \\ -Z(t) + f_M(Y(t)) + \mathbf{K}_2(\mathbf{\Gamma}_0) \\ \frac{1}{R}f_{MC}(X(t)) + Y(t) - \frac{Z(t)}{R} \end{pmatrix} \tag{11.37}$$

where $Z(t)$ is in (11.30) and $\mathbf{K}(\mathbf{\Gamma}_0) = (\mathbf{K}_1(\mathbf{\Gamma}_0), \mathbf{K}_2(\mathbf{\Gamma}_0))$ with

$$\mathbf{K}_1(\mathbf{\Gamma}_0) = -Gf_{MC}(\sigma_{MC_0}) + q_{MC_0} + Cv_{C_0} - q_{M_0}$$

$$\mathbf{K}_2(\mathbf{\Gamma}_0) = Li_{L_0} + f_M(q_{M_0}) - RCv_{C_0} + f_{MC}(\sigma_{MC_0}) + Rq_{M_0}.$$

The ICs in the (v, i)-domain act as constant inputs in the SEs in the (φ, q)-domain via the terms $\mathbf{K}_1(\mathbf{\Gamma}_0)$ and $\mathbf{K}_2(\mathbf{\Gamma}_0)$. It turns out that there coexist ∞^2 different reduced-order dynamics (one for each manifold $\mathcal{M}(\mathbf{K}_1(\mathbf{\Gamma}_0), \mathbf{K}_2(\mathbf{\Gamma}_0))$.

Suppose that the circuit parameters are $G = 2.1$, $R = 0.2083$, $C = 5.8823$, $L = 0.136$ and let $f_{MC}(\sigma_{MC}) = 0.7\sigma_{MC} + 0.5\sigma_{MC}^2$ and $f_M(q_M) = 0.01q_M - 0.05q_M^2$ (cf. [24]). Figure 11.8 reports the bifurcation diagram for (11.37), as a function of $\mathbf{K}_1$ and $\mathbf{K}_2/L$, numerically obtained by means of the software MatCont [25]. The black (resp., blue) solid curve corresponds to points where MatCont detects a supercritical (resp., subcritical) Hopf bifurcation. The black dashed line represents points of the parameter-space where a period-doubling bifurcation occurs. Finally,

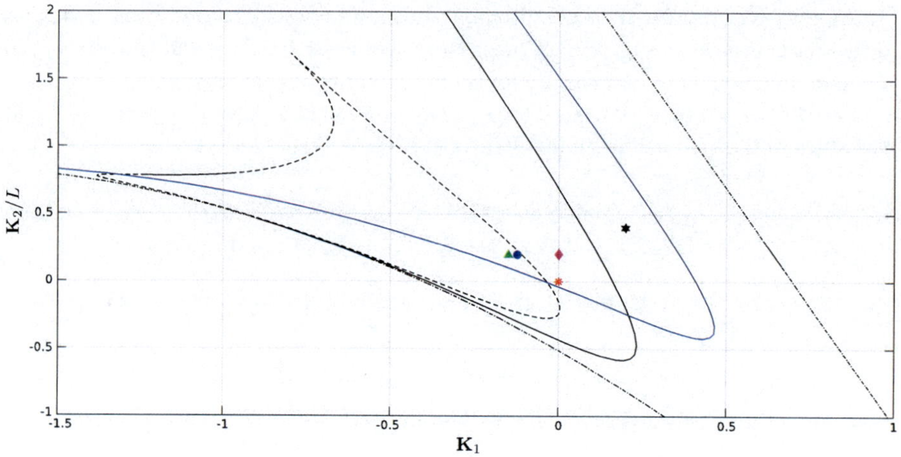

Fig. 11.8 Bifurcation diagram for a circuit with a memristor and a meminductor

the black dashed-dotted lines are points where a fold bifurcation is present. Within the region enclosed by the period-doubling line, complex dynamics are expected to occur. Note that these bifurcations are due to changing $\mathbf{K}_1(\boldsymbol{\Gamma}_0)$ and $\mathbf{K}_2(\boldsymbol{\Gamma}_0)$, i.e., due to varying the ICs for the state variables in the (v, i)-domain, but for a fixed set of circuit parameters and nonlinearities.

The bifurcation diagram in Fig. 11.8 is fundamental to drive, by means of the input sources $q_a(t; t_0)$ and $\varphi_e(t; t_0)$, solutions of the SEs (11.35) along different manifolds, and nonlinear dynamical behaviors. Let us consider a solution $\boldsymbol{\Gamma}(t; t_0, \boldsymbol{\Gamma}_0)$ starting at $t_0 = 0$ in a neighborhood of the origin $\boldsymbol{\Gamma}_0 = \mathbf{0}$ on the zero-manifold $\mathcal{M}(0, 0)$ (red asterisk in Fig. 11.8) and suppose the input is given by a sequence of rectangular voltage and current pulses with time duration $\Delta = 1$ (see Fig. 11.9). The bottom part of Fig. 11.9 also reports the corresponding variations of $\mathbf{K}_1$ and $\mathbf{K}_2$, according to (11.36), due to the inputs $a(t)$ and $e(t)$ shown in the upper part of the same figure. The time-domain complex behavior of $v_C(t)$ for this solution is depicted in Fig. 11.10. This behavior can be explained as follows. Each pulse causes a switch of the manifold according to the variations of $\mathbf{K}_1$ and $\mathbf{K}_2$, while in the time intervals between pulses each manifold is invariant. The solution first approaches a limit cycle attractor on manifold $\mathcal{M}(0, 0)$. The first pulse switches the solution to the manifold $\mathcal{M}(0, 0.2)$ (magenta diamond in Fig. 11.8) where there exists a different limit cycle attractor. The second pulse moves the solution to $\mathcal{M}(-0.12, 0.2)$ (blue ball in Fig. 11.8), passing through a line where the mem-circuit undergoes a period-doubling bifurcation. In such a case the mem-circuit exhibits an attractor that is a limit cycle with period 2. The subsequent pulse shifts the solution onto $\mathcal{M}(-0.14, 0.2)$ (green triangle in Fig. 11.8), where a complex behavior is displayed by the mem-circuit. The final pulse moves the solution to $\mathcal{M}(0.2, 0.4)$ (black asterisk in Fig. 11.8) where the mem-circuit has a stable EP. The

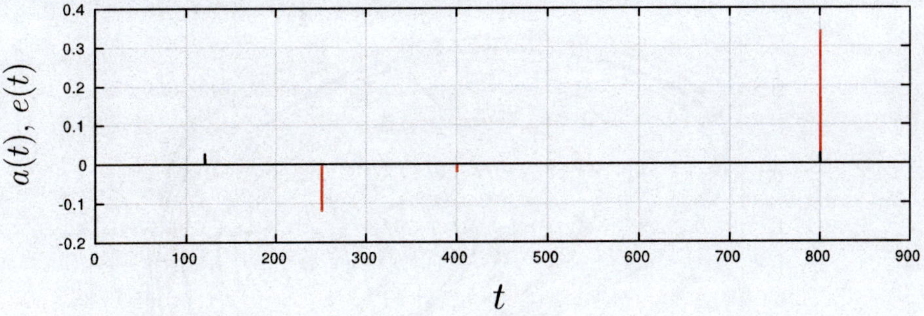

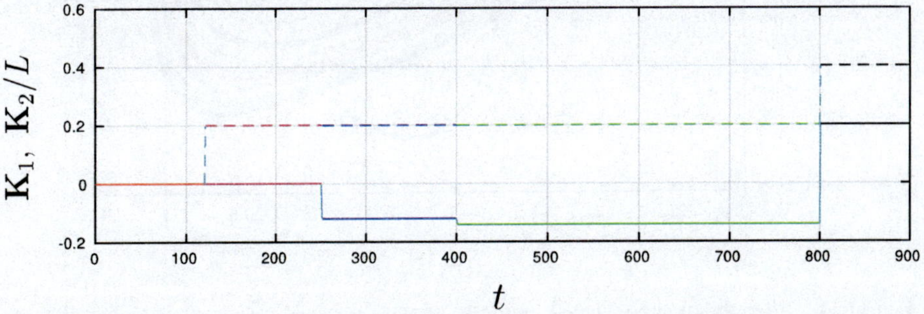

Fig. 11.9 **Upper figure: Input voltage** $e(t)$ (black) **and current** $a(t)$ (red) **given by a sequence of short (**$\Delta = 1$ **duration) rectangular pulses. Lower figure: corresponding behavior of** $\mathbf{K}_1$ (solid) and $\mathbf{K}_2/L$ (dashed)

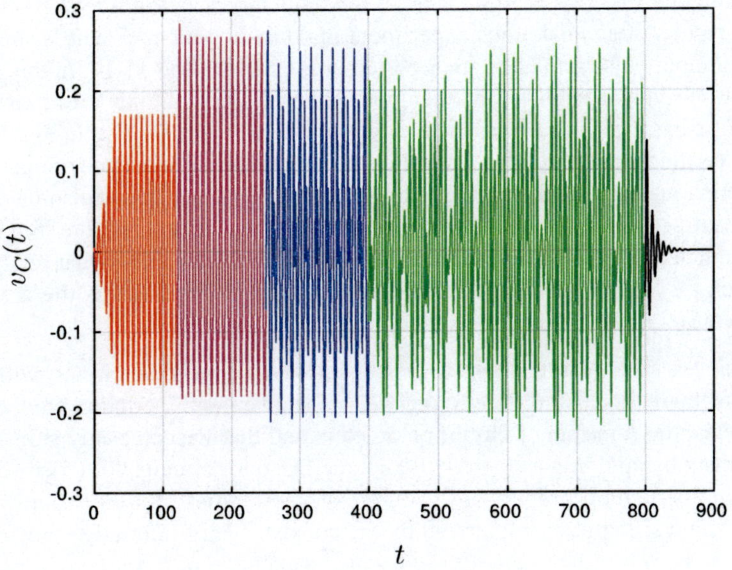

Fig. 11.10 Time-domain evolution of capacitor voltage $v_C(t)$

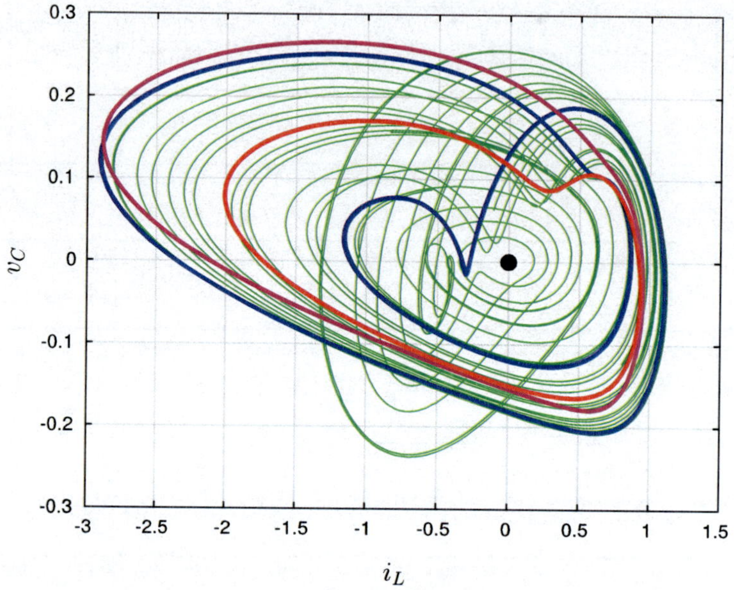

Fig. 11.11 Different attractors visited by the solution of a forced circuit, namely, two different limit cycles (red and magenta), a cycle with period 2 (blue), a complex attractor (green), and an EP (small black ball)

subsequent attractors experienced by the solutions in the intervals between pulses are reported in Fig. 11.11. A 3D view of the trajectory is in Fig. 11.12.

The results of an analogous experiment in which the mem-circuit is subject to a different input pulse sequence are reported in Figs. 11.13 and 11.14. In this case, the input pulse sequence consists of triangular pulses, each with time duration $\Delta = 20$. Each pulse has the same area as the corresponding rectangular pulse in Fig. 11.9. It is readily verified that the solution switches through the same sequence of manifolds as in the previous experiments with rectangular pulses. In fact, it can also be observed that the attractors encountered by the solution are the same as before. Such results confirm that, as predicted by the theory developed in this chapter, what really counts to switch a solution between different manifolds and dynamics is the area of the applied pulse, not the shape or its time duration.

Remark 11.3 The results thus obtained show that for this circuit with mem-elements there coexist *different regimes* as convergent, periodic, and complex regimes for the same set of circuit parameters and that we can easily shift between any of them by applying a single pulse source. Also for circuits with mem-elements it is then a structural property to feature extreme multistability, i.e., an extremely rich dynamic scenario characterized by the coexistence of infinitely many different attractors. The considered circuit represents an example of a source of controllable complex dynamics that can be implemented at nanoscale and is potentially useful for incorporation in a neuromorphic system along the lines in [14].

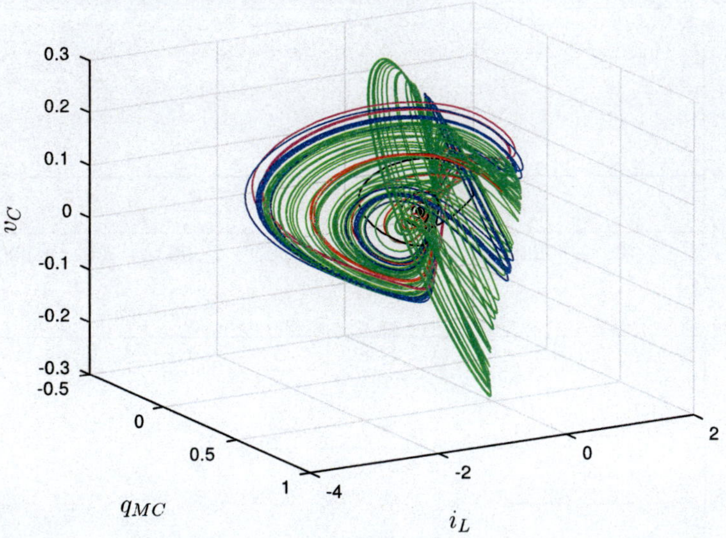

Fig. 11.12 3D view of the trajectory evolution

Example 11.1 Consider a mem-element circuit as in Fig. 11.15 containing a two-terminal element, given by the parallel connection of a σ-controlled memcapacitor and a flux-controlled memristor, and a linear inductor, connected to a two-port network with only linear resistors and flux and charge independent sources.

The two-terminal element MC_σ-M_φ is not included in those considered in the class $\mathcal{LME}$ (or in the class $\mathcal{LME}'$). This notwithstanding, we wish to show that it is still possible to derive the SE representation in the (φ, q)-domain.

Suppose there exists the hybrid representation of N_R

$$
\begin{aligned}
q_1(t;t_0) &= h_{11}\varphi_1(t;t_0) + h_{12}q_2(t;t_0) + q_a(t;t_0) \\
\varphi_2(t;t_0) &= h_{21}\varphi_1(t;t_0) + h_{22}q_2(t;t_0) + \varphi_e(t;t_0).
\end{aligned}
$$

KφL and KqL at port 1 yield

$$
\varphi_1(t;t_0) = \varphi_{MC}(t;t_0) = \varphi_M(t;t_0);\;\; q_1(t;t_0) + q_{MC}(t;t_0) + q_M(t;t_0) = 0
$$

and at port 2 we have

$$
q_2(t;t_0) = q_L(t;t_0);\, \varphi_2(t;t_0) = -\varphi_L(t;t_0).
$$

The CR of M_φ is

$$
q_M(t;t_0) = f_M(\varphi_M(t;t_0) + \varphi_{M_0}) - f_M(\varphi_{M_0})
$$

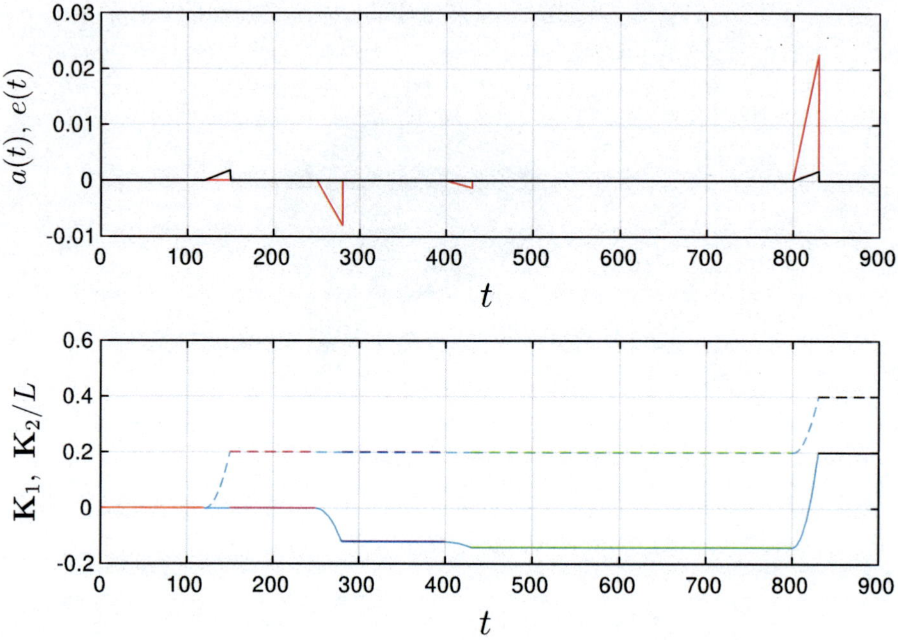

Fig. 11.13 Input voltage $e(t)$ (black) and current $a(t)$ (red) given by a sequence of long ($\Delta = 20$ duration) triangular pulses. Lower figure: corresponding behavior of $\mathbf{K}_1$ (solid) and $\mathbf{K}_2/L$ (dashed)

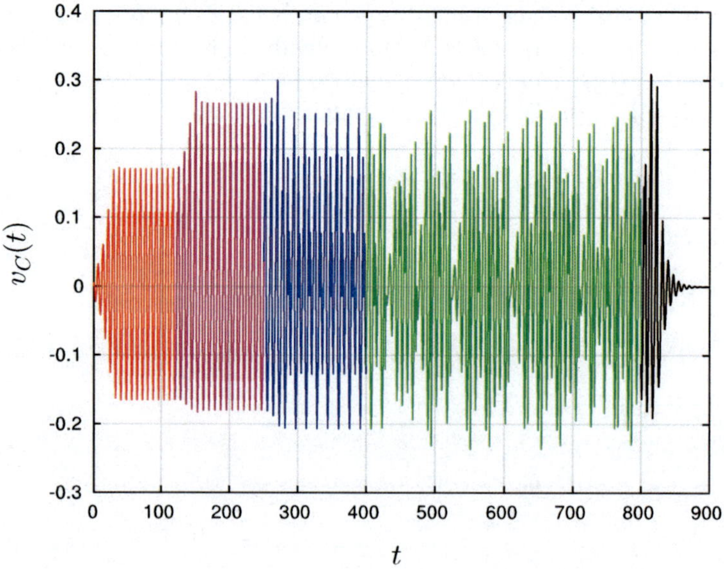

Fig. 11.14 Time-domain evolution of capacitor voltage $v_C(t)$

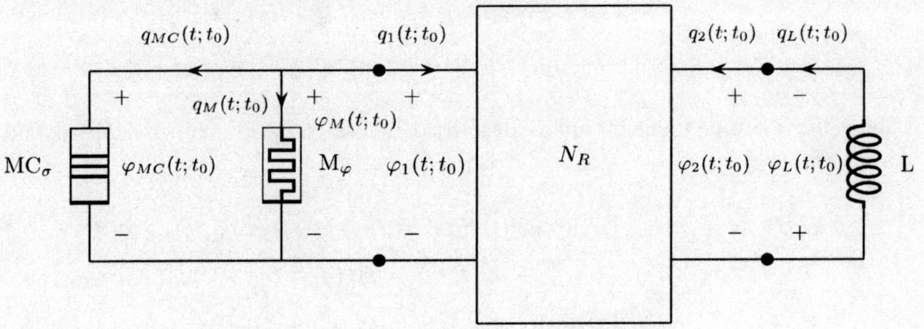

Fig. 11.15 Circuit with a σ-controlled memcapacitor and a flux-controlled memristor

that of MC_σ is given by

$$\varphi_{MC}(t;t_0) = f_{MC}(\sigma_{MC}(t;t_0) + \sigma_{MC_0}) - f_{MC}(\sigma_{MC_0})$$
$$\dot{\sigma}_{MC}(t;t_0) = q_{MC}(t;t_0) + q_{MC_0}$$

and, finally, the CR of L is

$$\varphi_L(t;t_0) = L\dot{q}_L(t;t_0) - \varphi_{L_0}.$$

The state variables of the circuit in the (φ, q)-domain are (according to Table 11.1) $\sigma_{MC}(t;t_0)$ and $q_L(t;t_0)$.

By substitution we obtain

$$\begin{aligned}
\dot{\sigma}_{MC}(t;t_0) &= q_{MC}(t;t_0) + q_{MC_0} \\
&= -q_1(t;t_0) - q_M(t;t_0) + q_{MC_0} \\
&= -h_{11}\varphi_{MC}(t;t_0) - h_{12}q_L(t;t_0) - q_a(t;t_0) \\
&\quad - f_M(\varphi_{MC}(t;t_0) + \varphi_{M_0}) + f_M(\varphi_{M_0}) + q_{MC_0} \\
&= -h_{11}[f_{MC}(\sigma_{MC}(t;t_0) + \sigma_{MC_0}) - f_{MC}(\sigma_{MC_0})] \\
&\quad -h_{12}q_L(t;t_0) - q_a(t;t_0) - f_M\big(f_{MC}(\sigma_{MC}(t;t_0) + \sigma_{MC_0}) \\
&\quad - f_{MC}(\sigma_{MC_0}) + \varphi_{M_0}\big) + f_M(\varphi_{M_0}) + q_{MC_0}.
\end{aligned}$$

Moreover,

$$\begin{aligned}
L\dot{q}_L(t;t_0) &= \varphi_L(t;t_0) + \varphi_{L_0} \\
&= \varphi_2(t;t_0) + \varphi_{L_0} \\
&= h_{21}\varphi_{MC}(t;t_0) + h_{22}q_L(t;t_0) + \varphi_e(t;t_0)
\end{aligned}$$

$$= h_{21}[f_{MC}(\sigma_{MC}(t;t_0) + \sigma_{MC_0}) - f_{MC}(\sigma_{MC_0})]$$
$$+h_{22}q_L(t;t_0) + \varphi_e(t;t_0).$$

Then, the mem-element circuit is described in the (φ, q)-domain by the second-order SE

$$\begin{aligned}
\dot{\sigma}_{MC}(t;t_0) &= -h_{11}[f_{MC}(\sigma_{MC}(t;t_0) + \sigma_{MC_0}) - f_{MC}(\sigma_{MC_0})] \\
&\quad -h_{12}q_L(t;t_0) - q_a(t;t_0) - f_M\big(f_{MC}(\sigma_{MC}(t;t_0) + \sigma_{MC_0}) \\
&\quad -f_{MC}(\sigma_{MC_0}) + \varphi_{M_0}\big) + f_M(\varphi_{M_0}) + q_{MC_0} \\
L\dot{q}_L(t;t_0) &= h_{21}[f_{MC}(\sigma_{MC}(t;t_0) + \sigma_{MC_0}) - f_{MC}(\sigma_{MC_0})] \\
&\quad +h_{22}q_L(t;t_0) + \varphi_e(t;t_0).
\end{aligned}$$

By time differentiation we can find the fourth-order SE in the state variables $\sigma_{MC}, q_{MC}, \varphi_M$, and i_L that describes the circuit dynamics in the (v, i)-domain. On this basis it is possible to study invariant manifolds, reduced-order dynamics on manifolds, and bifurcations without parameters. The mathematical details are left to the interested reader.

11.7 Examples Concerning Class $\mathcal{LME}'$

In this section we present a number of examples to illustrate how we can write an SE description in the (φ, q)-domain and in the (v, i)-domain for circuits with mem-elements belonging to the class $\mathcal{LME}'$. The examples enable one to write the SEs by inspection but a systematic procedure for writing the SEs may be developed along similar lines as for the class $\mathcal{LME}$. The reader is referred to [26] for more details.

The main new aspect of $\mathcal{LME}'$, with respect to $\mathcal{LME}$, is the presence of nonlinear charge-controlled capacitors or nonlinear flux-controlled inductors. Before illustrating the examples we then describe their CRs in the (φ, q)-domain and also in the (v, i)-domain.

11.7.1 Charge-Controlled Nonlinear Capacitor $\mathrm{C_q}$

A charge-controlled nonlinear capacitor $\mathrm{C_q}$ is defined by the algebraic CR $v_C(t) = f_C(q_C(t))$ in the (v, q)-domain, i.e., $\alpha = 0$ and $\beta = -1$. The corresponding CR in the (φ, q)-domain is

$$\dot{\varphi}_C(t;t_0) = f_C(q_C(t;t_0) + q_{C_0})$$

where $q_{MC_0} = q_{MC}(t_0)$. Hence, $O^{(-1,-1)}_{\mathrm{C}_q} = 1$, the state variable is $\varphi_C(t; t_0)$ and the IC is $\varphi_C(t_0; t_0) = 0$. It is worth noting that the CR depends upon q_{C_0}.

In the (v, i)-domain we have

$$v_C(t) = f_C(q_C(t))$$

where

$$\dot{q}_C(t) = i_C(t).$$

Hence, $O^{(0,0)}_{\mathrm{C_q}} = 1$, the state variable is $q_C(t)$ with corresponding ICs q_{C_0}. There is no reduction of order in passing from the (v, i)-domain to the (φ, q)-domain.

11.7.2 *Flux-Controlled Nonlinear Inductor* $\mathbf{L}_\varphi$

A flux-controlled nonlinear inductor L_φ is defined by the algebraic CR $i_L(t) = f_L(\varphi_L(t))$ in the (φ, i)-domain, i.e., $\alpha = -1$ and $\beta = 0$. The corresponding CR in the (φ, q)-domain is

$$\dot{q}_L(t; t_0) = f_L(\varphi_L(t; t_0) + \varphi_{L_0})$$

where $\varphi_{L_0} = \varphi_L(t_0)$. Hence, $O^{(-1,-1)}_{\mathrm{L}_\varphi} = 1$, the state variable is $q_L(t; t_0)$, and the IC is $q_L(t_0; t_0) = 0$. It is worth noting that the CR depends upon φ_{L_0}.

In the (v, i)-domain we have

$$i_L(t) = f_L(\varphi_L(t))$$

where

$$\dot{\varphi}_L(t) = v_L(t).$$

Hence, $O^{(0,0)}_{\mathrm{L}_\varphi} = 1$, the state variable is $\varphi_L(t)$ with corresponding IC φ_{L_0}. There is no reduction of order in passing from the (v, i)-domain to the (φ, q)-domain.

Remark 11.4 Note that the CR in the (φ, q)-domain depends upon the IC at t_0 for the state variable of the element in the (v, i)-domain. Since the state variable in the (φ, q)-domain is an incremental electric variable, then the IC at t_0 is by construction null.

Remark 11.5 Table 11.1 summarizes the results thus obtained for a nonlinear capacitor and inductor. Note that the chief result on order reduction in Remark 11.2 continues to hold even in the presence of nonlinear inductors and capacitors.

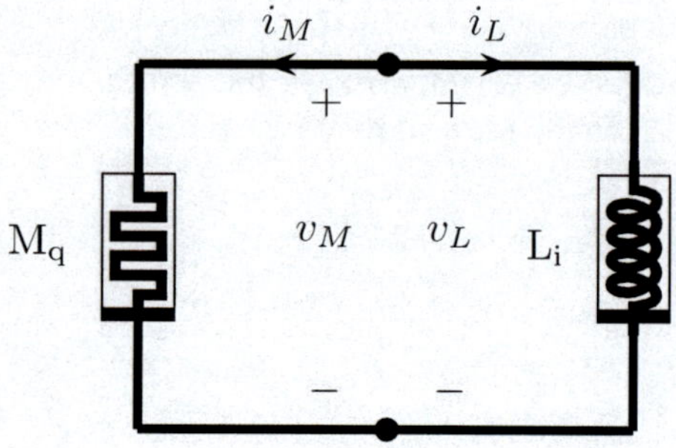

Fig. 11.16 Circuit with a nonlinear current-controlled inductor and a charge-controlled memristor

Example 11.2 Let us consider a simple circuit as in Fig. 11.16 given by a nonlinear current-controlled inductor and a charge-controlled memristor. The state variables in the (v, i)-domain are $i_L(t)$ and $q_M(t)$. The KVL and KCL are expressed as $v_M(t) = v_L(t)$ and $i_M(t) + i_L(t) = 0$, respectively, whereas the CRs are $v_M(t) = f'_M(q_M(t))i_M(t)$, with $\dot{q}_M(t) = i_M(t)$ and $v_L(t) = f'_L(i_L(t))(di_L(t)/dt)$. By substitution we obtain

$$f'_L(i_L(t))\frac{di_L(t)}{dt} = -f'_M(q_M(t))i_L$$

and

$$\dot{q}_M(t) = i_L(t).$$

To write the SEs we need to assume $f'_L(x) \neq 0$ for any $x \in \mathbb{R}$ (cf. Sect. 11.3), which implies f_L is globally invertible, hence the nonlinear inductor is also flux-controlled (we can write $i_L = (f_L)^{-1}(\varphi_L)$). Under such an assumption the SEs are given by the second-order system

$$\begin{aligned}\frac{di_L(t)}{dt} &= -\frac{f'_M(q_M(t))}{f'_L(i_L(t))}i_L(t)\\ \dot{q}_M(t) &= i_L(t)\end{aligned}$$

for $t \geq t_0$ and the ICs are $i_L(t_0) = i_{L_0}$ and $q_M(t_0) = q_{M_0}$.

The equivalent circuit in the (φ, q)-domain can be analyzed for $t \geq t_0$ via the KqL $q_L(t; t_0) + q_M(t; t_0) = 0$, the K$\varphi$L $\varphi_L(t; t_0) = \varphi_M(t; t_0)$ and the CR of the nonlinear inductor

$$\varphi_L(t; t_0) = f_L\big(\dot{q}_L(t; t_0)\big) - f_L(i_{L_0})$$

and memristor

$$\varphi_M(t; t_0) = f_M(q_M(t; t_0) + q_{M_0}) - f_M(q_{M_0}).$$

Since, by assumption, there exists f_L^{-1}, by substitution we obtain that in the (φ, q)-domain the circuit satisfies for $t \geq t_0$ the first-order SE

$$\dot{q}_L(t; t_0) = (f_L)^{-1}\big(f_M(-q_L(t; t_0) + q_{M_0}) - f_M(q_{M_0}) + f_L(i_{L0})\big)$$

in the state variable $q_L(t; t_0)$. The IC is $q_L(t_0; t_0) = 0$.

Passing from the (v, i)-domain to the (φ, q)-domain we have an order reduction for the SEs equal to 1. Note, however, that the right-hand side of the SE in the (φ, q)-domain depends upon the ICs i_{L0} and q_{M_0} for the state variable in the (v, i)-domain.

Let us show that the dynamics in the (v, i)-domain is characterized by the presence of *invariant manifolds*. From the KφL $\varphi_L(t; t_0) = \varphi_M(t; t_0)$ we obtain

$$f_L(i_L(t)) - f_M(q_M(t)) = f_L(i_{L0}) - f_M(q_{M_0}) = \Phi_0(i_{L0}, q_{M_0})$$

for any $t \geq t_0$, i.e., function $f_L(i_L(t)) - f_M(q_M(t))$ is an invariant of motion. This implies that the two-dimensional state space $(i_L(t), q_M(t))^T \in \mathbb{R}^2$ in the (v, i)-domain can be decomposed in ∞^1 1D manifolds, which are (positively) invariant for the dynamics in the (v, i)-domain, given by

$$\mathcal{M}(\Phi_0) = \{(i_L(t), q_M(t))^T \in \mathbb{R}^2 : f_L(i_L(t)) - f_M(q_M(t)) = \Phi_0\}$$

where $\Phi_0 \in \mathbb{R}$. On each manifold the dynamics is first order and can be described by the corresponding SE in the (φ, q)-domain. This also implies that *there coexist* ∞^1 *different first-order dynamics*, one for each manifold, in the (v, i)-domain.

Example 11.3 Let us consider the circuit with a current source $a(t)$, a voltage source $e(t)$, a σ-controlled memcapacitor, and a flux-controlled nonlinear inductor as in Fig. 11.17.

In the (v, i)-domain, omitting dependence on t, we have:

KVL: $v_{MC} = v_L + e$

KCL: $a = i_{MC} + i_L$

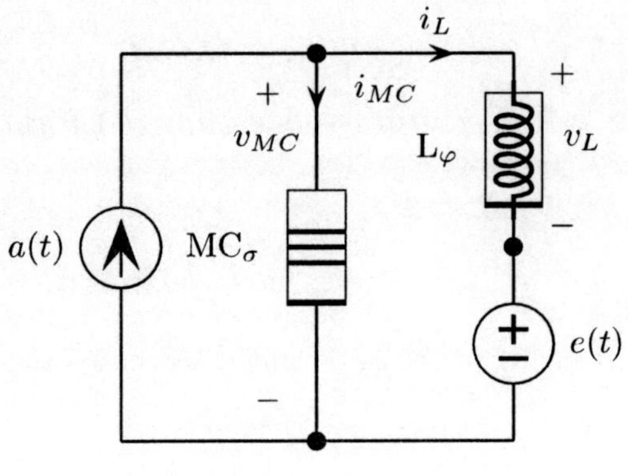

Fig. 11.17 Circuit with a current and a voltage source, a σ-controlled memcapacitor, and a flux-controlled nonlinear inductor

CR of MC_σ:

$$v_{MC} = f'_{MC}(\sigma_{MC})q_{MC}$$
$$\dot{\sigma}_{MC} = q_{MC}$$
$$\dot{q}_{MC} = i_{MC}.$$

CR of L_φ:

$$i_L = f_L(\varphi_L)$$

where

$$\dot{\varphi}_L = v_L.$$

By substitution, these yield the third-order SE in the (v, i)-domain

$$\begin{aligned}\dot{\varphi}_L &= f'_{MC}(\sigma_{MC})q_{MC} - e\\ \dot{\sigma}_{MC} &= q_{MC}\\ \dot{q}_{MC} &= -f_L(\varphi_L) + a\end{aligned}$$

in the state variables φ_L, σ_{MC} and q_{MC}.

In the (φ, q)-domain we can write:

KφL:

$$\varphi_{MC}(t; t_0) = \varphi_L(t; t_0) + \varphi_e(t; t_0)$$

KqL:

$$q_a(t; t_0) = q_{MC}(t; t_0) + q_L(t; t_0)$$

CR of MC_σ:

$$\begin{aligned}\varphi_{MC}(t; t_0) &= f_{MC}(\sigma_{MC}(t; t_0) + \sigma_{MC_0}) - f_{MC}(\sigma_{MC_0})\\ \dot{\sigma}_{MC}(t; t_0) &= q_{MC}(t; t_0) + q_{MC_0}\end{aligned}$$

CR of L_φ:

$$\dot{q}_L(t; t_0) = f_L(\varphi_L(t; t_0) + \varphi_{L_0}).$$

By substitution, we obtain the second-order SE in the (φ, q)-domain

$$\dot{q}_L(t; t_0) = f_L\big(f_{MC}(\sigma_{MC}(t; t_0) + \sigma_{MC_0}) - f_{MC}(\sigma_{MC_0}) - \varphi_e(t; t_0) + \varphi_{L_0}\big)$$
$$\dot{\sigma}_{MC}(t; t_0) = -q_L(t; t_0) + q_a(t; t_0) + q_{MC_0}$$

in the state variables $q_L(t; t_0)$ and $\sigma_{MC}(t; t_0)$.

Using KφL with the CR of the memcapacitor it can be checked that

$$\frac{d}{dt}[\varphi_L(t) - f_{MC}(\sigma_{MC}(t))] = e(t) \quad (11.38)$$

for any $t \geq t_0$.

Hence, the three-dimensional state space $(\varphi_L, \sigma_{MC}, q_{MC}) \in \mathbb{R}^3$ can be decomposed in ∞^1 1D manifolds, which are (positively) invariant for the dynamics in the (v, i)-domain, given by

$$\mathcal{M}(\Phi_0) = \{(\sigma_{MC}(t), q_{MC}(t), \varphi_L(t))^T \in \mathbb{R}^3 : \varphi_L(t) - f_{MC}(\sigma_{MC}(t)) = \Phi_0\}$$

where $\Phi_0 \in \mathbb{R}$. On each manifold the dynamics is second order and can be described by the corresponding SE in the (φ, q)-domain. This also implies that *there coexist* ∞^1 *different second-order dynamics*, one for each manifold, in the (v, i)-domain.

Due to (11.38), it is seen that manifolds can be programmed via the voltage source $e(t)$, while the current source $a(t)$ has no effect on manifold programming.

Example 11.4 **Consider a circuit with a charge-controlled memristor, a flux-controlled nonlinear inductor, and a σ-controlled memcapacitor as shown in Fig. 11.18.**

In the (v, i)-domain, omitting dependence on t, we have:
KVL: $v_M = v_L + v_{MC}$
KCLs: $i_M + i_{MC} = 0$, $i_L = i_{MC}$

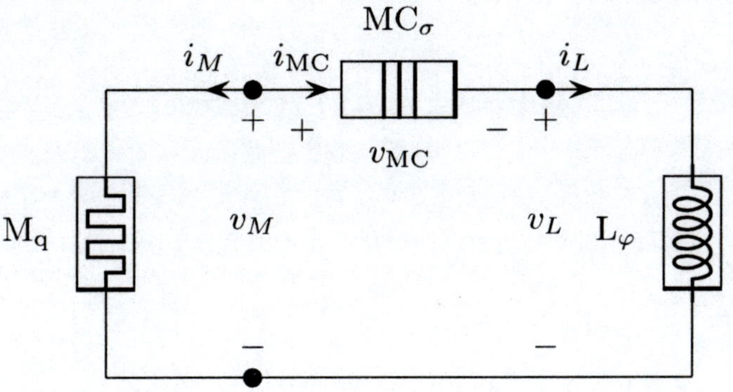

Fig. 11.18 Circuit with a charge-controlled memristor, a flux-controlled nonlinear inductor, and a σ-controlled memcapacitor

CR of MC_σ:

$$\begin{aligned} v_{MC} &= f'_{MC}(\sigma_{MC}) q_{MC} \\ \dot{\sigma}_{MC} &= q_{MC} \\ \dot{q}_{MC} &= i_{MC} \end{aligned}$$

CR of L_φ:

$$i_L = f_L(\varphi_L)$$

where

$$\dot{\varphi}_L = v_L$$

CR of M_φ:

$$v_M = f'_M(q_M) i_M$$

and

$$\dot{q}_M = i_M.$$

By substitution we obtain the fourth-order SE

$$\begin{aligned} \dot{\varphi}_L &= -f'_M(q_M) f_L(\varphi_L) - f'_{MC}(\sigma_{MC}) q_{MC} \\ \dot{q}_{MC} &= f_L(\varphi_L) \\ \dot{\sigma}_{MC} &= q_{MC} \\ \dot{q}_M &= -f_L(\varphi_L) \end{aligned}$$

in the state variables $\varphi_L, q_{MC}, \sigma_{MC}, q_M$.

In the (φ, q)-domain we have

KφL:

$$\varphi_M(t; t_0) = \varphi_L(t; t_0) + \varphi_{MC}(t; t_0)$$

KqL:

$$q_M(t; t_0) + q_L(t; t_0) = 0; \quad q_L(t; t_0) = q_{MC}(t; t_0)$$

CR of MC_σ:

$$\varphi_{MC}(t; t_0) = f_{MC}(\sigma_{MC}(t; t_0) + \sigma_{MC_0}) - f_{MC}(\sigma_{MC_0})$$

$$\dot{\sigma}_{MC}(t;t_0) = q_{MC}(t;t_0) + q_{MC_0}$$

CR of L_φ:

$$\dot{q}_L(t;t_0) = f_L(\varphi_L(t;t_0) + \varphi_{L_0})$$

CR of M_φ:

$$\varphi_M(t;t_0) = f_M(q_M(t;t_0) + q_{M_0}) - f_M(q_{M_0}).$$

By substitution we obtain the second-order SE

$$\begin{aligned}\dot{q}_L(t;t_0) &= f_L\big(f_M(q_L(t;t_0) + q_{M_0}) - f_M(q_{M_0})\\ &\qquad - f_{MC}(\sigma_{MC}(t;t_0) + \sigma_{MC_0}) + f_{MC}(\sigma_{MC_0}) + \varphi_{L_0}\big)\\ \dot{\sigma}_{MC}(t;t_0) &= -q_L(t;t_0) + q_{MC_0}\end{aligned}$$

in the state variables $q_L(t;t_0)$, $\sigma_{MC}(t;t_0)$.

To study the invariant manifolds for the dynamics in the (v,i)-domain, note that from KCL $i_M(t) + i_{MC}(t) = 0$ we obtain

$$\frac{d}{dt}(q_M(t) + q_{MC}(t)) = 0.$$

Moreover, KVL $v_M(t) = v_L(t) + v_{MC}(t)$ yields

$$\frac{d}{dt}\varphi_M(t) = \frac{d}{dt}\varphi_L(t) + f'_{MC}(\sigma_{MC}(t))q_{MC}(t)$$

and hence

$$\frac{d}{dt}(f_M(q_M(t)) = \frac{d}{dt}\varphi_L(t) + \frac{d}{dt}(f_{MC}(\sigma_{MC}(t))).$$

Then, we can identify the two invariants of motion

$$q_M(t) + q_{MC}(t)$$

and

$$f_M(q_M(t)) - \varphi_L(t) - f_{MC}(\sigma_{MC}(t))$$

which in turn permit to define ∞^2 two-dimensional invariant manifolds for the dynamics in the (v,i)-domain

$$\mathcal{M}(Q_0,\Phi_0) = \{(\varphi_L(t), q_{MC}(t), \sigma_{MC}(t), q_M(t))^T \in \mathbb{R}^4 :$$

$$q_M(t) + q_{MC}(t) = Q_0,$$
$$f_M(q_M(t)) - \varphi_L(t) - f_{MC}(\sigma_{MC}(t)) = \Phi_0\}$$

where $(\Phi_0, Q_0)^T \in \mathbb{R}^2$.

We leave to the reader the problem of inserting two independent sources in the circuit in order to programme the manifolds and reduced-order circuit dynamics.

11.8 Discussion

Nonlinear dynamic memory elements, as memristors, memcapacitors, and meminductors (a.k.a. mem-elements), are of paramount importance in conceiving neural networks, mem-computing machines, and reservoir computing systems with advanced computational primitives. This chapter has developed a systematic methodology for analyzing complex dynamics in nonlinear circuits with such emerging nanoscale mem-elements. The technique extends FCAM for nonlinear circuits with memristors to a broader class of nonlinear circuits $\mathcal{N}_e$ containing also memcapacitors and meminductors. FCAM has also been extended to deal with circuits containing mem-elements in combination with nonlinear capacitors and inductors.

After deriving the CR and equivalent circuit in the (φ, q)-domain of each two-terminal element in $\mathcal{N}_e$, the chapter has focused on relevant subclasses of $\mathcal{N}_e$ for which an SE description can be obtained. On this basis, salient features of the dynamics are highlighted and studied analytically: (1) the presence of invariant manifolds in the autonomous case; (2) the coexistence of infinitely many different reduced-order dynamics on manifolds; (3) the presence of bifurcations due to changing the ICs for a fixed set of parameters (bifurcations without parameters). Analytic formulas are also given to design non-autonomous circuits subject to pulses that drive trajectories through different manifolds and nonlinear reduced-order dynamics.

The results in the chapter provide a method for a comprehensive understanding of complex dynamical features and computational capabilities in nonlinear circuits with mem-elements, which is fundamental for a holistic approach in neuromorphic systems with such emerging nanoscale devices. The richness of the dynamical behavior highlighted via FCAM in mem-circuits and the possibility to tune via pulses the different dynamics make the circuits studied in the chapter valid candidates as sources of controllable complex dynamics to be incorporated in future neuromorphic architectures. Such architectures, which operate in the flux-charge domain, may be implemented at nanoscale exploiting the full variety of emerging mem-elements including memcapacitors and meminductors.

Appendix 1: Proof of Theorem 11.1

We have $\dot{\mathbf{x}}(t) = \mathbf{W}(t)$, $\dot{\mathbf{y}}(t) = \mathbf{T}(t)$, and $\dot{\mathbf{z}}(t) = \mathbf{R}(t)$. We can rewrite the SEs in the (φ, q)-domain as

$$\mathbf{M}\begin{pmatrix}\mathbf{W}(t)-\mathbf{W}_0\\ \mathbf{T}(t)-\mathbf{T}_0\\ \mathbf{R}(t)-\mathbf{R}_0\end{pmatrix} = -\mathbf{H}\begin{pmatrix}\mathbf{F}^a(\mathbf{X}(t))-\mathbf{F}^a(\mathbf{X}_0)\\ \mathbf{Y}(t)-\mathbf{Y}_0\\ \mathbf{z}(t)\end{pmatrix}$$
$$-\begin{pmatrix}\mathbf{0}\\ \mathbf{F}^b(\mathbf{Y}(t))-\mathbf{F}^b(\mathbf{Y}_0)\\ \mathbf{0}\end{pmatrix} - \mathbf{U}(t;t_0).$$

We have

$$\mathbf{z}(t) = \mathbf{H}_{22}^{-1}\Big[-\mathbf{M}_{22}(\mathbf{R}(t)-\mathbf{R}_0)$$
$$-\mathbf{H}_{21}\begin{pmatrix}\mathbf{F}^a(\mathbf{X}(t))-\mathbf{F}^a(\mathbf{X}_0)\\ \mathbf{Y}(t)-\mathbf{Y}_0\end{pmatrix} - \mathbf{U}_2(t;t_0)\Big].$$

Substituting $\mathbf{z}(t)$, we obtain

$$\mathbf{K}(\boldsymbol{\Gamma}_0) = \mathbf{M}_{11}\begin{pmatrix}\mathbf{W}(t)\\ \mathbf{T}(t)\end{pmatrix} + \mathbf{H}_{11}\begin{pmatrix}\mathbf{F}^a(\mathbf{X}(t))\\ \mathbf{Y}(t)\end{pmatrix}$$
$$+\begin{pmatrix}\mathbf{0}\\ \mathbf{F}^b(\mathbf{Y}(t))\end{pmatrix} + \mathbf{H}_{12}\mathbf{H}_{22}^{-1}\Big[-\mathbf{M}_{22}\mathbf{R}(t)$$
$$-\mathbf{H}_{21}\begin{pmatrix}\mathbf{F}^a(\mathbf{X}(t))\\ \mathbf{Y}(t)\end{pmatrix} - \mathbf{U}_2(t;t_0)\Big] + \mathbf{U}_1(t;t_0)$$

and the result follows. ∎

Appendix 2: Proof of Corollary 11.1

The property $\boldsymbol{\Gamma}(t;t_0,\boldsymbol{\Gamma}_0) \in \mathcal{M}(\mathbf{K}(\boldsymbol{\Gamma}_0))$ for any $t \geq t_0$ follows from Theorem 11.1 and (11.29). From (11.23), we obtain

$$\mathbf{M}_{11}\begin{pmatrix}\dot{\mathbf{X}}(t)\\ \dot{\mathbf{Y}}(t)\end{pmatrix} = -\mathbf{H}_{11}\begin{pmatrix}\mathbf{F}^a(\mathbf{X}(t))-\mathbf{F}^a(\mathbf{X}_0)\\ \mathbf{Y}(t)-\mathbf{Y}_0\end{pmatrix}$$
$$-\mathbf{H}_{12}\mathbf{z}(t) - \begin{pmatrix}\mathbf{0}\\ \mathbf{F}^b(\mathbf{Y}(t))-\mathbf{F}^b(\mathbf{Y}_0)\end{pmatrix}$$
$$+\mathbf{M}_{11}\begin{pmatrix}\mathbf{W}_0\\ \mathbf{T}_0\end{pmatrix} \tag{11.39}$$

and

$$\begin{aligned}
\mathbf{M}_{22}\dot{\mathbf{z}}(t) &= -\mathbf{H}_{21}\begin{pmatrix}\mathbf{F}^a(\mathbf{X}(t)) - \mathbf{F}^a(\mathbf{X}_0)\\ \mathbf{Y}(t) - \mathbf{Y}_0\end{pmatrix}\\
&\quad -\mathbf{H}_{22}\mathbf{z}(t) + \mathbf{M}_{22}\mathbf{R}_0\\
&= -\mathbf{H}_{21}\begin{pmatrix}\mathbf{F}^a(\mathbf{X}(t))\\ \mathbf{Y}(t)\end{pmatrix}\\
&\quad -\mathbf{H}_{22}\Big[\mathbf{z}(t) - \mathbf{H}_{22}^{-1}\mathbf{H}_{21}\begin{pmatrix}\mathbf{F}^a(\mathbf{X}_0)\\ \mathbf{Y}_0\end{pmatrix}\\
&\quad -\mathbf{H}_{22}^{-1}\mathbf{M}_{22}\mathbf{R}_0\Big]\\
&= -\mathbf{H}_{21}\begin{pmatrix}\mathbf{F}^a(\mathbf{X}(t))\\ \mathbf{Y}(t)\end{pmatrix} - \mathbf{H}_{22}\mathbf{Z}(t)
\end{aligned}$$

where we used (11.30). The corollary is proved by deriving $\mathbf{z}(t)$ from (11.30) and substituting in (11.39). ■

References

1. A. Wu, Z. Zeng, X. Zhu, J. Zhang, Exponential synchronization of memristor-based recurrent neural networks with time delays. Neurocomputing **74**(17), 3043–3050 (2011)
2. A. Wu, S. Wen, Z. Zeng, Synchronization control of a class of memristor-based recurrent neural networks. Inf. Sci. **183**(1), 106–116 (2012)
3. H. Kim, M. Sah, C. Yang, T. Roska, L.O. Chua, Memristor bridge synapses. Proc. IEEE **100**(6), 2061–2070 (2012)
4. Z. Guo, J. Wang, Z. Yan, Global exponential synchronization of two memristor-based recurrent neural networks with time delays via static or dynamic coupling. IEEE Trans. Syst. Man Cybern. Syst. **45**(2), 235–249 (2015)
5. Q. Xiao, Z. Huang, Z. Zeng, Passivity analysis for memristor-based inertial neural networks with discrete and distributed delays. IEEE Trans. Syst. Man Cybern. Syst. **49**(2), 375–385 (2019)
6. Z. Guo, J. Wang, Z. Yan, Attractivity analysis of memristor-based cellular neural networks with time-varying delays. IEEE Trans. Neural Netw. Learn. Syst. **25**(4), 704–717 (2014)
7. Y. Wan, J. Cao, Periodicity and synchronization of coupled memristive neural networks with supremums. Neurocomputing **159**(1), 137–143 (2015)
8. Q. Xu, Y. Lin, B. Bao, M. Chen, Multiple attractors in a non-ideal active voltage-controlled memristor based Chua's circuit. Chaos, Solitons Fractals **83**, 186–200 (2016)
9. X. Yang, D.W.C. Ho, Synchronization of delayed memristive neural networks: robust analysis approach. IEEE Trans. Cybern. **46**(12), 3377–3387 (2016)
10. S. Ding, Z. Wang, N. Rong, H. Zhang, Exponential stabilization of memristive neural networks via saturating sampled-data control. IEEE Trans. Cybern. **47**(10), 3027–3039 (2017)
11. H. Liu, Z. Wang, B. Shen, H. Dong, Delay-distribution-dependent H_∞ state estimation for discrete-time memristive neural networks with mixed time-delays and fading measurements. IEEE Trans. Cybern. 50(2):440-51 (2018)

12. M. Di Marco, M. Forti, L. Pancioni, Convergence and multistability of nonsymmetric cellular neural networks with memristors. IEEE Trans. Cybern. **47**(10), 2970–2983 (2017)
13. C. Chen, S. Zhu, Y. Wei, C. Yang, Finite-time stability of delayed memristor-based fractional-order neural networks. IEEE Trans. Cybern. **50**(4), 1607–1616 (2019)
14. S. Kumar, J.P. Strachan, R. Stanley Williams, Chaotic dynamics in nanoscale NbO_2 Mott memristors for analogue computing. Nature **548**(7667), 318 (2017)
15. C. Du, F. Cai, M.A. Zidan, W. Ma, S.H. Lee, W. D. Lu, Reservoir computing using dynamic memristors for temporal information processing. Nat. Commun. **8**(1), 2204 (2017)
16. F.L. Traversa, M. Di Ventra, Universal memcomputing machines. IEEE Trans. Neural Netw. Learn. Syst. **26**(11), 2702–2715 (2015)
17. J.S. Najem, M.S. Hasan, R. Stanley Williams, R.J. Weiss, G.S. Rose, G.J. Taylor, S.A. Sarles, C. Patrick Collier, Dynamical nonlinear memory capacitance in biomimetic membranes. Nat. Commun. **10**(1), 3239 (2019)
18. T. Driscoll, H.-T. Kim, B.-G. Chae, B.-J. Kim, Y.-W. Lee, N. Marie Jokerst, S. Palit, D.R. Smith, M. Di Ventra, D.N. Basov, Memory metamaterials. Science **325**(5947), 1518–1521 (2009)
19. J. Georgiou, K.M. Kossifos, M.A. Antoniades, A.H. Jaafar, N.T. Kemp, Chua mem-components for adaptive RF metamaterials, in *IEEE ISCAS 2018*, Florence, May 27–30 (2018), pp. 1–5
20. K. Rajagopal, S. Jafari, A. Karthikeyan, A. Srinivasan, B. Ayele, Hyperchaotic memcapacitor oscillator with infinite equilibria and coexisting attractors. Circuits Syst. Signal Proc. **37**, 3702–3724 (2018)
21. F. Yuan, G. Wang, P. Jin, X. Wang, G. Ma, Chaos in a meminductor-based circuit. Int. J. Bifurc. Chaos **26**(8), 1650130 (2016)
22. F. Yuan, G. Wang, Y. Shen, X. Wang, Coexisting attractors in a memcapacitor-based chaotic oscillator. Nonlinear Dyn. **86**(1), 37–50 (2016)
23. C. Li, C. Li, T. Huang, H. Wang, Synaptic memcapacitor bridge synapses. Neurocomputing **122**, 370–374 (2013)
24. G. Wang, S. Jiang, X. Wang, Y. Shen, F. Yuan, A novel memcapacitor model and its application for generating chaos. Math. Prob. Eng. **2016**, 3173696 (2016)
25. A. Dhooge, W. Govaerts, Yu. A. Kuznetsov, MATCONT: A MATLAB package for numerical bifurcation analysis of ODEs. ACM Trans. Math. Softw. **29**(2), 141–164 (2003)
26. M. Di Marco, M. Forti, F. Corinto, M. Gilli, State equations of memristor circuits with nonlinear lossless elements in the flux-charge domain, in *2019 IEEE International Symposium Circuits System (ISCAS)* (2019), pp. 1–5

Index

F. Corinto et al., *Nonlinear Circuits and Systems with Memristors*,
https://doi.org/10.1007/978-3-030-55651-8

N

O